Springer Monographs in Mathematics

Giuseppe Mastroianni · Gradimir V. Milovanović

Interpolation Processes

Basic Theory and Applications

Giuseppe Mastroianni
Università della Basilicata
Dipartimento di Matematica
Via dell'Ateneo Lucano
85100 Potenza, Italy
mastroianni@unibas.it

Gradimir V. Milovanović
Megatrend University
Faculty of Computer Sciences
Bulevar umetnosti 29
11070 Novi Beograd, Serbia
gvm@megatrend-edu.net

ISBN 978-3-540-68346-9   e-ISBN 978-3-540-68349-0

DOI 10.1007/978-3-540-68349-0

Springer Monographs in Mathematics ISSN 1439-7382

Library of Congress Control Number: 2008930793

Mathematics Subject Classification (2000): 33-xx, 41-xx, 42Axx, 45A05, 45B05, 45H05, 65B10, 65Dxx

© 2008 Springer-Verlag Berlin Heidelberg

This work is subject to copyright. All rights are reserved, whether the whole or part of the material is concerned, specifically the rights of translation, reprinting, reuse of illustrations, recitation, broadcasting, reproduction on microfilm or in any other way, and storage in data banks. Duplication of this publication or parts thereof is permitted only under the provisions of the German Copyright Law of September 9, 1965, in its current version, and permission for use must always be obtained from Springer. Violations are liable to prosecution under the German Copyright Law.

The use of general descriptive names, registered names, trademarks, etc. in this publication does not imply, even in the absence of a specific statement, that such names are exempt from the relevant protective laws and regulations and therefore free for general use.

Cover design: WMXDesign GmbH, Heidelberg

Printed on acid-free paper

9 8 7 6 5 4 3 2 1

springer.com

To Ida and Dobrila

Preface

Interpolation of functions is one of the basic part of *Approximation Theory*. There are many books on approximation theory, including interpolation methods that appeared in the last fifty years, but a few of them are devoted only to interpolation processes. An example is the book of J. Szabados and P. Vértesi: *Interpolation of Functions,* published in 1990 by World Scientific. Also, two books deal with a special interpolation problem, the so-called Birkhoff interpolation, written by G.G. Lorentz, K. Jetter, S.D. Riemenschneider (1983) and Y.G. Shi (2003).

The classical books on interpolation address numerous negative results, i.e., results on divergent interpolation processes, usually constructed over some equidistant system of nodes. The present book deals mainly with new results on convergent interpolation processes in uniform norm, for algebraic and trigonometric polynomials, not yet published in other textbooks and monographs on approximation theory and numerical mathematics. Basic tools in this field (orthogonal polynomials, moduli of smoothness, K-functionals, etc.), as well as some selected applications in numerical integration, integral equations, moment-preserving approximation and summation of slowly convergent series are also given.

The first chapter provides an account of basic facts on approximation by algebraic and trigonometric polynomials introducing the most important concepts on approximation of functions. Especially, in Sect. 1.4 we give basic results on interpolation by algebraic polynomials, including representations and computation of interpolation polynomials, Lagrange operators, interpolation errors and uniform convergence in some important classes of functions, as well as an account on the Lebesgue function and some estimates for the Lebesgue constant.

The second chapter is devoted to orthogonal polynomials on the real line and weighted polynomial approximation. For polynomials orthogonal on the real line we give the basic properties and introduce and discuss the associated polynomials, functions of the second kind, Stieltjes polynomials, as well as the Christoffel functions and numbers. The *classical orthogonal polynomials* as the most important class of orthogonal polynomials on the real line are treated in Sect. 2.3, and new results on *nonclassical orthogonal polynomials*, including methods for their numerical construction, are studied in Sect. 2.4. Introducing the weighted functional spaces, moduli of smoothness and K-functionals, the weighted best polynomial approximations on $(-1, 1)$, $(0, +\infty)$ and $(-\infty, +\infty)$ are treated in Sect. 2.5, as well as the weighted polynomial approximation of functions having interior isolated singularities.

Trigonometric approximation is considered in Chap. 3. Approximations by sums of Fourier and Fejér and de la Vallée Poussin means are given. Their discrete versions and the Lagrange trigonometric operator are also investigated. As a basic tool for studying approximating properties of the Lagrange and de la Vallée Poussin operators we consider the so-called Marcinkiewicz inequalities. Beside the uniform approximation we also investigate the Lagrange interpolation error in L^p-norm ($1 < p < +\infty$) and give some estimates in the L^1-Sobolev norm, including some weighted versions.

Chapter 4 treats algebraic interpolation processes $\{L_n(\mathcal{X})\}_{n \in \mathbb{N}}$ in the uniform norm, starting with the so-called optimal system of nodes $\mathcal{X}$, which provides Lebesgue constants of order $\log n$ and the convergence of the corresponding interpolation processes. Moreover, the error of such an approximation is near to the error of the best uniform approximation. Beside two classical examples of the well-known optimal system of nodes (zeros of the Jacobi polynomials $P_n^{(\alpha,\beta)}(x)$ ($-1 < \alpha, \beta \leq -1/2$) and the so-called Clenshaw's abscissas), we introduce more general results for constructing interpolation processes at nodes with an *arc sine distribution* having Lebsgue constants of order $\log n$. The so-called *additional nodes method with Jacobi zeros* is presented in Sect. 4.2.2. Some other optimal interpolation processes are analyzed in 4.2.3. The third section of this chapter is devoted to the weighted interpolation in the corresponding weighted spaces (Jacobi, Laguerre and Hermite cases). In addition, we consider the weighted interpolation of functions with internal isolated singularities.

The final chapter provides some selected applications in numerical analysis. In the first section on quadrature formulae we present some special Newton–Cotes rules, the Gauss–Christoffel, Gauss–Radau and Gauss–Lobatto quadratures, the so-called product integration rules, as well as a method for the numerical integration of periodic functions on the real line with respect to a rational weight function. Also, we include the error estimates of Gaussian rules for some classes of functions. The second section is devoted to methods for solving the Fredholm integral equations of the second kind. The methods are based on the so-called *Approximation and Polynomial Interpolation Theory* and lead to the construction of polynomial sequences converging to the exact solutions in some weighted uniform norms. In the third section we consider some kinds of moment-preserving approximations by polynomials and splines. In the last section of this chapter we consider two recent methods of summation of slowly convergent series based on integral representations of series and an application of the Gaussian quadratures. In the first method we assume that the general term of the series is expressible in terms of the Laplace transform (or its derivative) of a known function. It leads to the Gaussian quadrature formulas with respect to the Einstein and Fermi weight functions on $(0, +\infty)$. The second method is based on a contour integration over a rectangle in the complex plane, reducing then the summation of a slowly convergent series to a problem of Gaussian quadrature rules on $(0, +\infty)$ with respect to some hyperbolic weight functions.

Notation of this book is standard. If it is not defined in another way, throughout this book C, C_0, C_1, C_2, ... denote positive constants, which can take different values even in subsequent formulae. It will always be clear what indices and variables the constants are independent of. If we use the notation C_p, it means that this constant always depends on a parameter p. Sometimes, we will write $C \neq C(a, b, \ldots)$ in order to denote that the constant C is independent only of a, b, ..., but it can depend on parameters which are not mentioned in the list $(a, b, \ldots)$.

If A and B are two expressions depending on certain indices and variables, then we write

$$A \sim B \quad \text{if and only if} \quad 0 < C_1 \leq \left|\frac{A}{B}\right| \leq C_2$$

uniformly for the indices and variables considered.

Some five hundred references have been cited here. As a rule, we have studied the original sources and in some cases have retrieved some forgotten but useful results. At the end of the book we included an index, combined with subjects and names.

The book addresses researchers and students in mathematics, physics, and other computational and applied sciences.

We are indebted to several mathematicians who have supported us with valuable suggestions and useful comments: B.D. Bojanov, W. Gautschi, P. Junghanns, D.S. Lubinsky, E. Malkowsky, Th.M. Rassias, J. Szabados, V. Totik, and P. Vértesi.

Thanks also go to our collaborators: A.S. Cvetković, M.C. De Bonis, M.G. Russo, M. Stanić, and W. Themistoclakis, who helped us in the technical preparation of the manuscript, as well as to the national company *Electric Power Industry of Serbia* (Belgrade) for a financial support.

We are deeply grateful to the Springer-Verlag for including this book in the excellent series *Springer Monographs in Mathematics* and, in particular, to Mrs. Ellen Kattner for her continuing encouragement.

We dedicate this book to our wives, Ida and Dobrila, in appreciation of their patience and unwavering support.

Potenza/Niš,   *Giuseppe Mastroianni*
July 2008   *Gradimir V. Milovanović*

Contents

Preface vii

1 Constructive Elements and Approaches in Approximation Theory . 1
   1.1 Introduction to Approximation Theory 1
       1.1.1 Basic Notions 1
       1.1.2 Algebraic and Trigonometric Polynomials 4
       1.1.3 Best Approximation by Polynomials 7
       1.1.4 Chebyshev Polynomials 9
       1.1.5 Chebyshev Extremal Problems 14
       1.1.6 Chebyshev Alternation Theorem 17
       1.1.7 Numerical Methods 20
   1.2 Basic Facts on Trigonometric Approximation 24
       1.2.1 Trigonometric Kernels 24
       1.2.2 Fourier Series and Sums 30
       1.2.3 Moduli of Smoothness, Best Approximation and Besov
            Spaces 32
   1.3 Chebyshev Systems and Interpolation 38
       1.3.1 Chebyshev Systems and Spaces 38
       1.3.2 Algebraic Lagrange Interpolation 39
       1.3.3 Trigonometric Interpolation 40
       1.3.4 Riesz Interpolation Formula 44
       1.3.5 A General Interpolation Problem 46
   1.4 Interpolation by Algebraic Polynomials 48
       1.4.1 Representations and Computation of Interpolation
            Polynomials 48
       1.4.2 Interpolation Array and Lagrange Operators 51
       1.4.3 Interpolation Error for Some Classes of Functions 54
       1.4.4 Uniform Convergence in the Class of Analytic Functions . 56
       1.4.5 Bernstein's Example of Pointwise Divergence 61
       1.4.6 Lebesgue Function and Some Estimates for the Lebesgue
            Constant 63
       1.4.7 Algorithm for Finding Optimal Nodes 68

2 Orthogonal Polynomials and Weighted Polynomial Approximation . 75
   2.1 Orthogonal Systems and Polynomials 75
       2.1.1 Inner Product Space and Orthogonal Systems 75
       2.1.2 Fourier Expansion and Best Approximation 77
       2.1.3 Examples of Orthogonal Systems 79

		2.1.4	Basic Facts on Orthogonal Polynomials and Extremal Problems 89

- 2.1.4 Basic Facts on Orthogonal Polynomials and Extremal Problems 89
- 2.1.5 Zeros of Orthogonal Polynomials 93
- 2.2 Orthogonal Polynomials on the Real Line 95
  - 2.2.1 Basic Properties 95
  - 2.2.2 Asymptotic Properties of Orthogonal Polynomials 103
  - 2.2.3 Associated Polynomials and Christoffel Numbers 111
  - 2.2.4 Functions of the Second Kind and Stieltjes Polynomials . . 117
- 2.3 Classical Orthogonal Polynomials 121
  - 2.3.1 Definition of the Classical Orthogonal Polynomials 121
  - 2.3.2 General Properties of the Classical Orthogonal Polynomials 124
  - 2.3.3 Generating Function 128
  - 2.3.4 Jacobi Polynomials 131
  - 2.3.5 Generalized Laguerre Polynomials 140
  - 2.3.6 Hermite Polynomials 145
- 2.4 Nonclassical Orthogonal Polynomials 146
  - 2.4.1 Semi-classical Orthogonal Polynomials 146
  - 2.4.2 Generalized Gegenbauer Polynomials 147
  - 2.4.3 Generalized Jacobi Polynomials 148
  - 2.4.4 Sonin-Markov Orthogonal Polynomials 152
  - 2.4.5 Freud Orthogonal Polynomials 154
  - 2.4.6 Orthogonal Polynomials with Respect to Abel, Lindelöf, and Logistic Weights 159
  - 2.4.7 Strong Non-classical Orthogonal Polynomials 159
  - 2.4.8 Numerical Construction of Orthogonal Polynomials 160
- 2.5 Weighted Polynomial Approximation 166
  - 2.5.1 Weighted Functional Spaces, Moduli of Smoothness and K-functionals 166
  - 2.5.2 Weighted Best Polynomial Approximation on $[-1, 1]$ . . . 170
  - 2.5.3 Weighted Approximation on the Semi-axis 174
  - 2.5.4 Weighted Approximation on the Real Line 178
  - 2.5.5 Weighted Polynomial Approximation of Functions Having Isolated Interior Singularities 182

3 Trigonometric Approximation 193
- 3.1 Approximating Properties of Operators 193
  - 3.1.1 Approximation by Fourier Sums 193
  - 3.1.2 Approximation by Fejér and de la Vallée Poussin Means . . 195
- 3.2 Discrete Operators 197
  - 3.2.1 A Quadrature Formula 197
  - 3.2.2 Discrete Versions of Fourier and de la Vallée Poussin Sums 202
  - 3.2.3 Marcinkiewicz Inequalities 205

		3.2.4	Uniform Approximation 210

		3.2.4 Uniform Approximation 210

        3.2.4 Uniform Approximation 210
        3.2.5 Lagrange Interpolation Error in L^p 212
        3.2.6 Some Estimates of the Interpolation Errors in L^1-Sobolev Spaces 221
        3.2.7 The Weighted Case 224

4 Algebraic Interpolation in Uniform Norm 235
   4.1 Introduction and Preliminaries 235
        4.1.1 Interpolation at Zeros of Orthogonal Polynomials 235
        4.1.2 Some Auxiliary Results 239
   4.2 Optimal Systems of Nodes 248
        4.2.1 Optimal Systems of Knots on $[-1, 1]$ 248
        4.2.2 Additional Nodes Method with Jacobi Zeros 252
        4.2.3 Other "Optimal" Interpolation Processes 264
        4.2.4 Some Simultaneous Interpolation Processes 268
   4.3 Weighted Interpolation 271
        4.3.1 Weighted Interpolation at Jacobi Zeros 271
        4.3.2 Lagrange Interpolation in Sobolev Spaces 276
        4.3.3 Interpolation at Laguerre Zeros 278
        4.3.4 Interpolation at Hermite Zeros 287
        4.3.5 Interpolation of Functions with Internal Isolated Singularities 292

5 Applications 319
   5.1 Quadrature Formulae 319
        5.1.1 Introduction 319
        5.1.2 Some Remarks on Newton-Cotes Rules with Jacobi Weights 322
        5.1.3 Gauss-Christoffel Quadrature Rules 324
        5.1.4 Gauss-Radau and Gauss-Lobatto Quadrature Rules 328
        5.1.5 Error Estimates of Gaussian Rules for Some Classes of Functions 332
        5.1.6 Product Integration Rules 345
        5.1.7 Integration of Periodic Functions on the Real Line with Rational Weight 350
   5.2 Integral Equations 362
        5.2.1 Some Basic Facts 362
        5.2.2 Fredholm Integral Equations of the Second Kind 369
        5.2.3 Nyström Method 382
   5.3 Moment-Preserving Approximation 385
        5.3.1 The Standard L^2-Approximation 385
        5.3.2 The Constrained L^2-Polynomial Approximation 388
        5.3.3 Moment-Preserving Spline Approximation 389
   5.4 Summation of Slowly Convergent Series 397
        5.4.1 Laplace Transform Method 398

　　　　　5.4.2　Contour Integration Over a Rectangle 401
　　　　　5.4.3　Remarks on Some Slowly Convergent Power Series 411

References 415

Index 437

Chapter 1
Constructive Elements and Approaches in Approximation Theory

1.1 Introduction to Approximation Theory

1.1.1 Basic Notions

One of the main problems in approximation theory is how to find, for a given function f from a large space X, a simple function ϕ from some small subset Φ of X, such that ϕ is close in some sense to f. We say that ϕ is an *approximation* or an *approximant* to a given function f. Usually, X is a normed linear space of functions defined on a given set A. For example, A can be a compact interval $[a, b]$, the circle $\mathbb{T}$, etc. We use the circle $\mathbb{T}$ in the periodic case, when it represents the real line $\mathbb{R}$ with the identification of the points modulo 2π. The normed space can be the space of continuous functions $C(A)$, m-times continuous-differentiable functions $C^m(A)$, the space $L^p(A)$, and other Banach spaces. The space $L^p(A)$ is defined in the usual way

$$L^p(A) = \left\{ f \mid \|f\|_p := \left(\int_A |f(t)|^p\, dt \right)^{1/p} < +\infty \right\}, \quad 1 \le p < +\infty.$$

If f is defined everywhere on A and $\|f\|_\infty := \sup_{t \in A} |f(t)| < +\infty$ we write $f \in L^\infty(A)$.

When $A = \mathbb{T} \equiv [0, 2\pi)$ we simply write $L^p = L^p(\mathbb{T})$ and $L^\infty = L^\infty(\mathbb{T})$.

The distance between f and ϕ can be measured by the norm $\|f - \phi\|$. Then, the distance between $f \in X$ and Φ is determined by

$$E(f) = \inf_{\phi \in \Phi} \|f - \phi\|.$$

This infimum $E(f)$ is called the *error of best approximation* of the function f by elements from Φ in a given norm. If there exists an element $\phi^* \in \Phi$ such that

$$E(f) = \min_{\phi \in \Phi} \|f - \phi\| = \|f - \phi^*\|, \tag{1.1.1}$$

we say that ϕ^* is a *best approximation*. The question of existence and uniqueness of such an element is crucial. Also, algorithms for finding it are of great importance from numerical point of view.

If Φ is a finite dimensional subspace of X, then each $f \in X$ has a best approximant. Unfortunately the best approximation is not always unique. However, when

X is a strictly *normed space*,[1] then this element is unique. Some important special cases will be considered in Sect. 1.1.3. The proofs of general results can be found for example in [95, 235, 397].

Very often we approximate the function f by *algebraic polynomials* on $A = [a,b]$, i.e. we take $\Phi = \mathcal{P}_n$, where $\mathcal{P}_n$ is the set of all algebraic polynomials p_n of degree at most n,

$$p_n(x) = \sum_{k=0}^{n} a_k x^k \qquad (a_k \in \mathbb{R}). \tag{1.1.2}$$

If the coefficients $a_k \in \mathbb{C}$, the corresponding set will be denoted by $\mathcal{P}_n^{\mathbb{C}}$. If $a_n \neq 0$, the degree of p_n is strictly n. A polynomial is *monic* if $a_n = 1$.

For the circle $\mathbb{T}$, i.e., in the periodic case, we take $\Phi = \mathcal{T}_n$, where $\mathcal{T}_n$ is the set of all *trigonometric polynomials* t_n of degree at most n,

$$t_n(x) = \frac{a_0}{2} + \sum_{k=1}^{n} (a_k \cos kx + b_k \sin kx) \qquad (a_k, b_k \in \mathbb{R}). \tag{1.1.3}$$

If $|a_n| + |b_n| > 0$ the degree of t_n is strictly n.

There are two important particular cases of (1.1.3). Namely, if $b_1 = \cdots = b_n = 0$ we have a *cosine* polynomial

$$c_n(x) = a_0 + a_1 \cos x + \cdots + a_n \cos nx.$$

For $a_0 = a_1 = \cdots = a_n = 0$ it is a *sine* polynomial

$$s_n(x) = b_1 \sin x + \cdots + b_n \sin nx.$$

Defining the complex coefficients c_k ($|k| \leq n$) as

$$c_0 = \frac{a_0}{2}, \quad c_k = \bar{c}_{-k} = \frac{1}{2}(a_k - ib_k) \qquad (k = 1, \ldots, n),$$

we obtain the complex form of (1.1.3),

$$t_n(x) = \sum_{k=-n}^{n} c_k e^{ikx}. \tag{1.1.4}$$

Sometimes, we omit n in $p_n(x)$ and $t_n(x)$ and write simple $p(x)$ and $t(x)$ (or $P(x)$ and $T(x)$), respectively.

Evidently, $\mathcal{P}_n$ is a vector space of dimension $n+1$ over $\mathbb{R}$. Therefore, this space equipped with any norm is isomorphic to the Euclidean vector space $\mathbb{R}^{n+1}$, and these norms are equivalent to each other.

---

[1] If $\|f + g\| = \|f\| + \|g\|$ implies that $f = \alpha g$ ($\alpha \in \mathbb{R}$).

1.1 Introduction to Approximation Theory

Some typical norms of (1.1.2) on the space $\mathcal{P}_n$ are the *uniform* or *supremum norm* and L^p-*norm* ($p \geq 1$), i.e.,

$$\|p_n\|_A = \sup_{x \in A} |p_n(x)| \quad \text{and} \quad \|p_n\|_p = \left(\int_A |p_n(x)|^p \, dx\right)^{1/p}.$$

Similarly, $\mathcal{T}_n$ is a vector space of dimension $2n + 1$. The previous norms are also usual for trigonometric polynomials (1.1.3).

The main constructive elements in approximation theory are algebraic and trigonometric polynomials, rational functions and splines. In this book we deal only with polynomials. Two properties of polynomials are essential in approximation theory:

1° Each real continuous function on a finite closed interval can be uniformly approximated by polynomials.
2° Each polynomial $p_n \in \mathcal{P}_n$ and $t_n \in \mathcal{T}_n$ can be uniquely interpolated at $n + 1$ and $2n + 1$ points, respectively.

Concerning the first property there are two basic theorems of Weierstrass.

Theorem 1.1.1 *For each $f \in C[a, b]$ and each $\varepsilon > 0$ there is an algebraic polynomial p such that*

$$|f(x) - p(x)| < \varepsilon \quad (a \leq x \leq b).$$

Theorem 1.1.2 *For each function $f \in C(\mathbb{T})$ and each $\varepsilon > 0$ there is a trigonometric polynomial t such that*

$$|f(x) - t(x)| < \varepsilon \quad (x \in \mathbb{T}).$$

Theorem 1.1.1 was first proved in 1885 by Weierstrass (see [502]). There are several different proofs of these theorems and their extensions and ramifications (see Lubinsky [271] and Pinkus [400]). Theorem 1.1.1 can be interpreted in terms of the best approximation in the uniform (supremum) norm $\|f\|_{[a,b]} = \max_{a \leq x \leq b} |f(x)|$ ($f \in C[a, b]$). Let

$$E_n(f) = \inf_{p \in \mathcal{P}_n} \|f - p\|_{[a,b]}. \qquad (1.1.5)$$

Then, Theorem 1.1.1 asserts that

$$\lim_{n \to +\infty} E_n(f) = 0, \quad f \in C[a, b].$$

The property 2° of polynomials mentioned above enables another kind of approximation, which is known as the *interpolation of functions*. An important part of this book is devoted to the interpolation and interpolating processes in different spaces of functions.

1.1.2 Algebraic and Trigonometric Polynomials

Consider an algebraic polynomial $p_n(z)$ of degree n,

$$p_n(z) = \sum_{k=0}^{n} \alpha_k z^k \qquad (\alpha_k \in \mathbb{C}).$$

The following result is well-known as the *fundamental theorem of algebra* (cf. [359, p. 177]):

Theorem 1.1.3 *Every polynomial of degree n (≥ 1) with complex coefficients has exactly n zeros (counted with their multiplicities) in the complex plane.*

The zeros of a polynomial are continuous functions of the coefficients of the polynomial (see [359, pp. 177–178]).

Taking z on the circumference $|z| = 1$, i.e., $z = e^{i\theta}$, $p_n(z)$ becomes a trigonometric polynomial $t_n(\theta)$ of degree n,

$$t_n(\theta) = p_n(e^{i\theta}) = \frac{a_0}{2} + \sum_{k=1}^{n}(a_k \cos k\theta + b_k \sin k\theta), \qquad (1.1.6)$$

with complex coefficients in the general case.

The following important result is known as the *Haar property*.

Theorem 1.1.4 *An arbitrary trigonometric polynomial $t_n(\theta)$ of degree at most n, which is not identically zero, cannot have more than $2n$ distinct zeros in $\mathbb{T}$ (i.e., in any interval $[a, a + 2\pi)$, $a \in \mathbb{R}$).*

Proof Putting $z = e^{i\theta}$ and using Euler's formulas

$$\cos k\theta = \frac{1}{2}(e^{ik\theta} + e^{-ik\theta}), \qquad \sin k\theta = \frac{1}{2i}(e^{ik\theta} - e^{-ik\theta}),$$

we obtain

$$t_n(\theta) = \sum_{k=-n}^{n} c_{n+k} e^{ik\theta},$$

where

$$c_{n+k} = \begin{cases} \frac{1}{2}(a_{-k} + ib_{-k}), & k < 0, \\ \frac{1}{2}a_0, & k = 0, \\ \frac{1}{2}(a_k - ib_k), & k > 0. \end{cases}$$

1.1 Introduction to Approximation Theory

Thus we have
$$e^{in\theta} t_n(\theta) = q(z), \tag{1.1.7}$$
where q is an algebraic polynomial of degree at most $2n$. If $t_n(\theta) \not\equiv 0$, then $t_n(\theta)$ cannot have more than $2n$ distinct real zeros in $\mathbb{T}$. □

If the polynomial $q(z)$ in (1.1.7) is of degree strictly $2n$, then it has exactly $2n$ zeros in the complex plane (counted with their multiplicities). Denote them by $z_1, \ldots, z_{2n}$. In that case $t_n(\theta)$ has also exactly $2n$ zeros in any strip $a \le \text{Re}\,\theta < a + 2\pi$ ($a \in \mathbb{R}$) of the complex plane.

Using a factorization of $q(z)$ and putting $z_k = e^{i\theta_k}$ ($k = 1, \ldots, 2n$), we get

$$t_n(\theta) = c_{2n} e^{-in\theta} \prod_{k=1}^{2n} \left(e^{i\theta} - e^{i\theta_k} \right)$$

$$= c_{2n} \exp\left(\frac{i}{2} \sum_{k=1}^{2n} \theta_k \right) \prod_{k=1}^{2n} \left(e^{i(\theta-\theta_k)/2} - e^{-i(\theta-\theta_k)/2} \right),$$

i.e.,

$$t_n(\theta) = A \prod_{k=1}^{2n} \sin \frac{\theta - \theta_k}{2}, \tag{1.1.8}$$

where

$$A = (-1)^n 2^{2n} c_{2n} \exp\left(\frac{i}{2} \sum_{k=1}^{2n} \theta_k \right).$$

Thus, each $t_n \in \mathcal{T}_n \setminus \mathcal{T}_{n-1}$ can be represented in the form (1.1.8).

Consider now only real trigonometric polynomials.[2] If we put

$$c_0 = \frac{1}{2} a_0, \quad c_k = a_k - i b_k \quad (k = 1, \ldots, n),$$

where $a_k, b_k \in \mathbb{R}$, then we can represent a real trigonometric polynomial as the real part of an algebraic polynomial on the unit circle line $|z| = 1$. Namely,

$$t_n(\theta) = \text{Re}\left\{ \sum_{k=0}^{n} c_k z^k \right\}_{z=e^{i\theta}} = \text{Re}\left\{ \frac{1}{2} a_0 + \sum_{k=1}^{n} (a_k - i b_k) e^{ik\theta} \right\}$$

$$= \frac{1}{2} a_0 + \sum_{k=1}^{n} (a_k \cos k\theta + b_k \sin k\theta).$$

---
[2] Polynomials with real coefficients.

On the other hand, since
$$\operatorname{Re}\left\{\sum_{k=0}^{n} c_k z^k\right\} = \frac{1}{2}\left(\sum_{k=0}^{n} c_k z^k + \sum_{k=0}^{n} \bar{c}_k \bar{z}^k\right)$$

we have
$$t_n(\theta) = \frac{1}{2} z^{-n} \sum_{k=0}^{n} \left(c_k z^{n+k} + \bar{c}_k z^{n-k}\right)\bigg|_{z=e^{i\theta}},$$

i.e.,
$$t_n(\theta) = \frac{1}{2} e^{-in\theta} q(e^{i\theta}),$$

where
$$q(z) = \sum_{k=0}^{n} \left(c_k z^{n+k} + \bar{c}_k z^{n-k}\right)$$
$$= \bar{c}_n + \cdots + \bar{c}_1 z^{n-1} + 2c_0 z^n + c_1 z^{n+1} + \cdots + c_n z^{2n}.$$

Note that $q(z) = z^{2n} \bar{q}(1/z)$, i.e., q is a self-inversive polynomial of degree $2n$ (cf. [359, pp. 16–18]). According to the above we conclude that
$$|t_n(\theta)| = \frac{1}{2}|q(e^{i\theta})|.$$

Two simple, but important real trigonometric polynomials are
$$\cos n\theta \quad \text{and} \quad \frac{\sin(n+1)\theta}{\sin\theta}.$$

Both of them can be expressed in $\cos\theta$ as algebraic polynomials of degree n. Putting $x = \cos\theta$ we obtain the well-known Chebyshev polynomials of the first and second kind,
$$T_n(x) = T_n(\cos\theta) = \cos n\theta \quad \text{and} \quad U_n(x) = U_n(\cos\theta) = \frac{\sin(n+1)\theta}{\sin\theta},$$

respectively. Their algebraic representations for $|x| \leq 1$ are
$$T_n(x) = \cos(n\arccos x) \quad \text{and} \quad U_n(x) = \frac{\sin\big((n+1)\arccos x\big)}{\sqrt{1-x^2}}.$$

Remark 1.1.1 For Chebyshev polynomials the following relations
$$\frac{1}{2} + \sum_{k=1}^{n} T_{2k}(x) = \frac{1}{2} U_{2n}(x) \quad \text{and} \quad \sum_{k=1}^{n} T_{2k-1}(x) = \frac{1}{2} U_{2n-1}(x) \qquad (1.1.9)$$

hold.

Remark 1.1.2 If we put $y = \sin(\theta/2)$, then

$$\frac{T_{2n+1}(y)}{y} = \frac{\cos[(2n+1)\arccos y]}{y} = \frac{\cos[(2n+1)(\pi/2 - \arcsin y)]}{y}$$

$$= (-1)^n \frac{\sin[(2n+1)\arcsin y]}{y},$$

i.e.,

$$\frac{T_{2n+1}(y)}{y} = (-1)^n \frac{\sin\frac{(2n+1)\theta}{2}}{\sin\frac{\theta}{2}} = (-1)^n U_{2n}\left(\cos\frac{\theta}{2}\right). \quad (1.1.10)$$

According to (1.1.9) we get

$$\frac{T_{2n+1}(y)}{y} = (-1)^n \left(1 + 2\sum_{k=1}^n \cos k\theta\right), \quad y = \sin\frac{\theta}{2}, \quad (1.1.11)$$

because $T_{2k}(\cos(\theta/2)) = \cos k\theta$. Thus, (1.1.10) is an even trigonometric polynomial of degree n.

The Chebyshev polynomials will be treated in details later.

1.1.3 Best Approximation by Polynomials

In Sect. 1.1.1 we defined best approximation of the function $f \in X$ by elements from some subset Φ of X in a given norm. Here we deal with normed spaces $C[a, b]$ and $L^p[a, b]$ ($p \geq 1$) and with their subsets $\mathcal{P}_n$ and $\mathcal{T}_n$ (sets of algebraic and trigonometric polynomials of degree at most n, respectively).

Let $\Phi = \mathcal{P}_n$. Since $\mathcal{P}_n$ is a finite dimensional subspace of X ($X = C[a, b]$ or $X = L^p[a, b]$, $1 \leq p < +\infty$), the following result holds:

Theorem 1.1.5 *Let $f \in X$, where $X = C[a, b]$ or $X = L^p[a, b]$, $1 \leq p < +\infty$. Then for each $n \in \mathbb{N}$ there exists an algebraic polynomial P^* ($\in \mathcal{P}_n$) of best approximation in $\mathcal{P}_n$ for the function f.*

Usually we say that such a polynomial is *best uniform approximation* ($X = C[a, b]$) or *best L^p-approximation* ($X = L^p[a, b]$).

A similar situation appears in the periodic case, when we take $\Phi = \mathcal{T}_n$.

Theorem 1.1.6 *Let $f \in X$, where $X = C[0, 2\pi]$ or $X = L^p[0, 2\pi]$, $1 \leq p < +\infty$. Then for each $n \in \mathbb{N}$ there exists a trigonometric polynomial T^* ($\in \mathcal{T}_n$) of best approximation in $\mathcal{T}_n$ for the function f.*

Since the spaces $L^p[a,b]$ (in the periodic case $L^p[0,2\pi]$) for $1 < p < +\infty$ are strictly normed (cf. [384]), the polynomials of best L^p-approximations in such cases are unique. On the other hand, the spaces $L^1[a,b]$ and $C[a,b]$ are not strictly normed, so that in $L^1[a,b]$ we do not have uniqueness of best L^1-approximation. An illustration of this fact is the following example:

Example 1.1.1 Let
$$f(x) = \begin{cases} 0, & 0 \le x \le 1, \\ 1, & 1 < x \le 2. \end{cases}$$

Consider best L^1-approximation of this function in the set of all algebraic polynomials of degree zero. Since, for each c ($0 \le c \le 1$)
$$\int_0^2 |f(x) - c|\,dx = \int_0^1 c\,dx + \int_1^2 (1-c)\,dx = 1,$$

we conclude that every such constant c is best L^1-approximation in $\mathcal{P}_0$.

However, the situation is somewhat different in the space of continuous functions on $[a,b]$. Namely, as a consequence of the well-known *Chebyshev alternation theorem* (see Sect. 1.1.6), best uniform approximation P^* ($\in \mathcal{P}_n$) for a function $f \in C[a,b]$ is unique.

Instead of the term *uniform approximation*, we use also *Chebyshev approximation*. The concept of best approximation was introduced mainly by the work of the famous Russian mathematician Pafnutiĭ L'vovich Chebyshev (1821–1894), who studied properties of polynomials of least deviation from a given continuous function (see [58, 59]).

Let $f \in C[a,b]$ and $\|f\| = \|f\|_\infty = \|f\|_{[a,b]} = \max\limits_{a \le x \le b} |f(x)|$. As before, according to (1.1.1), i.e., (1.1.5), we put
$$E_n(f) = \min_{p \in \mathcal{P}_n} \|f - p\| = \|f - P^*\|. \tag{1.1.12}$$

In particular, for the function $f(x) = x^{n+1}$ on $[-1,1]$, it is clear that (1.1.12) becomes
$$E_n(x^{n+1}) = \min_{p \in \mathcal{P}_n} \|x^{n+1} - p(x)\| = \min_{q \in \hat{\mathcal{P}}_{n+1}} \|q\| = \|\hat{Q}^*_{n+1}\|, \tag{1.1.13}$$

where $\hat{\mathcal{P}}_{n+1}$ denotes the set of all monic polynomials of degree $n+1$ and $\hat{Q}^*_{n+1}$ is the monic polynomial of least uniform norm on $[-1,1]$ among all polynomials of degree $n+1$, with leading coefficient unity. In this way, with fixed leading coefficient, Chebyshev [58] introduced polynomials of least deviation from zero, which are known today as Chebyshev polynomials of the first kind $T_n(x)$. We mentioned such polynomials at the end of Sect. 1.1.2 in connection with a simple trigonometric polynomial. Precisely, Chebyshev showed that $\hat{Q}_n^*(x) = 2^{1-n} T_n(x)$. The notation $\cos(n \arccos x) - T_n(x)$ was introduced by Bernstein.[3]

The polynomials $T_n(x)$ appear prominently in various extremal problems with polynomials (cf. [359]). The following subsection deals with the Chebyshev polynomials.

1.1.4 Chebyshev Polynomials

1.1.4.1 Basic Properties

As we mentioned before, the Chebyshev polynomials of the first and second kind can be expressed for $|x| \leq 1$ in the forms

$$T_n(x) = \cos(n \arccos x) \quad \text{and} \quad U_n(x) = \frac{\sin((n+1) \arccos x)}{\sqrt{1-x^2}}, \quad (1.1.14)$$

respectively. It is easy to see that $T_0(x) = 1$, $T_1(x) = x$ and $U_0(x) = 1$, $U_1(x) = 2x$. Also, using

$$\cos(n+1)\theta + \cos(n-1)\theta = 2\cos\theta \cos n\theta,$$

with $x = \cos\theta$, we see that the polynomials T_n satisfy the recurrence relation

$$T_{n+1}(x) = 2x T_n(x) - T_{n-1}(x), \quad n = 1, 2, \ldots. \quad (1.1.15)$$

The same recurrence relation also holds for the polynomials of the second kind, i.e., $U_{n+1}(x) = 2x U_n(x) - U_{n-1}(x)$, $n \geq 1$. Starting from $T_0(x) = 1$ and $T_1(x) = x$ or $U_0(x) = 1$ and $U_1(x) = 2x$, we compute the both sequences of polynomials $\{T_n(x)\}_{n=0}^{+\infty}$ and $\{U_n(x)\}_{n=0}^{+\infty}$ very easily. For example, for $n = 0, 1, \ldots, 6$ we get

$T_0(x) = 1$, $U_0(x) = 1$,
$T_1(x) = x$, $U_1(x) = 2x$,
$T_2(x) = 2x^2 - 1$, $U_2(x) = 4x^2 - 1$,
$T_3(x) = 4x^3 - 3x$, $U_3(x) = 8x^3 - 4x$,
$T_4(x) = 8x^4 - 8x^2 + 1$, $U_4(x) = 16x^4 - 12x^2 + 1$,
$T_5(x) = 16x^5 - 20x^3 + 5x$, $U_5(x) = 32x^5 - 32x^3 + 6x$,
$T_6(x) = 32x^6 - 48x^4 + 18x^2 - 1$, $U_6(x) = 64x^6 - 80x^4 + 24x^2 - 1$.

The relation (1.1.15) is known as the *three-term recurrence relation*.

Since $x = \cos\theta$ and $\sqrt{1-x^2} = \sin\theta$, using

$$\cos n\theta = \frac{1}{2}\left(e^{in\theta} + e^{-in\theta}\right) = \frac{1}{2}\left[(\cos\theta + i\sin\theta)^n + (\cos\theta - i\sin\theta)^n\right],$$

---

[3] It was derived from another transliteration of the name Chebyshev in the form Tchebychev or related forms.

we can get the following expressions for all complex x

$$T_n(x) = \frac{1}{2}\left(\varrho^n + \varrho^{-n}\right), \qquad U_n(x) = \frac{\varrho^{n+1} - \varrho^{-n-1}}{\varrho - \varrho^{-1}}, \qquad (1.1.16)$$

where we put

$$\varrho = x + \sqrt{x^2 - 1} \qquad (x \in \mathbb{C}). \qquad (1.1.17)$$

The square root in (1.1.17) is such that $|x + \sqrt{x^2 - 1}| > 1$ whenever $x \in \mathbb{C} \setminus [-1, 1]$. For $x \in [-1, 1]$ these formulas reduce to (1.1.14).

In the general case, the explicit expressions for the Chebyshev polynomials of the first and second kind are

$$T_n(x) = \frac{n}{2} \sum_{k=0}^{[n/2]} \frac{(-1)^k (n-k-1)!}{k!(n-2k)!} (2x)^{n-2k} \qquad (n \geq 1)$$

and

$$U_n(x) = \sum_{k=0}^{[n/2]} \frac{(-1)^k (n-k)!}{k!(n-2k)!} (2x)^{n-2k},$$

respectively.

The Chebyshev polynomials $T_n(x)$ for $n = 0, 1, \ldots, 6$ are displayed in Fig. 1.1.1. Since $T_n(x) = \cos(n \arccos x)$ for $-1 \leq x \leq 1$, it is easy to see that $|T_n(x)| \leq 1$ for each $n \geq 0$ and $-1 \leq x \leq 1$ (see Fig. 1.1.1). Also, we have

$$|U_n(x)| \leq n + 1 \qquad \text{and} \qquad |\sqrt{1-x^2}\, U_n(x)| \leq 1$$

for $-1 \leq x \leq 1$. Some interesting values are:

$$T_n(\pm 1) = (\pm 1)^n, \qquad T_{2n}(0) = (-1)^n, \qquad T_{2n+1}(0) = 0,$$

$$T_n'(\pm 1) = (\pm 1)^n n^2, \qquad T_n^{(k)}(1) = \frac{n^2(n^2-1)\cdots(n^2-(k-1)^2)}{(2k-1)!!}.$$

1.1.4.2 Differential Equation

Differentiating $y = \cos(n \arccos x)$ we obtain the *differential equation of the Chebyshev polynomials*

$$(1-x^2)y'' - xy' + n^2 y = 0.$$

The second particular solution of this equation, $S_n(x) = \sin(n \arccos x)$ $(-1 \leq x \leq 1)$ can be expressed in terms of the Chebyshev polynomials of the second kind.

1.1 Introduction to Approximation Theory

Fig. 1.1.1 The graphs of $y = T_n(x)$ for $n = 0, 1, \ldots, 6$ and $-1 \le x \le 1$

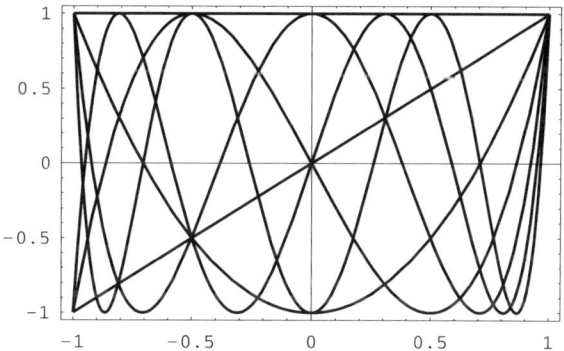

Namely, $S_n(x) = U_{n-1}(x)\sqrt{1-x^2}$. The corresponding *differential equation of the Chebyshev polynomials of the second kind* is

$$(1-x^2)y'' - 3xy' + n(n+2)y = 0.$$

1.1.4.3 Zeros and Extremal Points

According to (1.1.14) the *zeros of $T_n(x)$* can be expressed in an explicit form,

$$x_k = x_{n,k} = \cos\frac{(2k-1)\pi}{2n} \qquad (k = 1, \ldots, n). \tag{1.1.18}$$

The zeros x_k are real, distinct, and lie in $(-1, 1)$. In order to give a geometric interpretation of the zeros, we put $\theta_k = (2k-1)\pi/(2n)$, $k = 1, \ldots, n$. Now, it is clear that the zeros x_k are the projections onto the real line of the equally spaced points $\exp(i\theta_k)$ on the upper arc of the unit circle. Thus, the zeros x_k are more densely distributed around the endpoints ± 1 than in the interior of $(-1, 1)$. Precisely, using an idea from the theory of probability, we can introduce the *distribution* of zeros of the polynomials $T_n(x)$ when n tends to infinity, the so-called *limit distribution*.

Let $N_n(a, b)$ be the number of zeros of $T_n(x)$ in $[a, b] \subset [-1, 1]$, i.e., $N_n(a, b) = \{m \in \mathbb{Z}_n \mid a \le \cos\theta_k < b\}$, where $\mathbb{Z}_n = \{1, \ldots, n\}$. Then, the corresponding *density function* at the point $x \in (-1, 1)$ is given by

$$\psi(x) = \lim_{h \to 0}\frac{1}{h}\lim_{n \to +\infty}\frac{N_n(x, x+h)}{n}.$$

It is not difficult to see that $N_n(x, x+h) - (n/\pi)(\arccos x - \arccos(x+h))$ is equal to 0 or -1, so that

$$\lim_{n \to +\infty}\frac{N_n(x, x+h)}{n} = \frac{\arccos x - \arccos(x+h)}{\pi}.$$

This means that we have the so-called *arc sine distribution*, because the density function is given by

$$\psi(x) = \frac{1}{\pi} \frac{1}{\sqrt{1-x^2}} \quad (-1 < x < 1).$$

Thus, the limit distribution of zeros of the Chebyshev polynomials is $d\mu(x) = \psi(x)\,dx = \pi^{-1}(1-x^2)^{-1/2}\,dx$.

In order to ensure that the zeros are in increasing order, we often change k to $n-k+1$ in (1.1.18) so that

$$x_k = x_{n,k} = -\cos\frac{(2k-1)\pi}{2n} \quad (k=1,\ldots,n) \tag{1.1.19}$$

and $-1 < x_1 < x_2 < \cdots < x_n < 1$.

Other interesting points are the points where $T_n(x) = \pm 1$,

$$\xi_k = \xi_{n,k} = -\cos\frac{k\pi}{n} \quad (k=0,1,\ldots,n). \tag{1.1.20}$$

The points (1.1.20) are the *extremal points of $T_n(x)$*. Their limit distribution is also the arc sine distribution on $[-1, 1]$.

1.1.4.4 Chebyshev Polynomials in the Complex Plane

In order to investigate $T_n(z)$ for a complex z outside the interval $[-1, 1]$, we need the *Joukowski transformation*

$$z = \frac{1}{2}\left(w + \frac{1}{w}\right), \tag{1.1.21}$$

which maps $|w| > 1$ onto $\mathbb{C} \setminus [-1, 1]$ and maps the unit circle $|w| = 1$ onto the interval $[-1, 1]$. Taking $w = re^{i\theta}$ and $z = x + iy$, we get

$$x + iy = \frac{1}{2}\left(re^{i\theta} + \frac{1}{r}e^{-i\theta}\right) = \frac{1}{2}\left(r + \frac{1}{r}\right)\cos\theta + i\frac{1}{2}\left(r - \frac{1}{r}\right)\sin\theta,$$

i.e.,

$$x = \frac{1}{2}\left(r + \frac{1}{r}\right)\cos\theta, \qquad y = \frac{1}{2}\left(r - \frac{1}{r}\right)\sin\theta.$$

For a constant $r > 1$, these equations describe an ellipse E_r: $(x/a)^2 + (y/b)^2 = 1$, with semi-axes

$$a = \frac{1}{2}\left(r + \frac{1}{r}\right), \qquad b = \frac{1}{2}\left(r - \frac{1}{r}\right)$$

and foci ± 1 ($a^2 - b^2 = 1$). For $r = 1$, E_r reduces to the interval $[-1, 1]$.

1.1 Introduction to Approximation Theory

Thus, for a given $z \in \mathbb{C} \setminus [-1, 1]$ there exists exactly one ellipse E_r $(r > 1)$ passing through z, where r is determined by

$$r + \frac{1}{r} = 2a = |z+1| + |z-1|.$$

According to (1.1.16) and (1.1.21) we have

$$T_n(z) = \frac{1}{2}\left[\left(z + \sqrt{z^2-1}\right)^n + \left(z - \sqrt{z^2-1}\right)^n\right], \qquad (1.1.22)$$

where

$$z = \frac{1}{2}\left(w + w^{-1}\right) = \frac{1}{2}\left(re^{i\theta} + \frac{1}{r}e^{-i\theta}\right),$$

i.e., $z + \sqrt{z^2-1} = re^{i\theta}$, $z - \sqrt{z^2-1} = r^{-1}e^{-i\theta}$. Here, $|z + \sqrt{z^2-1}| = r > 1$ whenever $z \in \mathbb{C} \setminus [-1, 1]$.

Using (1.1.22) we find

$$|T_n(z)| = \frac{1}{2}\sqrt{r^{2n} + 2\cos 2n\theta + \frac{1}{r^{2n}}}, \qquad (1.1.23)$$

as well as

$$\operatorname{Re} T_n(z) = \frac{1}{2}\left(r^n + \frac{1}{r^n}\right)\cos n\theta, \qquad \operatorname{Im} T_n(z) = \frac{1}{2}\left(r^n - \frac{1}{r^n}\right)\sin n\theta.$$

As we know, the zeros of T_n are all inside $(-1, 1)$, but it is interesting to mention that for a fixed z, arbitrarily close to the interval $[-1, 1]$, the sequence $\{|T_n(z)|\}_{n \in \mathbb{N}}$ tends to infinity with geometric rate. Indeed, from (1.1.23) we find

$$\lim_{n \to +\infty} |T_n(z)|^{1/n} = r \qquad (z \in \mathbb{C} \setminus [-1, 1]). \qquad (1.1.24)$$

Based on the above, (1.1.24) holds for each $z \in E_r$ $(r > 1)$.

1.1.4.5 Some Other Relations

It is easy to prove the following relations:

$$U_n(x) = \frac{1}{n+1}T'_{n+1}(x) = \frac{1}{1-x^2}\left(xT_{n+1}(x) - T_{n+2}(x)\right),$$

$$T_n(x) = U_n(x) - xU_{n-1}(x) = xU_{n-1}(x) - U_{n-2}(x),$$

$$U_n(x) = \frac{1}{2(n+1)}\left(U'_{n+1}(x) - U'_{n-1}(x)\right),$$

$$U'_n(x) = \frac{1}{1-x^2}\left((n+1)U_{n-1}(x) - nxU_n(x)\right)$$

Also, for $2 \leq k \leq n$ we can check
$$T_n(x) = T_k(x)U_{n-k}(x) - T_{k-1}(x)U_{n-k-1}(x).$$
Indeed, using (1.1.16), the right hand side in this equality reduces to
$$\frac{1}{2}\bigl(U_n(x) - U_{n-2}(x)\bigr) = xU_{n-1}(x) - U_{n-2}(x) = T_n(x).$$

1.1.4.6 Orthogonality

The sequences $\{T_n(x)\}_{n=0}^{+\infty}$ and $\{U_n(x)\}_{n=0}^{+\infty}$ are *orthogonal* on $(-1, 1)$ in the following sense
$$\int_{-1}^{1} T_n(x)T_m(x)\frac{dx}{\sqrt{1-x^2}} = 0, \quad \int_{-1}^{1} U_n(x)U_m(x)\sqrt{1-x^2}\,dx = 0 \quad (n \neq m).$$

They are special cases of the so-called Jacobi polynomials $\{P_n^{(\alpha,\beta)}(x)\}_{n=0}^{+\infty}$, with parameters $\alpha, \beta > -1$. The Chebyshev polynomials of the first and second kind correspond to parameters $\alpha = \beta = -1/2$ and $\alpha = \beta = 1/2$, respectively. The Jacobi polynomials and other orthogonal polynomials will be treated in Chap. 2.

1.1.5 Chebyshev Extremal Problems

We start this subsection by considering the following extremal problem: *Among all polynomials of degree n, with leading coefficient unity, find the polynomial which deviates least from zero in a given norm* $\|\cdot\|$.

1.1.5.1 The Extremal Problem in the Uniform Norm

As we mentioned before, Chebyshev [58] solved the previous problem in the uniform norm $\|f\|_{[-1,1]} = \max\limits_{-1 \leq x \leq 1}|f(x)|$.

Let $\hat{T}_n(x)$ be the monic Chebyshev polynomial of the first kind of degree n, i.e.,
$$\hat{T}_0(x) = T_0(x) = 1, \qquad \hat{T}_n(x) = \frac{1}{2^{n-1}}T_n(x) \quad (k = 1, 2, \ldots),$$
where $T_n(x) = \cos(n \arccos x)$ for $-1 \leq x \leq 1$.

Theorem 1.1.7 *Let* $q(x) = \sum\limits_{\nu=0}^{n} a_\nu x^\nu$, *with* $a_n = 1$, *be an arbitrary monic polynomial of degree n. Then*
$$\|q\|_{[-1,1]} \geq \|\hat{T}_n\|_{[-1,1]} = \begin{cases} 2^{1-n}, & n > 0, \\ 1, & n = 0, \end{cases} \tag{1.1.25}$$

with equality only if $q(x) = \hat{Q}_n^*(x) = \hat{T}_n(x)$.

1.1 Introduction to Approximation Theory

Proof Replacing n by $n+1$ in (1.1.13) we see that for $n = 0$, the statement (1.1.25) is trivial. Therefore, we consider the case $n > 0$.

First, we note that the extremal points of $T_n(x)$, given by (1.1.20), are ordered, i.e.,

$$-1 = \xi_0 < \xi_1 < \cdots < \xi_n = 1.$$

Let $r(x) = \hat{T}_n(x) - q(x)$ be a polynomial of degree at most $n-1$. In order to prove (1.1.25) we suppose on the contrary that

$$Q = \|q\|_{[-1,1]} < \|\hat{T}_n\|_{[-1,1]} = 2^{1-n}. \tag{1.1.26}$$

Let $x \in [-1, 1]$. According to (1.1.26) we have

$$-2^{1-n} - q(x) \leq -2^{1-n} + Q < 0,$$
$$2^{1-n} - q(x) \geq 2^{1-n} - Q > 0,$$

from which we conclude that the polynomial $r(x)$ has alternatively positive and negative values at the extremal points ξ_k ($k = 0, 1, \ldots, n$), given by (1.1.20). Therefore, this polynomial must have at least n zeros, which is a contradiction, because $r \in \mathcal{P}_{n-1}$. □

From this important theorem we conclude (cf. Rivlin [414, 415]):

(a) The polynomial $P^* \in \mathcal{P}_{n-1}$ closest to the power function $x \mapsto f(x) = x^n$, where closeness is measured by $\|f - p\|_{[-1,1]}$ ($p \in \mathcal{P}_{n-1}$), is

$$P^*(x) = x^n - \hat{T}_n(x).$$

(b) Let $P(x) = \sum_{\nu=0}^{n} a_\nu x^\nu$ be an arbitrary algebraic polynomial of degree n and let $F: \mathcal{P}_n \to \mathbb{R}$ be a linear functional defined by

$$F(P) = a_n = \frac{1}{n!} P^{(n)}(0).$$

Among all $P \in \mathcal{P}_n$ satisfying $\|P\|_{[-1,1]} = 1$, the largest value of $|F(P)|$ is 2^{n-1}, and this value is attained only for $P(x) = \pm T_n(t)$. This fact can be expressed as an inequality for the leading coefficient of the polynomial $P \in \mathcal{P}_n$, i.e.,

$$|a_n| \leq 2^{n-1} \|P\|_{[-1,1]}. \tag{1.1.27}$$

This is the well-known *Chebyshev inequality*.

Remark 1.1.3 For polynomials $P \in \mathcal{P}_n$, van der Corput and Visser [487] proved that

$$\max_{-1 \leq t \leq 1} |P(t)|^2 - \min_{-1 \leq t \leq 1} |P(t)|^2 \geq \frac{|a_n|^2}{2^{2n-2}}.$$

This inequality contains the Chebyshev inequality (1.1.27).

It is interesting to mention that the Chebyshev polynomial $T_n(x)$ has also an extremal property in the following sense:

Theorem 1.1.8 *If $P \in \mathcal{P}_n$ such that $|P(x)| \le 1$ for $-1 \le x \le 1$, then*

$$|P(x)| \le |T_n(x)| \quad for \quad |x| \ge 1.$$

Consequently, the Chebyshev polynomial $T_n(x)$ is the fastest growing polynomial outside $[-1, 1]$ among all polynomials of degree n, with $\|P\|_{[-1,1]} \le 1$. More generally, we have (see Rivlin [415, p. 93]):

Theorem 1.1.9 *If $P \in \mathcal{P}_n$ and $\max\limits_{0 \le k \le n} |P(\xi_k)| \le 1$, where ξ_k are extremal points defined by (1.1.20), then for $m = 0, 1, \ldots, n$,*

$$|P^{(m)}(x)| \le |T_n^{(m)}(x)| \quad for \quad |x| \ge 1. \tag{1.1.28}$$

Equality is possible in (1.1.28) for $m \ge 1$ and $|x| > 1$ if and only if $P(x) = \pm T_n(x)$.

Further results in this direction can be found in [359, Chap. 5].

1.1.5.2 The Extremal Problem in L^1-norm

An analogous result to (1.1.25) in L^1-norm,

$$\|f\|_1 = \int_{-1}^{1} |f(x)| \, dx, \tag{1.1.29}$$

was proved by Korkin and Zolotarev [234]:

Theorem 1.1.10 *Let $q(x) = \sum\limits_{\nu=0}^{n} a_\nu x^\nu$, with $a_n = 1$, be an arbitrary monic polynomial of degree n. Then*

$$\|q\|_1 \ge \|\hat{U}_n\|_1 = 2^{1-n},$$

with equality only if $q(x) = \hat{U}_n(x)$, where $\hat{U}_n$ is the monic Chebyshev polynomial of the second kind of degree n, i.e., $\hat{U}_n(x) = 2^{-n} U_n(x)$.

Proof Using the norm (1.1.29), we define the functional $J \colon \mathcal{P}_n \to \mathbb{R}$ by

$$J(q) = \|q\|_1 = \int_{-1}^{1} |q(x)| \, dx.$$

Since

$$\int_{-1}^{1} x^k \operatorname{sgn} U_n(x) \, dx = \begin{cases} 0, & 0 \le k \le n-1, \\ 2^{1-n}, & k = n \end{cases}$$

(see [359, pp. 408–409]), for the monic polynomial $q(x)$, we have

$$\int_{-1}^{1} q(x) \operatorname{sgn} U_n(x)\, dx = \frac{1}{2^{n-1}},$$

from which it follows that

$$J(q) = \int_{-1}^{1} |q(x)|\, dx \geq \frac{1}{2^{n-1}}.$$

Also, for the monic polynomial $\hat{U}_n(x) = 2^{-n} U_n(x)$, we have

$$\int_{-1}^{1} |\hat{U}_n(x)|\, dx = \int_{-1}^{1} \frac{1}{2^n} U_n(x) \operatorname{sgn} U_n(x)\, dx = \frac{1}{2^{n-1}}.$$

Thus, the polynomial $\hat{U}_n(x)$ minimizes the functional $J(q) = \|q\|_1$.

It remains to prove that this is the only polynomial which minimizes the functional J. For this let us suppose that there exists another monic polynomial, say $R(x)$, of degree n, such that $J(R) = 2^{1-n}$. Then, it follows that $R(x) \equiv \hat{U}_n(x)$, which is a contradiction. $\square$

Similar extremal problems can be considered also in other norms (cf. [359, Chap. 5]).

1.1.6 Chebyshev Alternation Theorem

Let $f \in C[a,b]$. As before, we put $\|f\| = \|f\|_\infty = \|f\|_{[a,b]} = \max_{a \leq x \leq b} |f(x)|$ and

$$E_n(f) = \min_{p \in \mathcal{P}_n} \|f - p\| = \|f - P^*\|. \tag{1.1.30}$$

The polynomial $P^*(x)$ of best uniform approximation can be characterized by the *alternation theorem*, which was proved by Chebyshev [58] in the case when f is a differentiable function. A complete proof of this important theorem was given independently by Blichfeldt [42] and Kirchberger [229] at the beginning of the 20th century.

Theorem 1.1.11 *If $f \in C[a,b]$, then $P^* \in \mathcal{P}_n$ is the polynomial of best uniform approximation to f if and only if there exist $n+2$ points $x_0, x_1, \ldots, x_{n+1}$ ($a \leq x_0 < x_1 < \cdots < x_{n+1} \leq b$) such that*

$$f(x_k) - P^*(x_k) = \varepsilon(-1)^k \|f - P^*\| \quad (k = 0, 1, \ldots, n+1), \tag{1.1.31}$$

where the constant ε is $+1$ or -1.

We say that P^* realizes the *Chebyshev alternation* for f in $[a, b]$ if the conditions (1.1.31) are satisfied. The points x_k in (1.1.31) are called *alternation points* for the approximating polynomial (approximant) P^*. Theorem 1.1.11, which is sometimes called as the *equioscillation theorem*, holds also for an arbitrary real Haar space (see Definition 1.3.1 in Sect. 1.3.1), taking it instead of $\mathcal{P}_n$. The proof of this theorem (in original or generalized form) can be found in several books (cf. DeVore and Lorentz [95], Feinerman and Newman [124], Korneichuk [235], Meinardus [315], Petrushev and Popov [397]).

As we mentioned before, the uniqueness of best uniform approximation is a consequence of the Chebyshev alternation theorem.

Theorem 1.1.12 *The polynomial $P^* \in \mathcal{P}_n$ of best uniform approximation to a given function $f \in C[a, b]$ is unique.*

Proof Assume that for a function $f \in C[a, b]$ there are two polynomials of best approximation in $\mathcal{P}_n$, P^* and $\tilde{P}^*$:

$$\|f - P^*\| = \|f - \tilde{P}^*\| = E_n(f).$$

Then $Q^* = (P^* + \tilde{P}^*)/2$ ($\in \mathcal{P}_n$) is also a polynomial of best uniform approximation for f. This means that there exist $n + 2$ points x_k in $[a, b]$ (see Theorem 1.1.11) such that

$$f(x_k) - Q^*(x_k) = \frac{1}{2}[(f(x_k) - P^*(x_k)) + (f(x_k) - \tilde{P}^*(x_k))] = \varepsilon(-1)^k E_n(f)$$

for $k = 0, 1, \ldots, n + 1$, where $\varepsilon = \pm 1$. Since

$$|f(x_k) - P^*(x_k)| \leq E_n(f) \quad \text{and} \quad |f(x_k) - \tilde{P}^*(x_k)| \leq E_n(f),$$

the previous equality can be fulfilled only if

$$f(x_k) - P^*(x_k) = f(x_k) - \tilde{P}^*(x_k) \qquad (k = 0, 1, \ldots, n + 1),$$

i.e., if $P^*(x_k) = \tilde{P}^*(x_k)$ for each $k = 0, 1, \ldots, n + 1$. This means that $P^* \equiv \tilde{P}^*$, i.e., the polynomial of best uniform approximation for a continuous function on $[a, b]$ is unique. □

Another useful application of the Chebyshev alternation theorem is the following lower estimate of $E_n(f)$, obtained by de la Vallée Poussin [85, p. 85].

Theorem 1.1.13 *Let $f \in C[a, b]$, $P \in \mathcal{P}_n$, and $n + 2$ points $x_0, x_1, \ldots, x_{n+1}$ be such that $a \leq x_0 < x_1 < \cdots < x_{n+1} \leq b$. If $f(x_k) - P(x_k) = \varepsilon(-1)^k \mu_k$, with $\varepsilon = \pm 1$ and $\mu_k > 0$, $k = 0, 1, \ldots, n + 1$, then*

$$E_n(f) \geq \mu = \min_{0 \leq k \leq n+1} \mu_k.$$

1.1 Introduction to Approximation Theory

Proof Let $P^* \in \mathcal{P}_n$ be the polynomial of best uniform approximation to f and suppose that $\|f - P^*\| = E_n(f) < \mu$. Then, for the polynomial $Q(x) = P(x) - P^*(x)$ ($Q \in \mathcal{P}_n$) we must have

$$\text{sgn } Q(x_k) = \text{sgn}\left[\left(f(x_k) - P^*(x_k)\right) - \left(f(x_k) - P(x_k)\right)\right] = \varepsilon(-1)^k,$$

which means that the polynomial $Q(x)$ has alternate signs at $n+2$ points x_k, $k = 0, 1, \ldots, n+1$. Thus, $Q(x)$ must have at least $n+1$ different zeros in $[a, b]$, which is a contradiction, because $Q \in \mathcal{P}_n$. Therefore, $E_n(f) \geq \mu$. □

This theorem is very useful in numerical methods for finding the polynomial of best uniform approximation.

1.1.6.1 Some Classical Special Cases

We consider now two special cases when the polynomial of the best approximation and the corresponding quantity $E_n(f)$ can be determined in an explicit form.

1. The first one is the well-known Chebyshev example $f(x) = x^{n+1}$ on $[-1, 1]$. Taking the extremal points of $T_{n+1}(x)$ as the alternation points x_k in (1.1.31), i.e.,

$$x_k = \xi_k^{(n+1)} = -\cos\frac{k\pi}{n+1} \quad (k = 0, 1, \ldots, n+1), \tag{1.1.32}$$

according to Theorems 1.1.11 and 1.1.12, we must have $x^{n+1} - P^*(x) = \hat{T}_{n+1}(x)$, i.e.,

$$P^*(x) = x^{n+1} - \frac{1}{2^n} T_{n+1}(x) \quad \text{and} \quad E_n(x^{n+1}) = \frac{1}{2^n}.$$

Thus, we have just obtained the Chebyshev result given by Theorem 1.1.7.

2. The second example is best uniform approximation of the function

$$f(x) = \frac{1}{x - a} \quad (a > 1)$$

on the interval $[-1, 1]$ by polynomials in $\mathcal{P}_n$. It was first solved by Chebyshev. This problem was also investigated by Bernstein [35, pp. 120–121], who gave the trigonometric representation

$$\frac{1}{x - a} - P^*(x) = \frac{(a - \sqrt{a^2 - 1})^n}{a^2 - 1} \cos(n\varphi + \delta),$$

where $x = \cos\varphi$ and $(ax - 1)/(x - a) = \cos\delta$. It is clear that

$$E_n\left(\frac{1}{x - a}\right) = \frac{(a - \sqrt{a^2 - 1})^n}{a^2 - 1}.$$

A more transparent solution was given by Ahieser [6, pp. 69–71] in the following form:

$$\frac{1}{x-a} - P^*(x) = \frac{M}{2}\left\{v^n \frac{\alpha-v}{1-\alpha v} + v^{-n}\frac{1-\alpha v}{\alpha-v}\right\},$$

where

$$x = \frac{1}{2}\left(v + \frac{1}{v}\right), \quad a = \frac{1}{2}\left(\alpha + \frac{1}{\alpha}\right) \quad (|v|=1,\ |\alpha|<1)$$

and

$$M = E_n\left(\frac{1}{x-a}\right) = \frac{4\alpha^{n+2}}{(1-\alpha^2)^2} = \frac{(a-\sqrt{a^2-1})^n}{a^2-1},$$

because $\alpha = a - \sqrt{a^2-1}$. Some details of the solution can be found in [315, pp. 34–36].

1.1.7 Numerical Methods

Several methods for numerically computing the best uniform polynomial approximation to a given continuous function on $[a,b]$ are described in Meinardus [315, pp. 105–130]. In this subsection we mention only the so-called Remez algorithms [408, 409], which can be applied also for generalized cases of arbitrary real Haar subspaces (see Sect. 1.3.1), instead of $\mathcal{P}_n$. Precisely, we will describe a variant of these algorithms which is known as the *second Remez algorithm*.

Without loss of generality, we consider the approximation problem for a continuous function on $[-1,1]$. For a such function $f \in C[-1,1]$ and an algebraic polynomial $P \in \mathcal{P}_n$, i.e., $P(x) = \sum_{v=0}^{n} a_v x^v$, we put

$$\delta_n(x) = \delta_n(x;\mathbf{a}) = f(x) - \sum_{v=0}^{n} a_v x^v,$$

where $\mathbf{a} = (a_0, a_1, \ldots, a_n) \in \mathbb{R}^{n+1}$. According to the Chebyshev alternation theorem, we formulate the following algorithm:

1° *Start with $n+2$ selected points $\{x_k\}_{k=0}^{n+1}$, such that*

$$-1 \le x_0 < x_1 < \cdots < x_{n+1} \le 1;$$

2° *Find $a_0, a_1, \ldots, a_n, E$ from the linear system of equations*

$$\delta_n(x_k;\mathbf{a}) = f(x_k) - \sum_{v=0}^{n} a_v x_k^v = (-1)^k E \quad (k=0,1,\ldots,n+1); \quad (1.1.33)$$

1.1 Introduction to Approximation Theory

Because of (1.1.33) in each of the intervals $[x_k, x_{k+1}]$ the function $x \mapsto \delta_n(x)$ possesses at least one zero z_k ($x_k < z_k < x_{k+1}$).

3° *Determine the points* $\{z_k\}_{k=-1}^{n+1}$ *such that*

$$z_{-1} = -1, \quad \delta_n(z_k) = 0 \ (k = 0, 1, \ldots, n), \quad z_{n+1} = 1;$$

4° *Select the points* $\hat{x}_k \in [z_{k-1}, z_k]$, $k = 0, 1, \ldots, n+1$, *such that*

$$(\operatorname{sgn} \delta_n(x_k))\delta_n(\hat{x}_k) = \max_{z_{k-1} \leq x \leq z_k} \left\{ \operatorname{sgn} \delta_n(x_k)\delta_n(x) \right\};$$

5° *If* $\|\delta_n(\cdot, \mathbf{a})\| > \max\limits_{0 \leq k \leq n+1} |\delta_n(\hat{x}_k; \mathbf{a})|$ *then there exists a point* $\hat{x} \in [-1, 1]$ *such that* $|\delta_n(\hat{x}; \mathbf{a})| = \|\delta_n(\cdot, \mathbf{a})\|$. *In that case, put the point* $\hat{x}$ *in place of some point in* $\{\hat{x}_k\}_{k=0}^{n+1}$ *so that the function* $x \mapsto \delta_n(x, \mathbf{a})$ *would preserve the alternating signs on the new set of points* $\{\hat{x}_k\}_{k=0}^{n+1}$ *obtained in this way;*

6° *For a given tolerance* ε, *check the condition* $|\|\delta_n(\cdot; \mathbf{a})\| - |E|| < \varepsilon$. *If this condition is satisfied then stop; otherwise, put* $x_k := \hat{x}_k$ ($k = 0, 1, \ldots, n+1$) *and go to* 2°.

Thus, as the best polynomial approximation to a given continuous function we take the algebraic polynomial which satisfies the "tolerance condition" in the last step of the algorithm. The algorithm is not essentially sensitive to the choice of the initial points $\{x_k\}_{k=0}^{n+1}$. Very often it is convenient to take the extremal points of $T_{n+1}(x)$, i.e., (1.1.32).

A modified version of the second Remez algorithm for polynomial approximation of differentiable functions was given by Murnaghan and Wrench [372] (see also [371]).

The linear convergence of the second Remez algorithm can be proved for each continuous function (cf. [397, pp. 13–15]):

Theorem 1.1.14 *Let* $f \in C[-1, 1]$ *and* $P^* \in \mathcal{P}_n$ *be its polynomial of best uniform approximation. The polynomial* P_ν ($\in \mathcal{P}_n$) *generated at the* ν-*th step by the second Remez algorithm satisfies the condition*

$$\|P_\nu - P^*\| \leq C\varrho^\nu,$$

where $0 < \varrho < 1$ *and* C *is a constant independent of* ν.

Under some restrictions on the smoothness of the function f it is possible to prove the quadratic convergence of this algorithm (cf. Meinardus [315, pp. 111–113]).

In order to illustrate this Remez algorithm we give two examples.

Example 1.1.2 Consider a continuous function defined on $[-1, 1]$ by

$$f(x) := \sqrt{3 + 2x + 4x^2}.$$

Table 1.1.1 The polynomial coefficients and the quantity $E^{(\nu)}$ in the second Remez algorithm

ν	$a_0^{(\nu)}$	$a_1^{(\nu)}$	$a_2^{(\nu)}$	$a_3^{(\nu)}$	$E^{(\nu)}$
1	1.7510913738	0.5217341087	0.8859831812	−0.1397680974	−0.0190405662
2	1.7494089696	0.5261703529	0.8893792012	−0.1442043415	−0.0207541821
3	1.7494113495	0.5261479622	0.8893942621	−0.1441819510	−0.0207716229
4	1.7494113492	0.5261479629	0.8893942627	−0.1441819516	−0.0207716232

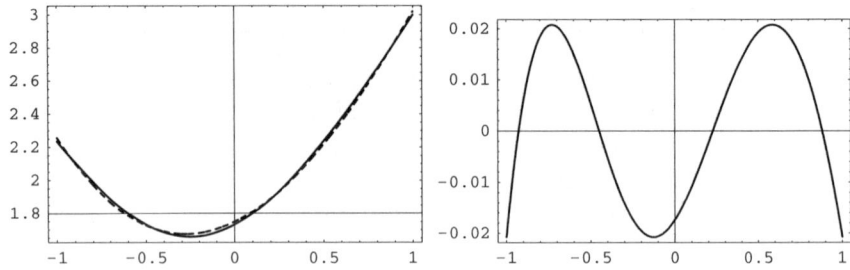

Fig. 1.1.2 The graphics of $x \mapsto f(x) = \sqrt{3 + 2x + 4x^2}$ (*solid line*) and $x \mapsto P^*(x)$ (*dashed line*) for $n = 3$ (*left*) and the deviation $\delta_3^*(x) = f(x) - P^*(x)$ (*right*)

For this simple function we want to find its best polynomial approximation in the set of all polynomials of degree at most three. Thus, in the ν-th iteration we have
$\delta_3^{(\nu)}(x) = f(x) - (a_0^{(\nu)} + a_1^{(\nu)}x + a_2^{(\nu)}x^2 + a_3^{(\nu)}x^3)$.

Starting from the extremal points of $T_4(x)$, i.e., $\{-1, -\sqrt{2}/2, 0, \sqrt{2}/2, 1\}$, the second Remez algorithm generates the sequences of polynomial coefficients, given in Table 1.1.1. Also, the corresponding quantity $E^{(\nu)}$ is presented in the last column of this table.

Thus, as an approximate solution (rounded to 8 decimals) we can take

$$P^*(x) = 1.74941135 + 0.52614796x + 0.88939426x^2 - 0.14418195x^3.$$

The alternation points are: $x_0 = 1$, $x_1 = -0.72898482$, $x_2 = -0.12747162$, $x_3 = 0.58607094$, $x_4 = 1$, and

$$E_3(f) = \|f - P^*\| \approx 2.077 \times 10^{-2}.$$

The corresponding graphics are displayed in Fig. 1.1.2.

Example 1.1.3 Consider now the function defined by $f(x) := \sqrt{|\sin(\pi x/2)|}$ on $[-1, 1]$. For its best polynomial approximation in the set $\mathcal{P}_6$ we get

Fig. 1.1.3 The graphics of functions
$x \mapsto f(x) = \sqrt{|\sin \pi x/2|}$
and $x \mapsto P^*(x)$ for $n = 6$
(*dashed line*) and $n = 12$
(*solid line*)

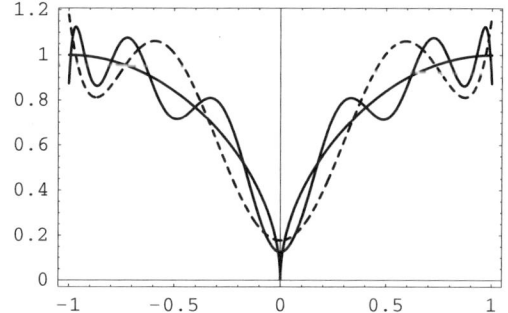

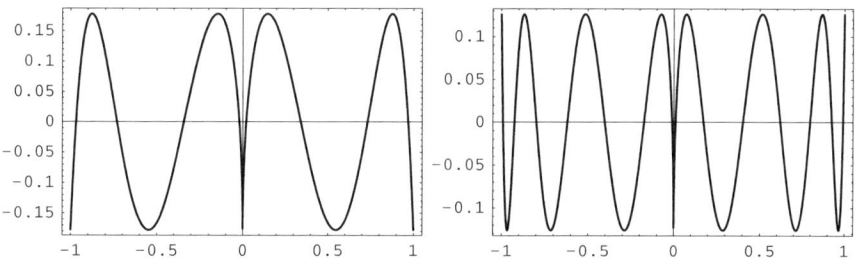

Fig. 1.1.4 The deviation $\delta_n^*(x) = f(x) - P^*(x)$ for $n = 6$ (*left*) and $n = 12$ (*right*)

$$P^*(x) = 0.17787718 + 5.93869742x^2 - 12.34268353x^4 + 7.40398610x^6$$

and $E_6(f) = \|f - P^*\| \approx 0.1779$. Similarly, in the set $\mathcal{P}_{12}$ we obtain

$$P^*(x) = 0.126045547094028 + 16.9552165863717x^2 - 149.188517382840x^4$$
$$+ 573.700153068282x^6 - 1048.54708556994x^8$$
$$+ 901.727265823760x^{10} - 293.899123618993x^{12},$$

with $E_{12}(f) \approx 0.1260$. The corresponding graphics are presented in Figs. 1.1.3 and 1.1.4.

Remark 1.1.4 Some details on the *first Remez algorithm* can be found in [397, pp. 10–12]. Several modifications of this algorithm for solving linear and nonlinear Chebyshev approximation problems on compact $B \subset \mathbb{R}^s$, as well as their convergence, were studied by Reemtsen [407].

1.2 Basic Facts on Trigonometric Approximation

1.2.1 Trigonometric Kernels

First, we introduce two very important trigonometric sums

$$D_n(\theta) = \frac{1}{2} + \sum_{k=1}^{n} \cos k\theta \tag{1.2.1}$$

and

$$F_n(\theta) = \frac{1}{n+1} \sum_{k=0}^{n} D_k(\theta), \tag{1.2.2}$$

which are known as the *Dirichlet kernel* and the *Fejér kernel*, respectively.

As we can see the Dirichlet and Fejér kernels are even trigonometric polynomials of degree n. A simple form of these kernels can be found by using the following

Proposition 1.2.1 *For each $\theta \neq 2\nu\pi$ ($\nu \in \mathbb{Z}$) and $n \in \mathbb{N}$ we have*

$$\sum_{k=0}^{n} \cos k\theta = \frac{\sin((n+1)\theta/2) \cos(n\theta/2)}{\sin(\theta/2)}, \tag{1.2.3}$$

$$\sum_{k=1}^{n} \sin k\theta = \frac{\sin((n+1)\theta/2) \sin(n\theta/2)}{\sin(\theta/2)}, \tag{1.2.4}$$

$$\sum_{k=0}^{n} \sin(2k+1)\frac{\theta}{2} = \frac{\sin^2((n+1)\theta/2)}{\sin(\theta/2)}. \tag{1.2.5}$$

Proof Putting $z = e^{i\theta}$ in

$$1 + z + z^2 + \cdots + z^n = \frac{z^{n+1} - 1}{z - 1} \quad (z \neq 1),$$

we find

$$\sum_{k=0}^{n} e^{ik\theta} = \frac{e^{i(n+1)\theta} - 1}{e^{i\theta} - 1} = e^{in\theta/2} \frac{e^{i(n+1)\theta/2} - e^{-i(n+1)\theta/2}}{e^{i\theta/2} - e^{-i\theta/2}},$$

i.e.,

$$\sum_{k=0}^{n} \cos k\theta + i \sum_{k=1}^{n} \sin k\theta = \frac{\sin((n+1)\theta/2)}{\sin(\theta/2)} e^{in\theta/2}, \tag{1.2.6}$$

1.2 Basic Facts on Trigonometric Approximation

Fig. 1.2.1 The Dirichlet kernel for $n = 7$

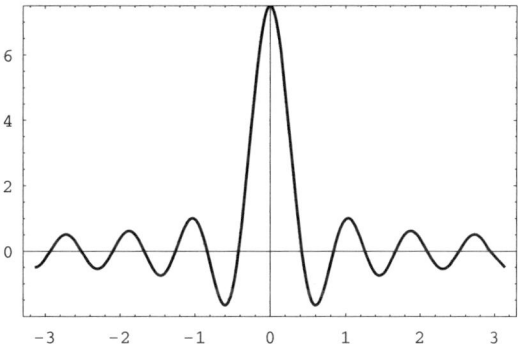

from which we get (1.2.3) and (1.2.4). Finally if we multiply (1.2.6) by $e^{i\theta/2}$, we obtain

$$\sum_{k=0}^{n} e^{i(2k+1)\theta/2} = \frac{\sin((n+1)\theta/2)}{\sin(\theta/2)} e^{i(n+1)\theta/2},$$

whose imaginary part gives (1.2.5). □

We have for $\theta \neq 2\nu\pi$ ($\nu \in \mathbb{Z}$) by (1.2.3)

$$D_n(\theta) = \mathrm{Re}\left\{\frac{1}{2} + \sum_{k=1}^{n} e^{ik\theta}\right\} = \frac{\sin((2n+1)\theta/2)}{2\sin(\theta/2)}. \qquad (1.2.7)$$

Moreover since the zeros of the Dirichlet kernel in $[0, 2\pi)$ are

$$\theta_k = \frac{2k\pi}{2n+1} \qquad (k = 1, \ldots, 2n),$$

according to (1.1.8), we can write

$$D_n(\theta) = A \prod_{k=1}^{2n} \sin\frac{\theta - \theta_k}{2},$$

where using $D_n(0) = n + 1/2$ we determine the constant A, so that

$$D_n(\theta) = \frac{2n+1}{2} \prod_{k=1}^{2n} \frac{\sin\dfrac{\theta - \theta_k}{2}}{\sin\dfrac{\theta_k}{2}}.$$

Figure 1.2.1 shows the Dirichlet kernel for $n = 7$ and $-\pi < \theta < \pi$. In this interval $D_7(\theta)$ has 14 real zeros.

Fig. 1.2.2 The Fejér kernel for $n = 7$

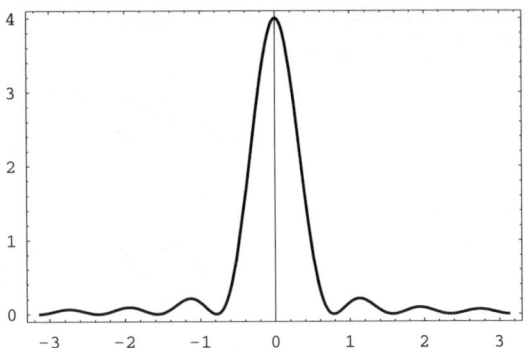

By (1.2.7) and (1.2.5), the Fejér kernel (1.2.2) can be expressed in the form

$$F_n(\theta) = \frac{1}{2(n+1)} \left(\frac{\sin((n+1)\theta/2)}{\sin(\theta/2)} \right)^2.$$

Thus, this kernel is nonnegative and it can also be given by (cf. [359, p. 311])

$$F_n(\theta) = \frac{1}{2} + \sum_{k=1}^{n} \frac{n-k+1}{n+1} \cos k\theta.$$

The case $n = 7$ is displayed in Fig. 1.2.2.

Notice that more generally

$$\left(\frac{\sin(n\theta/2)}{\sin(\theta/2)} \right)^{2k} \qquad (n, k \in \mathbb{N})$$

is a trigonometric polynomial of degree $\nu = k(n-1)$. Indeed, it is the k-th power of the Fejér kernel $F_{n-1}(\theta)$ up to a multiplicative constant. As an example of such a polynomial, we mention the *Jackson kernel* [220]

$$J_n(\theta) = \frac{3}{2n(2n^2+1)} \left(\frac{\sin(n\theta/2)}{\sin(\theta/2)} \right)^4,$$

which is an even trigonometric polynomial of degree $2(n-1)$.

Using Chebyshev polynomials of the first kind and equality (1.1.10), we can get the following other expressions

$$D_n(\theta) = (-1)^n \frac{T_{2n+1}(x)}{2x}, \qquad F_n(\theta) = \frac{1 + (-1)^n T_{2n+2}(x)}{4(n+1)x^2},$$

of the Dirichlet and Fejér kernels, where $x = \sin(\theta/2)$.

Now it is easy to prove that

$$\frac{1}{\pi} \int_0^{2\pi} D_n(\theta) \, d\theta = \frac{1}{\pi} \int_0^{2\pi} F_n(\theta) \, d\theta = 1$$

1.2 Basic Facts on Trigonometric Approximation

holds, however it is more difficult to calculate the integral

$$\Lambda_n := \frac{1}{\pi} \int_0^{2\pi} |D_n(\theta)|\, d\theta = \frac{2}{\pi} \int_0^{\pi} |D_n(\theta)|\, d\theta. \tag{1.2.8}$$

Theorem 1.2.1 *For each $n \in \mathbb{N}$ we have*

$$\Lambda_n = \frac{4}{\pi^2} \log n + r_n, \tag{1.2.9}$$

where $|r_n| \leq 3$.

Proof At first, we write $D_n(\theta)$ in the form

$$D_n(\theta) = \frac{\sin n\theta}{\theta} + \frac{1}{2}\left[\cot\frac{\theta}{2} - \frac{2}{\theta}\right]\sin n\theta + \frac{1}{2}\cos n\theta. \tag{1.2.10}$$

The function

$$\varphi(\theta) = \begin{cases} \cot\dfrac{\theta}{2} - \dfrac{2}{\theta}, & 0 < \theta \leq \pi, \\ 0, & \theta = 0, \end{cases}$$

is bounded on $[0, \pi]$ and $-2/\pi \leq \varphi(\theta) \leq 0$ holds. Further, we have

$$\int_0^\pi |\sin n\theta|\, d\theta = \int_0^\pi |\cos n\theta|\, d\theta = n \int_0^{\pi/n} \sin n\theta\, d\theta = 2. \tag{1.2.11}$$

Since

$$\frac{2}{\pi} \int_0^\pi \left|\frac{\sin n\theta}{\theta}\right| d\theta = \frac{2}{\pi} \sum_{k=0}^{n-1} \int_{k\pi/n}^{(k+1)\pi/n} \left|\frac{\sin n\theta}{\theta}\right| d\theta,$$

we can estimate this integral in the following way. First, we have the inequality

$$\frac{2}{\pi} \int_0^\pi \left|\frac{\sin n\theta}{\theta}\right| d\theta \leq \frac{2}{\pi} \int_0^{\pi/n} \frac{\sin n\theta}{\theta}\, d\theta + \frac{2}{\pi} \sum_{k=1}^{n-1} \frac{n}{k\pi} \int_{k\pi/n}^{(k+1)\pi/n} |\sin n\theta|\, d\theta$$

$$= \frac{2}{\pi} \int_0^\pi \frac{\sin t}{t}\, dt + \frac{2}{\pi} \sum_{k=1}^{n-1} \frac{n}{k\pi} \cdot \frac{2}{n}$$

$$= \frac{2}{\pi} \operatorname{Si}(\pi) + \frac{4}{\pi^2} \sum_{k=1}^{n-1} \frac{1}{k},$$

where

$$\operatorname{Si}(\pi) = \int_0^\pi \frac{\sin t}{t}\, dt = 1.8519370\ldots.$$

On the other hand,

$$\frac{2}{\pi}\int_0^\pi \left|\frac{\sin n\theta}{\theta}\right| d\theta \geq \frac{2}{\pi}\sum_{k=0}^{n-1}\frac{n}{(k+1)\pi}\int_{k\pi/n}^{(k+1)\pi/n}|\sin n\theta|\,d\theta = \frac{4}{\pi^2}\sum_{k=1}^n \frac{1}{k}$$

holds. Then because of the inequalities

$$-\frac{1}{n} < \sum_{k=1}^{n-1}\frac{1}{k} - \log n - \gamma < 0,$$

where $\gamma = 0.5772156649\ldots$ (Euler's constant), we get

$$\frac{2}{\pi}\int_0^\pi \left|\frac{\sin n\theta}{\theta}\right| d\theta = \frac{4}{\pi^2}\log n + \frac{1}{\pi}\left(\mathrm{Si}\,(\pi) + \frac{4\gamma}{\pi}\right) + q_n, \qquad (1.2.12)$$

where $|q_n| < \mathrm{Si}\,(\pi)/\pi$.

Now, using (1.2.10) and (1.2.11), we obtain

$$\frac{2}{\pi}\int_0^\pi |D_n(\theta)|\,d\theta = \frac{2}{\pi}\int_0^\pi \left|\frac{\sin n\theta}{\theta}\right| d\theta + d_n,$$

where

$$|d_n| \leq \frac{2}{\pi}\left\{\frac{1}{2}\cdot\frac{2}{\pi}\int_0^\pi |\sin n\theta|\,d\theta + \frac{1}{2}\int_0^\pi |\cos n\theta|\,d\theta\right\} = \frac{2}{\pi}\left(\frac{2}{\pi}+1\right).$$

So, according to (1.2.12), we conclude that for the integral in (1.2.8) the following estimate

$$\Lambda_n = \frac{4}{\pi^2}\log n + r_n$$

holds, with

$$|r_n| \leq \frac{1}{\pi}\left(\mathrm{Si}\,(\pi) + \frac{4\gamma}{\pi}\right) + |q_n| + |d_n| \leq \frac{2}{\pi}\left[\mathrm{Si}\,(\pi) + \frac{2(\gamma+1)}{\pi} + 1\right] \leq 3. \qquad \square$$

Remark 1.2.1 A better approximation of Λ_n is

$$\Lambda_n = \frac{4}{\pi^2}(\log n + 3) + \varepsilon_n,$$

where $|\varepsilon_n| < 1/10$ for each $n \geq 5$.

In 1919 de la Vallée-Poussin [85] introduced the kernel $V_m^n(\theta)$ for arbitrary integers n, m $(0 \leq m \leq n)$ by

$$V_m^n(\theta) = \frac{1}{n-m+1}\Big[D_m(\theta) + D_{m+1}(\theta) + \cdots + D_n(\theta)\Big], \qquad (1.2.13)$$

1.2 Basic Facts on Trigonometric Approximation

Fig. 1.2.3 The de la Vallée-Poussin kernel $V_3^7(\theta)$

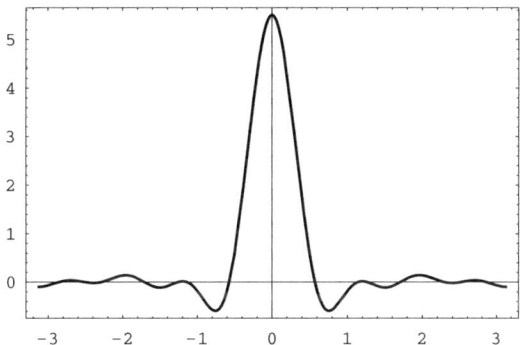

which today is known as the *de la Vallée-Poussin kernel*. Notice that

$$V_n^n(\theta) = D_n(\theta), \qquad V_0^n(\theta) = F_n(\theta).$$

Since

$$D_{m+k}(\theta) = D_m(\theta) + \cos(m+1)\theta + \cdots + \cos(m+k)\theta,$$

we obtain

$$V_m^n(\theta) = D_m(\theta) + \sum_{k=m+1}^{n} \frac{n-k+1}{n-m+1} \cos k\theta.$$

Also, in terms of the Fejér kernel, we have

$$V_m^n(\theta) = \frac{1}{n-m+1}\bigl[(n+1)F_n(\theta) - m F_{m-1}(\theta)\bigr]$$

$$= \frac{1}{n-m+1}\left[\frac{\sin^2((n+1)\theta/2)}{2\sin^2(\theta/2)} - \frac{\sin^2(m\theta/2)}{2\sin^2(\theta/2)}\right],$$

which gives

$$V_m^n(\theta) = \frac{\cos m\theta - \cos(n+1)\theta}{4(n-m+1)\sin^2(\theta/2)}. \tag{1.2.14}$$

Moreover, in terms of the Chebyshev polynomials, we find

$$V_m^n(\theta) = \frac{(-1)^n T_{2n+2}(x) + (-1)^m T_{2m}(x)}{4(n-m+1)x^2}, \qquad x = \sin\frac{\theta}{2}.$$

The kernel $V_3^7(\theta)$ is shown in Fig. 1.2.3.

Finally, the following result holds:

Theorem 1.2.2 *For each pair of integers $0 \leq m < n$ we have*

$$\frac{1}{\pi}\int_0^{2\pi} |V_m^n(\theta)|\,d\theta = \frac{4}{\pi^2}\log\frac{n}{n-m+1} + O(1). \tag{1.2.15}$$

The proof of this result can be found in [235, p. 100].

1.2.2 Fourier Series and Sums

Using the trigonometric formulas

$$\sin nx \cos mx = \frac{1}{2}\big[\sin(n+m)x + \sin(n-m)x\big],$$

$$\sin nx \sin mx = \frac{1}{2}\big[\cos(n-m)x - \cos(n+m)x\big],$$

$$\cos nx \cos mx = \frac{1}{2}\big[\cos(n+m)x + \cos(n-m)x\big],$$

it is easy to prove that

$$\int_0^{2\pi} \sin nx \cos mx\, dx = 0,$$

$$\int_0^{2\pi} \cos nx \cos mx\, dx = \begin{cases} 0, & n \neq m, \\ \pi, & n = m \neq 0, \\ 2\pi, & n = m = 0, \end{cases} \quad (1.2.16)$$

$$\int_0^{2\pi} \sin nx \sin mx\, dx = \begin{cases} 0, & n \neq m, \\ \pi, & n = m \neq 0, \\ 0, & n = m = 0. \end{cases} \quad (1.2.17)$$

This means that the *trigonometric system*

$$T = \{1, \cos x, \sin x, \cos 2x, \sin 2x, \ldots, \cos nx, \sin nx, \ldots\} \quad (1.2.18)$$

is an orthogonal system in the Hilbert space $L^2(\mathbb{T})$ with the inner product defined by

$$(u, v) = \int_{\mathbb{T}} u(x)v(x)\, dx = \int_0^{2\pi} u(x)v(x)\, dx. \quad (1.2.19)$$

By the Weierstrass theorem, the trigonometric system T is also dense in the Hilbert space $L^2(\mathbb{T})$. So it forms an orthogonal basis in $L^2(\mathbb{T})$ and each function $f \in L^2(\mathbb{T})$ can be represented as the sum of its *Fourier series*

$$\frac{1}{2}a_0 + \sum_{k=1}^{+\infty}(a_k \cos kx + b_k \sin kx), \quad (1.2.20)$$

1.2 Basic Facts on Trigonometric Approximation

with the *Fourier coefficients* given by

$$a_k = a_k(f) := \frac{1}{\pi} \int_0^{2\pi} f(x) \cos kx\, dx, \tag{1.2.21}$$

$$b_k = b_k(f) := \frac{1}{\pi} \int_0^{2\pi} f(x) \sin kx\, dx. \tag{1.2.22}$$

In view of this classical result, it is natural to consider the $(n+1)$-th partial sums of the Fourier series (1.2.20), i.e.

$$S_n f(x) := \frac{a_0}{2} + \sum_{k=1}^n (a_k \cos kx + b_k \sin kx), \tag{1.2.23}$$

in order to approximate a generic 2π-periodic integrable function $f \in L^1(\mathbb{T})$ by trigonometric polynomials.

The partial sums (1.2.23) are called the *Fourier sums* and by (1.2.21), (1.2.22) they can be represented in the form

$$S_n f(x) = \frac{1}{2} a_0 + \sum_{k=1}^n (a_k \cos kx + b_k \sin kx)$$

$$= \frac{1}{2\pi} \int_0^{2\pi} f(t)\, dt + \frac{1}{\pi} \sum_{k=1}^n \int_0^{2\pi} f(t)\big(\cos kt \cos kx + \sin kt \sin kx\big)\, dt$$

$$= \frac{1}{\pi} \int_0^{2\pi} f(t) \left[\frac{1}{2} + \sum_{k=1}^n \cos k(x-t)\right] dt,$$

i.e.

$$S_n f(x) = \frac{1}{\pi} \int_0^{2\pi} D_n(x-t) f(t)\, dt, \tag{1.2.24}$$

where D_n is the Dirichlet kernel defined by (1.2.1). Moreover (1.2.24) can be also represented as

$$S_n f(x) = \frac{1}{\pi} \int_0^{2\pi} D_n(t) f(x-t)\, dt = \frac{1}{\pi} \int_0^{2\pi} D_n(t) f(x+t)\, dt,$$

or

$$S_n f(x) = \frac{1}{\pi} \int_0^{\pi} D_n(t) \big[f(x+t) + f(x-t)\big] dt,$$

since we recall that we have for 2π-periodic functions $f, g \in L^1$ and for each $x \in \mathbb{R}$

$$\int_x^{x+2\pi} f(t)\, dt = \int_0^{2\pi} f(t)\, dt = \int_0^{2\pi} f(t+x)\, dt$$

and
$$\int_0^{2\pi} g(x-t)f(t)\,dt = \int_0^{2\pi} g(t)f(x-t)\,dt.$$

Also, the arithmetic means of the Fourier sums

$$\sigma_n f(x) = \frac{1}{n+1}\sum_{k=0}^n S_k f(x) \quad \text{and} \quad V_m^n f(x) = \frac{1}{n-m+1}\sum_{k=m}^n S_k f(x) \quad (1.2.25)$$

play an important role in the theory of Fourier series. The means $\sigma_n f$ and $V_m^n f$ are known as *Fejér sums* and *de la Vallée Poussin sums*, respectively. They can be written in the following explicit form

$$\sigma_n f(x) = \frac{a_0}{2} + \sum_{k=1}^n \frac{n-k+1}{n+1}(a_k \cos kx + b_k \sin kx),$$

$$V_m^n f(x) = \frac{a_0}{2} + \sum_{k=1}^n \mu_k(a_k \cos kx + b_k \sin kx), \quad (1.2.26)$$

where a_k, b_k are the Fourier coefficients given by (1.2.21) and (1.2.22), respectively, and

$$\mu_k := \begin{cases} 1 & \text{if } 1 \le k \le m, \\ \dfrac{n-k+1}{n-m+1} & \text{if } m < k \le n. \end{cases}$$

Alternatively, using the Fejér kernel (1.2.2) and de la Vallée Poussin kernel (1.2.13), as well as (1.2.24), the previous sums can be expressed by the following integrals:

$$\sigma_n f(x) = \frac{1}{\pi}\int_0^{2\pi} F_n(x-t)f(t)\,dt, \quad V_m^n f(x) = \frac{1}{\pi}\int_0^{2\pi} V_m^n(x-t)f(t)\,dt.$$

The approximating properties of the sums $S_n f$, $\sigma_n f$, and $V_m^n f$ will be studied in Chap. 3.

1.2.3 Moduli of Smoothness, Best Approximation and Besov Spaces

The aim of this subsection is to introduce some basic tools that are useful to study the trigonometric polynomial approximation of a generic 2π-periodic function $f \in L^p$, $1 \le p \le +\infty$.

At first we introduce the *finite forward differences* of the first order of a function $x \mapsto f(x)$,

$$\vec{\Delta}_h f(x) := f(x+h) - f(x), \qquad h > 0. \quad (1.2.27)$$

1.2 Basic Facts on Trigonometric Approximation

For all $k \in \mathbb{N}$, the forward differences of order k are defined by $\vec{\Delta}_h^k := \vec{\Delta}_h \vec{\Delta}_h^{k-1}$ or in an explicit form by

$$\vec{\Delta}_h^k f(x) := \sum_{i=0}^{k} (-1)^i \binom{k}{i} f(x+kh-ih). \tag{1.2.28}$$

A measure of the smoothness of f is given by the *k-th modulus of smoothness* of f,

$$\omega^k(f,t)_p := \sup_{0<h\leq t} \begin{cases} \left(\int_0^{2\pi-kh} |\vec{\Delta}_h^k f(x)|^p dx \right)^{1/p}, & 1 \leq p < +\infty, \\ \max_{x \in [0, 2\pi-hk]} |\vec{\Delta}_h^k f(x)|, & p = +\infty, \end{cases}$$

where $k \in \mathbb{N}$ and $t \in \mathbb{R}^+$. For brevity, we set $\omega := \omega^1$.

The main properties of such a modulus of smoothness are:

1° $\omega^k(f,\theta)_p \leq \omega^k(f,t)_p$ $(\theta \leq t)$;

2° $\omega^k(f,t)_p/t^k \leq 2^k \omega^k(f,\theta)_p/\theta^k$ $(\theta \leq t)$;

3° $\omega^k(f+g,\theta)_p \leq \omega^k(f,\theta)_p + \omega^k(g,\theta)_p$ $(f, g \in L^p)$;

4° $\omega(f, t+\theta)_p \leq \omega(f,t)_p + \omega(f,\theta)_p$ $(t, \theta \in \mathbb{R}^+)$;

5° $\omega^k(f,nt)_p \leq n^k \omega^k(f,t)_p$ $(n \in \mathbb{N})$;

6° $\omega^k(f,\lambda t)_p \leq ([\lambda]+1)^k \omega^k(f,t)_p$ $(\lambda \in \mathbb{R}^+)$;

7° $\omega^k(f,t)_p \leq 2\omega^{k-1}(f,t)_p$ $(k \geq 2)$;

8° $\omega^k(f,t)_p \leq t\omega^{k-1}(f',t)_p$ $(k \geq 2,\ f \in AC,\ f' \in L^p)$;

9° $\omega^k(f,t)_p \leq t^k \|f^{(k)}\|_p$ $(f^{(k)} \in L^p)$,

where here and in the sequel AC denotes the set of all absolutely continuous functions and $f^{(k)}$ is the k-th derivative of f (in the sense distributions when $f \in L^p$). For the proofs and more details on such properties, the interested reader may consult [397] and [475, pp. 93–110].

Some subspaces of 2π-periodic functions in L^p can be characterized by using also the *K-functional* defined as

$$K^k(f,t)_p := \inf_{g^{(k-1)} \in AC} \left\{ \|f-g\|_p + t^k \|g^{(k)}\|_p \right\}. \tag{1.2.29}$$

This K-functional was introduced by Peetre [388]. It is an interpolatory functional between the L^p-spaces and the *Sobolev spaces*

$$W_r^p := \left\{ f \in L^p \ \Big|\ f^{(r-1)} \in AC \text{ and } \|f^{(r)}\| < +\infty \right\}, \quad r \geq 1,\ 1 \leq p \leq +\infty,$$

with

$$\|f\|_{W_r^p} := \|f\|_p + \|f^{(r)}\|_p.$$

Indeed $\omega^k(f,t)_p$ and $K^k(f,t)_p$ are "equivalent" tools for characterizing the smoothness of functions $f \in L^p$. In fact, the following result is valid (see [397, p. 66]):

Theorem 1.2.3 *For all 2π-periodic functions $f \in L^p$, $1 \leq p \leq +\infty$, any $t \in \mathbb{R}^+$ and for each $k \in \mathbb{N}$, there exists a constant C, depending only on k, such that*

$$C^{-1}\omega^k(f,t)_p \leq K^k(f,t)_p \leq C\omega^k(f,t)_p \qquad (1.2.30)$$

holds.

Remark 1.2.2 In proving the upper bound of (1.2.30) the generalized Steklov function[4] is used and, for all $t > 0$, a function $g = g_t$ (depending on t) is found such that

$$\|f - g\|_p + t^k \|g^{(k)}\|_p \leq C\omega^k(f,t)_p \qquad (1.2.31)$$

holds. Here, C is a constant which does not depend on f and t. In such cases, very often, we write $C \neq C(f,t)$.

As we will see in the sequel, the connection between the Peetre K-functional and the modulus of smoothness turns out to be useful in several contexts.

Alternatively, we can write the following equivalence (see [100])

$$\omega^k(f,t)_p \sim \min_{T \in \mathcal{T}_{[1/t]}} \left(\|f - T\|_p + t^k \|T^{(k)}\|_p \right). \qquad (1.2.32)$$

Another basic notion in the approximation of 2π-periodic functions $f \in L^p$, $1 \leq p \leq +\infty$, is the error of best L^p-approximation by trigonometric polynomials of degree at most n defined as

$$E_n^*(f)_p := \inf_{T \in \mathcal{T}_n} \|f - T\|_p.$$

The best approximation and the moduli of smoothness are strictly related by the following well-known theorem of Jackson (cf. [397]):

---

[4]For $f \in L$ and $h > 0$, the Steklov function f_h is defined by $f_h(x) = (1/h)\int_0^h f(x+t)\,dt$, and the generalized Steklov function as

$$f_{k,h}(x) = \frac{(-1)^k}{h^k} \int_0^h \cdots \int_0^h \left\{ \sum_{\nu=1}^k (-1)^\nu \binom{k}{\nu-1} f\left(x + \frac{k-\nu+1}{k}(t_1 + \cdots + t_k)\right) \right\} dt_1 \cdots dt_k.$$

1.2 Basic Facts on Trigonometric Approximation

Theorem 1.2.4 *Let $f \in L^p$, $1 \le p \le +\infty$, and $k \in \mathbb{N}$. Then there exists a constant C, depending only on $k \subset \mathbb{N}$, such that for each $n \in \mathbb{N}$ and $n > k$, we have*

$$E_n^*(f)_p \le C\, \omega^k\left(f, \frac{1}{n}\right)_p. \tag{1.2.33}$$

Using (1.2.33) it is possible to prove another useful inequality, known as the *Favard inequality* (cf. [475]):

Theorem 1.2.5 *Let $f \in AC$, $f' \in L^p$, $1 \le p \le +\infty$, and $n \in \mathbb{N}$. Then*

$$E_n^*(f)_p \le \frac{c}{n} E_n^*(f')_p \tag{1.2.34}$$

holds, where c is an absolute positive constant.

Note that the inequality (1.2.33) cannot be inverted. Indeed, if for instance we take $f(x) = \cos x$, $k = 1$ and $p = +\infty$, then for all $m > 1$ we have $E_m^*(f)_\infty = 0$, but $\omega(f, t)_\infty = t$. Nevertheless it is possible to prove a weak "inverse" of the Jackson theorem by using the well-known *Bernstein inequality* (cf. [359, p. 584])

$$\|T'\|_p \le n \|T\|_p \qquad (1 \le p \le +\infty), \tag{1.2.35}$$

which holds for each $T \in \mathcal{T}_n$. A proof of the Bernstein inequality will be given later in Sect. 1.3.4.

Theorem 1.2.6 *Let $f \in L^p$, $1 \le p \le \infty$. Then for each $n, k \in \mathbb{N}$, with $n \ge k$,*

$$\omega^k\left(f, \frac{1}{n}\right)_p \le \frac{C}{n^k} \sum_{j=0}^{n} (1+j)^{k-1} E_j^*(f)_p \tag{1.2.36}$$

holds, where C is a positive constant depending only on k.

The inequality (1.2.36) is known as the *Salem-Stechkin inequality* [424, 453].

By virtue of the Jackson and Salem-Stechkin inequalities it is possible to link the smoothness of a function f with the *rate of convergence* of its best approximation. For instance, for all $k > \alpha > 0$, we have

$$\sup_{t>0} \frac{\omega^k(f,t)_p}{t^\alpha} < +\infty \quad \text{if and only if} \quad \sup_n n^\alpha E_n^*(f)_p < +\infty.$$

This equivalence can be generalized as follows. Define the following two norms for $0 < r \in \mathbb{R}$, $1 \leq p \leq +\infty$, and $1 \leq q \leq +\infty$

$$\|f\|_{B^p_{r,q}} := \|f\|_p + \begin{cases} \left(\int_0^1 \left[\dfrac{\omega^k(f,t)_p}{t^{r+1/q}}\right]^q dt\right)^{1/q}, & 1 \leq q < +\infty, \\ \sup_{t>0} \dfrac{\omega^k(f,t)_p}{t^r}, & q = +\infty, \end{cases}$$

for $k > r$, and

$$\|f\|_{E^p_{r,q}} := \|f\|_p + \begin{cases} \left(\sum_{i=1}^{+\infty} \dfrac{[(1+i)^r E^*_i(f)_p]^q}{1+i}\right)^{1/q}, & 1 \leq q < +\infty, \\ \sup_{i \geq 1} i^r E^*_i(f)_p, & q = +\infty. \end{cases}$$

Then using the Jackson and Salem-Stechkin inequalities, it is simple to prove that (cf. [99])

$$\|f\|_{B^p_{r,q}} \sim \|f\|_{E^p_{r,q}} \tag{1.2.37}$$

holds. This means we can use any of these two norms in order to define the so-called *Besov spaces* given by

$$B^p_{r,q} := \left\{ f \in L^p \;\middle|\; \|f\|_{B^p_{r,q}} < +\infty \right\},$$

or equivalently

$$E^p_{r,q} := \left\{ f \in L^p \;\middle|\; \|f\|_{E^p_{r,q}} < +\infty \right\}.$$

These spaces were introduced and extensively investigated in [41]. A special case of Besov spaces occurs for $1 \leq p \leq +\infty$ and $q = +\infty$. In this case we have the L^p-Zygmund-Hölder spaces (see [402])

$$B^p_{r,\infty} = \left\{ f \in L^p \;\middle|\; \sup_{t>0} \dfrac{\omega^k(f,t)_p}{t^r} < +\infty, \; k > r \right\},$$

or equivalently

$$E^p_{r,\infty} = \left\{ f \in L^p \;\middle|\; \sup_{i \geq 1} i^r E^*_i(f)_p < +\infty \right\}.$$

Another interesting case, which is useful in several applications, is the case when $p = q = 2$. Let $f \in L^2$ be an arbitrary 2π-periodic function and

$$c_0 = \dfrac{a_0}{2}, \quad c_i^2 = c_i^2(f) = a_i^2 + b_i^2 \quad (i = 1, 2, \ldots),$$

1.2 Basic Facts on Trigonometric Approximation

where a_i and b_i are the Fourier coefficients of the function f in the trigonometric system (see (1.2.21) and (1.2.21), respectively). Then, for all $r \geq 0$, we define the space

$$L_r^2 := \left\{ f \in L^2 \ \bigg| \ \sum_{i=0}^{+\infty} c_i^2 (1+i)^{2r} < +\infty \right\},$$

equipped with the norm

$$\|f\|_{L_r^2} := \left(\sum_{i=0}^{+\infty} c_i^2 (1+i)^{2r} \right)^{1/2}.$$

If $r = 0$, the *Parseval identity* gives

$$\|f\|_{L_r^2} := \left(\sum_{i=0}^{+\infty} c_i^2 \right)^{1/2} = \|f\|_2.$$

Moreover, if $r > 0$ is an integer, from the equality $c_k(f)^2 = c_k(f')^2/k^2$ and the Parseval identity, we get

$$\left(\sum_{i=0}^{+\infty} c_i^2 (1+i)^{2r} \right)^{1/2} \sim \|f\|_2 + \|f^{(r)}\|_2,$$

that is the ordinary *trigonometric Sobolev norm*. Finally, if $r > 0$ is an arbitrary real number, then setting $E_0^*(f)_2 = \|f\|_2$ and recalling that $E_k^*(f)_2^2 = \sum_{i>k} c_i^2$ holds, we can write

$$\|f\|_{E_{r,2}^2}^2 = \|f\|_2^2 + \sum_{k=1}^{+\infty} (1+k)^{2r-1} E_k^*(f)_2^2 = \sum_{i=0}^{+\infty} c_i^2 + \sum_{k=1}^{+\infty} (1+k)^{2r-1} \sum_{i>k} c_i^2$$

$$= \sum_{i=0}^{+\infty} c_i^2 + \sum_{i=1}^{+\infty} c_i^2 \sum_{k=1}^{i-1} (1+k)^{2r-1} \sim \sum_{i=0}^{+\infty} c_i^2 + \sum_{i=1}^{+\infty} c_i^2 i^{2r-1}$$

$$\sim \sum_{i=0}^{+\infty} c_i^2 (1+i)^{2r}.$$

Hence by (1.2.37) we can conclude with the following well-known equivalence

$$\|f\|_{L_r^2} \sim \begin{cases} \|f\|_2 & \text{if } r=0, \\ \|f\|_{W_r^2} & \text{if } r \in \mathbb{N}, \\ \|f\|_{B_{r,2}^2} & \text{if } r > 0. \end{cases}$$

1.3 Chebyshev Systems and Interpolation

1.3.1 Chebyshev Systems and Spaces

We start with the set $\mathcal{P}_n$ of all algebraic polynomials of degree at most n defined on $A = [a, b]$. In Sect. 1.1.1 we mentioned two essential properties of polynomials in approximation theory. One of them is that *each algebraic polynomial of degree at most $n - 1$ can be uniquely interpolated at n points* (note that n is replaced by $n - 1$).

This property is equivalent to the fact that any $p \in \mathcal{P}_{n-1}$ that vanishes at n points is identically zero. It can be extended to other finite-dimensional subspaces of $C(A)$, where A is a Hausdorff topological space.

Definition 1.3.1 Let $\phi_k : A \to \mathbb{R}$ ($k = 1, \ldots, n$) be continuous functions. The set $H = \{\phi_1, \ldots, \phi_n\}$ is called a *Chebyshev system* or *Haar system* of dimension n on A if $X_n = \mathrm{span}\{\phi_1, \ldots, \phi_n\}$ over $\mathbb{R}$ is an n-dimensional subspace of $C(A)$ and any function of X_n that has n distinct zeros in A is identically zero. In that case X_n is called a *Chebyshev space* or *Haar space*.

Of course, it is clear that A in the previous definition must contain at least n points. Usually, we use $A = [a, b]$ and $A = \mathbb{T}$. Since the functions of a Haar system are linearly independent, we can conclude that any other basis $\{\psi_1, \ldots, \psi_n\}$ of X_n is also a Haar system. For many details on these systems see [47, 225].

Using well-known facts from linear algebra we can formulate a number of equivalences:

Proposition 1.3.1 *Let $H = \{\phi_1, \ldots, \phi_n\}$ be a Haar system on A. The following statements are equivalent:*

(a) *Each $\phi = a_1 \phi_1 + \cdots + a_n \phi_n \neq 0$ has at most $n - 1$ distinct zeros in A.*
(b) *If $x_1, \ldots, x_n$ are distinct points of A, then*

$$D = D(x_1, \ldots, x_n) = \begin{vmatrix} \phi_1(x_1) & \cdots & \phi_n(x_1) \\ \vdots & & \\ \phi_1(x_n) & & \phi_n(x_n) \end{vmatrix} \neq 0. \qquad (1.3.1)$$

(c) *If $x_1, \ldots, x_n$ are distinct points of A and $f_1, \ldots, f_n$ are arbitrary numbers, then there exists a unique $\phi = \sum_{k=1}^{n} a_k \phi_k \in X_n = \mathrm{span}(H)$ such that*

$$\phi(x_k) = f_k \qquad (k = 1, \ldots, n). \qquad (1.3.2)$$

Notice that (1.3.2) is a system of linear equations

$$a_1 \phi_1(x_k) + \cdots + a_n \phi_n(x_k) = f_k \qquad (k = 1, \ldots, n), \qquad (1.3.3)$$

1.3 Chebyshev Systems and Interpolation

which has a unique solution for the coefficients $a_1, \ldots, a_n$. It is equivalent to (1.3.1).

We call the elements of X_n *polynomials* and ϕ in (1.3.2) an *interpolation polynomial* with prescribed values f_k at the points x_k. The statement (c) in Proposition 1.3.1 means that an interpolation polynomial ϕ exists uniquely. The points x_k are called *interpolation nodes*.

1.3.2 Algebraic Lagrange Interpolation

Take $A = [a, b]$ and $\phi_k(x) = x^{k-1}$ ($k = 1, \ldots, n$). Then (1.3.1) becomes the well-known Vandermonde determinant

$$D = V(x_1, \ldots, x_n) = \begin{vmatrix} 1 & x_1 & \cdots & x_1^{n-1} \\ \vdots & & & \\ 1 & x_n & & x_n^{n-1} \end{vmatrix} = \prod_{j<k}^n (x_k - x_j). \tag{1.3.4}$$

Assuming the nodes are distinct, we have $D \neq 0$ and therefore $\{1, x, \ldots, x^{n-1}\}$ and $\mathcal{P}_{n-1}$ are a Haar system and Haar space, respectively. Consequently there is a unique solution of (1.3.3), i.e., a unique algebraic interpolation polynomial $\phi(x) = \sum_{k=1}^n a_k x^{k-1} \in \mathcal{P}_{n-1}$. In the case when the prescribed values f_k at the nodes x_k are the values assumed by a certain function f, i.e., $f_k = f(x_k)$, we say $\phi(x)$ is the *Lagrange algebraic interpolation polynomial* of f and denote it by $L_n f(x)$.

An explicit form of such a polynomial can be obtained by considering the so-called *fundamental Lagrange polynomials* defined as follows

$$\ell_{n,k}(x) := \prod_{\substack{\nu=1 \\ \nu \neq k}}^n \frac{x - x_\nu}{x_k - x_\nu}, \qquad k = 1, \ldots, n \tag{1.3.5}$$

or equivalently by

$$\ell_{n,k}(x) := \frac{q_n(x)}{q_n'(x_k)(x - x_k)} \tag{1.3.6}$$

where q_n is the polynomial of degree n whose zeros are the knots $x_1, \ldots, x_n$, i.e., $q_n(x) = \prod_{\nu=1}^n (x - x_\nu)$.

Of course, $\ell_{n,k}(x)$ are algebraic polynomials of degree $n-1$ and

$$\ell_{n,k}(x_\nu) = \begin{cases} 1 & \text{if } \nu = k, \\ 0 & \text{if } \nu \neq k, \end{cases}$$

holds for each $v, k = 1, \ldots, n$. Consequently, we can write the Lagrange polynomial $L_n f(x)$ in the following form

$$L_n f(x) = \sum_{k=1}^{n} \ell_{n,k}(x) f(x_k) \tag{1.3.7}$$

which is known as the *Lagrange interpolation formula*.

Example 1.3.1 Let $\lambda_1 < \cdots < \lambda_n$. The system of functions $\{x^{\lambda_1}, \ldots, x^{\lambda_n}\}$ is a Haar system on $A = (0, +\infty)$. For distinct points $x_1, \ldots, x_n \in A$ the determinant

$$\begin{vmatrix} x_1^{\lambda_1} & \cdots & x_1^{\lambda_n} \\ \vdots & & \\ x_n^{\lambda_1} & & x_n^{\lambda_n} \end{vmatrix}$$

is different from zero. Such systems of generalized polynomials are known as *Müntz systems* and generalized polynomials as the *Müntz polynomials* (cf. [47]). For $\lambda_1 = 0$ this system is a Haar system on $A = [0, +\infty)$. For integer exponents $\lambda_k = k - 1$ ($k = 1, \ldots, n$), it reduces to the standard polynomial system considered before, and it is a Haar system on $A = \mathbb{R}$.

Remark 1.3.1 The Müntz systems can be considered also for complex sequences $\{\lambda_1, \ldots, \lambda_n\}$ (cf. [47]), where we have the following definition for x^λ:

$$x^\lambda = e^{\lambda \log x}, \qquad x \in (0, +\infty), \ \lambda \in \mathbb{C},$$

and the value at $x = 0$ to be the limit of x^λ as $x \to 0$ from $(0, +\infty)$ whenever the limit exists.

1.3.3 Trigonometric Interpolation

Let $A = \mathbb{T}$ and

$$H = \{1, \cos x, \sin x, \cos 2x, \sin 2x, \ldots, \cos nx, \sin nx\}. \tag{1.3.8}$$

If x_k ($k = 0, 1, \ldots, 2n$) are distinct points in $\mathbb{T}$, for example

$$0 \leq x_0 < x_1 < \cdots < x_{2n} < 2\pi, \tag{1.3.9}$$

then, we have to evaluate the determinant

$$D = \begin{vmatrix} 1 & \cos x_0 & \sin x_0 & \cos 2x_0 & \sin 2x_0 & \cdots & \cos nx_0 & \sin nx_0 \\ 1 & \cos x_1 & \sin x_1 & \cos 2x_1 & \sin 2x_1 & & \cos nx_1 & \sin nx_1 \\ \vdots & & & & & & & \\ 1 & \cos x_{2n} & \sin x_{2n} & \cos 2x_{2n} & \sin 2x_{2n} & & \cos nx_{2n} & \sin nx_{2n} \end{vmatrix}.$$

1.3 Chebyshev Systems and Interpolation

At first, we transform its elements to the complex form. Multiplying the 3rd, 5th, ... columns by the imaginary unit i and then adding them to the 2nd, 4th, ... columns, respectively, we get

$$D = \begin{vmatrix} 1 & e^{ix_0} & \sin x_0 & e^{i2x_0} & \sin 2x_0 & \cdots & e^{inx_0} & \sin nx_0 \\ 1 & e^{ix_1} & \sin x_1 & e^{i2x_1} & \sin 2x_1 & & e^{inx_1} & \sin nx_1 \\ \vdots & & & & & & & \\ 1 & e^{ix_{2n}} & \sin x_{2n} & e^{i2x_{2n}} & \sin 2x_{2n} & & e^{inx_{2n}} & \sin nx_{2n} \end{vmatrix}.$$

In the second step, we multiply the 3th, 5th, ... columns by $-2i$ and add to them the 2nd, 4th, ... columns, respectively. Then, we obtain

$$(-2i)^n D = \begin{vmatrix} 1 & e^{ix_0} & e^{-ix_0} & e^{i2x_0} & e^{-i2x_0} & \cdots & e^{inx_0} & e^{-inx_0} \\ 1 & e^{ix_1} & e^{-ix_1} & e^{i2x_1} & e^{-i2x_1} & & e^{inx_1} & e^{-inx_1} \\ \vdots & & & & & & & \\ 1 & e^{ix_{2n}} & e^{-ix_{2n}} & e^{i2x_{2n}} & e^{-i2x_{2n}} & & e^{inx_{2n}} & e^{-inx_{2n}} \end{vmatrix},$$

whence, by interchanging the columns,

$$(-2i)^n (-1)^{n(n+1)} D = \begin{vmatrix} e^{-inx_0} & e^{-i(n-1)x_0} & \cdots & e^{i(n-1)x_0} & e^{inx_0} \\ e^{-inx_1} & e^{-i(n-1)x_1} & & e^{i(n-1)x_1} & e^{inx_1} \\ \vdots & & & & \\ e^{-inx_{2n}} & e^{-i(n-1)x_{2n}} & & e^{i(n-1)x_{2n}} & e^{inx_{2n}} \end{vmatrix}.$$

Finally, multiplying the ν-th row by e^{inx_ν} ($\nu = 0, 1, \ldots, 2n$) we obtain the following determinant of Vandermonde type (see (1.3.4))

$$(-2i)^n (-1)^{n(n+1)} e^{in(x_0+x_1+\cdots+x_{2n})} D = \begin{vmatrix} 1 & e^{ix_0} & e^{i2x_0} & \cdots & e^{i2nx_0} \\ 1 & e^{ix_1} & e^{i2x_1} & & e^{i2nx_1} \\ \vdots & & & & \\ 1 & e^{ix_{2n}} & e^{i2x_{2n}} & & e^{i2nx_{2n}} \end{vmatrix},$$

i.e.,

$$D = (-1)^{n(n+1)} e^{in(x_0+x_1+\cdots+x_{2n})} \left(\frac{i}{2}\right)^n \prod_{j<k} \left(e^{ix_k} - e^{ix_j}\right).$$

Similarly as in getting the factorization (1.1.8), we find

$$D = \frac{1}{2^n} \prod_{j<k} \left(2 \sin \frac{x_k - x_j}{2}\right).$$

Because of (1.3.9), we conclude that $D \neq 0$, and therefore, (1.3.8) is a Haar system and $X_{2n+1} = \mathcal{T}_n$ is the corresponding Haar space of dimension $2n+1$. Hence, the existence and uniqueness of a trigonometric interpolation polynomial are guaranteed. In order to find an explicit expression of this polynomial $\Phi(x)$, we introduce the polynomials $\ell_k^*(x)$ by

$$\ell_k^*(x) := \prod_{\substack{\nu=0 \\ \nu \neq k}}^{2n} \frac{\sin \dfrac{x - x_\nu}{2}}{\sin \dfrac{x_k - x_\nu}{2}} \qquad (k = 0, 1, \ldots, 2n). \tag{1.3.10}$$

It is easy to see that $\ell_k^*(x)$ are trigonometric polynomials of order n and

$$\ell_k^*(x_\nu) = \begin{cases} 1 & \text{if } \nu = k, \\ 0 & \text{if } \nu \neq k, \end{cases} \tag{1.3.11}$$

holds for each $\nu, k = 0, 1, \ldots, 2n$. Therefore, taking $2n + 1$ arbitrary values $f_0, f_1, \ldots, f_{2n}$, we can express the polynomial $\Phi(x)$, which satisfies the conditions

$$\Phi(x_k) = f_k \qquad (k = 0, 1, \ldots, 2n), \tag{1.3.12}$$

in the form

$$\Phi(x) = \sum_{k=0}^{2n} \ell_k^*(x) f_k. \tag{1.3.13}$$

If f_k are the values of a continuous function f at the nodes x_k, i.e., $f_k \equiv f(x_k)$ ($k = 0, 1, \ldots, 2n$), we say that $\Phi(x)$ is the *trigonometric interpolation polynomial* of f and we denote it by $L_n^* f(x)$ or $L_n^*(f; x)$.

The formula (1.3.13) is known as the *trigonometric interpolation formula* (in the Lagrange form).

The form (1.3.10) of the polynomials $\ell_k^*(x)$ corresponds to (1.3.5). We can also state for such polynomials an expression similar to (1.3.6). In fact, setting

$$u(x) := \prod_{\nu=0}^{2n} \sin \frac{x - x_\nu}{2} \qquad \text{and} \qquad u_k(x) := \frac{u(x)}{\sin \dfrac{x - x_k}{2}},$$

we have

$$\prod_{\substack{\nu=0 \\ \nu \neq k}}^{2n} \sin \frac{x_k - x_\nu}{2} = u_k(x_k) = 2u'(x_k),$$

and then we obtain

$$\ell_k^*(x) = \frac{u_k(x)}{u_k(x_k)} = \frac{u(x)}{2u'(x_k) \sin \dfrac{x - x_k}{2}}. \tag{1.3.14}$$

Notice that $u(x)$ is not a trigonometric polynomial.

1.3 Chebyshev Systems and Interpolation

In a particular, but very important case of equidistant nodes

$$x_k := \frac{2k\pi}{2n+1} \quad (k=0,1,\ldots,2n), \tag{1.3.15}$$

we can give a simple form of (1.3.14). According to (1.1.9) and (1.1.11),

$$D_n(x) = \frac{\sin(2n+1)\frac{x}{2}}{\sin\frac{x}{2}}$$

is a trigonometric polynomial of degree n with zeros at the points (1.3.15) except $x_0 = 0$, i.e.,

$$D_n(x_0) = D_n(0) = 2n+1, \qquad D_n(x_k) = 0 \quad (k=1,\ldots,2n).$$

Defining

$$\ell_k^*(x) := \frac{1}{2n+1} D_n(x - x_k) = \frac{1}{2n+1} \frac{\sin(2n+1)\frac{x - x_k}{2}}{\sin\frac{x - x_k}{2}} \quad (k=0,1,\ldots,2n),$$

we see that these polynomials satisfy the conditions (1.3.11), so that the interpolating polynomial at the equidistant points (1.3.15) has the following representation

$$L_n^* f(x) = \sum_{k=0}^{2n} \frac{\sin(2n+1)\frac{x - x_k}{2}}{(2n+1)\sin\frac{x - x_k}{2}} f(x_k). \tag{1.3.16}$$

Example 1.3.2 The system of functions $\{1, \cos x, \cos 2x, \ldots, \cos nx\}$ is a Haar system on $A = [0, \pi)$. The corresponding determinant is given by (see [481, p. 129])

$$\begin{vmatrix} 1 & \cos x_0 & \cos 2x_0 & \cdots & \cos nx_0 \\ 1 & \cos x_1 & \cos 2x_1 & & \cos nx_1 \\ \vdots & & & & \\ 1 & \cos x_n & \cos 2x_n & & \cos nx_n \end{vmatrix} = 2^{n(n-1)/2} \prod_{j<k} (\cos x_j - \cos x_k).$$

Example 1.3.3 The system of functions $\{\sin x, \sin 2x, \ldots, \sin nx\}$ is a Haar system on $A = (0, \pi)$. For $0 < x_k < \pi$ $(k=1,\ldots,n)$ we have

$$\begin{vmatrix} \sin x_0 & \sin 2x_0 & \cdots & \sin nx_0 \\ \sin x_1 & \sin 2x_1 & & \sin nx_1 \\ \vdots & & & \\ \sin x_n & \sin 2x_n & & \sin nx_n \end{vmatrix} = 2^{n(n-1)/2} \left(\prod_{i=1}^n \sin x_i \right) \prod_{j<k} (\cos x_j - \cos x_k).$$

1.3.4 Riesz Interpolation Formula

Consider now an arbitrary trigonometric polynomial of degree n given by (1.1.6), i.e.,

$$t_n(x) = \frac{a_0}{2} + \sum_{k=1}^{n} (a_k \cos kx + b_k \sin kx) \qquad (1.3.17)$$

and take $2n$ equidistant points

$$\xi_k := \frac{(2k-1)\pi}{2n} \qquad (k = 1, \ldots, 2n).$$

Following [384] we can prove:

Proposition 1.3.2 *The polynomial (1.3.17) can be expressed in the form*

$$t_n(x) = a_n \cos nx + \frac{\cos nx}{2n} \sum_{k=1}^{2n} (-1)^k \cot \frac{x - \xi_k}{2} t_n(\xi_k). \qquad (1.3.18)$$

Proof Since ξ_k ($k = 1, \ldots, 2n$) are zeros of $\cos nx$ we have (see (1.1.8))

$$\cos nx = A \prod_{m=1}^{2n} \sin \frac{x - \xi_m}{2}, \qquad (1.3.19)$$

where A is a constant. Notice also that for an arbitrary $k \in \{1, \ldots, 2n\}$

$$\cot \frac{x - \xi_k}{2} = \frac{\cos \frac{x - \xi_k}{2}}{\sin \frac{x - \xi_k}{2}} = -\frac{\sin \frac{x - (\pi + \xi_k)}{2}}{\sin \frac{x - \xi_k}{2}},$$

so that we can conclude that

$$P_k^*(x) := \frac{(-1)^k}{2n} \cos nx \cot \frac{x - \xi_k}{2}$$

is a trigonometric polynomial of degree n. This polynomial has the same form as (1.3.19), except that ξ_k is replaced by $\pi + \xi_k$. It is clear that

$$P_k^*(\xi_m) = \begin{cases} 1 & \text{if } k = m, \\ 0 & \text{if } k \neq m. \end{cases}$$

Therefore,

$$t_n^*(x) = \frac{\cos nx}{2n} \sum_{k=1}^{2n} (-1)^k \cot \frac{x - \xi_k}{2} t_n(\xi_k)$$

1.3 Chebyshev Systems and Interpolation

is a trigonometric polynomial of degree n which interpolates $t_n(x) - \alpha \cos nx$ at the nodes ξ_k ($k = 1, \ldots, 2n$). Now, because of Theorem 1.1.4, we must have $t_n^*(x) = t_n(x) - \alpha \cos nx$, where α is a constant which can be determined in the following way. Since

$$\int_{-\pi}^{\pi} \cos nx \, t_n^*(x) \, dx = \frac{1}{2n} \sum_{k=1}^{2n} (-1)^k t_n(\xi_k) \int_{-\pi}^{\pi} \cos^2 nx \cot \frac{x - \xi_k}{2} \, dx$$

and

$$\int_{-\pi}^{\pi} \cos^2 nx \cot \frac{x - \xi_k}{2} \, dx = \int_{-\pi}^{\pi} \cos^2 n(\theta + \xi_k) \cot \frac{\theta}{2} \, d\theta$$

$$= \int_{-\pi}^{\pi} \sin^2 n\theta \cot \frac{\theta}{2} \, d\theta = 0,$$

we conclude from the condition

$$0 = \int_{-\pi}^{\pi} \cos nx \, t_n^*(x) \, dx = \int_{-\pi}^{\pi} \cos nx \, [t_n(x) - \alpha \cos nx] \, dx$$

that $\alpha = a_n$, which proves the result. $\square$

Differentiating (1.3.18) and putting $x = 0$, we get

$$t'(0) = \frac{1}{4n} \sum_{k=1}^{2n} (-1)^{k+1} \frac{t_n(\xi_k)}{\sin^2 \frac{\xi_k}{2}}. \tag{1.3.20}$$

This formula holds for every trigonometric polynomial of degree at most n. In a special case we can take the polynomial in θ, $t_n(\theta + x)$, where x is a fixed number and apply (1.3.20). In that case we obtain

$$t_n'(x) = \frac{1}{4n} \sum_{k=1}^{2n} \frac{(-1)^{k+1}}{\sin^2 \frac{\xi_k}{2}} t_n(x + \xi_k). \tag{1.3.21}$$

This formula is known as the *Riesz interpolation formula*.

Using (1.3.21) for $t_n(x) = \sin nx$ and setting $x = 0$ (see [384] or [397]) we obtain

$$n = \frac{1}{4n} \sum_{k=1}^{2n} \frac{1}{\sin^2 \frac{\xi_k}{2}}. \tag{1.3.22}$$

Taking now L^p-norm ($1 \leq p \leq +\infty$) over $A = \mathbb{T} = [0, 2\pi)$ in (1.3.21) we get

$$\|t_n'\|_p \leq \frac{1}{4n} \sum_{k=1}^{2n} \frac{1}{\sin^2 \frac{\xi_k}{2}} \|t_n\|_p.$$

Finally, using the equality (1.3.22) we obtain the well-known *Bernstein inequality*.

Theorem 1.3.1 *If $t_n \in \mathcal{T}_n$ then*
$$\|t_n'\|_p \leq n\|t_n\|_p \qquad (1 \leq p \leq +\infty),$$
with equality only for $t_n(x) = \gamma \sin n(x - x_0)$, where γ and x_0 are constants.

Remark 1.3.2 Notice that we can put here $L^\infty[0, 2\pi] = C[0, 2\pi]$.

Remark 1.3.3 The Bernstein inequality holds also for $0 < p < 1$ (cf. [359, pp. 584–585]).

1.3.5 A General Interpolation Problem

In this subsection, instead of (1.3.2), we consider a moregeneral interpolation problem. Let ϑ_k ($k = 1, \ldots, n$) be n given linear functionals defined on a Haar space X_n. The problem is finding an element $\phi \in X_n$ such that

$$\vartheta_k(\phi) = c_k \qquad (k = 1, \ldots, n). \tag{1.3.23}$$

The vector space of all linear functionals defined on X_n is denoted by X_n^*.

Notice that we have the case (1.3.2) if the functionals ϑ_k are defined as values of ϕ at distinct points x_k, i.e., $\vartheta_k(\phi) = \phi(x_k)$.

Proposition 1.3.3 *Let ϑ_k ($k = 1, \ldots, n$) be n linear functionals on X_n. The interpolation problem (1.3.23) has a unique solution for arbitrary values c_k ($k = 1, \ldots, n$) if and only if the functionals $\vartheta_1, \ldots, \vartheta_n$ are independent in X_n^*.*

The proof of this statement as well as several examples can be found in [75].

Let $X_n = \text{span}\{\phi_1, \ldots, \phi_n\}$. Taking $\phi = \sum_{\nu=1}^{n} a_\nu \phi_\nu \in X_n$, we get from (1.3.23) the following system of linear equations

$$\vartheta_k(\phi) = \vartheta_k\left(\sum_{\nu=1}^{n} a_\nu \phi_\nu\right) = \sum_{\nu=1}^{n} a_\nu \vartheta_k(\phi_\nu) = c_k \qquad (k = 1, \ldots, n),$$

whose matrix has the form

$$G = \begin{bmatrix} \vartheta_1(\phi_1) & \cdots & \vartheta_1(\phi_n) \\ \vdots & & \\ \vartheta_n(\phi_1) & & \vartheta_n(\phi_n) \end{bmatrix}. \tag{1.3.24}$$

An interesting case leads to the *Hermite interpolation*.

1.3 Chebyshev Systems and Interpolation

Let x_j $(j = 1, \ldots, m)$ be distinct points in A, which are equipped with multiplicities $k_j > 0$ $(j = 1, \ldots, m)$, with $k_1 + \cdots + k_m = n$. Defining the index $\alpha(i, j)$ by

$$\alpha(i, j) = i + \sum_{\nu=1}^{j-1} k_\nu \qquad (i = 0, 1, \ldots, k_j - 1;\ j = 1, \ldots, m),$$

we suppose n functionals in the forms

$$\vartheta_{\alpha(i,j)}(\phi) = \phi^{(i)}(x_j) \qquad (i = 0, 1, \ldots, k_j - 1;\ j = 1, \ldots, m). \qquad (1.3.25)$$

Then, the matrix (1.3.24) can be represented as a block matrix

$$G = \begin{bmatrix} G_1^T & \cdots & G_m^T \end{bmatrix}^T, \qquad (1.3.26)$$

where its blocks are

$$G_j = \begin{bmatrix} \phi_1(x_j) & \cdots & \phi_n(x_j) \\ \phi_1'(x_j) & & \phi_n'(x_j) \\ \vdots & & \\ \phi_1^{(k_j-1)}(x_j) & & \phi_n^{(k_j-1)}(x_j) \end{bmatrix} \qquad (j = 1, \ldots, m).$$

If $\det G \neq 0$, the existence and uniqueness of an interpolation polynomial $\phi = \sum_{\nu=1}^{n} a_\nu \phi_\nu \in X_n$, which satisfies the conditions

$$\phi^{(i)}(x_j) = f_{i,j} \qquad (i = 0, 1, \ldots, k_j - 1;\ j = 1, \ldots, m),$$

are guaranteed for any choice of the numbers $f_{i,j}$.[5] Such ϕ is known as the *Hermite interpolation polynomial*.

Remark 1.3.4 In the extreme case when $k_j = 1$ for each j, the Hermite interpolation reduces to the basic interpolation problem treated in Proposition 1.3.1. In that case, the determinant of the matrix (1.3.26) becomes (1.3.1).

Another extreme case is when we have only one point $x_1 = \alpha \in A$, i.e., when $m = 1$ and $k_1 = n$. If we take $X_n = \mathcal{P}_{n-1}$ and $c_i = f_{i,1} = f^{(i)}(\alpha)$ $(i = 0, 1, \ldots, n-1)$, we get the *Taylor interpolation*, i.e.,

$$\phi(x) = f(\alpha) + f'(\alpha)(x - \alpha) + \cdots + \frac{1}{(n-1)!} f^{(n-1)}(\alpha)(x - \alpha)^{n-1}.$$

---

[5] To avoid indexing difficulties, we use the numbers $f_{i,j}$ instead of c_k.

Here, it is convenient to take $\phi_k(x) = (x-\alpha)^{k-1}$ ($k = 1, \ldots, n-1$). Then, the matrix G reduces to a diagonal matrix $G_0 = \text{diag}\,(0!, 1!, 2!, \ldots, (n-1)!)$ and consequently

$$\det G_0 = 1! \cdot 2! \cdots (n-1)! \neq 0.$$

This means that the Taylor polynomial exists and is unique.

Example 1.3.4 In the case $X_n = \mathcal{P}_{n-1}$, $\phi_k(x) = x^{k-1}$ ($k = 1, \ldots, n$), $A = [a, b]$, and $a \leq x_1 < \cdots < x_m \leq b$ equipped with multiplicities $k_j > 0$ ($j = 1, \ldots, m$) such that $k_1 + \cdots + k_m = n$, the determinant of the matrix (1.3.26) is given by

$$\det G = \prod_{j=1}^{m} \left(\prod_{\nu=0}^{k_j-1} \nu! \right) \prod_{j<k} (x_k - x_j) \neq 0.$$

Thus, there exists a unique *Hermite algebraic interpolation polynomial* for any choice of the numbers $f_{i,j}$.

Example 1.3.5 Let $X_{2n+1} = \mathcal{T}_n$, $A = \mathbb{T}$,

$$\phi_0(x) = 1, \quad \phi_{2k-1}(x) = \cos kx, \quad \phi_{2k}(x) = \sin kx \quad (k = 1, \ldots, n),$$

and let x_j ($j = 1, \ldots, m$) be distinct points in $\mathbb{T}$, which are equipped with multiplicities $k_j > 0$ ($j = 1, \ldots, m$), with $k_1 + \cdots + k_m = 2n+1$. Then, the determinant of the matrix (1.3.26) can be expressed in the form [401]

$$\det G = \frac{1}{2^n} \prod_{j=1}^{m} \left(\prod_{\nu=0}^{k_j-1} \nu! \right) \prod_{j<k} \left(2 \sin \frac{x_k - x_j}{2} \right).$$

Since $\det G \neq 0$, the *Hermite trigonometric interpolation polynomial* exists uniquely.

Remark 1.3.5 If some of m sequences in (1.3.25) are discontinuous (do not consist terms for each $i = 0, 1, \ldots, k_j - 1$) we have the *lacunary interpolation* or *Birkhoff interpolation* (cf. [268] and [431]).

1.4 Interpolation by Algebraic Polynomials

1.4.1 Representations and Computation of Interpolation Polynomials

In Sect. 1.3.2 we have already introduced a polynomial of minimal degree which interpolates a function f at n fixed points $x_1, x_2, \ldots, x_n$. Such a polynomial is known

1.4 Interpolation by Algebraic Polynomials

as the Lagrange interpolation polynomial and it was discovered in 1795 by Joseph Louis Lagrange. Otherwise, interpolation by polynomials is a very old subject in mathematics. The first formula with equally spaced points $\{x_k\}$ was found in the 1670's by Isaac Newton.

Defining the so-called *divided differences* of f recursively by

$$[x_1; f] = f(x_1), \quad [x_1, \ldots, x_k; f] = \frac{[x_2, \ldots, x_k; f] - [x_1, \ldots, x_{k-1}; f]}{x_k - x_1},$$

the *Newton interpolation formula* can be given also for non-equally spaced nodes $\{x_k\}$ in the form

$$L_n f(x) = f(x_1) + \sum_{k=2}^{n} (x - x_1) \cdots (x - x_{k-1})[x_1, \ldots, x_k; f]. \tag{1.4.1}$$

The computation of divided differences requires $n(n-1)$ additions and $\frac{1}{2}n(n-1)$ divisions(cf. Gautschi [163, p. 92]). Adding another data point $(x_{n+1}, f(x_{n+1}))$ requires the generation of only the next difference $[x_1, \ldots, x_{n+1}; f]$ and adding to (1.4.1) the term $(x - x_1) \cdots (x - x_n)[x_1, \ldots, x_{n+1}; f]$, but this formula is unstable.

The Lagrange interpolation formula (1.3.7) is very attractive for many theoretical purposes, but its application for practical computational work is not so efficient, especially when we want to add a new node x_{n+1}. However, the formula

$$L_n f(x) = \sum_{k=1}^{n} \ell_{n,k}(x) f(x_k) \tag{1.4.2}$$

can be rewritten in a form that makes the Lagrange interpolation formula computationally efficient and also allows additional nodes to be added.

Introducing the auxiliary quantities

$$\lambda_k^{(n)} = \prod_{\substack{\nu=1 \\ \nu \neq k}}^{n} \frac{1}{x_k - x_\nu}, \quad k = 1, \ldots, n, \tag{1.4.3}$$

we can express the fundamental Lagrange polynomials (1.3.5), i.e., (1.3.6), in the form $\ell_{n,k}(x) = \lambda_k^{(n)} q_n(x)/(x - x_k)$.

Indeed, dividing (1.4.2) through $\sum_{\nu=1}^{n} \ell_{n,\nu}(x) \equiv 1$, we get

$$L_n f(x) = \frac{\sum_{k=1}^{n} \ell_{n,k}(x) f(x_k)}{\sum_{\nu=1}^{n} \ell_{n,\nu}(x)} = \frac{\sum_{k=1}^{n} \frac{\lambda_k^{(n)}}{x - x_k} q_n(x) f(x_k)}{\sum_{\nu=1}^{n} \frac{\lambda_\nu^{(n)}}{x - x_\nu} q_n(x)},$$

i.e.,
$$L_n f(x) = \frac{\sum_{k=1}^{n} \frac{\lambda_k^{(n)}}{x - x_k} f(x_k)}{\sum_{\nu=1}^{n} \frac{\lambda_\nu^{(n)}}{x - x_\nu}} \quad (x \notin \{x_1, \ldots, x_n\}). \quad (1.4.4)$$

This formula is called the *barycentric formula*. Comparing (1.4.4) and (1.4.2) we obtain

$$\ell_{n,k}(x) = \frac{\frac{\lambda_k^{(n)}}{x - x_k}}{\sum_{\nu=1}^{n} \frac{\lambda_\nu^{(n)}}{x - x_\nu}}, \quad k = 1, \ldots, n.$$

An efficient algorithm for computing the required quantities $\lambda_k^{(n)}$ was suggested by Werner [503] (see also [163, p. 96] and [164]):

Starting with $\lambda_1^{(1)} = 1$, for $k = 2, \ldots, n$ do

$$\lambda_\nu^{(k)} = \frac{\lambda_\nu^{(k-1)}}{x_\nu - x_k} \quad (\nu = 1, \ldots, k-1); \qquad \lambda_k^{(k)} = -\sum_{\nu=1}^{k-1} \lambda_\nu^{(k)}.$$

The first set of equations in the k-loop is a direct consequence of (1.4.3), while the second equation follows from the identity

$$1 \equiv \sum_{\nu=1}^{k} \ell_{k,\nu}(x) = \sum_{\nu=1}^{k} \lambda_\nu^{(k)} \prod_{\substack{i=1 \\ i \neq \nu}}^{k} (x - x_i)$$

by comparing the leading coefficients on the left and right side. It gives $\sum_{\nu=1}^{k} \lambda_\nu^{(k)} = 0$.

The previous algorithm requires exactly $(n-1)^2$ additions and $\frac{1}{2}n(n-1)$ divisions for computing the n quantities (1.4.3). In order to include a new data point $(x_{n+1}, f(x_{n+1}))$, we must only extend the k-loop in the previous algorithm through $n+1$.

Recently, Higham [211] has given an error analysis of the evaluation of the interpolating polynomial using two more computationally attractive forms: a modified Lagrange form and a barycentric form. His analysis shows that barycentric Lagrange interpolation should be the polynomial interpolation method of choice. Also, Berrut and Trefethen in a nice survey paper [39] have introduced, in a general way, barycentric Lagrange interpolation, including the barycentric formula, convergence rates, numerical stability, etc.

1.4.2 Interpolation Array and Lagrange Operators

Without loss of generality, we consider interpolation on the interval $A = [-1, 1]$. Suppose now that we have a continuous function $f:[-1,1] \to \mathbb{R}$ and a sequence of monic polynomials $\{q_n\}_{n \in \mathbb{N}}$ ($q_n \in \hat{\mathcal{P}}_n$) such that for each $n \geq 1$, the polynomial q_n has n distinct zeros $x_{n,k}$ ($k = 1, \ldots, n$) in $[-1, 1]$, i.e.,

$$-1 \leq x_{n,1} < x_{n,2} < \cdots < x_{n,n} \leq 1. \tag{1.4.5}$$

Let $\mathcal{X}$ be the corresponding infinite triangular array of these zeros

$$\mathcal{X} := \begin{Bmatrix} x_{1,1} & & & \\ x_{2,1} & x_{2,2} & & \\ \vdots & & \ddots & \\ x_{n,1} & x_{n,2} & \cdots & x_{n,n} \\ \vdots & & & & \ddots \end{Bmatrix}. \tag{1.4.6}$$

Thus, the n-th row of this array $\mathcal{X}$ consists of the zeros of the polynomial q_n. According to this fact, sometimes we write $\mathcal{Q}$ instead of $\mathcal{X}$. For example, taking the zeros of the Chebyshev polynomials T_n, we use the notation $\mathcal{T}$ for the corresponding triangular array.

Now, we can associate to $\{q_n\}_{n \in \mathbb{N}}$, i.e., to the array $\mathcal{X}$, a sequence of Lagrange polynomials $\{L_n(\mathcal{X}, f)\}_{n \in \mathbb{N}}$, defined by

$$L_n(\mathcal{X}, f)(x_{n,k}) = L_n(\mathcal{X}, f; x_{n,k}) = f(x_{n,k}), \quad k = 1, \ldots, n.$$

Note that obviously $L_n(\mathcal{X}, f) \in \mathcal{P}_{n-1}$ and the index n in $L_n(\mathcal{X}, f)$ denotes the number of knots. The triangular $\mathcal{X}$ is called the *interpolation array* or the *system of interpolation nodes*. In the notation $L_n(\mathcal{X}, f)(x)$ we aim to emphasize the dependence on the system of knots $\mathcal{X}$.

Referring to (1.3.7), the Lagrange interpolation formula is

$$L_n(\mathcal{X}, f)(x) = L_n(\mathcal{X}, f; x) = \sum_{k=1}^{n} \ell_{n,k}(\mathcal{X}; x) f(x_k), \tag{1.4.7}$$

where $\ell_{n,k}(\mathcal{X}, x) = \ell_{n,k}(x)$ ($\in \mathcal{P}_{n-1}$) are the fundamental Lagrange polynomials given by

$$\ell_{n,k}(\mathcal{X}; x) := \prod_{\substack{\nu=1 \\ \nu \neq k}}^{n} \frac{x - x_{n,\nu}}{x_{n,k} - x_{n,\nu}}, \quad k = 1, \ldots, n, \tag{1.4.8}$$

or equivalently by

$$\ell_{n,k}(\mathcal{X}; x) := \frac{q_n(x)}{q_n'(x_{n,k})(x - x_{n,k})}, \quad k = 1, \ldots, n. \tag{1.4.9}$$

Usually, for a fixed n, we simplify the notation putting x_k instead of $x_{n,k}$ and omitting $\mathcal{X}$ (as in Sect. 1.3.2).

For a given interpolation array $\mathcal{X}$, according to (1.4.7) we define a sequence of operators $L_n(\mathcal{X}): C[-1, 1] \to \mathcal{P}_{n-1}$ ($n = 1, 2, \ldots$) such that $L_n(\mathcal{X})f = L_n(\mathcal{X}, f)$. The sequence $\{L_n(\mathcal{X})\}_{n\in\mathbb{N}}$ defines an *interpolatory process*. For each $n \in \mathbb{N}$, $L_n(\mathcal{X})$ is a linear map and $L_n(\mathcal{X}, f) = f$ holds when f is a polynomial of degree at most $n-1$ ($f \in \mathcal{P}_{n-1}$).

The main question is the convergence $L_n(\mathcal{X}, f) \to f$ when $n \to +\infty$. In other words, what kind of the interpolation array $\mathcal{X}$ provides this convergence. Here, we consider the space $C^0 = C[-1, 1]$ of all continuous functions f equipped with the uniform norm

$$\|f\| = \|f\|_\infty = \max_{|x|\leq 1} |f(x)|$$

and consider $L_n(\mathcal{X})$ as a map from C^0 into itself, i.e., $L_n(\mathcal{X}): C^0 \to C^0$. It is easy to compare the *interpolation error* $\|f - L_n(\mathcal{X}, f)\|$ with the error of best uniform approximation $E_{n-1}(f)$, given by

$$E_{n-1}(f) = \min_{p\in\mathcal{P}_{n-1}} \|f - p\| = \|f - P^*\|,$$

where P^* ($\in \mathcal{P}_{n-1}$) is the polynomial of best uniform approximation to the function $f \in C^0$. Since, by the Chebyshev alternation theorem, this polynomial interpolates f in at least n points, there exists, for each $f \in C^0$, an interpolation array $\mathcal{Y}$ for which

$$\|f - L_n(\mathcal{Y}, f)\| = E_{n-1}(f)$$

tends to zero as $n \to +\infty$. Note that for a given $f \in C^0$ the related array $\mathcal{Y}$ is unknown. Moreover, for the whole class C^0, the situation is much less favorable.

Taking the polynomial of best approximation P^* ($\in \mathcal{P}_{n-1}$) to $f \in C^0$, we get

$$|f(x) - L_n(\mathcal{X}, f; x)| \leq |f(x) - P^*(x)| + |P^*(x) - L_n(\mathcal{X}, f; x)|$$

$$\leq E_{n-1}(f) + |L_n(\mathcal{X}, f - P^*; x)|$$

$$\leq \left(1 + \sum_{k=1}^{n} |\ell_{n,k}(\mathcal{X}; x)|\right) E_{n-1}(f),$$

i.e.,

$$|f(x) - L_n(\mathcal{X}, f; x)| \leq (1 + \lambda_n(\mathcal{X}; x)) E_{n-1}(f), \qquad (1.4.10)$$

where we introduced the *Lebesgue function*

$$\lambda_n(\mathcal{X}; x) := \sum_{k=1}^{n} |\ell_{n,k}(\mathcal{X}; x)|. \qquad (1.4.11)$$

1.4 Interpolation by Algebraic Polynomials

Using the norm of the operator $L_n(\mathcal{X}): C^0 \to C^0$, given by

$$\|L_n(\mathcal{X})\| := \sup_{f \in C^0} \frac{\|L_n(\mathcal{X}, f)\|}{\|f\|} = \sup_{\|f\|=1} \|L_n(\mathcal{X}, f)\|, \qquad (1.4.12)$$

it is easy to see that this norm is exactly the norm of the Lebesgue function (1.4.11), and therefore it is called the *Lebesgue constant* and is denoted by $\Lambda_n(\mathcal{X})$. Thus,

$$\Lambda_n(\mathcal{X}) = \|L_n(\mathcal{X})\| = \|\lambda_n(\mathcal{X}; x)\| = \max_{-1 \le x \le 1} |\lambda_n(\mathcal{X}; x)|. \qquad (1.4.13)$$

Finally, taking the maximum in (1.4.10) over $[-1, 1]$, we get

$$\|f - L_n(\mathcal{X}, f)\| \le (1 + \Lambda_n(\mathcal{X})) E_{n-1}(f). \qquad (1.4.14)$$

According to (1.4.10) and (1.4.14), the behaviour of (1.4.11) and (1.4.13) plays an important role in the study of the convergence of the Lagrange polynomials.

In 1914 Faber [123] proved that

$$\Lambda_n(\mathcal{X}) \ge \frac{1}{12} \log n, \qquad n \ge 1, \qquad (1.4.15)$$

for any interpolation array $\mathcal{X}$. Based on this result Faber [123] obtained:

Theorem 1.4.1 *For any fixed interpolation array $\mathcal{X}$ there exists a function $f \in C^0$ for which the interpolation polynomials $L_n(\mathcal{X}, f)$ do not converge uniformly to f, in fact,*

$$\limsup_{n \to +\infty} \|L_n(\mathcal{X}, f)\| = +\infty. \qquad (1.4.16)$$

Grünwald [196, 197] and Marcinkiewicz [282] obtained independently the following result:

Theorem 1.4.2 *If $\mathcal{T}$ is the Chebyshev array of nodes, then there exists a continuous function f_0 for which*

$$\limsup_{n \to +\infty} |L_n(\mathcal{T}, f_0; x)| = +\infty, \qquad \text{for any } x \in [-1, 1].$$

Finally, in 1980 Erdős and Vértesi [120] proved the following statement on the divergence of the Lagrange interpolation processes:

Theorem 1.4.3 *For any interpolation array $\mathcal{X}$ on $[-1, 1]$ one can find a function $f \in C^0$ such that*

$$\limsup_{n \to +\infty} \|L_n(\mathcal{X}, f)\| = +\infty \quad \text{a.e. in } [-1, 1].$$

Moreover, the divergence set is of second category on $[-1, 1]$.

A survey about many results in this direction can be found in [112, 498], as well as in the monograph [465], written by Szabados and Vértesi.

In the sequel, we will give some estimates for the Lebesgue constant of some interpolation arrays. But first, we come back to some classical results on the interpolation error. Also, we give a historical review of some interesting interpolatory processes.

1.4.3 Interpolation Error for Some Classes of Functions

In order to be able to estimate the *error of interpolation*,

$$R_n(x) = R_n(\mathcal{X}, f; x) = f(x) - L_n(\mathcal{X}, f; x), \qquad (1.4.17)$$

for any $x \neq x_k$ in $[-1, 1]$ we need some additional assumptions about the function $f \in C[-1, 1]$. Since the nodes x_k, $k = 1, \ldots, n$, are zeros of the polynomial q_n and $R_n(x_k) = 0$ for each x_k, it is clear that we can write, in general,

$$R_n(x) = \Phi(x; f, n, \mathcal{X}) q_n(x),$$

but $\Phi(x; f, n, \mathcal{X})$ depends, above all, on the structural properties of the function f. Very often, the error of interpolation (1.4.17) is called the *remainder term*.

1.4.3.1 The Error in the Class of Continuous-Differentiable Functions

By restricting the class $C[-1, 1]$ to the class of n times continuous-differentiable functions $C^n[-1, 1]$, the following well-known estimate for (1.4.17) is valid.

Theorem 1.4.4 *Let $f \in C^n[-1, 1]$ and $q_n(x) = \prod_{k=1}^{n}(x - x_k)$. Then, for each $x \in [-1, 1]$, there exists a point $\xi = \xi(x) \in (-1, 1)$ such that*

$$R_n(x) = f(x) - L_n(f)(x) = \frac{f^{(n)}(\xi)}{n!} q_n(x). \qquad (1.4.18)$$

Repeatedly applying Rolle's theorem, Augustin Luis Cauchy gave an elegant proof of (1.4.18), which today can be found in every standard book on classical analysis.

An alternative estimate of $R_n(x)$ can be given in terms of divided differences (cf. [186, p. 74])

$$R_n(x) = f(x) - L_n(f)(x) = q_n(x)[x; x_1, \ldots, x_n; f]. \qquad (1.4.19)$$

The monic polynomial $q_n(x) = (x - x_1) \cdots (x - x_n)$ is called the *node polynomial*.

1.4 Interpolation by Algebraic Polynomials

Taking $M_n = \max_{|x| \leq 1} |f^{(n)}(x)|$ and using (1.4.18) we obtain the following estimate

$$|R_n(x)| = |f(x) - L_n(f)(x)| \leq \frac{M_n}{n!} |q_n(x)| \qquad (-1 \leq x \leq 1). \qquad (1.4.20)$$

For an arbitrary system of knots $\mathcal{X}$, $|q_n(x)| \leq 2^n$ holds for all $|x| \leq 1$ and we get a crude estimate

$$|R_n(x)| \leq \frac{2^n M_n}{n!} \qquad (-1 \leq x \leq 1).$$

But, using this estimate for a restricted class of functions and for arbitrary systems of nodes, we can easily prove the uniform convergence of $L_n(\mathcal{X}, f)$ to f in $[-1, 1]$. Namely, if we take functions f having uniformly bounded derivatives of any order, i.e.,

$$\sup_n M_n = \sup_n \left(\max_{|x| \leq 1} |f^{(n)}(x)| \right) = M < +\infty,$$

we get

$$\sup_{\mathcal{X}} \max_{|x| \leq 1} |f(x) - L_n(\mathcal{X}, f; x)| \leq \frac{2^n M}{n!}. \qquad (1.4.21)$$

Since the right side in (1.4.21) tends to zero when $n \to +\infty$, we have the uniform convergence of $L_n(\mathcal{X}, f)$ to f in $[-1, 1]$ for any $\mathcal{X}$.

According to Theorem 1.1.7, the Chebyshev nodes (1.1.19) are often a good choice of interpolation nodes. In that case, (1.4.20) becomes

$$|R_n(x)| = |f(x) - L_n(f)(x)| \leq \frac{M_n}{2^{n-1} n!} |T_n(x)| \qquad (-1 \leq x \leq 1).$$

Such an interpolation is called the *Chebyshev interpolation*. The node polynomials with equally spaced and Chebyshev nodes are displayed in Fig. 1.4.1.

1.4.3.2 The Error in the Class of Analytic Functions

For analytic functions we can give an explicit formula for the remainder term

$$R_n(z) = f(z) - L_n(f)(z),$$

using the well-known Cauchy residue theorem.

Theorem 1.4.5 *Let Γ be a simple closed contour in $\mathbb{C}$ and let the nodes $x_k = x_{n,k} \in \text{int } \Gamma$. For an analytic function $z \mapsto f(z)$ on and inside Γ, we have*

$$R_n(z) = \frac{1}{2\pi i} \oint_\Gamma \frac{q_n(z)}{q_n(\zeta)} \frac{f(\zeta)}{\zeta - z} d\zeta \qquad (z \in \text{int } \Gamma), \qquad (1.4.22)$$

where $q_n(z) = \prod_{k=1}^n (z - x_k)$.

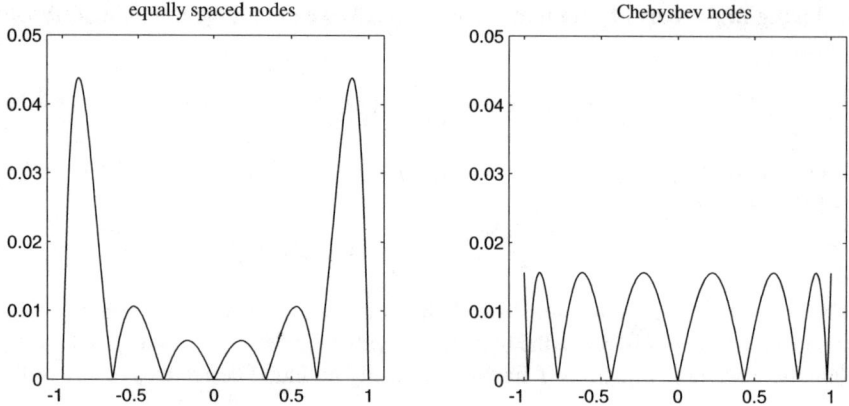

Fig. 1.4.1 Graph of $|q_n(x)|$ for $n = 7$, with equally spaced nodes (*left*) and Chebyshev nodes (*right*)

Proof Define the function $\zeta \mapsto F(\zeta)$ by

$$F(\zeta) = \frac{q_n(z)}{q_n(\zeta)} \frac{f(\zeta)}{\zeta - z},$$

which has simple poles at z and x_k, $k = 1, \ldots, n$, with residues $f(z)$ and $f(x_k)q_n(x_k)/((x_k - z)\pi'_n(x_k)) = -\ell_{n,k}(z)f(x_k)$, $k = 1, \ldots, n$, respectively. Here, $\ell_{n,k}(z)$ are the fundamental Lagrange polynomials.

An application of the residue theorem gives

$$\frac{1}{2\pi i} \oint_\Gamma F(\zeta)\, d\zeta = f(z) - \sum_{k=1}^n \ell_{n,k}(z) f(x_k) = f(z) - L_n(f)(z),$$

i.e., (1.4.22). □

Remark 1.4.1 The restriction (1.4.5) for the nodes x_k is omitted in Theorem 1.4.5. The nodes can be different complex numbers inside the contour Γ.

1.4.4 Uniform Convergence in the Class of Analytic Functions

Interpolation polynomials were widely used in the nineteenth century, but without a rigorous analysis of their convergence. Perhaps the first significant, but negative result, is due to Charles Méray in 1884, in the complex plane. He observed that the interpolation polynomial for the function $z \mapsto f(z) = 1/z$ at the nth roots of unity is $L_n(f)(z) = z^{n-1}$, and clearly this does not converge to f as $n \to +\infty$, on the unit circle or off. Of course, in this example f is not an analytic function in $|z| \leq 1$

1.4 Interpolation by Algebraic Polynomials

(because of the pole at 0), so that no sequence of polynomials (interpolatory or not) can converge to f uniformly on the unit circle (cf. [272]).

For some distributions of nodes $\mathcal{X}$ and functions analytic in a sufficiently large region, using Theorem 1.4.5, it is possible to prove the uniform convergence of the corresponding interpolatory process in a sub-region. In other words, the location of the interpolation points and the analyticity of f play an important role in this subject. The key to convergence is contained in the asymptotic behaviour of $|q_n(z)|^{1/n}$, and therefore we investigate convergence supposing the existence of the limit

$$A(z) = \lim_{n \to +\infty} |q_n(z)|^{1/n} = \lim_{n \to +\infty} \left(\prod_{k=1}^n |z - x_{n,k}| \right)^{1/n} \tag{1.4.23}$$

on certain regions in the complex plane and uniform convergence there.

Theorem 1.4.6 *Let the conditions of Theorem 1.4.5 be satisfied, let γ be a simple closed contour inside Γ, and*

$$\delta = \min_{\substack{\zeta \in \Gamma \\ z \in D}} |\zeta - z|, \qquad D = \overline{\operatorname{int} \gamma} \subset \operatorname{int} \Gamma.$$

If there exist the constants $\alpha, \beta, 0 \leq \alpha < \beta$, such that

$$(\forall z \in D) \quad A(z) \leq \alpha \qquad \text{and} \qquad (\forall \zeta \in \Gamma) \quad A(\zeta) \geq \beta, \tag{1.4.24}$$

where A is given by (1.4.23), then the interpolatory process $\{L_n(\mathcal{X}, f)\}_{n \in \mathbb{N}}$ converges to f uniformly in D.

Proof Let $L = \ell(\Gamma)$ be the length of Γ and $M = \max_{\zeta \in \Gamma} |f(\zeta)|$. Then, according to Theorem 1.4.5 we have

$$|R_n(z)| \leq \frac{1}{2\pi} \oint_\Gamma \frac{|q_n(z)|}{|q_n(\zeta)|} \frac{|f(\zeta)|}{|\zeta - z|} |d\zeta| \leq \frac{ML}{2\pi\delta} \cdot \frac{\max_{z \in D} |q_n(z)|}{\min_{\zeta \in \Gamma} |q_n(\zeta)|}$$

for each $z \in D$. Based on (1.4.23) and using (1.4.24) we can conclude that for a sufficiently small ε there exist two integers $n_1, n_2 \in \mathbb{N}$ such that for $n > \max(n_1, n_2)$ the following inequalities

$$(\forall z \in D) \quad |q_n(z)| \leq (\alpha + \varepsilon)^n \qquad \text{and} \qquad (\forall \zeta \in \Gamma) \quad |q_n(\zeta)| \geq (\beta - \varepsilon)^n$$

hold. Thus,

$$\frac{\max_{z \in D} |q_n(z)|}{\min_{\zeta \in \Gamma} |q_n(\zeta)|} \leq q^n \qquad \text{and} \qquad |R_n(z)| \leq \frac{ML}{2\pi\delta} q^n,$$

where $q = (\alpha + \varepsilon)/(\beta - \varepsilon)$. The last estimate holds uniformly for $z \in D$.

Taking, $\varepsilon < (\beta - \alpha)/2$ we see that $q < 1$ and, therefore, $\lim_{n \to +\infty} R_n(z) = 0$ uniformly in $D = \text{int}\,\Gamma$. □

In the sequel, we will restrict our consideration to nodes distributed in $[-1, 1]$ like (1.4.5), with a given limit distribution $d\mu(x)$, and investigate the limit $A(z)$, given by (1.4.23). Precisely, in terms of potential theory, we are interested in the limit case of the logarithmic potential

$$U_n(z) = \frac{1}{n} \log \frac{1}{|q_n(z)|} = \log \frac{1}{|q_n(z)|^{1/n}},$$

induced by the discrete measure that has mass $1/n$ at every zero of q_n.

In particular, we will analyze two cases:

1° *Uniformly distributed nodes over* $[-1, 1]$. For the equally spaced nodes

$$x_{n,k} = -1 + \frac{2(k-1)}{n-1}, \qquad k = 1, \ldots, n \quad (n \geq 2).$$

we have $d\mu(x) = dx/2$ and

$$U(z) = \lim_{n \to +\infty} U_n(z) = \frac{1}{2} \int_{-1}^{1} \log \frac{1}{|z-x|}\, dx,$$

for all z outside $[-1, 1]$. Indeed, from

$$e^{-U_n(z)} = |q_n(z)|^{1/n} = 2 \prod_{k=1}^{n} \left| \frac{z+1}{2} - \frac{k-1}{n-1} \right|^{1/n},$$

i.e.,

$$U_n(z) = -\log 2 - \frac{1}{n} \sum_{k=1}^{n} \log \left| \frac{z+1}{2} - \frac{k-1}{n-1} \right|,$$

it follows (by the definition of the integral)

$$U(z) = -\log 2 - \int_0^1 \log\left| \frac{z+1}{2} - t\right| dt = \frac{1}{2}\int_{-1}^1 \log\frac{1}{|z-x|}\,dx.$$

With a little work, we get

$$U(z) = 1 - \frac{1}{2} \text{Re}\{(z+1)\log(z+1) - (z-1)\log(z-1)\}.$$

Notice that $U(\pm 1) = 1 - \log 2$ and $U(0) = 1$.

In the general case, for a given limit distribution $d\mu(x)$, the logarithmic potential can be expressed in the form

$$U(z) = \int_{-1}^{1} \log \frac{1}{|z-x|}\, d\mu(x). \qquad (1.4.25)$$

1.4 Interpolation by Algebraic Polynomials

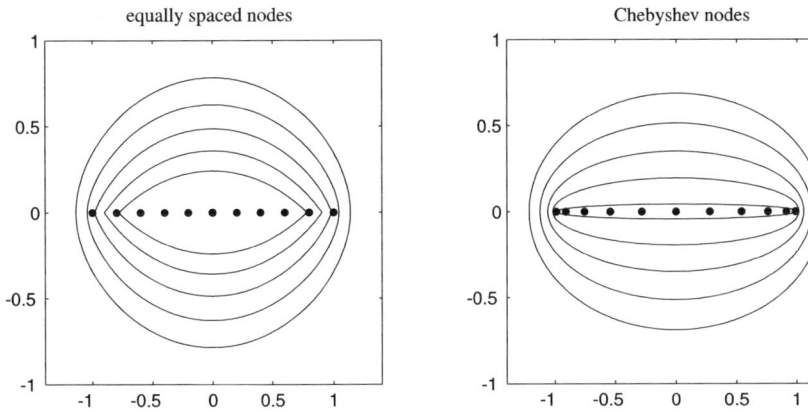

Fig. 1.4.2 The equipotential curves for equally spaced nodes (*left*) and Chebyshev nodes (*right*)

$2°$ *Arc sine distribution of nodes over* $[-1, 1]$. Since the limit measure is given by $d\mu(x) = \pi^{-1}(1 - x^2)^{-1/2} dx$, based on (1.4.25), we have

$$U(z) = \frac{1}{\pi} \int_{-1}^{1} \log \frac{1}{|z - x|} \frac{dx}{\sqrt{1 - x^2}}.$$

After some computation (cf. Saff and Totik [423, pp. 45–46]) we get

$$U(z) = \log \frac{2}{|z + \sqrt{z^2 - 1}|}, \qquad (1.4.26)$$

where $\sqrt{z^2 - 1}$ denotes the branch that behaves like z near infinity. For example, this holds for the Chebyshev nodes (see Sect. 1.1.4)

$$x_{n,k} = -\cos \frac{(2k - 1)\pi}{2n}, \qquad k = 1, \ldots, n. \qquad (1.4.27)$$

Notice that the *minus* sign in (1.4.27) is taken to ensure the ordering (1.4.5). Then, the potential (1.4.26) can be obtained directly from (1.1.24) and (1.4.23), $U(z) = \log(2/r)$, where $r = |z + \sqrt{z^2 - 1}|$. Notice that $U(z) = \log 2 \approx 0.69\ldots$, when $z \in [-1, 1]$ ($r = 1$).

In Fig. 1.4.2 we displayed the equipotential curves for the uniformly distributed nodes and Chebyshev nodes (arc sine distribution). The potential levels $U(z) = c$ were taken between 0.05 and 0.65 with an equidistant step 0.05. In general, as c increases, the level curves "shrink" towards the interval $[-1, 1]$. Putting $\gamma_c = \{z \in \mathbb{C} \mid U(z) = c\}$, it is clear that $\operatorname{int} \gamma_{c_2} \subset \operatorname{int} \gamma_{c_1}$ for $c_2 > c_1$. Of course, $A(z) = \exp(-U(z))$.

The case of the arc sine distribution of nodes is much nicer. The curves are ellipses $\gamma_c = E_r$, where $c = \log(2/r) < \log 2$, i.e., $r = 2e^{-c} > 1$. The *limit ellipse* is just the interval $[-1, 1]$ for $r = 1$ (see Sect. 1.1.4), and also $[-1, 1] = E_1 \subset$

int $E_r \subset$ int E_R for $r < R$. In this case, according to the previous discussion and Theorem 1.4.6, we can conclude that the Lagrange interpolatory process $\{L_n(\mathcal{T}, f)\}_{n \in \mathbb{N}}$ converges uniformly on $[-1, 1]$ ($= E_1$) if f is *analytic on* $[-1, 1]$, i.e., analytic in any region int E_r ($r > 1$), no matter how thin. Precisely, the following result holds:

Theorem 1.4.7 *If f is analytic on the closed ellipse $D = \overline{\text{int } E_r}$ ($r > 1$), then the Lagrange interpolatory process $\{L_n(\mathcal{T}, f)\}_{n \in \mathbb{N}}$ converges uniformly on D.*

In the case of equally spaced nodes the equipotential curve γ passing through ± 1 (ends of the interval $[-1, 1]$) is determined by $U(z) = U(\pm 1) = \log(e/2)$, and its interior D by $U(z) \geq \log(e/2)$, i.e., $A(z) \leq \alpha = 2/e$. Thus, if f is analytic on D, then the corresponding Lagrange interpolatory process converges uniformly on $[-1, 1]$ and, of course, on the complex region D. However, if f has any singular point ($\neq 0$) inside D, it is possible to find an equipotential curve γ^*, determined by $U(z) = \log(1/A(z)) = c^*$, where $1 - \log 2 < c^* < 1$, so that the function f be analytic inside $D^* = \text{int } \gamma^*$. In that case, the convergence is only on D^* and, in particular, on the corresponding central part $(-x^*, x^*)$ of the interval $[-1, 1]$. Thus, for analytic functions on the real line, interpolation at equally spaced points can be a bad idea.

The following Runge's example from 1901 gives an excellent explanation of the previous fact. Namely, let

$$f_a(x) := \frac{1}{1 + (x/a)^2} \qquad (x \in [-1, 1])$$

and let the interpolation nodes be equally spaced points in $[-1, 1]$. This function is analytic on the whole real line, but its continuation $z \mapsto f_a(z)$ to the complex plane does have poles at $\pm ia$, which can be quite close to the interval $[-1, 1]$.

Runge showed that for a sufficiently small a (e.g., $|a| \leq 1/5$)

$$e(n) = \|f_a - L_n(f_a)\| = \max_{-1 \leq x \leq 1} |f_a(x) - L_n(f_a)(x)| \to +\infty, \qquad n \to +\infty.$$

Precisely, according to the previous investigation, the critical value of a can be determined from the equation $U(ia) = 1 - \log 2$, which gives $a^* \approx 0.5255$. Thus, for $|a| > a^*$, $e(n) \to 0$ as $n \to +\infty$. The cases $a = 1$ and $a = 1/4$ with $n = 11$ nodes are displayed in Fig. 1.4.3.

For a small value of the parameter a there exists x^* (depending on a), such that

$$\limsup_{n \to +\infty} |f_a(x) - L_n(f_a)(x)| = \begin{cases} 0 & \text{if } |x| < x^*, \\ +\infty & \text{if } |x| > x^*. \end{cases}$$

Thus, we have pointwise convergence in the central zone $(-x^*, x^*)$ of the interval $[-1, 1]$ and divergence in the lateral zones. The bound x^* as a function on a is given in Fig. 1.4.4. For example, $x^* = 0.7942\ldots$ for $a = 1/4$.

1.4 Interpolation by Algebraic Polynomials

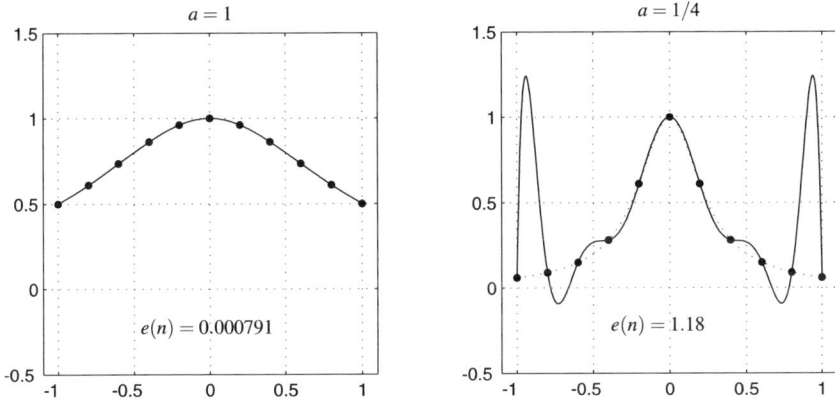

Fig. 1.4.3 The Runge's example for $n = 11$ equally spaced nodes, when $a = 1$ (*left*) and $a = 1/4$ (*right*)

Fig. 1.4.4 The bound x^* as a function of the parameter a in Runge's example

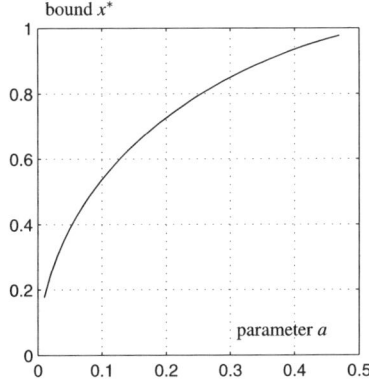

1.4.5 Bernstein's Example of Pointwise Divergence

The following interesting example of pointwise divergence properties of the Lagrange interpolation on equidistant nodes was discovered in 1916 by Bernstein [33] (see also [34]). Namely, for $g_1(x) := |x|$ on $[-1, 1]$, and equidistant interpolation nodes $x_{n,k} = -1 + 2(k-1)/(n-1)$, $k = 1, \ldots, n$, he proved

$$\limsup_{n \to +\infty} |g_1(x) - L_n(g_1)(x)| = +\infty \quad \text{for every } x \in [-1, 1] \tag{1.4.28}$$

except at $x = \pm 1$ and $x = 0$. Notice that $x = \pm 1$ are interpolation nodes, where the error is zero, so that this fact is trivial for these points. The same is true for the point $x = 0$ when n is odd, but not if n is even. Thus, for the point $x = 0$ the situation is more complicated. In 1939 D. L. Berman proved in his thesis that the Lagrange polynomials at zero converge to its true function value, and S. M. Lozinskiĭ established an upper bound for the approximation error and showed that the error tends to

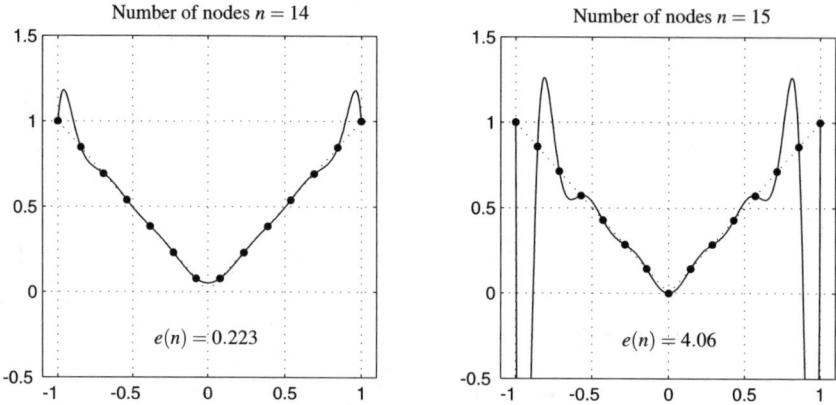

Fig. 1.4.5 Bernstein's example for $n = 14$ (*left*) and $n = 15$ (*right*)

0 with $O(1/n)$. The cases with $n = 14$ and $n = 15$ nodes are presented in Fig. 1.4.5.

Recently Byrne, Mills, and Smith [54] proved

$$\limsup_{n \to +\infty} \frac{1}{n} \log \big| |x| - L_n(g_1)(x) \big| = \frac{1}{2}[(1+x)\log(1+x) + (1-x)\log(1-x)] \quad (1.4.29)$$

if $0 < |x| < 1$. This result shows that for each x with $0 < |x| < 1$, there exists a subsequence of $\{L_n(g_1)(x)\}_{n=2,3,\ldots}$, say $\{L_n(g_1)(x)\}_{n=n_1,n_2,\ldots}$, whose rate of divergence is geometrically fast, but it seems that the sequence $\{n_1, n_2, \ldots\}$ should depend on x. Li and Mohapatra [264] showed that there actually exists a subsequence that works for almost all x. Their interesting result is the following:

Theorem 1.4.8 *For all* $x \in \mathbb{R}$, *we have*

$$\lim_{n \to +\infty} \left| \frac{|x| - L_n(g_1)(x)}{q_n(x)} \right|^{1/n} = e,$$

where $q_n(x) = \prod_{k=1}^{n}(x - x_{n,k})$.

The extension of (1.4.29) to $g_3(x) = |x|^3$ was given in [410].

In [411] Revers showed that (1.4.28) holds true for $g_\alpha(x) = |x|^\alpha$, $\alpha \in (0, 1)$ (see also [413]). For this function, he also established the surprising formula (see [412])

$$\lim_{n \to +\infty} (2n)^\alpha L_{2n}(g_\alpha)(0) = 2\left(\frac{2}{\pi}\right)^{\alpha+1} \sin\frac{\pi\alpha}{2} \int_0^{+\infty} \frac{t^{\alpha-1}}{e^t + e^{-t}} \, dt, \quad (1.4.30)$$

where $\alpha \in (0, 1]$. Motivated by numerical calculations and by aesthetic reasons [410], Revers [411] conjectured that relations (1.4.28), (1.4.29), and (1.4.30) remain valid for all $\alpha > 0$ (except when α is an even integer). Such conjectures, as well as certain strong asymptotics for $g_\alpha(x) - L_n(g_\alpha)(x)$, $-1 < x < 1$, have recently been proved by Ganzburg [141]. An interesting approach for studying asymptotics of errors of the Lagrange interpolation to the function g_α was also given by Lubinsky [274]. It was based on some consideration for the Runge function f_α.

Taking $q_n(x) = \prod_{k=1}^n (x - x_{n,k})$, with distinct real zeros $x_{n,k}$, and using an idea from [274], Kubayi and Lubinsky [241] presented a representation for the error $f - L_n(f)$ of the Lagrange interpolation polynomial at the zeros of q_n, involving the Hilbert transform. Namely, they showed that $f - L_n(f) = -q_n H_e[H[f]/q_n]$, where H denotes the Hilbert transform, and H_e is an extension of it. Using this fact, they proved the convergence of the Lagrange interpolation for certain functions analytic in $(-1, 1)$ that are not assumed analytic in any ellipse with foci at ± 1, for example, $f(x) = (1 - x^2)^{-\alpha}$, $x \in (-1, 1)$, $0 < \alpha < 1$.

1.4.6 Lebesgue Function and Some Estimates for the Lebesgue Constant

For a given interpolation array $\mathcal{X}$, in Sect. 1.4.2 we defined the Lebesgue function $x \mapsto \lambda_n(\mathcal{X}; x)$ and the Lebesgue constant $\Lambda_n(\mathcal{X})$ by (1.4.11) and (1.4.13), respectively, and pointed out their importance in the convergence of interpolation polynomials.

We start this section with the formulation of elementary properties of the Lebesgue function $\lambda_n(\mathcal{X}; x)$ for an arbitrary array $\mathcal{X}$ given by (1.4.6) (cf. [52, 279]):

1° The function $\lambda_n(\mathcal{X}; x)$ is a piecewise polynomial satisfying $\lambda_n(\mathcal{X}; x) \geq 1$ and $\lambda_n(\mathcal{X}; x) = 1$ if and only if $x = x_{n,k}$ ($k = 1, \ldots, n$);
2° Between the consecutive nodes $x_{n,k-1}$ and $x_{n,k}$ ($k = 2, \ldots, n$) the function $\lambda_n(\mathcal{X}; x)$ has a single maximum, which will be denoted by $\mu_k(\mathcal{X})$,

$$\mu_k(\mathcal{X}) = \max_{x_{n,k-1} \leq x \leq x_{n,k}} \lambda_n(\mathcal{X}; x);$$

3° In the intervals $(-1, x_{n,1})$ and $(x_{n,n}, 1)$ the Lebesgue function is convex and monotone decreasing and increasing, respectively. If $x_{n,1} > -1$ and $x_{n,n} < 1$ the values $\lambda_n(\mathcal{X}, -1)$ and $\lambda_n(\mathcal{X}, 1)$ will be denoted by $\mu_1(\mathcal{X})$ and $\mu_{n+1}(\mathcal{X})$.

We put

$$m_n(\mathcal{X}) = \min_{1 \leq k \leq n+1} \mu_k(\mathcal{X}) \quad \text{and} \quad M_n(\mathcal{X}) = \max_{1 \leq k \leq n+1} \mu_k(\mathcal{X}).$$

It is clear that $\Lambda_n(\mathcal{X}) = M_n(\mathcal{X})$.

In the sequel we discuss some special interpolation arrays. The cases of equidistant nodes and Chebyshev nodes for $n = 7$ are presented in Fig. 1.4.6.

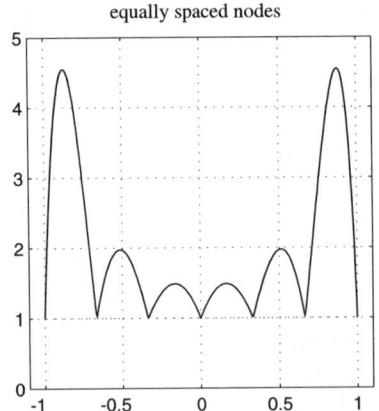 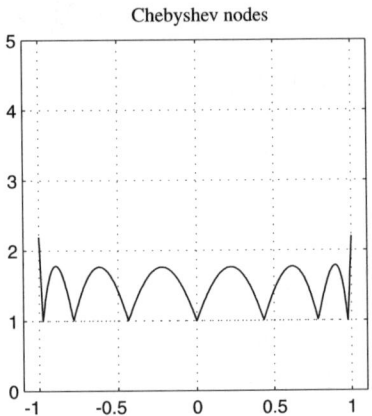

Fig. 1.4.6 Graph of the Lebesgue function for $n = 7$, with equally spaced nodes (*left*) and Chebyshev nodes (*right*)

1.4.6.1 Equidistant Nodes

For equally spaced nodes on $[-1, 1]$,

$$x_{n,k} = -1 + 2\frac{k-1}{n-1}, \qquad k = 1, \ldots, n,$$

we denote the corresponding array by $\mathcal{E}$ (starting with the two nodes $x_{2,1} = -1$ and $x_{2,2} = 1$). As we mentioned before, such a choice is usually bad (see, for instance, the examples of Runge and Bernstein).

In 1917 Tietze [474] proved that the relative maxima $\mu_k(\mathcal{E})$ of the Lebesgue function are strictly decreasing from the outside towards the middle of the interval. For $M_n(\mathcal{E}) = \Lambda_n(\mathcal{E})$ Tietze obtained a rather conservative lower bound (for details see the survey of Brutman [52]). Tureckiĭ [480], [481, Problem XX] rediscovered the monotone behaviour of the local maxima $\mu_k(\mathcal{E})$ and found the following asymptotic expression

$$\Lambda_n(\mathcal{E}) \sim \frac{2^n}{e n \log n} \qquad (n \to +\infty). \tag{1.4.31}$$

He also investigated the behaviour of $m_n(\mathcal{E})$ and showed that it tends asymptotically to $\log n / \pi$. Schönhage [428] found an expression which is a little bit more precise than (1.4.31), and Mills and Smith [324] improved it by finding the following asymptotic expansion

$$\log \Lambda_{n+1}(\mathcal{E}) = (n+1)\log 2 - \log n - \log\log n - 1 + \sum_{k=1}^{m} \frac{A_k}{(\log n)^k} + O\left(\frac{1}{(\log n)^{m+1}}\right),$$

1.4 Interpolation by Algebraic Polynomials

where $A_1 = -\gamma$ ($\gamma = 0.577\ldots$ is Euler's constant), $A_2 = \gamma^2/2 - \pi^2/12$, etc. We mention also the two-sided estimate

$$\frac{2^{n-2}}{n^2} < \Lambda_{n+1}(\mathcal{E}) < \frac{2^{n+3}}{n} \quad (n \geq 1)$$

obtained by Trefethen and Weideman [478].

1.4.6.2 Chebyshev Nodes

Taking the Chebyshev nodes $x_{n,k} = -\cos(2k-1)\pi/(2n)$, $k = 1, \ldots, n$, Bernstein [34] established an asymptotic behaviour of $\Lambda_n(\mathcal{T})$ in the form

$$\Lambda_n(\mathcal{T}) \sim \frac{2}{\pi} \log n \quad (n \to +\infty). \tag{1.4.32}$$

There are several estimates for $\Lambda_n(\mathcal{T})$. Some of them were obtained by using the fact that

$$\Lambda_n(\mathcal{T}) = \lambda_n(\mathcal{T}, 1) = \frac{1}{n} \sum_{k=1}^{n} \cot \frac{(2k-1)\pi}{4n}.$$

For example, Rivlin [416] proved that

$$a_0 + \frac{2}{\pi} \log n < \Lambda_n(\mathcal{T}) < 1 + \frac{2}{\pi} \log n,$$

where

$$a_0 = \frac{2}{\pi}\left(\gamma + \log \frac{8}{\pi}\right) = 0.9625\ldots,$$

and Shivakumar and Wong [433], Dzyadyk and Ivanov [105], and Günttner [202] independently established the asymptotic formula

$$\Lambda_n(\mathcal{T}) \sim \frac{2}{\pi} \log n + a_0 + \sum_{k=1}^{+\infty} \frac{a_k}{n^{2k}} \quad (n \to +\infty),$$

where

$$a_k = \frac{(-1)^{k+1}(2^{2k-1}-1)^2 \pi^{2k-1} B_{2k}^2}{4^{k-1} k (2k)!} \quad (k \in \mathbb{N}),$$

and B_k are the Bernoulli numbers. For the smallest local maximum Brutman [51] obtained

$$m_n(\mathcal{T}) > \frac{2}{\pi} \log n + \chi \quad \left(\chi = \frac{2}{\pi}\left(\gamma + \log \frac{4}{\pi}\right) = 0.52125\ldots\right), \tag{1.4.33}$$

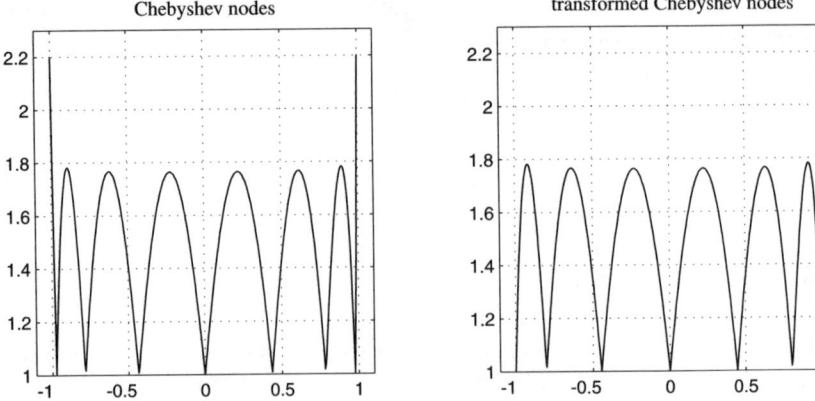

Fig. 1.4.7 Graph of the Lebesgue function for $n = 7$, with the Chebyshev nodes (*left*) and the transformed Chebyshev nodes to $[-1, 1]$ (*right*)

and Günttner [204] improved this result as follows

$$m_n(T) > \frac{2}{\pi} \log n + \chi + \frac{\pi}{18n^2} - \frac{49\pi^3}{10800 n^4}.$$

A slightly smaller Lebesgue constant can be obtained for the Chebyshev nodes transformed from $[x_{n,1}, x_{n,n}]$ to $[-1, 1]$, i.e., taking

$$\tilde{x}_{n,k} = -\frac{\cos \frac{(2k-1)\pi}{2n}}{\cos \frac{\pi}{2n}} \qquad (k = 1, \ldots, n).$$

Brutman [51] proved that

$$\Lambda_n(\hat{T}) < \frac{2}{\pi} \log n + 0.7219,$$

and Günttner [203] found the expansion

$$\Lambda_n(\hat{T}) = \frac{2}{\pi} \log n + \sum_{k=1}^{m} \frac{b_k}{(\log n)^k} + O\left(\frac{1}{(\log n)^{m+1}}\right) \qquad (m = 0, 1, \ldots),$$

where $b_0 = a_0 - 4/3\pi = 0.5381\ldots$, $b_1 = 16/81\pi^3 = 0.006371\ldots$, etc. In Fig. 1.4.7 we display the cases with original the Chebyshev nodes and the transformed Chebyshev nodes.

There are also estimates of the Lebesgue constant taking the Chebyshev extremal points given by (1.1.20), as well as other systems of nodes. A review of that was given by Brutman [52].

1.4 Interpolation by Algebraic Polynomials

It is an open problem to get the exact value of the optimal Lebesgue constants

$$\Lambda_n^* = \Lambda_n(\mathcal{X}^*) = \min_{\mathcal{X}} \Lambda_n(\mathcal{X}). \tag{1.4.34}$$

However, the Faber estimate (1.4.15) and Bernstein result (1.4.32) for $\Lambda_n(\mathcal{T})$ give the order $\log n$ of the optimal Lebesgue constant Λ_n^*.

In 1931 Bernstein [38] returned to the problem (1.4.34) and found that

$$\Lambda_n^* \sim \frac{2}{\pi} \log n \qquad (n \to +\infty).$$

Also, he conjectured that the greatest of the relative maxima of $\lambda_n(\mathcal{X}; x)$ (i.e., the Lebesgue constant $\Lambda_n(\mathcal{X})$) would be minimized if all these maxima are mutually equal. In 1961 Erdős [115], using a very delicate argument, obtained the estimate

$$\frac{2}{\pi} \log n - c_1 \leq \Lambda_n^* \leq \frac{2}{\pi} \log n + c_2 \qquad (n \geq 2). \tag{1.4.35}$$

Without loss of generality, the mentioned conjecture can be formulated only for *canonical arrays* $\mathcal{X}$, i.e., when $x_{n,1} = -1$ and $x_{n,n} = 1$. Erdős [113–115] conjectured that there is a unique canonical array $\mathcal{X}^*$ for which all maxima in the Lebesgue function are equal, as well as that for any array $\mathcal{X}$,

$$m_n(\mathcal{X}) = \min_{2 \leq k \leq n} \mu_k(\mathcal{X}) \leq \Lambda_n(\mathcal{X}^*) \leq \max_{2 \leq k \leq n} \mu_k(\mathcal{X}) = M_n(\mathcal{X}). \tag{1.4.36}$$

In 1978 Kilgore [228] and de Boor and Pinkus [83] proved this, the so-called *Bernstein-Erdős conjecture*:

Theorem 1.4.9 *Let* $n \geq 3$. *Then there exists a unique optimal canonical array* $\mathcal{X}^*$ *such that*

$$\mu_2(\mathcal{X}^*) = \cdots = \mu_n(\mathcal{X}^*). \tag{1.4.37}$$

Moreover, for arbitrary array $\mathcal{X}$ *the inequalities* (1.4.36) *hold.*

Using this theorem Vértesi [497] improved the Erdős estimate (1.4.35) and proved the following estimate

$$\Lambda_n^* = \frac{2}{\pi} \log n + \chi + o(1) \qquad (n \to +\infty), \tag{1.4.38}$$

where the constant χ is the same as in (1.4.33). This is the first estimate with the corresponding array which gives the value of Λ_n^* with the error $o(1)$.

At the end of this subsection we mention a recent result of Brutman, Gopengauz, and Toledano [53]. They studied the asymptotic behaviour of the integral of the Lebesgue function induced by interpolation at the Chebyshev nodes and determined explicitly two leading terms in the corresponding asymptotic expansion. Namely,

$$\int_{-1}^{1} \lambda_n(\mathcal{T}; x) \, dx = \frac{8}{\pi^2} \log n + A + o(1), \qquad \text{as } n \to +\infty, \tag{1.4.39}$$

where

$$A = \frac{8}{\pi^2}\left(\gamma + \int_0^{\pi/2} \frac{\sin x}{x}\,dx - \int_0^{\pi/2} \frac{1-\cos x}{x}\,dx + \int_0^{\pi/2} \frac{\cot x(x-\sin x)}{(1-2x/\pi)x}\,dx\right),$$

and γ is Euler's constant. The numerical value of the previous constant is $A \cong$ 1.41701860548.

The problem of finding a set of nodes $\mathcal{X}$ minimizing the value of the integral of the Lebesgue function

$$I_n(\mathcal{X}) = \int_{-1}^1 \lambda_n(\mathcal{X};x)\,dx \tag{1.4.40}$$

was posed by Erdős in 1961 (see [115]). It was announced there (with an indication of a possible method of proof) that

$$I_n^* = \min_{\mathcal{X}} I_n(\mathcal{X}) > C_1 \log n. \tag{1.4.41}$$

This estimate was generalized by Erdős and Szabados [117] for an arbitrary interval $[a,b] \subseteq [-1,1]$ in the form

$$\int_a^b \lambda_n(\mathcal{X};x)\,dx \geq C_2 \log n, \quad C_2 = \frac{b-a}{40}, \quad n > N(a,n).$$

(See also a recent estimate given in [118].)

In contrast to the estimate of Λ_n^* in (1.4.38), the best constant in (1.4.40) is unknown. As to the behaviour of the integral (1.4.40) for special sets of interpolation nodes, Erdős [116] conjectured that asymptotically the minimum in (1.4.41) is attained for the zeros of the Chebyshev polynomial $T_n(x)$. The results of some extensive numerical computations of I_n^* and $I_n(T)$ strongly suggest that this conjecture is true, namely that $I_n^* - I_n(T) = o(1)$, when $n \to +\infty$ (see [53]). According to (1.4.39) it would be

$$I_n^* = \frac{8}{\pi^2} \log n + A + o(1), \quad n \to +\infty,$$

where the constant A was defined earlier.

1.4.7 Algorithm for Finding Optimal Nodes

In this section we present an algorithm for finding optimal nodes for polynomial interpolation on $[-1,1]$.

For a fixed n, we take nodes in an increasing arrangement as in (1.4.5),

$$-1 = x_0 < x_1 < x_2 < \cdots < x_n < x_{n+1} = 1. \tag{1.4.42}$$

1.4 Interpolation by Algebraic Polynomials

Fig. 1.4.8 The Lebesgue function for six nodes: $\{-0.9, -0.45, -0.15, 0.2, 0.5, 0.85\}$

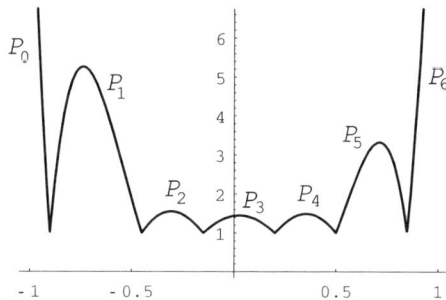

According to the property $1°$ (mentioned at the beginning of the previous section), the Lebesgue function $\lambda_n(x)$ is a piecewise polynomial, i.e.,

$$\lambda_n(x) = \sum_{k=1}^{n} |\ell_{n,k}(x)| = P_i(x), \quad x \in [x_i, x_{i+1}] \quad (i = 0, 1, \ldots, n).$$

Evidently, for $i = 0, 1, \ldots, n$, this formula defines $n + 1$ polynomials P_i such that $P_i \in \mathcal{P}_{n-1}$ $(i = 0, 1, \ldots, n)$ and

$$P_i(x_k) = \begin{cases} (-1)^{i+k}, & 1 \le k \le i, \\ (-1)^{i+k-1}, & i+1 \le k \le n. \end{cases}$$

The case of $n = 6$ nodes is presented in Fig. 1.4.8.

These polynomials can be expressed using the Lagrange interpolation formula in n points of our mesh (1.4.42). Thus, we have

$$P_i(t) = \sum_{j=1}^{n} P_i(x_j)\varphi_j(t) = \sum_{j \le i}(-1)^{i+j}\varphi_j(t) + \sum_{j \ge i+1}(-1)^{i+j-1}\varphi_j(t), \quad (1.4.43)$$

where

$$\varphi_j(t) = \prod_{\substack{\nu=1 \\ \nu \ne j}}^{n} \frac{t - x_\nu}{x_j - x_\nu}, \quad j = 1, \ldots, n. \quad (1.4.44)$$

In the sequel, we need

$$\frac{\partial \varphi_j(t)}{\partial x_j} = -\varphi_j(t) \sum_{\substack{\nu=1 \\ \nu \ne j}}^{n} \frac{1}{x_j - x_\nu}, \quad (1.4.45)$$

$$\frac{\partial \varphi_j(t)}{\partial x_k} = \frac{t - x_j}{(t - x_k)(x_j - x_k)} \varphi_j(t), \quad k \ne j. \quad (1.4.46)$$

For an arbitrarily selected system of nodes (1.4.42), we should determine the points of local maxima on $[-1, 1]$, denoted by ξ_i, $i = 0, 1, \ldots, n$. Evidently, $\xi_0 = -1$ and $\xi_n = 1$, while $\xi_i \in (x_i, x_{i+1})$, $i = 1, \ldots, n-1$.

These extremum points can be determined by solving the equation

$$P_i'(t) = \sum_{j \leq i}(-1)^{i+j}\varphi_j'(t) + \sum_{j \geq i+1}(-1)^{i+j-1}\varphi_j'(t) = 0$$

by some numerical methods. For example, we can use the bisection method on each of the intervals $[x_i, x_{i+1}]$, $i = 1, \ldots, n-1$. Alternatively, an application of the standard Newton method

$$\xi_i^{(m+1)} = \xi_i^{(m)} - \frac{P_i'(\xi_i^{(m)})}{P_i''(\xi_i^{(m)})}, \quad m = 0, 1, \ldots, \tag{1.4.47}$$

provides the quadratic convergence of the iterations $\xi_i^{(m)}$ to ξ_i, when $m \to +\infty$ ($i = 1, \ldots, n-1$). Here,

$$P_i'(t) = \sum_{j \leq i}(-1)^{i+j}\varphi_j(t)\alpha_j(t) + \sum_{j \geq i+1}(-1)^{i+j-1}\varphi_j(t)\alpha_j(t),$$

$$P_i''(t) = \sum_{j \leq i}(-1)^{i+j}\varphi_j(t)\beta_j(t) + \sum_{j \geq i+1}(-1)^{i+j-1}\varphi_j(t)\beta_j(t),$$

and

$$\alpha_j(t) = \sum_{\nu \neq j}\frac{1}{t - x_\nu}, \quad \beta_j(t) = \left(\sum_{\nu \neq j}\frac{1}{t - x_\nu}\right)^2 - \sum_{\nu \neq j}\frac{1}{(t - x_\nu)^2}.$$

As a starting value for (1.4.47) one can take the midpoint of the interval $[x_i, x_{i+1}]$, i.e.,

$$\xi_i^{(0)} = \frac{1}{2}(x_i + x_{i+1}), \quad i = 1, \ldots, n-1.$$

The corresponding maxima are

$$P_0(-1), \ P_1(\xi_1), \ P_2(\xi_2), \ \ldots, \ P_{n-1}(\xi_{n-1}), \ P_n(1).$$

According to Theorem 1.4.9, the Lebesgue constant will be minimized if all these maxima are mutually equal, i.e., when

$$P_0(-1) = P_1(\xi_1) = P_2(\xi_2) = \cdots = P_{n-1}(\xi_{n-1}) = P_n(1).$$

In order to find the optimal nodes, we consider this system of equations in the following form

$$\mathbf{F}(\mathbf{x}) = \mathbf{0}, \tag{1.4.48}$$

1.4 Interpolation by Algebraic Polynomials

where

$$\mathbf{x} = \begin{bmatrix} x_1 \\ x_2 \\ \vdots \\ x_n \end{bmatrix}, \quad \mathbf{F}(\mathbf{x}) = \begin{bmatrix} f_1(\mathbf{x}) \\ f_2(\mathbf{x}) \\ \vdots \\ f_n(\mathbf{x}) \end{bmatrix} = \begin{bmatrix} P_0(-1) - P_1(\xi_1) \\ P_1(\xi_1) - P_2(\xi_2) \\ \vdots \\ P_{n-1}(\xi_{n-1}) - P_n(1) \end{bmatrix}$$

and $P_0, P_1, \ldots, P_n$ are defined by (1.4.43). Now, by a linearization of (1.4.48) at $\mathbf{x} = \mathbf{x}^{(0)} = [x_1^{(0)} \ x_2^{(0)} \ \ldots \ x_n^{(0)}]^T$, where $\mathbf{x}^{(0)}$ is an appropriate starting vector of nodes, we get

$$\mathbf{F}(\mathbf{x}^{(0)}) + W(\mathbf{x}^{(0)})(\mathbf{x} - \mathbf{x}^{(0)}) = \mathbf{0}, \quad (1.4.49)$$

where W is the *Jacobian matrix* given by

$$W(\mathbf{x}) = \left[\frac{\partial f_i}{\partial x_j} \right]_{n \times n} = \begin{bmatrix} \frac{\partial f_1}{\partial x_1} & \frac{\partial f_1}{\partial x_2} & \cdots & \frac{\partial f_1}{\partial x_n} \\ \frac{\partial f_2}{\partial x_1} & \frac{\partial f_2}{\partial x_2} & \cdots & \frac{\partial f_2}{\partial x_n} \\ \vdots & & & \\ \frac{\partial f_n}{\partial x_1} & \frac{\partial f_n}{\partial x_2} & \cdots & \frac{\partial f_n}{\partial x_n} \end{bmatrix}.$$

Under the condition $\det W(\mathbf{x}^{(0)}) \neq 0$, the solution of (1.4.49) gives a new approximation of nodes

$$\mathbf{x}^{(1)} = \mathbf{x}^{(0)} - W^{-1}(\mathbf{x}^{(0)})\mathbf{F}(\mathbf{x}^{(0)}). \quad (1.4.50)$$

The functions f_i ($i = 1, \ldots, n$) and their partial derivatives $\partial f_i/\partial x_k$ ($i, k = 1, \ldots, n$) can be calculated very easily by

$$f_{i+1}(\mathbf{x}) = \sum_{j \leq i} (-1)^{i+j} \big(\varphi_j(\xi_i) + \varphi_j(\xi_{i+1})\big) + \varphi_{i+1}(\xi_i) - \varphi_{i+1}(\xi_{i+1})$$

$$+ \sum_{j \geq i+2} (-1)^{i+j-1} \big(\varphi_j(\xi_i) + \varphi_j(\xi_{i+1})\big) \quad (i = 0, 1, \ldots, n-1)$$

and

$$\frac{\partial f_{i+1}(\mathbf{x})}{\partial x_k} = \sum_{j \leq i} (-1)^{i+j} \left(\frac{\partial \varphi_j(\xi_i)}{\partial x_k} + \frac{\partial \varphi_j(\xi_{i+1})}{\partial x_k} \right) + \frac{\partial \varphi_{i+1}(\xi_i)}{\partial x_k} - \frac{\partial \varphi_{i+1}(\xi_{i+1})}{\partial x_k}$$

$$+ \sum_{j \geq i+2} (-1)^{i+j-1} \left(\frac{\partial \varphi_j(\xi_i)}{\partial x_k} + \frac{\partial \varphi_j(\xi_{i+1})}{\partial x_k} \right) \quad \begin{pmatrix} i = 0, 1, \ldots, n-1 \\ k = 1, 2, \ldots, n \end{pmatrix},$$

where the values of the functions φ_j and their partial derivatives are given by (1.4.44) and (1.4.45)–(1.4.46), respectively.

According to the results of the previous section, it is convenient to take the Chebyshev nodes as starting values, i.e.,

$$x_k^{(0)} = -\cos\frac{(2k-1)\pi}{2n}, \qquad k=1,\ldots,n. \qquad (1.4.51)$$

Based on the previous considerations we can state the following algorithm:

1° *Start with $v:=-1$ and n Chebyshev points $\{x_k^{(0)}\}_{k=1}^n$, given by (1.4.51);*
2° *Put $v:=v+1$ and $x_k:=x_k^{(v)}$, $k=1,\ldots,n$;*
3° *Determine the points of the local maxima ξ_i ($\in (x_i,x_{i+1})$), $i=1,\ldots,n-1$, using Newton's method (1.4.47);*
4° *Determine $x^{(v+1)}$ by (1.4.50), i.e.,*

$$\mathbf{x}^{(v+1)} = \mathbf{x}^{(v)} - W^{-1}(\mathbf{x}^{(v)})\mathbf{F}(\mathbf{x}^{(v)});$$

5° *For a given tolerance ε, check the condition $\|\mathbf{x}^{(v+1)} - \mathbf{x}^{(v)}\| < \varepsilon$. If this condition is satisfied then stop; otherwise, go to 2°.*

As we can see, this algorithm describes the well-known Newton-Kantorovič method, which is applied, in this case, to the special system of nonlinear equations given by (1.4.48). The convergence is quadratic. Notice that $\|\cdot\|$ is a norm in $\mathbb{R}^n$.

Example 1.4.1 Here we give some numerical examples with $n=6, 21, 51$, and 100 nodes. All computations are performed in double precision arithmetic with machine

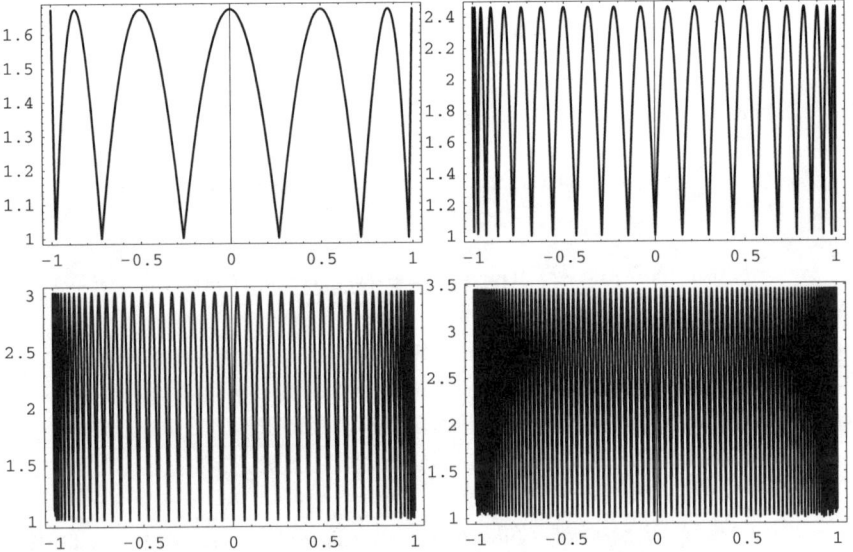

Fig. 1.4.9 Optimal Lebesgue functions for $n=6, 21, 51$, and 100 nodes

1.4 Interpolation by Algebraic Polynomials

precision (m.p. $\approx 2.22 \times 10^{-16}$). In each case only a few iterations (four or five) are enough to get results with machine precision, which are provided by very good starting values (Chebyshev nodes) and by the quadratic convergence of the method.

For example, in the case $n = 6$, the optimal nodes are:

$$-x_1 = x_6 = 0.9778619853559459, \quad -x_2 = x_5 = 0.7178745602803507,$$

$$-x_3 = x_4 = 0.2629539766769217.$$

Optimal Lebesgue functions are displayed in Fig. 1.4.9. The corresponding Lebesgue constants are:

$$\Lambda_6^* = 1.672210365022631, \quad \Lambda_{21}^* = 2.460787748789093,$$
$$\Lambda_{51}^* = 3.024619144410691, \quad \Lambda_{100}^* = 3.453082266076696.$$

Another method based on a nonlinear Remez search was presented in [14].

Remark 1.4.2 An analytic solution for $n = 3$ can be found in [38]. The optimal nodes are $-x_1 = x_3 = 2\sqrt{2}/3$, $x_2 = 0$, and the corresponding optimal Lebesgue constant is $\Lambda_3^* = 5/4$.

Chapter 2
Orthogonal Polynomials and Weighted Polynomial Approximation

2.1 Orthogonal Systems and Polynomials

2.1.1 Inner Product Space and Orthogonal Systems

Suppose that X is a complex linear space of functions with an inner product $(f, g): X^2 \to \mathbb{C}$ such that

(a) $(f + g, h) = (f, h) + (g, h)$ (Linearity),
(b) $(\alpha f, g) = \alpha(f, g)$ (Homogeneity),
(c) $(f, g) = \overline{(g, f)}$ (Hermitian Symmetry),
(d) $(f, f) \geq 0$, $(f, f) = 0 \iff f = 0$ (Positivity),

where $f, g, h \in X$ and α is a complex scalar. The bar in the above line denotes the complex conjugate. The space X is called an *inner product space*.

If X is a real linear space, then the inner product $(f, g): X^2 \to \mathbb{R}$ is such that the condition (c) is reduced to

(c') $(f, g) = (g, f)$ (Symmetry).

An important inequality for the inner product is the *Cauchy-Schwarz-Buniakowsky inequality* (cf. [328, p. 87])

$$|(f, g)| \leq \|f\| \|g\| \qquad (f, g \in X), \tag{2.1.1}$$

where the *norm* of f is defined by $\|f\| = \sqrt{(f, f)}$.

A system S of elements of an inner product space is called *orthogonal* if $(f, g) = 0$ for every $f \neq g$ $(f, g \in S)$. An orthogonal system S is called *orthonormal* if $(f, f) = 1$ for all $f \in S$.

Suppose that $U = \{g_0, g_1, g_2, \ldots\}$ is a system of linearly independent functions in a complex inner product space X. Starting from this system of elements and using the well-known *Gram-Schmidt orthogonalizing process* we can construct the corresponding orthogonal (orthonormal) system $S = \{\varphi_0, \varphi_1, \varphi_2, \ldots\}$, where φ_n is, in fact, a linear combination of the functions $g_0, g_1, \ldots, g_n$, such that $(\varphi_n, \varphi_k) = 0$ for $n \neq k$.

Using the functions g_n and *Gram matrix* of order $n+1$,

$$G_{n+1} = \begin{bmatrix} (g_0, g_0) & (g_0, g_1) & \cdots & (g_0, g_n) \\ (g_1, g_0) & (g_1, g_1) & & (g_1, g_n) \\ \vdots & & & \\ (g_n, g_0) & (g_n, g_1) & & (g_n, g_n) \end{bmatrix}$$

an explicit expression for the orthogonal functions φ_n can be obtained. Notice that this matrix is non-singular. Namely, it is well-known that $\Delta_{n+1} = \det G_{n+1} \neq 0$ if and only if the system of functions $\{g_0, g_1, g_2, \ldots, g_n\}$ is linearly independent. Moreover, we can prove that $\det G_{n+1} > 0$.

Firstly, the matrix G_{n+1} is Hermitian, because of the property (c) of the inner product, i.e., $(g_i, g_j) = \overline{(g_j, g_i)}$. Putting $\mathbf{x} = [x_0 \ x_1 \ \cdots \ x_n]^T$ and $\psi_n = \sum_{k=0}^{n} \overline{x}_k g_k$, we can see that the Gram matrix is also positive definite. Namely, then

$$\mathbf{x}^* G_{n+1} \mathbf{x} = \sum_{i=0}^{n} \sum_{j=0}^{n} (g_i, g_j) \overline{x}_i x_j$$

can be expressed in the form

$$\mathbf{x}^* G_{n+1} \mathbf{x} = (\psi_n, \psi_n) = \|\psi_n\|^2,$$

which is positive, except if $\psi_n = 0$ (i.e., $\mathbf{x} = \mathbf{0}$). This means that $\Delta_{n+1} = \det G_{n+1} > 0$.

Theorem 2.1.1 *The orthonormal functions φ_n are given by*

$$\varphi_n(z) = \frac{1}{\sqrt{\Delta_n \Delta_{n+1}}} \begin{vmatrix} (g_0, g_0) & (g_0, g_1) & \cdots & (g_0, g_{n-1}) & g_0(z) \\ (g_1, g_0) & (g_1, g_1) & & (g_1, g_{n-1}) & g_1(z) \\ \vdots & & & & \\ (g_n, g_0) & (g_n, g_1) & & (g_n, g_{n-1}) & g_n(z) \end{vmatrix} \quad (2.1.2)$$

where $\Delta_n = \det G_n$ and $\Delta_0 = 1$.

Proof For the proof of this statement it is enough to prove that φ_n given by (2.1.2) satisfies the orthogonality condition $(\varphi_n, g_k) = 0$ for each $k = 0, 1, \ldots, n-1$.
Since

$$(\varphi_n, g_k) = \frac{1}{\sqrt{\Delta_n \Delta_{n+1}}} \begin{vmatrix} (g_0, g_0) & (g_0, g_1) & \cdots & (g_0, g_{n-1}) & (g_0, g_k) \\ (g_1, g_0) & (g_1, g_1) & & (g_1, g_{n-1}) & (g_1, g_k) \\ \vdots & & & & \\ (g_n, g_0) & (g_n, g_1) & & (g_n, g_{n-1}) & (g_n, g_k) \end{vmatrix},$$

2.1 Orthogonal Systems and Polynomials

we see immediately that this determinant is equal to zero for each $k = 0, 1, \ldots, n-1$, and for $k = n$ we have $(\varphi_n, g_n) = \sqrt{\Delta_{n+1}/\Delta_n}$. Expanding the determinant from (2.1.2) along the last column, we obtain an expansion

$$\varphi_n(z) = \frac{1}{\sqrt{\Delta_n \Delta_{n+1}}} \bigl(c_0 g_0(z) + c_1 g_1(z) + \cdots + c_n g_n(z)\bigr),$$

in terms of g_k, where $c_n = \Delta_n$. Therefore,

$$(\varphi_n, \varphi_n) = \frac{c_n}{\sqrt{\Delta_n \Delta_{n+1}}} (g_n, \varphi_n) = 1.$$

$\square$

2.1.2 Fourier Expansion and Best Approximation

Taking an orthonormal system of functions $S = \{\varphi_0, \varphi_1, \varphi_2, \ldots\}$, it is easy to construct the corresponding *Fourier expansion* for a given function $f \in X$,

$$f(z) \sim \sum_{k=0}^{+\infty} f_k \varphi_k(z). \qquad (2.1.3)$$

The *Fourier coefficients* f_k are given by

$$f_k = (f, \varphi_k) \qquad (k = 0, 1, \ldots), \qquad (2.1.4)$$

which follows directly from (2.1.3). This sequence of coefficients is bounded. Indeed, applying the Cauchy-Schwarz-Buniakowsky inequality (2.1.1) to (2.1.4), we obtain

$$|f_k| = |(f, \varphi_k)| \le \|f\| \|\varphi_k\| = \|f\|.$$

The partial sums of (2.1.3), i.e.,

$$s_n(z) = \sum_{k=0}^{n} f_k \varphi_k(z), \qquad (2.1.5)$$

play a very important role in approximation theory.

Let X_n be a subspace of X spanned by $S_n = \{\varphi_0, \varphi_1, \ldots, \varphi_n\}$ ($\dim X_n = n+1$), i.e., $X_n = \operatorname{span} S_n$. The following theorem shows that s_n is the closest element to $f \in X$ among all elements of the subspace X_n with respect to the metric induced by the given norm. Thus, the partial sum s_n is the *best approximation* to $f \in X$ in the subspace X_n.

Theorem 2.1.2 *Let $f \in X$ and X_n be a subspace of X spanned by $\{\varphi_0, \varphi_1, \ldots\}$. Then*

$$\min_{\varphi \in X_n} \|f - \varphi\|^2 = \|f - s_n\|^2 = \|f\|^2 - \sum_{k=0}^{n} |f_k|^2, \qquad (2.1.6)$$

where s_n is given by (2.1.5).

Proof Let $f \in X$ and let s_n be given by (2.1.5). An arbitrary element of X_n can be expressed as a linear combination of the orthonormal functions $\varphi_0, \varphi_1, \ldots, \varphi_n$, i.e., $\varphi = \sum_{k=0}^{n} a_k \varphi_k$. Then

$$\|f - \varphi\|^2 = (f - \varphi, f - \varphi) = (f, f) - (f, \varphi) - (\varphi, f) + (\varphi, \varphi).$$

Since

$$(f, \varphi) = \sum_{k=0}^{n} \overline{a}_k (f, \varphi_k) = \sum_{k=0}^{n} \overline{a}_k f_k, \quad (\varphi, f) = \sum_{k=0}^{n} a_k \overline{f}_k, \quad (\varphi, \varphi) = \sum_{k=0}^{n} \overline{a}_k a_k,$$

we get

$$\|f - \varphi\|^2 = \|f\|^2 - \sum_{k=0}^{n} |f_k|^2 + \sum_{k=0}^{n} \left(f_k \overline{f}_k - \overline{a}_k f_k - a_k \overline{f}_k + a_k \overline{a}_k \right)$$

$$= \|f\|^2 - \sum_{k=0}^{n} |f_k|^2 + \sum_{k=0}^{n} |f_k - a_k|^2.$$

This expression attains the minimal value for $a_k = f_k$ $(k = 0, 1, \ldots, n)$, i.e., when $\varphi = s_n$, and the minimum is given by (2.1.6). □

Since $(f, s_n) = (s_n, s_n) = \sum_{k=0}^{n} |f_k|^2$, we see that the error $e_n = f - s_n$ in the best approximation is orthogonal to s_n, i.e.,

$$(f - s_n, s_n) = 0.$$

Also, for each $k = 0, 1, \ldots, n$,

$$(f - s_n, \varphi_k) = \left(f - \sum_{v=0}^{n} f_v \varphi_v, \varphi_k \right) = (f, \varphi_k) - \sum_{v=0}^{n} f_v (\varphi_v, \varphi_k) = 0, \quad (2.1.7)$$

In other words, the error e_n is orthogonal to each φ_k, i.e., e_n is orthogonal to the subspace X_n.

Based on (2.1.6) we conclude that

$$\|f - s_0\| \geq \|f - s_1\| \geq \|f - s_2\| \geq \cdots,$$

which also follows directly from the fact that $X_0 \subset X_1 \subset X_2 \subset \cdots$.

Notice also that (2.1.6) implies the *Bessel inequality*

$$\sum_{k=0}^{n} |f_k|^2 \leq \|f\|^2,$$

2.1 Orthogonal Systems and Polynomials

which holds for every $n \in \mathbb{N}$. When $n \to +\infty$, this becomes

$$\sum_{k=0}^{+\infty} |f_k|^2 \leq \|f\|^2.$$

Thus, the series on the left hand side in this limit inequality converges, which implies that

$$\lim_{k \to +\infty} f_k = \lim_{k \to +\infty} (f, \varphi_k) = 0.$$

Therefore, we conclude that the Fourier coefficients of any function $f \in X$ approach zero.

The limit Bessel inequality reduces to an equality (*Parseval's equality*) in the case when $\text{span}\{\varphi_0, \varphi_1, \varphi_2, \ldots\}$ is dense in X.

2.1.3 Examples of Orthogonal Systems

In this section we mention several interesting orthogonal systems.

2.1.3.1 Trigonometric System

In Sect. 1.2.2 we have seen that the trigonometric system (1.2.18), i.e.,

$$\{1, \cos x, \sin x, \cos 2x, \sin 2x, \ldots, \cos nx, \sin nx, \ldots\}$$

is orthogonal with respect to the inner product defined by (1.2.19). According to (1.2.16) and (1.2.17), the corresponding orthonormal system is

$$\left\{ \frac{1}{\sqrt{2\pi}}, \frac{\cos x}{\sqrt{\pi}}, \frac{\sin x}{\sqrt{\pi}}, \frac{\cos 2x}{\sqrt{\pi}}, \frac{\sin 2x}{\sqrt{\pi}}, \ldots, \frac{\cos nx}{\sqrt{\pi}}, \frac{\sin nx}{\sqrt{\pi}}, \ldots \right\}.$$

2.1.3.2 Chebyshev Polynomials

Let

$$(f, g) = \int_{-1}^{1} f(x)g(x)(1 - x^2)^{\lambda - 1/2} dx, \quad \lambda > -1/2. \tag{2.1.8}$$

The Chebyshev polynomials of the first kind $\{T_n\}_{n \in \mathbb{N}_0}$ and of the second kind $\{U_n\}_{n \in \mathbb{N}_0}$ are orthogonal on $[-1, 1]$ with respect to the inner product (2.1.8) for $\lambda = 0$ and $\lambda = 1$, respectively (see Sect. 1.1.4). The corresponding orthonormal systems are

$$\left\{ \frac{1}{\sqrt{\pi}}, \sqrt{\frac{2}{\pi}} T_1, \sqrt{\frac{2}{\pi}} T_2, \ldots \right\} \quad \text{and} \quad \left\{ \sqrt{\frac{2}{\pi}}, \sqrt{\frac{2}{\pi}} U_1, \sqrt{\frac{2}{\pi}} U_2, \ldots \right\}, \tag{2.1.9}$$

respectively.

2.1.3.3 Orthogonal Polynomials on the Unit Circle

The system of monomials $\{z^n\}_{n\in\mathbb{N}_0}$ is orthonormal with respect to the inner product

$$(f,g) = \frac{1}{2\pi}\int_{-\pi}^{\pi} f(e^{i\theta})\overline{g(e^{i\theta})}\, v(\theta)\, d\theta, \qquad (2.1.10)$$

where $v(\theta) = 1$. This is the simplest case of polynomials orthogonal on the unit circle with respect to (2.1.10). Such polynomials were introduced and studied by Szegő ([468, 469]) and Smirnov ([445, 446]). A more general case was considered by Achieser and Kreĭn [7], Geronimus ([177, 178]), Nevai ([378, 379]), Simon [436], etc. (see also surveys [9] and [334], as well as in a very impressive new book of Barry Simon in two volumes [437, 438]). These polynomials are linked to many questions in the theory of time series, digital filters, statistics, image processing, scattering theory, control theory, etc.

The general theory of orthogonality on a union of circular arcs was also considered by several authors (cf. Peherstorfer and Steinbauer [394–396], Simon [437, 438]). For the zeros of such polynomials see Lukashov and Peherstorfer [277], Simon [439, 440], etc.

2.1.3.4 Orthogonal Polynomials on the Unit Disk

The system of monomials $p_n(z) = \sqrt{(n+1)/\pi}\, z^n$, $n = 0, 1, \ldots$, is orthonormal with respect to the inner product defined by the following double integral

$$(f,g) = \iint_{|z|\le 1} f(z)\overline{g(z)}\, dx\, dy.$$

2.1.3.5 Orthogonal Polynomials on the Ellipse

Let E_r $(r > 1)$ denote the ellipse with its foci at ± 1 and such that the sum of its semi-axes is r.

The Chebyshev polynomials of the first kind $\{T_n\}_{n\in\mathbb{N}_0}$ are orthogonal with respect to the inner product defined by the following contour integral

$$(f,g) = \oint_{E_r} \frac{f(z)\overline{g(z)}}{|1-z^2|^{1/2}}\, ds \qquad (ds^2 = dx^2 + dy^2).$$

Namely, we have

$$(T_n, T_n) = \begin{cases} \dfrac{\pi}{2}\left(r^{2n} + r^{-2n}\right), & n > 0, \\ 2\pi, & n = 0. \end{cases}$$

2.1 Orthogonal Systems and Polynomials

However, the polynomials

$$p_n(z) = 2\sqrt{\frac{n+1}{\pi}}\left(r^{2n+2} - r^{-2n-2}\right)U_n(z), \quad n = 0, 1, \ldots,$$

where $U_n(z)$ are the Chebyshev polynomials of the second kind, are orthonormal with respect to the inner product

$$(f, g) = \iint_{\text{int } E_r} f(z)\overline{g(z)}\, dx\, dy.$$

2.1.3.6 Malmquist-Takenaka System of Rational Functions

Let $\{a_\nu\}_{\nu \in \mathbb{N}_0}$ be a sequence of complex numbers such that $|a_\nu| < 1$ ($\nu = 0, 1, \ldots$). The system of rational functions

$$w_n(z) = \frac{\prod_{\nu=0}^{n-1}(z - a_\nu)}{\prod_{\nu=0}^{n}(z - 1/\bar{a}_\nu)}, \quad n \in \mathbb{N}_0, \tag{2.1.11}$$

considered by Malmquist [281], Takenaka [472], Walsh [501, Sects. 9.1 and 10.7], Djrbashian [102], etc., is orthogonal with respect to the inner product defined by (2.1.10), where again $v(\theta) = 1$. Thus, the orthogonal functions (2.1.11), which are known as the *Malmquist-Takenaka functions (basis)*, generalize Szegő's orthogonal polynomials. Here,

$$\|w_n\|^2 = (w_n, w_n) = \frac{|a_0 a_1 \cdots a_n|^2}{1 - |a_n|^2}$$

(cf. [345, 362]).

2.1.3.7 Polynomials Orthogonal on the Radial Rays

The following sequence of polynomials

$$1,\ z,\ z^2,\ z^3,\ z^4 - \frac{1}{3},\ z^5 - \frac{5}{11}z,\ z^6 - \frac{7}{13}z^2,\ z^7 - \frac{3}{5}z^3,\ z^8 - \frac{14}{17}z^4 + \frac{21}{221},\ \ldots \tag{2.1.12}$$

is orthogonal with respect to the inner product

$$(f, g) = \int_0^1 \left[f(x)\overline{g(x)} + f(ix)\overline{g(ix)} + f(-x)\overline{g(-x)} + f(-ix)\overline{g(-ix)}\right]\omega(x)\, dx,$$

where $\omega(x) = (1-x^4)^{1/2}x^2$. This is a special case of polynomials orthogonal on the radial rays introduced by Milovanović in [336]. For details and several properties of such polynomials see [333, 337, 341, 360, 361]. It is interesting that the polynomial sequence (2.1.12) contains a subsequence which was obtained from a physical problem connected to a non-linear diffusion equation (cf. [448]).

2.1.3.8 Müntz Orthogonal Polynomials

Let $\Lambda = \{\lambda_0, \lambda_1, \ldots\}$ be a given sequence of complex numbers such that $\text{Re}(\lambda_k) > -1/2$, $k \in \mathbb{N}_0$, and let $\Lambda_n = \{\lambda_0, \lambda_1, \ldots, \lambda_n\}$. As we mentioned in Chap. 1 (Remark 1.3.1), Müntz systems can be considered also for the complex sequences. According to [23, 473], and [48], we can introduce the so-called *Müntz-Legendre polynomials* on $(0, 1]$ by

$$P_n(x) = P_n(x; \Lambda_n) = \frac{1}{2\pi i} \oint_\Gamma W_n(z) x^z \, dz, \qquad n = 0, 1, \ldots, \qquad (2.1.13)$$

where the simple contour Γ surrounds all poles of the rational function

$$W_n(z) = \prod_{\nu=0}^{n-1} \frac{z + \bar{\lambda}_\nu + 1}{z - \lambda_\nu} \cdot \frac{1}{z - \lambda_n} \qquad (n \in \mathbb{N}_0).$$

An empty product for $n = 0$ should be taken to be equal to 1.

The polynomials (2.1.13) are orthogonal with respect to the inner product $(f, g) = \int_0^1 f(x) \overline{g(x)} \, dx$.

The corresponding orthonormal polynomials are $P_n^*(x) = (1 + \lambda_n + \bar{\lambda}_n)^{1/2} P_n(x)$. In the simplest case when $\lambda_\nu \neq \lambda_\mu$ ($\nu \neq \mu$) it is easy to show that the polynomials $P_n(x)$ can be expressed in a power form $P_n(x) = \sum_{k=0}^n c_{n,k} x^{\lambda_k}$, where

$$c_{n,k} = \frac{\prod_{\nu=0}^{n-1}(1 + \lambda_k + \bar{\lambda}_\nu)}{\prod_{\substack{\nu=0 \\ \nu \neq k}}^{n}(\lambda_k - \lambda_\nu)}, \qquad k = 0, 1, \ldots, n.$$

An important special case of the Müntz-Legendre polynomials when

$$\lambda_{2k} = \lambda_{2k+1} = k, \qquad k = 0, 1, \ldots,$$

was considered in [338]. Namely, we put $\lambda_{2k} = k$ and $\lambda_{2k+1} = k + \varepsilon$, $k = 0, 1, \ldots$, where ε decreases to zero. The corresponding limit process leads to the orthogonal Müntz polynomials with logarithmic terms,

$$P_n(x) = R_n(x) + S_n(x) \log x, \qquad n = 0, 1, \ldots, \qquad (2.1.14)$$

2.1 Orthogonal Systems and Polynomials

where $R_n(x)$ and $S_n(x)$ are algebraic polynomials of degree $\left[\frac{n}{2}\right]$ and $\left[\frac{n-1}{2}\right]$, respectively, i.e.,

$$R_n(x) = \sum_{\nu=0}^{[n/2]} a_\nu^{(n)} x^\nu, \qquad S_n(x) = \sum_{\nu=0}^{[(n-1)/2]} b_\nu^{(n)} x^\nu. \qquad (2.1.15)$$

Notice that $P_n(1) = R_n(1) = 1$. The first few Müntz polynomials (2.1.14) are:

$$\begin{aligned}
P_0(x) &= 1, \\
P_1(x) &= 1 + \log x, \\
P_2(x) &= -3 + 4x - \log x, \\
P_3(x) &= 9 - 8x + 2(1 + 6x) \log x, \\
P_4(x) &= -11 - 24x + 36x^2 - 2(1 + 18x) \log x, \\
P_5(x) &= 19 + 276x - 294x^2 + 3(1 + 48x + 60x^2) \log x, \\
P_6(x) &= -21 - 768x + 390x^2 + 400x^3 - 3(1 + 96x + 300x^2) \log x.
\end{aligned}$$

The explicit expressions for the coefficients of the polynomials (2.1.15) for arbitrary n are given in [338]. These Müntz polynomials can be used in the proof of the irrationality of $\zeta(3)$ and of other familiar numbers (see [47, pp. 372–381] and [486]).

Similarly, if we take

$$\lambda_{3k} = \lambda_{3k+1} = \lambda_{3k+2} = k, \qquad k = 0, 1, \ldots,$$

i.e., $\lambda_{3k} = k - \varepsilon$, $\lambda_{3k+1} = k$, $\lambda_{3k+2} = k + \varepsilon$, $k = 0, 1, \ldots$, where ε tends to zero, we get the corresponding orthogonal Müntz polynomials:

$$\begin{aligned}
P_0(x) &= 1, \\
P_1(x) &= 1 + \log x, \\
P_2(x) &= 1 + 2\log x + \frac{1}{2} \log^2 x, \\
P_3(x) &= -7 + 8x - 4\log x - \frac{1}{2} \log^2 x, \\
P_4(x) &= 29 - 28x + (11 + 24x)\log x + \log^2 x, \\
P_5(x) &= -97 + 98x - 4(7 + 15x)\log x + (36x - 2)\log^2 x, \\
P_6(x) &= 127 - 342x + 216x^2 + (32 - 108x)\log x + (2 - 108x)\log^2 x.
\end{aligned}$$

These polynomials have the form

$$P_n(x) = R_n(x) + S_n(x) \log x + T_n(x) \log^2 x,$$

where $R_n(x)$, $S_n(x)$, and $T_n(x)$ are algebraic polynomials of degree $\left[\frac{n}{3}\right]$, $\left[\frac{n-1}{3}\right]$, and $\left[\frac{n-2}{3}\right]$, respectively. Notice that $P_n(1) = R_n(1) = 1$.

A direct evaluation of the Müntz polynomials $P_n(x)$ in the power form can be problematic in finite arithmetic, especially when n is a large number and x is close to 1. A numerical method for a stable evaluation of Müntz polynomials and some applications were given in [338] (see also [339]).

2.1.3.9 Müntz Orthogonal Polynomials of the Second Kind

In [72] and [362] an external operation for the Müntz polynomials from $M(\Lambda)$ and the corresponding inner product were defined. Namely, at first an operation $\odot$ for monomials was introduced in the following way:

$$x^\alpha \odot x^\beta = x^{\alpha\beta} \qquad (x \in (0, +\infty),\ \alpha, \beta \in \mathbb{C}),$$

and then it was extended to the Müntz polynomials $P \in M_n(\Lambda)$ and $Q \in M_m(\Lambda)$, i.e.,

$$P(x) = \sum_{i=0}^{n} p_i x^{\lambda_i} \quad \text{and} \quad Q(x) = \sum_{j=0}^{m} q_j x^{\lambda_j},$$

as

$$(P \odot Q)(x) = \sum_{i=0}^{n}\sum_{j=0}^{m} p_i q_j x^{\lambda_i \lambda_j}.$$

Under the restrictions that for each i and j, $|\lambda_i| > 1$ and $\text{Re}(\lambda_i \overline{\lambda_j} - 1) > 0$, the following inner product can be defined ([362])

$$[P, Q] = \int_0^1 (P \odot \overline{Q})(x)\, \frac{dx}{x^2}, \qquad (2.1.16)$$

as well as the Müntz polynomials

$$Q_n(x) = \frac{1}{2\pi i} \oint_\Gamma W_n(z) x^z\, dz, \qquad n = 0, 1, \ldots, \qquad (2.1.17)$$

where

$$W_n(s) = \frac{\prod_{\nu=0}^{n-1}(s - 1/\overline{\lambda_\nu})}{\prod_{\nu=0}^{n}(s - \lambda_\nu)}, \qquad n = 0, 1, \ldots, \qquad (2.1.18)$$

and the simple contour Γ surrounds all the points λ_ν, $\nu = 0, 1, \ldots, n$.

We note that the rational functions (2.1.18) form a Malmquist-Takenaka system. Indeed, putting $a_\nu = 1/\overline{\lambda_\nu}$ $\nu = 0, 1, \ldots$, these functions reduce to (2.1.11).

2.1 Orthogonal Systems and Polynomials

Under the previous conditions on the sequence Λ, the Müntz polynomials $Q_n(x)$, $n = 0, 1, \ldots$, defined by (2.1.17), are orthogonal with respect to the inner product (2.1.16). Furthermore, this orthogonality is connected to the orthogonality of the Malmquist-Takenaka system (2.1.11) (see ([362]) and

$$[Q_n, Q_m] = \frac{\delta_{n,m}}{(|\lambda_n|^2 - 1)|\lambda_0 \lambda_1 \cdots \lambda_{n-1}|^2}.$$

2.1.3.10 Generalized Exponential Polynomials

Let $A = \{\alpha_0, \alpha_1, \alpha_2, \ldots\}$ be a complex sequence such that $\operatorname{Re} \alpha_k > 0$. For each k ($k \geq 0$) denote by $m_k \geq 1$ the multiplicity of the appearance of the numbers α_k in the set $A_k = \{\alpha_0, \alpha_1, \alpha_2, \ldots, \alpha_k\}$. With the sequence A we associate the sequence of functions $\{t^{m_k-1} e^{-\alpha_k t}\}_{k \in \mathbb{N}_0}$, which can be orthogonalized with respect to the inner product

$$(f, g) = \int_0^{+\infty} f(t)\overline{g(t)}\, dt, \qquad (2.1.19)$$

for example, using the well-known Gram-Schmidt method. Such an orthonormal system $\{g_k(t)\}_{k \in \mathbb{N}_0}$ is unique up to a multiplicative constant of the form $e^{i\gamma_k}$, with $\operatorname{Im} \gamma_k = 0$.

For example, if we take $A = \{1/2, 1, 1, 2, 5/2, \ldots\}$, for which $m_0 = m_1 = 1$, $m_2 = 2$, $m_3 = m_4 = 1, \ldots$, using an orthogonalizing process we get the exponential functions (generalized exponential polynomials)

$$g_0(t) = e^{-t/2},$$
$$g_1(t) = \sqrt{2}\, e^{-t}(3 - 2e^{t/2}),$$
$$g_2(t) = \sqrt{2}\, e^{-t}(5 - 6e^{t/2} + 6t),$$
$$g_3(t) = 2e^{-2t}(15 - 8e^t - 6e^{3t/2} + 12te^t),$$
$$g_4(t) = \tfrac{1}{2}\sqrt{5}\, e^{-5t/2}(147 - 240e^{t/2} + 80e^{3t/2} + 15e^{2t} - 48te^{3t/2}), \ldots,$$

which are orthonormal with respect to the inner product (2.1.19) (cf. [339]).

2.1.3.11 Discrete Chebyshev Polynomials

Let

$$(f, g) = f(0)g(0) + f(1)g(1) + f(2)g(2) + f(3)g(3).$$

Starting from monomials and using the Gram-Schmidt orthogonalizing process we get orthogonal polynomials with respect to this inner product:

$$g_0(x) = 1, \quad g_1(x) = x - \frac{1}{3}, \quad g_2(x) = x^2 - 3x + 1, \quad g_3(x) = x^3 - \frac{9}{2}x^2 + \frac{47}{10}x - \frac{3}{10}.$$

The corresponding orthonormal polynomials are

$$\frac{1}{2}, \quad \frac{2x-3}{2\sqrt{5}}, \quad \frac{x^2-3x+1}{2}, \quad \frac{10x^3-45x^2+47x-3}{6\sqrt{5}}.$$

Continuing this process we find $g_4(x) = x(x-1)(x-2)(x-3)$ and, as we can see, $g_4(x) = 0$ for $x = 0, 1, 2, 3$. Therefore, $\|g_4\| = 0$.

This is a special case of the so-called *discrete Chebyshev orthogonal polynomials* (cf. Szegő [470, pp. 33–34]) which can be expressed (except for constant factors) in the form

$$t_n(x) = n! \Delta^n \left\{ \binom{x}{n} \binom{x-N}{n} \right\}, \quad n = 0, 1, \ldots, N-1, \tag{2.1.20}$$

where the inner product is given by

$$(f, g) = \sum_{k=0}^{N-1} f(k)g(k) \tag{2.1.21}$$

and N is a given natural number. In (2.1.20) the symbol Δ is the forward difference operator with unit spacing acting on the variable x. The inner product (2.1.21) can be rewritten in the integral form

$$(f, g) = \int_{\mathbb{R}} f(x)g(x)d\mu(x),$$

with the distribution $d\mu(x) = \sum_{k=0}^{N-1} \delta(x-k)dx$, where $\mu(x)$ is a step function with jumps of one unit at the points $x = k, k = 0, 1, \ldots, N-1$. Here, δ is the Dirac delta function. Notice that

$$\|t_n\|^2 = (t_n, t_n) = \frac{N(N^2-1^2)(N^2-2^2)\cdots(N^2-n^2)}{2n+1}.$$

Chebyshev also considered the case when the set of equidistant points $\{0, 1, \ldots, N-1\}$ is replaced by an arbitrary set of N distinct points. Finally, we mention that there are several other cases of discrete polynomials when the jumps in $\mu(x)$ are different from one unit (see [29, pp. 221–227], [382], [470, pp. 34–37]).

2.1.3.12 Formal Orthogonal Polynomials with Respect to a Moment Functional

Let a complex valued linear functional $\mathcal{L}$ be given on the linear space of all algebraic polynomials $\mathcal{P}$. The values of the linear functional $\mathcal{L}$ at the set of monomials are called moments and they are denoted by μ_k. Thus, $\mathcal{L}[x^k] = \mu_k$, $k \in \mathbb{N}_0$. A sequence of polynomials $\{P_n(x)\}_{n=0}^{\infty}$ is called a *formal orthogonal polynomial sequence* with respect to a *moment functional* $\mathcal{L}$ provided for all nonnegative integers k and n,

2.1 Orthogonal Systems and Polynomials

1° $P_n(x)$ is a polynomial of degree n,
2° $\mathcal{L}[P_n(x)P_k(x)] = 0$ for $k \neq n$,
3° $\mathcal{L}[P_n(x)^2] \neq 0$.

If a sequence of orthogonal polynomial exists for a given linear functional $\mathcal{L}$, then $\mathcal{L}$ is called *quasi-definite linear functional*. Under the condition $\mathcal{L}[P_n^2(x)] > 0$, the functional $\mathcal{L}$ is called positive definite. For details see Chihara [60, pp. 6–17].

The necessary and sufficient conditions for the existence of a sequence of orthogonal polynomials with respect to the linear functional $\mathcal{L}$ are that for each $n \in \mathbb{N}$ the Hankel determinants

$$\Delta_n = \begin{vmatrix} \mu_0 & \mu_1 & \cdots & \mu_{n-1} \\ \mu_1 & \mu_2 & & \mu_n \\ \vdots & & & \\ \mu_{n-1} & \mu_n & & \mu_{2n-2} \end{vmatrix} \neq 0. \quad (2.1.22)$$

When $\mathcal{L}$ is positive definite, we can define $(p,q) := \mathcal{L}[p(x)\overline{q(x)}]$ for all algebraic polynomials $p(x)$ and $q(x)$, so that the orthogonality with respect to the moment functional $\mathcal{L}$ is consistent with the standard definition of orthogonality with respect to an inner product. On the other hand, there are several interesting quasi-definite cases when the moments are non-real. We mention here two cases:

(a) *Orthogonality with respect to an oscillatory weight.* Let $w(x) = xe^{im\pi x}$, where $x \in [-1, 1]$ and $m \in \mathbb{Z}$. Putting

$$\mathcal{L}[p] := \int_{-1}^{1} p(x)w(x)\,dx \quad (p \in \mathcal{P}), \quad (2.1.23)$$

i.e.,

$$\mu_k = \frac{(-1)^{m+k}(k+1)!}{(im\pi)^{k+1}} \sum_{\nu=0}^{k} \frac{(1+(-1)^\nu)(-im\pi)^\nu}{(\nu+1)!},$$

in [346] it was proved that for every integer m ($\neq 0$), the sequence of formal orthogonal polynomials with respect to the functional $\mathcal{L}$ exists uniquely. Such orthogonal polynomials have several interesting properties and they can be applied in numerical integration of highly oscillatory functions (see [346]). The corresponding case with the Chebyshev weight, i.e., when $w(x) = x(1-x^2)^{-1/2}e^{i\zeta x}$, $\zeta \in \mathbb{R}$, was recently investigated in [347].

According to (2.1.23) we can define

$$(p,q) := \int_{-1}^{1} p(x)q(x)w(x)\,dx \quad (p,q \in \mathcal{P}),$$

but, as we can see, (p,q) is not Hermitian and not positive definite. Namely, the properties (c) and (d) of the inner product do not hold!

(b) *Orthogonality on the semicircle*. Polynomials orthogonal on the semicircle $\Gamma = \{z \in \mathbb{C} \mid z = e^{i\theta}, \ 0 \leq \theta \leq \pi\}$ with respect to the moment functional

$$\mathcal{L}[z^k] = \mu_k = \int_0^\pi e^{ik\theta} d\theta = \begin{cases} \pi, & k = 0, \\ \dfrac{2i}{k}, & k \text{ odd}, \\ 0, & k \neq 0 \text{ even}, \end{cases}$$

have been introduced by Gautschi and Milovanović (see [170] and [171]). The corresponding non-Hermitian inner product is given by

$$(p,q) = \int_\Gamma p(z)q(z)(iz)^{-1} dz = \int_0^\pi p(e^{i\theta})q(e^{i\theta}) d\theta.$$

A few first polynomials of this orthogonal system are

$$\pi_0(x) = 1,$$

$$\pi_1(x) = x - \frac{2i}{\pi},$$

$$\pi_2(x) = x^2 - \frac{\pi i}{6}x - \frac{1}{3},$$

$$\pi_3(x) = x^3 - \frac{8i}{5\pi}x^2 - \frac{3}{5}x + \frac{8i}{15\pi},$$

$$\pi_4(x) = x^4 - \frac{9\pi i}{56}x^3 - \frac{6}{7}x^2 + \frac{27\pi i}{280}x + \frac{3}{35},$$

$$\pi_5(x) = x^5 - \frac{128i}{81\pi}x^4 - \frac{10}{9}x^3 + \frac{256i}{189\pi}x^2 + \frac{5}{21}x - \frac{128i}{945\pi}.$$

The general case of complex polynomials orthogonal with respect to a *complex weight function* was considered by Gautschi, Landau and Milovanović [175]. For some properties and applications see [153, 326, 327, 329]. A generalization of such polynomials on a circular arc was given by de Bruin [50], and further investigations were done by Milovanović and Rajković [354].

In this chapter we mainly consider those polynomial orthogonal systems, when an inner product is defined on some lines or on a curve in the complex plane $\mathbb{C}$. Furthermore, starting from Sect. 2.2, we only consider polynomials orthogonal on the real line.

By $\mathcal{P}_n$ we denote the set of all algebraic polynomials (with complex coefficients) of degree at most n. Further, let $\hat{\mathcal{P}}_n$ be the set of all monic polynomials of degree n, i.e.,

$$\hat{\mathcal{P}}_n = \{z^n + q(z) \mid q(z) \in \mathcal{P}_{n-1}\}.$$

2.1 Orthogonal Systems and Polynomials

A system of polynomials $\{p_n\}$, where

$$p_n(z) = \gamma_n z^n + \text{lower degree terms}, \quad \gamma_n > 0,$$
$$(p_n, p_m) = \delta_{nm}, \quad n, m \geq 0, \tag{2.1.24}$$

is called a system of *orthonormal polynomials* with respect to the inner product $(\cdot, \cdot)$. In many considerations and applications we use the *monic orthogonal polynomials*

$$\pi_n(z) = \frac{p_n(z)}{\gamma_n} = z^n + \text{lower degree terms}. \tag{2.1.25}$$

Sometimes, we also use the notation $\hat{p}_n(z)$ for monic orthogonal polynomials instead of $\pi_n(z)$.

2.1.4 Basic Facts on Orthogonal Polynomials and Extremal Problems

Let $d\mu$ be a finite positive Borel measure in the complex plane $\mathbb{C}$, with an infinite set as its support, and let $L^2(d\mu)$ denote the Hilbert space of measurable functions f for which $\int |f(z)|^2 d\mu(z) < +\infty$. Finally, let the inner product $(\cdot, \cdot)$ be defined by

$$(f, g) = \int f(z)\overline{g(z)} d\mu(z) \qquad (f, g \in L^2(d\mu)). \tag{2.1.26}$$

Starting from the system of monomials $U = \{1, z, z^2, \ldots\}$, by the Gram-Schmidt orthogonalization process, we can obtain the unique orthonormal polynomials (2.1.24). In order to emphasize the orthogonality with respect to the given measure $d\mu$, we write

$$p_n(z) = p_n(d\mu; z) = \gamma_n z^n + \text{lower degree terms}, \quad \gamma_n = \gamma_n(d\mu) > 0.$$

Also, the monic orthogonal polynomials (2.1.25) are unique.

The following extremal property characterizes orthogonal polynomials:

Theorem 2.1.3 *The polynomial* $\pi_n(z) = p_n(z)/\gamma_n = z^n + \cdots$ *is the unique monic polynomial of degree n of minimal $L^2(d\mu)$-norm, i.e.,*

$$\min_{p \in \hat{\mathcal{P}}_n} \int |p(z)|^2 d\mu(z) = \int |\pi_n(z)|^2 d\mu(z) = \frac{1}{\gamma_n^2}. \tag{2.1.27}$$

Proof Using the polynomials $\{p_k(z)\}$, orthonormal with respect to the measure $d\mu$, an arbitrary monic polynomial $p(z) \in \hat{\mathcal{P}}_n$ can be expressed in the form

$$p(z) = \sum_{k=0}^{n-1} c_k p_k(z) + \frac{1}{\gamma_n} p_n(z).$$

Then, we have

$$\|p\|^2 = \sum_{k=0}^{n-1} |c_k|^2 + \frac{1}{\gamma_n^2} \geq \frac{1}{\gamma_n^2},$$

with equality if and only if $c_0 = c_1 = \cdots = c_{n-1} = 0$, i.e.,

$$p(z) = p^*(z) = \frac{1}{\gamma_n} P_n(z) = \pi_n(z).$$

□

This extremal property is completely equivalent to orthogonality. Namely, many questions regarding orthogonal polynomials can be answered by using only this extremal property (cf. [477] and [452]).

Notice also that the previous theorem gives the polynomial of the best approximation to the monomial z^n in the class $\mathcal{P}_{n-1}$. Indeed, according to Theorem 2.1.2, the best L^2-approximation to $f(z) = z^n$ is given by $s_{n-1}(z) = \sum_{k=0}^{n-1} f_k p_k(z)$, where $f_k = (z^n, p_k)$, $k = 0, 1, \ldots, n-1$. But, by Theorem 2.1.3, $f(z) - s_{n-1}(z) = \pi_n(z)$, so that the polynomial of the best approximation in this case can be expressed in the form $s_{n-1}(z) = z^n - \pi_n(z)$.

Now, we define the function $(z, t) \mapsto K_n(z, t)$ by

$$K_n(z, t) = \sum_{k=0}^{n} p_k(z) \overline{p_k(t)} \qquad (n \geq 0), \tag{2.1.28}$$

which plays a fundamental role in the integral representation of the partial sums of the orthogonal expressions.

For a function $f \in L^2(d\mu)$ we can determine its Fourier coefficients f_k with respect to the inner product (2.1.26). Thus, using (2.1.4) and the orthonormal polynomials $\{p_k(z)\}$, we have

$$f_k = (f, p_k) = \int f(t) \overline{p_k(t)} \, d\mu(t).$$

Then, the partial sums (2.1.5) can be expressed in an integral form

$$s_n(z) = \sum_{k=0}^{n} f_k p_k(z) = \sum_{k=0}^{n} (f, p_k) p_k(z) = \int f(t) K_n(z, t) \, d\mu(t).$$

Suppose that f is an arbitrary polynomial of degree at most n, i.e., $f(z) = P(z)$ ($P(z) \in \mathcal{P}_n$). Then, the corresponding partial sum s_n coincides with f and we obtain

$$P(z) = \int P(t) K_n(z, t) \, d\mu(t) \qquad (P(z) \in \mathcal{P}_n). \tag{2.1.29}$$

2.1 Orthogonal Systems and Polynomials

Because of that, the function K_n is very often called the *reproducing kernel*. Notice that $\overline{K_n(z,t)} = K_n(t,z)$ and

$$K_n(z,z) = \sum_{k=0}^{n} |p_k(z)|^2 \geq |p_0(z)|^2 = \gamma_0^2 > 0$$

for each $z \in \mathbb{C}$ and $n \geq 0$.

The reciprocal of this function is known as the *Christoffel function*,

$$\lambda_n(z) = \lambda_n(d\mu; z) = \frac{1}{K_{n-1}(z,z)} = \left(\sum_{k=0}^{n-1} |p_k(z)|^2\right)^{-1}. \tag{2.1.30}$$

The following extremal problem is related to the reproducing kernel (cf. [359] and [485]):

Theorem 2.1.4 *For every $P(z) \in \mathcal{P}_n$ such that $P(t) = 1$, we have*

$$\int |P(z)|^2 d\mu(z) \geq \lambda_{n+1}(d\mu; t), \tag{2.1.31}$$

with equality only for

$$P(z) = P^*(z) = \frac{K_n(z,t)}{K_n(t,t)}.$$

Proof Let t be a fixed complex number and $P(z) \in \mathcal{P}_n$. In order to find the minimum of the integral on the left hand side in (2.1.31) under the constraint $P(t) = 1$, we represent $P(z)$ as a linear combination of the orthonormal polynomials $p_k(z) = p_k(d\mu; z)$, i.e., $P(z) = \sum_{k=0}^{n} c_k p_k(z)$. Then, we have

$$F(P) = \int |P(z)|^2 d\mu(z) = \sum_{k=0}^{n} |c_k|^2.$$

Since $P(t) = \sum_{k=0}^{n} c_k p_k(t) = 1$, using the Cauchy inequality for the complex sequences $\mathbf{c} = \{c_k\}_{k=0}^{n}$ and $\mathbf{p} = \{p_k(t)\}_{k=0}^{n}$ (see Mitrinović [364, p. 32]), we have

$$1 = \left|\sum_{k=0}^{n} c_k p_k(t)\right|^2 \leq \left(\sum_{k=0}^{n} |c_k|^2\right)\left(\sum_{k=0}^{n} |p_k(t)|^2\right) = F(P) K_n(t,t), \tag{2.1.32}$$

which implies $F(P) \geq 1/K_n(t,t) = \lambda_{n+1}(d\mu; t)$, i.e., (2.1.31).

In the case of equality in (2.1.32), which is attained only if the sequences $\mathbf{c}$ and $\overline{\mathbf{p}}$ are proportional, i.e., when $c_k = \gamma \, \overline{p_k(t)}$ ($k = 0, 1, \ldots, n$), with some complex constant λ, we find that

$$P(t) = \gamma \sum_{k=0}^{n} |p_k(t)|^2 = \gamma K_n(t,t) = 1.$$

Thus, $\gamma = 1/K_n(t,t)$ and the extremal polynomial is given by

$$P(z) = P^*(z) = \gamma \sum_{k=0}^{n} \overline{p_k(t)}\, p_k(z) = \frac{K_n(z,t)}{K_n(t,t)}.$$

□

According to this theorem, the Christoffel function can also be expressed in the form

$$\lambda_n(d\mu;t) = \min_{\substack{P \in \mathcal{P}_{n-1} \\ P(t)=1}} \int |P(z)|^2 \, d\mu(z). \qquad (2.1.33)$$

Using the previous theorem we can also prove:

Theorem 2.1.5 *Let t be a fixed complex constant, and let $P(z)$ be an arbitrary polynomial of degree at most n, normalized by the condition*

$$\|P\|^2 = \int |P(z)|^2 \, d\mu(z) = 1.$$

The maximum of $|P(t)|^2$ taken over all such polynomials is attained for

$$P(z) = \gamma \frac{K_n(z,t)}{\sqrt{K_n(t,t)}} \qquad (|\gamma|=1).$$

The maximum itself is $K_n(t,t)$.

Proof Taking $Q(z) = P(z)/P(t)$ we see that $Q(t) = 1$ and, according to Theorem 2.1.4,

$$\int |Q(z)|^2 \, d\mu(z) = \frac{1}{|P(t)|^2} \geq \lambda_{n+1}(d\mu;t),$$

with equality case for $Q(z) = P(z)/P(t) = K_n(z,t)/K_n(t,t)$.

Thus, $|P(t)|^2 \leq K_n(t,t)$, with equality only for

$$P(z) = P(t)Q(z) = \gamma \sqrt{K_n(t,t)} \frac{K_n(z,t)}{K_n(t,t)} = \gamma \frac{K_n(z,t)}{\sqrt{K_n(t,t)}},$$

where $|\gamma| = 1$. □

Under certain conditions, the extremal property from Theorem 2.1.3, can be extended to $L^r(d\mu)$-norm ($1 < r < +\infty$), so that the unique monic polynomial $p_n^*(z) = z^n + \cdots$ of the minimal $L^r(d\mu)$-norm exists, i.e.,

$$\min_{p \in \hat{\mathcal{P}}_n} \int |p(z)|^r \, d\mu(z) = \int |p_n^*(z)|^r \, d\mu(z). \qquad (2.1.34)$$

For measures with support on the real line, an interesting special case $r = 2s+2$, where $s \in \mathbb{N}_0$, leads to a case of the *power orthogonality*. Then, the extremal (monic) polynomials in (2.1.34), denoted by $p_n^*(x) = \pi_{n,s}(x) = \pi_{n,s}(x; d\mu)$, exist uniquely and they are known as *s-orthogonal polynomials* (for more details see [146, 179, 180, 340, 387]). These polynomials must satisfy the "orthogonality conditions" (cf. Ghizzetti and Ossicini [180], Milovanović [340])

$$\int_\mathbb{R} \pi_{n,s}(x)^{2s+1} x^k \, d\mu(x) = 0, \quad k = 0, 1, \ldots, n-1. \tag{2.1.35}$$

In the case $s = 0$, the s-orthogonal polynomials reduce to the standard orthogonal polynomials, $\pi_{n,0} = \pi_n$.

Also, the *generalized Christoffel function* can be defined for $0 < r < +\infty$, by

$$\lambda_n(d\mu, r; t) = \min_{\substack{P \in \mathcal{P}_{n-1} \\ P(t)=1}} \int |P(z)|^r \, d\mu(z). \tag{2.1.36}$$

Notice that (2.1.36) for $r = 2$ reduces to (2.1.33), i.e., $\lambda_n(d\mu, 2; t) = \lambda_n(d\mu; t)$.

Several properties of the generalized Christoffel functions for measures on $\mathbb{R}$ can be found in Nevai [375, pp. 106–123].

2.1.5 Zeros of Orthogonal Polynomials

Now, we study some basic properties of the zeros of orthogonal polynomials. According to the fundamental theorem of algebra, we know that any polynomial of degree n has exactly n zeros, counting multiplicities. The zeros of orthogonal polynomials play a very important role in interpolation theory, quadrature formulas, etc.

Using Theorem 2.1.3 it is easy to prove a general result on the location of zeros. This result is connected with the support of the measure supp $(d\mu)$, which is a closed set. Firstly, we need some definitions:

Definition 2.1.1 A set $A \subset \mathbb{C}$ is *convex* if for each pair of points $z, t \in A$ the line connecting z and t is a subset of A.

Definition 2.1.2 The *convex hull* Co (B) of a set $B \subset \mathbb{C}$ is the smallest convex set containing B.

Definition 2.1.3 Let D_∞ be the connected component of the complement of E that contains the point ∞, then D_∞ is open and

$$\text{Pc}(E) = \mathbb{C} \setminus D_\infty$$

is the *polynomial convex hull* of E.

It is clear that Co (supp $(d\mu)$) is the intersection of all closed half-planes containing supp $(d\mu)$. Also,

$$\text{supp}(d\mu) \subset \text{Pc}(\text{supp}(d\mu)) \subset \text{Co}(\text{supp}(d\mu)).$$

The following result is due to Fejér (see [422] and [485]).

Theorem 2.1.6 *All the zeros of the (monic) polynomial $\pi_n(d\mu; z)$ lie in the convex hull of the support $E = \text{supp}(d\mu)$.*

Proof Suppose that ζ is a zero of $\pi_n(d\mu; z) = \pi_n(z)$ such that $\zeta \notin \text{Co}(E)$, where $E = \text{supp}(d\mu)$. Then

$$\pi_n(z) = (z - \zeta)q(z) \qquad (q(z) \in \hat{\mathcal{P}}_n).$$

Since $\zeta \notin \text{Co}(E)$, there exists a line L separating E and ζ. Let $\hat{\zeta}$ be the orthogonal projection of ζ on L. Then, for every $z \in E$, we have $|z - \hat{\zeta}| < |z - \zeta|$, i.e.,

$$|(z - \hat{\zeta})q(z)| < |(z - \zeta)q(z)| = |\pi_n(z)|,$$

from which it follows that

$$\int |(z - \hat{\zeta})q(z)|^2 d\mu(z) < \int |\pi_n(z)|^2 d\mu(z),$$

which is a contradiction to Theorem 2.1.3. Thus, we conclude that there are no zeros outside Co (E). $\square$

An improvement of this theorem was given by Saff [422]:

Theorem 2.1.7 *If Co (supp $(d\mu)$) is not a line segment, then all the zeros of the polynomial $\pi_n(d\mu; z)$ lie in the interior of Co (supp $(d\mu)$).*

For example, if C is the unit circle $|z| = 1$ and supp $(d\mu) \subset C$, then Theorem 2.1.7 asserts that all the zeros of $\pi_n(d\mu; z)$ must lie in the open unit disk $|z| < 1$. This is a classical result of Szegő [470, p. 292] for orthogonal polynomials on the unit circle.

An interesting question is related with a number of zeros of $\pi_n(d\mu; z)$ which are outside $E = \text{supp}(d\mu)$ (i.e., in Co $(E) \setminus E$). If the set E has holes, it is possible that all the zeros are in the holes, as in the case of polynomials orthogonal on the unit circle. Here, we mention a result of Widom [505] (see Saff [422] for the proof).

Theorem 2.1.8 *Let $E = \text{supp}(d\mu)$ and A be a closed set such that $A \cap \text{Pc}(E) = \emptyset$. Then the number of zeros of $\pi_n(d\mu; z)$ on A is uniformly bounded in n.*

2.2 Orthogonal Polynomials on the Real Line

2.2.1 Basic Properties

One of the most important classes of orthogonal polynomials is the class of orthogonal polynomials on the real line. In this section we consider the basic facts that hold for such polynomials. Thus, we suppose here that the support of a positive Borel measure $d\mu$ is on the real line, i.e.,

$$\operatorname{supp}(d\mu) = \{x \in \mathbb{R} \mid \mu(x+\varepsilon) - \mu(x-\varepsilon) > 0 \text{ for every } \varepsilon > 0\},$$

as well as that all moments $\mu_k = \int_\mathbb{R} x^k \, d\mu(x)$, $k = 0, 1, \ldots$, exist and are finite. Also, we suppose that $\operatorname{supp}(d\mu)$ contains infinitely many points, i.e., that the distribution function $\mu \colon \mathbb{R} \to \mathbb{R}$ is a non-decreasing function with infinitely many points of increase. We are now working with real-valued functions so that the inner product (2.1.26) reduces to

$$(f, g) = \int_\mathbb{R} f(x) g(x) \, d\mu(x). \tag{2.2.1}$$

As before, orthonormal and monic polynomials will be denoted by $p_n(x)$ and $\pi_n(x)$, respectively (see (2.1.24) and (2.1.25)). Taking $U = \{1, x, x^2, \ldots\}$, these polynomials can be obtained by the Gram-Schmidt orthogonalizing process. The corresponding Gram matrix can be expressed in terms of the moments μ_k ($k = 0, 1, \ldots, 2n$) in the form

$$G_{n+1} = \begin{bmatrix} \mu_0 & \mu_1 & \cdots & \mu_n \\ \mu_1 & \mu_2 & & \mu_{n+1} \\ \vdots & & & \\ \mu_n & \mu_{n+1} & & \mu_{2n} \end{bmatrix}, \tag{2.2.2}$$

and the orthonormal polynomial of degree n as

$$p_n(x) = p_n(d\mu; x) = \frac{1}{\sqrt{\Delta_n \Delta_{n+1}}} \begin{vmatrix} \mu_0 & \mu_1 & \cdots & \mu_{n-1} & 1 \\ \mu_1 & \mu_2 & & \mu_n & x \\ \vdots & & & & \\ \mu_n & \mu_{n+1} & & \mu_{2n-1} & x^n \end{vmatrix},$$

where $\Delta_n = \det G_n$ and $\Delta_0 = 1$. The leading coefficient is given by $\gamma_n = \gamma_n(d\mu) = \sqrt{\Delta_n / \Delta_{n+1}}$. The matrices of the form (2.2.2) are known as *Hankel matrices* and the corresponding determinants as *Hankel determinants* (see (2.1.22)).

If μ is an absolutely continuous function, then we say that $\mu'(x) = w(x)$ is a *weight function*. In that case, the measure $d\mu$ can be expressed as $d\mu(x) = w(x) \, dx$, where the weight function $x \mapsto w(x)$ is non-negative and measurable in Lebesgue's sense for which all moments exist and $\mu_0 > 0$. Then, instead of $p_n(d\mu; \cdot)$, $\gamma_n(d\mu)$, supp $(d\mu), \ldots,$ we usually write $p_n(w; \cdot)$, $\gamma_n(w)$, supp $(w), \ldots,$ respectively. If supp $(w) = [a, b]$, where $-\infty < a < b < +\infty$, we say that $\{p_n\}$ is a system of orthonormal polynomials in a finite interval $[a, b]$. For (a, b) we say that it is an *interval of orthogonality*.

Here we will not consider the case of a *discrete measure* when $d\mu(x)$ is concentrated on a finite number of points. In the general case, the function μ can be written in the form $\mu = \mu_{ac} + \mu_s + \mu_j$, where μ_{ac} is absolutely continuous, μ_s is singular, and μ_j is a jump function.

Now we give a few basic properties of orthogonal polynomials on the real line.

2.2.1.1 Three-Term Recurrence Relations

The following statement gives a fundamental relation for polynomials orthogonal with respect to the inner product (2.2.1):

Theorem 2.2.1 *The system of orthonormal polynomials* $\{p_n(x)\}$, *associated with the measure* $d\mu(x)$, *satisfy a three-term recurrence relation*

$$xp_n(x) = b_{n+1}p_{n+1}(x) + a_n p_n(x) + b_n p_{n-1}(x) \qquad (n \geq 0), \qquad (2.2.3)$$

where $p_{-1}(x) = 0$ *and the coefficients* $a_n = a_n(d\mu)$ *and* $b_n = b_n(d\mu)$ *are given by*

$$a_n = \int_{\mathbb{R}} xp_n(x)^2 d\mu(x) \quad \text{and} \quad b_n = \int_{\mathbb{R}} xp_{n-1}(x)p_n(x) d\mu(x) = \frac{\gamma_{n-1}}{\gamma_n}.$$

Proof The three-term recurrence relation is a consequence of the inner product (2.1.6) that $(xf, g) = (f, xg)$. Indeed, expanding $xp_n(x)$ in terms of the orthonormal polynomials $p_k(x)$ $(k = 0, 1, \ldots, n+1)$,

$$xp_n(x) = \sum_{k=0}^{n+1} c_k^{(n)} p_k(x),$$

where the Fourier coefficients are given by

$$c_k^{(n)} = \int_{\mathbb{R}} xp_n(x)p_k(x)d\mu(x) = (xp_n, p_k) = (p_n, xp_k) \qquad (k = 0, 1, \ldots, n+1),$$

we conclude that $c_k^{(n)} = 0$ for $k + 1 < n$. Then, putting

$$c_{n-1}^{(n)} = (p_n, xp_{n-1}) = b_n \quad \text{and} \quad c_n^{(n)} = (xp_n, p_n) = a_n,$$

we find that $c_{n+1}^{(n)} = (xp_n, p_{n+1}) = b_{n+1}$. By comparing the leading coefficients on both sides of (2.2.3) we get $\gamma_n = b_{n+1}\gamma_{n+1}$, i.e., $b_n = \gamma_{n-1}/\gamma_n$. □

2.2 Orthogonal Polynomials on the Real Line

Since $p_0(x) = \gamma_0 = 1/\sqrt{\mu_0}$ and $\gamma_{n-1} = b_n \gamma_n$ we have that $\gamma_n = \gamma_0/(b_1 b_2 \cdots b_n)$. Notice that $b_n > 0$ for each n.

Conversely, for two given real sequences $\{a_n\}_{n \in \mathbb{N}_0}$ and $\{b_n\}_{n \in \mathbb{N}}$, where $b_n > 0$ for each $n \in \mathbb{N}$, one can construct a sequence of polynomials using the three-term recurrence relation (2.2.3), starting with the initial values $p_{-1}(x) = 0$ and $p_0(x) = 1$. It is well-known by Favard's theorem (cf. Chihara [60]) that there exists a positive measure $d\sigma(x)$ on $\mathbb{R}$ such that

$$\int_{\mathbb{R}} p_n(x) p_m(x) \, d\sigma(x) = C_n \delta_{nm}, \qquad C_n > 0, \ n, m \geq 0.$$

The measure $d\sigma(x)$ is not unique which depending on the fact of whether or not the Hamburger moment problem is determined (see Sect. 2.2.3). A sufficient condition for a unique measure is the Carleman condition given by $\sum_{n=1}^{+\infty} (1/b_n) = +\infty$. Evidently, it holds if $\{b_n\}_{n \in \mathbb{N}}$ is a bounded sequence.

For the monic orthogonal polynomials $\pi_n(x) = p_n(x)/\gamma_n$ the following result holds:

Theorem 2.2.2 *The monic orthogonal polynomials* $\{\pi_n(x)\}$ *satisfy the following three-term recurrence relation*

$$\pi_{n+1}(x) = (x - \alpha_n) \pi_n(x) - \beta_n \pi_{n-1}(x), \qquad n = 0, 1, 2, \ldots, \tag{2.2.4}$$

where $\alpha_n = a_n$ *and* $\beta_n = b_n^2 > 0$.

Because of the orthogonality, we have

$$\alpha_n = \frac{(x\pi_n, \pi_n)}{(\pi_n, \pi_n)} \quad (n \geq 0), \qquad \beta_n = \frac{(\pi_n, \pi_n)}{(\pi_{n-1}, \pi_{n-1})} \quad (n \geq 1).$$

The coefficient β_0 which is multiplied by $\pi_{-1}(x) = 0$ in the three-term recurrence relation (2.2.4) may be arbitrary. Sometimes, it is convenient to define it by $\beta_0 = \mu_0 = \int_{\mathbb{R}} d\mu(x)$. Then the norm of π_n can be expressed in the form

$$\|\pi_n\| = \sqrt{(\pi_n, \pi_n)} = \sqrt{\beta_0 \beta_1 \cdots \beta_n}. \tag{2.2.5}$$

Remark 2.2.1 The recursion coefficients α_n and β_n in (2.2.4) can be expressed in terms of the Hankel determinants

$$\Delta_n = \begin{vmatrix} \mu_0 & \mu_1 & \cdots & \mu_{n-1} \\ \mu_1 & \mu_2 & & \mu_n \\ \vdots & & & \\ \mu_{n-1} & \mu_n & & \mu_{2n-2} \end{vmatrix} \quad \text{and} \quad \Delta'_n = \begin{vmatrix} \mu_0 & \mu_1 & \cdots & \mu_{n-2} & \mu_n \\ \mu_1 & \mu_2 & & \mu_{n-1} & \mu_{n+1} \\ \vdots & & & & \\ \mu_{n-1} & \mu_n & & \mu_{2n-3} & \mu_{2n-1} \end{vmatrix},$$

where $\Delta_0 = 1$ and $\Delta'_0 = 0$. Namely,

$$\alpha_k = \frac{\Delta'_{k+1}}{\Delta_{k+1}} - \frac{\Delta'_k}{\Delta_k} \quad (k \geq 0), \qquad \beta_k = \frac{\Delta_{k-1} \Delta_{k+1}}{\Delta_k^2} \quad (k \geq 1).$$

2.2.1.2 Christoffel's Formulae

The function K_n, defined earlier in (2.1.28), now reduces to

$$K_n(x,t) = \sum_{k=0}^{n} p_k(x) p_k(t) \qquad (n \geq 0), \tag{2.2.6}$$

Notice that $K_n(t,x) = K_n(x,t)$. Using the three-term recurrence relation (2.2.3) we can prove:

Theorem 2.2.3 Let $K_n(x,t)$ be defined by (2.2.6). Then

$$K_n(x,t) = b_{n+1} \frac{p_{n+1}(x) p_n(t) - p_n(x) p_{n+1}(t)}{x - t}, \tag{2.2.7}$$

where b_{n+1} is defined in Theorem 2.2.1.

Formula (2.2.7) is known as the *Christoffel-Darboux identity*.
Letting $t \to x$ we find the confluent form of (2.2.7),

$$K_n(x,x) = \sum_{k=0}^{n} p_k(x)^2 = b_{n+1} \big(p'_{n+1}(x) p_n(x) - p'_n(x) p_{n+1}(x) \big). \tag{2.2.8}$$

The following result is known as Christoffel's formula (cf. Szegő [470, pp. 29–30]):

Theorem 2.2.4 Let $\{p_n(x)\}$ be a system of orthonormal polynomials associated with the measure $d\mu(x)$ on $[a,b]$ and let

$$\varrho(x) = c(x-\xi_1)(x-\xi_2)\ldots(x-\xi_m), \qquad c \neq 0, \tag{2.2.9}$$

be a nonnegative polynomial $[a,b]$, with distinct zeros ξ_ν $(\nu = 1,\ldots,m)$ outside (a,b). Then the orthogonal polynomials $\{q_n(x)\}$, associated with the measure $d\sigma(x) = \varrho(x) d\mu(x)$, can be expressed in the form

$$\varrho(x) q_n(x) = \begin{vmatrix} p_n(x) & p_{n+1}(x) & \cdots & p_{n+m}(x) \\ p_n(\xi_1) & p_{n+1}(\xi_1) & & p_{n+m}(\xi_1) \\ \vdots & & & \\ p_n(\xi_m) & p_{n+1}(\xi_m) & & p_{n+m}(\xi_m) \end{vmatrix}. \tag{2.2.10}$$

Remark 2.2.2 In case of a zero ξ_k, of multiplicity s (> 1), the corresponding rows of (2.2.10) should be replaced by the derivatives of order $0, 1, \ldots, s-1$ of the polynomials $p_n(x), p_{n+1}(x), \ldots, p_{n+m}(x)$ at $x = \xi_k$.

2.2.1.3 Zeros

The three-term recurrence relation (2.2.3) suggests us to study an infinite, symmetric, tridiagonal matrix, known as *Jacobi matrix*,

$$J = J(d\mu) = \begin{bmatrix} a_0 & b_1 & & & O \\ b_1 & a_1 & b_2 & & \\ & b_2 & a_2 & b_3 & \\ & & \ddots & \ddots & \ddots \\ O & & & & \end{bmatrix}. \qquad (2.2.11)$$

Assuming both sequences $\{a_n\}_{n\in\mathbb{N}_0}$ and $\{b_n\}_{n\in\mathbb{N}}$ to be uniformly bounded, the associated Jacobi matrix (2.2.11) can be understood as a linear operator J acting on ℓ^2, the space of all complex square-summable sequences, where the value of the operator J at the vector x is a product of the infinite vector $\mathbf{x}$ and the infinite matrix J in the matrix sense. In the case when the sequences $\{a_n\}_{n\in\mathbb{N}_0}$ and $\{b_n\}_{n\in\mathbb{N}}$ are not uniformly bounded, an operator acting on ℓ^2 cannot be defined that easily. Additional properties of the sequence of orthogonal polynomials are needed in order to be able to uniquely define the operator.

Suppose now that supp $(d\mu)$ is bounded and denote by $\Delta(d\mu)$ the smallest closed interval containing supp $(d\mu)$. As a corollary of Theorem 2.1.6 we have that all the zeros of $p_n(d\mu; x)$ ($n \geq 1$) lie in $\Delta(d\mu)$. Furthermore, we can prove that they are mutually different.

Theorem 2.2.5 *All zeros of $p_n(d\mu; x)$, $n \geq 1$, are real and distinct and are located in the interior of the interval $\Delta(d\mu)$.*

Proof Suppose $p_n(d\mu; x)$ has m distinct zeros $x_1, \ldots, x_m$ in the interior of the interval $\Delta(d\mu)$ that are of odd order and let $q(x) := (x - x_1) \cdots (x - x_m)$. Then, for each $x \in \Delta(d\mu)$, we have that $p_n(d\mu; x)q(x) \geq 0$, which implies

$$\int_\mathbb{R} p_n(x)q(x)d\mu(x) > 0.$$

On the other hand, if $m < n$ this integral is equal to zero because of orthogonality. This contradiction implies that $m = n$, which means that all zeros are simple and are located in the interior of the interval $\Delta(d\mu)$. ☐

Let $x_{n,1} < x_{n,2} < \cdots < x_{n,n}$ denote the zeros of $p_n(d\mu; x)$ in increasing order.

Theorem 2.2.6 *The zeros of $p_n(d\mu; x)$ and $p_{n+1}(d\mu; x)$ interlace, i.e.,*

$$x_{n+1,k} < x_{n,k} < x_{n+1,k+1} \qquad (k = 1, \ldots, n;\ n \in \mathbb{N}).$$

The proof of this interlacing property can be given using the inequality

$$p'_{n+1}(x)p_n(x) - p'_n(x)p_{n+1}(x) > 0 \qquad (x \in \mathbb{R}),$$

which follows from (2.2.8) (cf. [328, p. 105]).

Consider now the three-term recurrence relation (2.2.3), in which n is substituted by k. Then, taking this relation for $k = 0, 1, \ldots, n-1$, one can obtain the following system of equations

$$x\mathbf{p}_n(x) = J_n(d\mu)\mathbf{p}_n(x) + b_n p_n(x)\mathbf{e}_n, \qquad (2.2.12)$$

where

$$J_n(d\mu) = \begin{bmatrix} a_0 & b_1 & & & O \\ b_1 & a_1 & b_2 & & \\ & b_2 & a_2 & \ddots & \\ & & \ddots & \ddots & b_{n-1} \\ O & & & b_{n-1} & a_{n-1} \end{bmatrix}, \quad \mathbf{p}_n(x) = \begin{bmatrix} p_0(x) \\ p_1(x) \\ p_2(x) \\ \vdots \\ p_{n-1}(x) \end{bmatrix}, \quad \mathbf{e}_n = \begin{bmatrix} 0 \\ 0 \\ 0 \\ \vdots \\ 1 \end{bmatrix}.$$

The tridiagonal matrix $J_n = J_n(d\mu)$ is the $n \times n$ leading principal minor matrix of the infinite Jacobi matrix (2.2.11). It is clear that $p_n(x) = 0$ if and only if

$$x\mathbf{p}_n(x) = J_n \mathbf{p}_n(x),$$

i.e., the zeros $x_{n,k}$ of $p_n(x)$ are the same as the eigenvalues of the Jacobi matrix J_n. Also, notice that the monic polynomial $\pi_n(x)$ can be expressed in the following determinant form

$$\pi_n(x) = \det(xI_n - J_n),$$

where I_n is the identity matrix of order n.

Using bounds for the eigenvalues of the Jacobi matrices we can obtain certain bounds for the zeros of orthogonal polynomials. For example, by Gershgorin's theorem we have

$$x_{n,k} \in \bigcup_{i=0}^{n-1} [a_i - b_i - b_{i+1}, a_i + b_i + b_{i+1}] \qquad (b_0 = 0),$$

i.e.,

$$\min_{0 \le i \le n-1} (a_i - b_i - b_{i+1}) \le x_{n,k} \le \max_{0 \le i \le n-1} (a_i + b_i + b_{i+1}). \qquad (2.2.13)$$

Some sharper bounds can be found in [181–183, 254, 488].

Let $x_i = x_{n,i}$, $i = 1, \ldots, n$, be the zeros of the orthonormal polynomial $p_n(d\mu; x)$. Putting $x = x_i$ and $t = x_j$ in (2.2.6) we get

$$K_n(x_i, x_j) = \sum_{k=0}^{n-1} p_k(x_i) p_k(x_j) = \langle \mathbf{p}_n(x_i), \mathbf{p}_n(x_j) \rangle = \mathbf{p}_n(x_j)^T \mathbf{p}_n(x_i), \qquad (2.2.14)$$

2.2 Orthogonal Polynomials on the Real Line

where $\mathbf{p}_n(x) = [p_0(x)\ p_1(x)\ \cdots\ p_{n-1}(x)]^T$ and $\langle \cdot, \cdot \rangle$ is the usual inner product in the Euclidean space $\mathbb{R}^n$.

Theorem 2.2.7 *We have for the inner products in* (2.2.14)

$$K_n(x_i, x_j) = \begin{cases} 0, & \text{if } i \ne j, \\ b_n p_{n-1}(x_i) p'_n(x_i), & \text{if } i = j, \end{cases} \quad (2.2.15)$$

where b_n is defined in Theorem 2.2.1.

Proof According to (2.2.7) and (2.2.8) we have $K_n(x_i, x_j) = 0$ $(i \ne j)$ and

$$K_n(x_i, x_i) = -b_{n+1} p'_n(x_i) p_{n+1}(x_i),$$

respectively. Using the three-term recurrence relation (2.2.3) at $x = x_i$ we get $0 = b_{n+1} p_{n+1}(x_i) + b_n p_{n-1}(x_i)$, so that the previous formula reduces to the desired result. □

An important question in interpolation theory is the distance between consecutive zeros of the orthonormal polynomial $p_n(d\mu; x)$.

Theorem 2.2.8 *Let $d\mu(x) = w(x)dx$, where $w(x)$ is a weight function on $[-1, 1]$, bounded from zero, i.e., $w(x) \ge c > 0$, and let $x_\nu = -\cos\theta_\nu$ $(0 < \theta_\nu < \pi;\ \nu = 1, \ldots, n)$ be the zeros of $p_n(w; x)$. Putting $\theta_0 = 0$ and $\theta_{n+1} = \pi$, we have*

$$\theta_{\nu+1} - \theta_\nu < K \frac{\log n}{n}, \quad \nu = 0, 1, \ldots, n,$$

where the constant K depends only on c.

Theorem 2.2.9 *If $A \le (1-x^2)^{1/2} w(x) \le B$ on $[-1, 1]$, where A and B are positive constants, then for the zeros $x_\nu = -\cos\theta_\nu$ $(0 < \theta_\nu < \pi;\ \nu = 1, \ldots, n)$ of $p_n(w; x)$, we have*

$$\theta_{\nu+1} - \theta_\nu < \frac{4\pi B}{A} \cdot \frac{1}{n}, \quad \nu = 0, 1, \ldots, n,$$

where $\theta_0 = 0$ and $\theta_{n+1} = \pi$.

The proofs of these results can be found in Szegő [470, pp. 112–115].

2.2.1.4 Some Special Weights

First we give a result for polynomials orthogonal with respect to a monotonic weight function on $[a, b]$ (see Szegő [470, p. 163]).

Theorem 2.2.10 *Let $x \mapsto w(x)$ be a non-decreasing weight function in $[a,b]$, where b is finite. If $\{q_n(x)\}$ is the corresponding system of orthogonal polynomials, then the functions $x \mapsto \sqrt{w(x)}|q_n(x)|$ attain their maximum in $[a,b]$ for $x = b$.*

A corresponding statement also holds for any subinterval $[x_0, b] \subset [a, b]$ where $x \mapsto w(x)$ is a non-decreasing function.

At the end of this section we mention some results for the polynomials $\{q_n(x)\}_{n \in \mathbb{N}_0}$ orthogonal with respect to an even weight function $x \mapsto w(x)$ on a symmetric interval $[-a, a]$. First, we have that $q_n(-x) = (-1)^n q_n(x)$, i.e., $q_n(x)$ is an even or odd polynomial depending on the parity of n. The monic polynomials $\{q_n(x)\}$ satisfy the recurrence relation (2.2.4) with $\alpha_n = 0$, i.e.,

$$q_{n+1}(x) = x q_n(x) - \beta_n q_{n-1}(x), \quad n = 0, 1, \ldots, \tag{2.2.16}$$

with $q_0(x) = 1$ and $q_{-1} = 0$.

Theorem 2.2.11 *Let $\{q_n(x)\}_{n \in \mathbb{N}_0}$ be a system of polynomials orthogonal with respect to an even weight $w(x)$ on $(-a, a)$. Then,*

(a) $\{p_n^{(1)}(t)\}_{n \in \mathbb{N}_0} = \{q_{2n}(\sqrt{t})\}_{n \in \mathbb{N}_0}$ *is a system of polynomials orthogonal on $[0, a^2]$ with respect to the weight $w_1(t) = w(\sqrt{t})/\sqrt{t}$;*

(b) $\{p_n^{(2)}(t)\}_{n \in \mathbb{N}_0} = \{q_{2n+1}(\sqrt{t})/\sqrt{t}\}_{n \in \mathbb{N}_0}$ *is a system of polynomials orthogonal on $[0, a^2]$ with respect to the weight $w_2(t) = \sqrt{t}\, w(\sqrt{t})$.*

Proof (a) Let $n \neq k$. Since

$$\int_{-a}^{a} q_{2n}(x) q_{2k}(x) w(x)\, dx = 2 \int_0^a q_{2n}(x) q_{2k}(x) w(x)\, dx = 0,$$

we have, by a change of variables $x^2 = t$,

$$\int_0^{a^2} q_{2n}(\sqrt{t}) q_{2k}(\sqrt{t}) \frac{w(\sqrt{t})}{\sqrt{t}}\, dt = 0 \quad (n \neq k).$$

Thus, the system of polynomials $\{q_{2n}(\sqrt{t})\}_{n \in \mathbb{N}_0}$ is orthogonal on $[0, a^2]$ with respect to the weight $w_0(t) = w(\sqrt{t})/\sqrt{t}$.

The proof of the statement (b) is analogous. □

Theorem 2.2.12 *If the orthogonal polynomial system $\{q_n(x)\}_{n \in \mathbb{N}_0}$ satisfies the three-term recurrence relation (2.2.16), then the systems $\{p_n^{(\nu)}(t)\}_{n \in \mathbb{N}_0}$, $\nu = 1, 2$, from Theorem 2.2.12, satisfy the recurrence relations*

$$p_{n+1}^{(\nu)}(t) = (t - \alpha_n^{(\nu)}) p_n^{(\nu)}(t) - \beta_n^{(\nu)} p_{n-1}^{(\nu)}(t), \quad n = 0, 1, \ldots,$$

with $p_0^{(\nu)}(t) = 1$ and $p_{-1}^{(\nu)}(t) = 0$, where $\alpha_0^{(1)} = \beta_1$, $\alpha_0^{(2)} = \beta_1 + \beta_2$, and for

2.2 Orthogonal Polynomials on the Real Line

$n \in \mathbb{N}$

$$\alpha_n^{(1)} = \beta_{2n} + \beta_{2n+1}, \qquad \beta_n^{(1)} = \beta_{2n-1}\beta_{2n}$$

and

$$\alpha_n^{(2)} = \beta_{2n+1} + \beta_{2n+2}, \qquad \beta_n^{(2)} = \beta_{2n} + \beta_{2n+1}.$$

Proof According to (2.2.16) and Theorem 2.2.12, it is easy to see that

$$p_{n+1}^{(1)}(t) + \beta_{2n+1} p_n^{(1)}(t) = t p_n^{(2)}(t)$$

and

$$p_{n+1}^{(2)}(t) + \beta_{2n+2} p_n^{(1)}(t) = p_{n+1}^{(1)}(t).$$

Combining these relations we get the stated results. □

2.2.2 Asymptotic Properties of Orthogonal Polynomials

In this section we deal with asymptotic properties of orthogonal polynomials when the degree of these polynomials tends to infinity. Many more results in this direction are known for polynomials orthogonal on a finite interval, than for ones on unbounded intervals. In order to have some precise asymptotic results we need certain restrictions on the corresponding weight functions. The first progress in this subject is due to Szegő (cf. [470, pp. 296–312]).

Definition 2.2.1 A weight function w defined on $(-1, 1)$ belongs to *Szegő's class* ($w \in \mathcal{S}$) if

$$\int_{-1}^{1} \frac{\log w(x)}{\sqrt{1-x^2}} \, dx > -\infty. \tag{2.2.17}$$

Definition 2.2.2 For a given function $v : [-\pi, \pi] \to \mathbb{R}^+$ satisfying $\log v \in L^1[-\pi, \pi]$ the *Szegő function* $D(z) = D(v; z)$ is defined by

$$D(v; z) = \exp\left\{ \frac{1}{4\pi} \int_{-\pi}^{\pi} \log v(\theta) \frac{e^{i\theta} + z}{e^{i\theta} - z} \, d\theta \right\}, \qquad |z| < 1. \tag{2.2.18}$$

The function (2.2.18) has the following properties:

1° Almost everywhere in $-\pi \leq \theta \leq \pi$,

$$\lim_{r \to 1-} D(v; re^{i\theta}) = D(v; e^{i\theta}) \quad \text{exists, and} \quad |D(v; e^{i\theta})|^2 = v(\theta);$$

2° $D(v; z) \neq 0$ in the unit disk $|z| < 1$;
3° $D(v; 0) > 0$.

Let $\{p_n(w;x)\}$ be a system of orthonormal polynomials with respect to the weight function $w \in \mathcal{S}$ and let

$$v(\theta) := w(\cos\theta)|\sin\theta|, \quad \theta \in [-\pi, \pi]. \quad (2.2.19)$$

Also, we define $\varrho: \mathbb{C} \setminus [-1, 1] \to \mathbb{C}$ by

$$\varrho(z) = z + \sqrt{z^2 - 1}, \quad (2.2.20)$$

where we take that branch of $\sqrt{z^2 - 1}$ for which $|\varrho(z)| > 1$ whenever $z \in \mathbb{C} \setminus [-1, 1]$. Notice that

$$\lim_{z \to \infty} \varrho(z) = \infty \quad \text{and} \quad \lim_{z \to \infty} |z^{-1}\varrho(z)| = 2.$$

Theorem 2.2.13 *Under condition (2.2.17), i.e., $w \in \mathcal{S}$, Szegő's asymptotic property*

$$p_n(w; z) = \frac{1}{\sqrt{2\pi}} \varrho(z)^n D(v; 1/\varrho(z))^{-1}(1 + o(1)), \quad n \to +\infty, \quad (2.2.21)$$

holds for each $z \in \mathbb{C} \setminus [-1, 1]$, where v is given by (2.2.19).

This asymptotic (known also as *strong* asymptotic) was firstly proved by Szegő [469] and later in a slightly more general form by Bernstein [36] and [37]. Conversely, Geronimus [178] proved that Szegő's asymptotic or even the uniform boundedness of $p_n(w; z)\varrho(z)^{-n}$ implies (2.2.17).

In the case of the Chebyshev polynomials of the first kind ($w(x) = 1/\sqrt{1-x^2}$, $v(\theta) = 1$), the Szegő function (2.2.18) becomes $D(v; z) = 1$, so that (2.2.21) reduces to

$$p_n(w; z) = \frac{1}{\sqrt{2\pi}} \varrho(z)^n (1 + o(1)), \quad n \to +\infty. \quad (2.2.22)$$

Of course, this result follows directly from (1.1.16) and (2.1.9). Namely, we have

$$p_n(w; z) = \sqrt{\frac{2}{\pi}} T_n(z) = \sqrt{\frac{2}{\pi}} \cdot \frac{1}{2}\left(\varrho(z)^n + \varrho(z)^{-n}\right), \quad n \to +\infty,$$

i.e., (2.2.22).

Taking $z = \infty$, (2.2.21) gives the behavior of the leading coefficient γ_n in $p_n(w; z) = \gamma_n z^n + \cdots$,

$$\lim_{n \to +\infty} \frac{\gamma_n}{2^n} = \frac{1}{\sqrt{2\pi}} D(v; 0)^{-1} = \frac{1}{\sqrt{\pi}} \exp\left\{-\frac{1}{2\pi} \int_{-1}^{1} \frac{\log w(x)}{\sqrt{1-x^2}} dx\right\}. \quad (2.2.23)$$

In terms of asymptotics on $[-1, 1]$, (2.2.21) and (2.2.23) are essentially equivalent to the mean asymptotic (see [273])

$$\lim_{n \to +\infty} \int_0^\pi \left|\sqrt{v(\theta)}\, p_n(\cos\theta) - \sqrt{\frac{2}{\pi}} \cos\left(n\theta + \arg D(v; e^{i\theta})\right)\right|^2 d\theta = 0.$$

2.2 Orthogonal Polynomials on the Real Line

If we want to have an asymptotic that holds uniformly for $x = \cos\theta$ in a compact subinterval of $(-1, 1)$, instead of the previous one, we need some stronger conditions than $w \in \mathcal{S}$. For example, the following result holds which is due to Szegő holds (cf. [470, pp. 297–298]):

Theorem 2.2.14 *Let $w(x)$ be a weight function on $[-1, 1]$, such that the function $v(\theta)$, defined by (2.2.19), satisfies the Lipschitz-Dini condition*

$$|v(\theta+\delta) - v(\theta)| < L|\log\delta|^{-1-\lambda},$$

where L and λ are fixed positive constants, and $m \le v(\theta) \le M$. Putting

$$\operatorname{sgn} D(v; e^{i\theta}) = |D(v; e^{i\theta})|^{-1} D(v; e^{i\theta}) = e^{i\gamma(\theta)},$$

we have uniformly on the segment $-1 \le x \le 1$ (or $0 \le \theta \le \pi$, $x = \cos\theta$),

$$(1-x^2)^{1/4} w(x)^{1/2} p_n(x) = \sqrt{\frac{2}{\pi}} \cos[n\theta + \gamma(\theta)] + O[(\log n)^{-\lambda}].$$

The constant factor in the O-term depends only on the constants L, λ, m and M.

Some extensions of Szegő's theory to weights with support on several intervals were given by Widom [506] and others (e.g. see Peherstorfer [392]).

Remark 2.2.3 A nice connection between orthonormal polynomials $p_n(w; x)$ on $[-1, 1]$ and orthonormal polynomials $\varphi_n(v; z)$ ($= k_n z^n +$ lower degree terms) on the unit circle with respect to the inner product (2.1.10), with $v(\theta)$ given by (2.2.19) can be established for each $n \ge 1$ in the form (cf. [470, p. 294])

$$p_n(w; x) = \frac{C_n}{\sqrt{2\pi}} \left(z^{-n} \varphi_{2n}(v; z) + z^n \varphi_{2n}(v; z^{-1}) \right),$$

where $x = (z + z^{-1})/2$ and $C_n = (1 + \varphi_{2n}(v; 0)/k_{2n})^{-1/2}$.

In 1940 Erdős and Turán [119] introduced another type of asymptotic for weights w that are positive almost everywhere on $(-1, 1)$:

$$\lim_{n \to +\infty} p_n(w; z)^{1/n} = \varrho(z), \quad z \in \mathbb{C} \setminus [-1, 1].$$

This nth root asymptotic was recently intensively studied by Stahl and Totik [452].

Also, we mention the *ratio* asymptotic

$$\lim_{n \to +\infty} \frac{p_{n+1}(w; z)}{p_n(w; z)} = \varrho(z), \quad z \in \mathbb{C} \setminus [-1, 1] \qquad (2.2.24)$$

which is closely connected to the three-term recurrence relation (2.2.3).

The behavior of the recursion coefficients in the three-term relation (2.2.3) determines an important class of the measures (cf. Nevai [375, p. 10]):

Definition 2.2.3 Let $a \in \mathbb{R}$ and $b \in [0, +\infty)$. The distribution function μ belongs to the Nevai class $\mathcal{N}(a, b)$ if[1]

$$\lim_{n \to +\infty} a_n = a \quad \text{and} \quad \lim_{n \to +\infty} b_n = \lim_{n \to +\infty} \frac{\gamma_{n-1}}{\gamma_n} = \frac{b}{2}.$$

According to this definition, one can prove that (2.2.24) is equivalent to $\mu \in \mathcal{N}(0, 1)$, where $\mu' = w$.

The following results were proved in [375, pp. 33–37]:

Theorem 2.2.15 *Let $\mu \in \mathcal{N}(a, b)$ and $z \in \mathbb{C} \setminus \mathrm{supp}\,(d\mu)$. Then*

$$\lim_{n \to +\infty} \frac{p_{n-1}(d\mu; z)}{p_n(d\mu; z)} = \begin{cases} 0 & \text{for } b = 0, \\ \varrho((z-a)/b)^{-1} & \text{for } b > 0. \end{cases} \quad (2.2.25)$$

Theorem 2.2.16 *Suppose that $a \in \mathbb{R}$ and $b > 0$. If $\mu \in \mathcal{N}(a, b)$, then for every $x \in \mathrm{supp}\,(d\mu) \setminus [a-b, a+b]$*

$$\lim_{n \to +\infty} \frac{p_{n-1}(d\mu; x)}{p_n(d\mu; x)} = \varrho\left(\frac{x-a}{b}\right).$$

The same limit as in (2.2.25) also holds for the quotient $p'_{n-1}(d\mu; z)/p'_n(d\mu; z)$.

Nevai also proved conversely that if (2.2.25) holds for all $z \in \mathbb{C} \setminus \mathbb{R}$, then $\mu \in \mathcal{N}(a, b)$. Recently, Simon [435] studied ratio asymptotics, i.e., existence of the limit of the quotient of monic orthogonal polynomials $\pi_{n+1}(d\mu; z)/\pi_n(d\mu; z)$ and the existence of weak limits of $[p_n(d\mu; x)]^2 d\mu$ as $n \to +\infty$. He proved that the existence of ratio asymptotics at a single point $z_0 \in \mathbb{C} \setminus \mathbb{R}$ with $\mathrm{Im}(z_0) \neq 0$ implies that μ is in a Nevai class.

The investigation of polynomials orthogonal on unbounded intervals with respect to general weights begun with Géza Freud in the 1960's (for details see Freud [134], Nevai [378], as well as recent monographs written by Mhaskar [316] and by Levin and Lubinsky [258]).

For example, for polynomials orthogonal with respect to a weight function on $\mathbb{R}$ with certain regular behavior at infinity the following result holds (see Rakhmanov [405]):

Theorem 2.2.17 *Let $w(x)$ be a weight function on $\mathbb{R}$ such that for some $\lambda > 1$*

$$\lim_{|x| \to +\infty} \frac{\log w(x)}{|x|^\lambda} = -1.$$

---

[1] The class $\mathcal{N}(a, b)$ is denoted by $\mathcal{M}(a, b)$ in [375]. Without loss of generality we can always assume that $a = 0, b = 1$ or $a = 0, b = 0$.

2.2 Orthogonal Polynomials on the Real Line

Then, for the orthonormal polynomials $\{p_n(w;z)\}_{n\in\mathbb{N}_0}$ *the following limit relation is valid uniformly with respect to z in any compact subset of* $\mathbb{C}\setminus\mathbb{R}$

$$\lim_{n\to+\infty}\frac{\log|p_n(w;z)|}{n^{1-1/\lambda}}=G(\lambda)|\operatorname{Im} z|,$$

where

$$G(\lambda)=\frac{\lambda}{\lambda-1}\left(\frac{1}{\sqrt{\pi}}\frac{\Gamma((\lambda+1)/2)}{\Gamma(\lambda/2)}\right)^{1/\lambda}.$$

The case of a weight function on $(0,+\infty)$ with the same kind of behavior at infinity can be obtained from the previous theorem. Using these results Van Assche [483] obtained the weighted zero distribution of polynomials orthogonal on $(-\infty,+\infty)$ or $(0,+\infty)$.

An important and nice survey article on the asymptotic behavior of orthogonal polynomials was recently written by Lubinsky [273] (see also his previous surveys [269] and [270]). In this survey all kinds of asymptotics for orthogonality on finite and unbounded intervals are discussed, as well as their developments and their history. A special treatment is given for the so-called Freud weights on $\mathbb{R}$ (see Sect. 2.4.5 for details).

The second part of Lubinsky's paper [273] is devoted to certain identities for special weights that are useful for general classes of weights. The approach is the following: suppose that we know an explicit expression for the n-th orthonormal polynomial $p_n(v;x)=\gamma_n(v)x^n+\cdots$ with respect to the special weight v on I and that we wish to determine the behavior of the orthonormal polynomials $p_n(w;x)$ as $n\to+\infty$, where w is another weight also given on I, which is close to v. If we have a sufficiently good approximation $w\cong v$, then we can expect that $p_n(w;x)\approx p_n(v;x)$, but in many cases this cannot easily be justified. As a useful tool, Lubinsky [273] mentions the Korous' identity, based on the reproducing kernel (2.2.6) for the polynomials $p_k(v;\cdot)$, i.e.,

$$K_{n-1}(v;x,t)=\sum_{k=0}^{n-1}p_k(v;x)p_k(v;t).$$

Let

$$r(x)=p_n(w;x)-\frac{\gamma_n(w)}{\gamma_n(v)}p_n(v;x)\quad\text{and}\quad R(t,x)=\frac{1-[v(x)/w(x)][w(t)/v(t)]}{x-t},$$

where $r\in\mathcal{P}_{n-1}$. Then, according to (2.1.29) and the orthogonality, we have

$$r(x)=\int_I K_{n-1}(v;x,t)r(t)v(t)\,dt=\int_I K_{n-1}(v;x,t)p_n(w;t)v(t)\,dt,$$

i.e.,

$$r(x)=\int_I K_{n-1}(v;x,t)p_n(w;t)\left[v(t)-\frac{v(x)}{w(x)}w(t)\right]dt.$$

Finally, using the Christoffel-Darboux formula (2.2.7) for $K_{n-1}(v;x,t)$, we obtain

$$p_n(w;x) - \frac{\gamma_n(w)}{\gamma_n(v)}p_n(v;x)$$

$$= \frac{\gamma_{n-1}(v)}{\gamma_n(v)}\left[p_n(v;x)\int_I p_{n-1}(v;t)p_n(w;t)R(t,x)v(t)\,dt\right.$$

$$\left.-p_{n-1}(v;x)\int_I p_n(v;t)p_n(w;t)R(t,x)v(t)\,dt\right].$$

Now for a sufficiently good approximation $w \approx v$, $R(t,x)$ is small in some sense. Knowing the bounds on $p_n(v;x)$ and $\gamma_{n-1}(v)/\gamma_n(v)$ for the special weight v, we may then apply the Cauchy-Schwarz inequality to the integrals on the right side in the previous equality and use the orthonormality of $p_n(w;\cdot)$ in order to show that $p_n(w;x) - (\gamma_n(w)/\gamma_n(v))p_n(v;x)$ is small. The estimates for v/w are essential in this approach, but they are not always available. Therefore, a choice of the special weights is very important. In [273] Lubinsky considers in detail three classes of identities (the so-called Bernstein-Szegő identities, the Fokas-Its-Kitaev identity (or Riemann-Hilbert), and Rakhmanov's projection identity) and gives a comparison of their applications in asymptotics of some special class of orthogonal polynomials.

In the sequel of this section we briefly mention them.

2.2.2.1 Bernstein-Szegő Identities

A weight function of the form

$$w(x) = \frac{\sqrt{1-x^2}}{S(x)}, \qquad x \in (-1,1), \qquad (2.2.26)$$

where $S(x)$ is a polynomial that is positive on $[-1,1]$, except possibly for simple zeros at ± 1, is known as the *Bernstein-Szegő weight*. For example, if $S(x)$ is a polynomial of degree m, then it can be represented in the form

$$S(\cos\theta) = |h(e^{i\theta})|^2, \qquad \theta \in [-1,1],$$

where h is an algebraic polynomial of degree m, with $h(0) > 0$, and with all zeros in $\{z \in \mathbb{C}\,|\,|z| > 1\}$, except possibly for simple zeros at ± 1 corresponding to the zeros of $S(x)$ at ± 1 (cf. [359, pp. 22–26]). In this case, the corresponding orthogonal polynomials and all related quantities can be explicitly written down. For example, if $x = \cos\theta$ and $z = e^{i\theta}$, then for $n > m/2$,

$$p_n(x) = \sqrt{\frac{2}{\pi}}(\sin\theta)^{-1}\mathrm{Im}\{z^{n+1}\overline{h(z)}\}, \qquad \gamma_n = \sqrt{\frac{2}{\pi}}h(0)2^n.$$

2.2 Orthogonal Polynomials on the Real Line

A traditional way of proving asymptotic formulas for some general weights is based on an approximation of such weights with a weight of the form (2.2.26). Then, we use the explicit expression for the orthogonal polynomials with respect to (2.2.26) and use some error analysis to get the desired asymptotics. The catch is that for a good approximation we need high a degree of the polynomial S, but then the exact expressions will also be valid only for high degrees of the orthogonal polynomials. Several applications in this direction, including the weights given in (2.2.19), were presented in [273].

2.2.2.2 The Fokas-Its-Kitaev (Riemann-Hilbert) Identity

This kind of identities, based on the *Sokhotski-Plemelj formulae*, emerged in the theory of orthogonal polynomials in the early 1990's (see [129, 130]).

We start with the *Cauchy* or (*Hilbert*) *transform* of a certain function φ,

$$\Phi(z) = C[\varphi](z) = \frac{1}{2\pi i}\int_\mathbb{R} \frac{\varphi(t)}{t-z} dt, \quad z \in \mathbb{C} \setminus \mathbb{R}.$$

The boundary values from the upper and lower half planes can be defined as

$$\Phi_+(x) = \lim_{\varepsilon\to 0+} C[\varphi](x+i\varepsilon), \quad \Phi_-(x) = \lim_{\varepsilon\to 0+} C[\varphi](x-i\varepsilon),$$

whenever the limits exist. In this case the *Sokhotski-Plemelj formulae* (cf. [373, p. 43])

$$\Phi_+(x) - \Phi_-(x) = \varphi(x), \quad \Phi_+(x) + \Phi_-(x) = \frac{1}{\pi i} \text{P.V.} \int_\mathbb{R} \frac{\varphi(t)}{t-x} dt$$

hold. For example, this is true a.e. if a measurable function $\varphi : \mathbb{R} \to \mathbb{R}$ satisfies the condition $\int_\mathbb{R} |\varphi(t)|/(1+|t|)\, dt < +\infty$ (cf. [273]). However, if φ satisfies a Lipschitz condition of some positive order on an interval, then the Plemelj formulae hold in the interior of that interval.

In the Riemann-Hilbert problems the difference of boundary values is replaced by their ratio, i.e., for a given function φ, we look for a function G analytic in $\mathbb{C} \setminus \mathbb{R}$ satisfying

$$G_+(x) = \varphi(x)G_-(x), \quad x \in \mathbb{R},$$

and some normalization condition.

Let w be a given weight function on the real line $\mathbb{R}$, for which all moments $\mu_k = \int_\mathbb{R} x^k w(x)\, dx$, $k = 0, 1, 2, \ldots$, exist and are finite, $\mu_0 > 0$, and assume that $t^k w(t)$ satisfies a Lipschitz condition of some positive order throughout $\mathbb{R}$ for each $k \geq 0$.

Let $\pi_n(x) = \pi_n(w; x) = p_n(x)/\gamma_n = x^n + \cdots$ denote the n-th monic orthogonal polynomial with respect to the measure $w(x)dx$. If we now have a two-dimensional vector-row, defined by $\mathbf{v} = [\pi_n(z) \ C[w\pi_n](z)]$, then it satisfies the following three properties (cf. [92]):

$\mathbf{v}: \mathbb{C} \setminus \mathbb{R} \to \mathbb{C}^2$ is analytic;

$$\mathbf{v}_+(x) = \mathbf{v}_-(x) \begin{bmatrix} 1 & w(x) \\ 0 & 1 \end{bmatrix}, \quad x \in \mathbb{R};$$

$$\mathbf{v}(z) \begin{bmatrix} z^{-n} & 0 \\ 0 & z^n \end{bmatrix} \to [1 \ 0], \quad |z| \to +\infty,$$

from which we can formally conclude some properties of the vector $\mathbf{v}$. This simple observation shows how the orthogonal polynomials can be formulated through a Riemann-Hilbert problem. The formulation involves a 2×2 complex valued matrix function $Y(z) \in \mathbb{C}^{2 \times 2}$ instead of a two-dimensional vector-row. This connection between a Riemann-Hilbert problem and orthogonal polynomials appeared for the first time in the papers [129] and [130].

Now we can consider the following matrix Riemann-Hilbert problem:

(I) $Y: \mathbb{C} \setminus \mathbb{R} \to \mathbb{C}^{2 \times 2}$ is analytic;

(II) $Y_+(x) = Y_-(x) \begin{bmatrix} 1 & w(x) \\ 0 & 1 \end{bmatrix}, \quad x \in \mathbb{R};$

(III) $Y(z) \begin{bmatrix} z^{-n} & 0 \\ 0 & z^n \end{bmatrix} = \begin{bmatrix} 1 & 0 \\ 0 & 1 \end{bmatrix} + O\left(\frac{1}{|z|}\right), \quad |z| \to +\infty.$

Theorem 2.2.18 *The Riemann-Hilbert problem* (I)–(III) *has a unique solution, given by*

$$Y(z) = \begin{bmatrix} \pi_n(z) & C[w\pi_n](z) \\ -2\pi i \gamma_{n-1}^2 \pi_{n-1}(z) & -2\pi i \gamma_{n-1}^2 C[w\pi_{n-1}](z) \end{bmatrix}. \tag{2.2.27}$$

Furthermore, there exist matrices $P, Q \in \mathbb{C}^{2 \times 2}$ *such that*

$$Y(z) \begin{bmatrix} z^{-n} & 0 \\ 0 & z^n \end{bmatrix} = I + \frac{P}{z} + \frac{Q}{z^2} + O\left(\frac{1}{|z|^3}\right), \quad |z| \to +\infty.$$

The leading coefficient γ_n *and the coefficients* α_n *and* β_n *in the three-term recurrence relation* (2.2.4) *for monic polynomials can be expressed in terms of the elements of the matrices* $P = [p_{ij}]_{2 \times 2}$ *and* $Q = [q_{ij}]_{2 \times 2}$,

$$\gamma_n = \sqrt{\frac{1}{-2\pi i \, p_{12}}}, \quad \alpha_n = p_{11} + \frac{q_{12}}{p_{12}}, \quad \beta_n = p_{12} p_{21}.$$

2.2 Orthogonal Polynomials on the Real Line

For a proof of this theorem and several applications to specific weight functions see [86, 89–92, 130, 240, 273]. Since some Riemann-Hilbert problems can be solved explicitly, the identity obtained in this way allows one to deduce very sharp asymptotic formulae for orthogonal polynomials. Moreover, combining the identities with a series of transformations, as well as with some variants of the steepest descent method (cf. [87]), this approach has produced very spectacular results for analytic weights (see Sect. 2.4).

2.2.2.3 Rakhmanov's Identity

This approach appeared in 1992 in Rakhmanov's paper [406] on strong asymptotics for orthogonal polynomials associated with exponential weights on $\mathbb{R}$. Rakhmanov's projection identity identifies the orthogonal polynomials as the solution of some fairly explicit extremal problem, where the extremal error gives the error between the polynomial and an explicit expression. Practically, if we can estimate the extremal error, then we get an asymptotic formula with the given error. A nice presentation of this approach for an exponential weight on $[-1, 1]$ was given in [273, Theorem 2.13].

2.2.3 Associated Polynomials and Christoffel Numbers

Let $d\mu(x)$ be a given nonnegative measure on the real line $\mathbb{R}$, with compact or unbounded support, for which all moments $\mu_k = \int_{\mathbb{R}} x^k d\mu(x)$, $k = 0, 1, 2, \ldots$, exist and are finite, and $\mu_0 > 0$. Let $\{\pi_n(x)\}$ be the corresponding system of monic orthogonal polynomials with respect to the measure $d\mu(x)$. These polynomials satisfy the three-term recurrence relation (2.2.4), i.e.,

$$\pi_{n+1}(x) = (x - \alpha_n)\pi_n(x) - \beta_n \pi_{n-1}(x), \qquad n = 0, 1, 2, \ldots, \qquad (2.2.28)$$

with $\pi_0(x) = 1$ and $\pi_{-1}(x) = 0$. As before, we put $\beta_0 = \mu_0 = \int_{\mathbb{R}} d\mu(x)$.

2.2.3.1 Associated Polynomials

We introduce another system of polynomials $\{\sigma_n(x)\}$, the so-called *associated polynomials*, by

$$\sigma_n(x) = \int_{\mathbb{R}} \frac{\pi_n(x) - \pi_n(t)}{x - t} d\mu(t) \qquad (n \geq 0). \qquad (2.2.29)$$

Notice that $\sigma_0(x) = 0$, $\sigma_1(x) = \int_{\mathbb{R}} d\mu(t) = \mu_0$, and $\deg(\sigma_n(x)) = n - 1$ $(n \geq 1)$.

It is easily seen that these polynomials also satisfy the recurrence relation (2.2.28). Namely, starting from (2.2.28) we have

$$\frac{\pi_{n+1}(x) - \pi_{n+1}(t)}{x-t} = \pi_n(t) + (x - \alpha_n)\frac{\pi_n(x) - \pi_n(t)}{x-t} - \beta_n \frac{\pi_{n-1}(x) - \pi_{n-1}(t)}{x-t}.$$

Now, integrating with respect to $d\mu(t)$ and using the orthogonality $\int_{\mathbb{R}} \pi_n(t)\,d\mu(t) = 0$ for each $n \geq 1$, we conclude that, for $n \geq 1$, the associated polynomials satisfy the same three-term recurrence relation as the polynomials $\{\pi_n(x)\}$. Defining $\sigma_{-1}(x) = -1$ and assuming again $\beta_0 = \mu_0$, this relation also holds for $n = 0$. Thus, we have

$$\sigma_{n+1}(x) = (x - \alpha_n)\sigma_n(x) - \beta_n \sigma_{n-1}(x), \quad \sigma_0(x) = 0, \ \sigma_{-1}(x) = -1, \quad (2.2.30)$$

for each $n \geq 0$.

Sometimes for the monic associated polynomials $\hat{\sigma}_{n+1}(x)$ we use the notation $\pi_n^{[1]}(x)$ ($\deg \pi_n^{[1]} = n$), i.e.,

$$\pi_n^{[1]}(x) = \frac{1}{\beta_0} \int_{\mathbb{R}} \frac{\pi_{n+1}(x) - \pi_{n+1}(t)}{x-t}\,d\mu(t) \quad (n \geq 0), \quad (2.2.31)$$

where $\beta_0 = \mu_0 = \int_{\mathbb{R}} d\mu(t)$. Then, from (2.2.30) we have

$$\pi_{n+1}^{[1]}(x) = (x - \alpha_{n+1})\pi_n^{[1]}(x) - \beta_{n+1}\pi_{n-1}^{[1]}(x), \quad \pi_0^{[1]}(x) = 1, \ \pi_{-1}^{[1]}(x) = 0.$$

This enables us to introduce the monic *associated polynomials* (or *numerator polynomials*) $\pi_n^{[\nu]}(x)$ *of order* ν in the following way

$$\pi_{n+1}^{[\nu]}(x) = (x - \alpha_{n+\nu})\pi_n^{[\nu]}(x) - \beta_{n+\nu}\pi_{n-1}^{[\nu]}(x), \quad \pi_0^{[\nu]}(x) = 1, \ \pi_{-1}^{[\nu]}(x) = 0.$$

Notice that $\pi_n^{[0]}(x) = \pi_n(x)$ and $\pi_n^{[1]}(x) = \sigma_{n+1}(x)/\beta_0 = -(1/\beta_0)(\partial \pi_{n+1}/\partial \alpha_0)$. In general, we have

$$\pi_n^{[\nu]}(x) = \frac{(-1)^\nu}{\beta_0 \beta_1 \ldots \beta_{\nu-1}} \frac{\partial^\nu \pi_{n+\nu}(x)}{\partial \alpha_0 \partial \alpha_1 \ldots \partial \alpha_{\nu-1}} \quad (\nu = 0, 1, \ldots).$$

According to Favard's theorem, the set of (monic) associated polynomials $\{\pi_n^{[\nu]}(x)\}$ is orthogonal with respect to some measure $d\mu_\nu(x)$. Such kind of polynomials were considered by many authors (cf. [28, 60, 64, 194, 195, 261, 299, 377, 430, 484]).

Supposing that $d\mu(x) = w(x)\,dx$ and $[a,b] = \mathrm{supp}(d\mu)$, Grosjean [194] developed a theory for finding an explicit expression for the measure $d\mu_1(x) = W_1(x)\,dx$ and a procedure for obtaining its interval of orthogonality $[c,d] = \mathrm{supp}(d\mu_1) \subset \mathrm{supp}(d\mu)$. These results are obtained for an arbitrary weight $w(x)$ being piecewise continuous as well as containing discrete mass points. Thus, if $w(x)$ is a piecewise weight function on $[a,b]$, then $[c,d] = [a,b]$ and

$$W(x) = W_1(x) = \frac{\beta_0 w(x)}{\left(\mathrm{P.V.}\int_a^b \frac{w(t)\,dt}{t-x}\right)^2 + (\pi w(x))^2}, \quad a < x < b, \quad (2.2.32)$$

2.2 Orthogonal Polynomials on the Real Line

where $\beta_0 = \mu_0 = \int_a^b w(x)\,dx$ and $\int_a^b W(x)\,dx = \beta_1$. In the case $w(x) = 1$ on $[-1, 1]$, (2.2.32) reduces to

$$W(x) = \frac{2}{\left(\log \dfrac{1+x}{1-x}\right)^2 + \pi^2}, \quad -1 < x < 1. \tag{2.2.33}$$

Following (2.2.31) and taking the normalization $\int_{\mathbb{R}} d\mu_\nu(x) = \beta_\nu$, the polynomials $\pi_n^{[\nu+1]}(x)$ can be expressed in the form

$$\pi_n^{[\nu+1]}(x) = \frac{1}{\beta_\nu} \int_{\mathbb{R}} \frac{\pi_{n+1}^{[\nu]}(x) - \pi_{n+1}^{[\nu]}(t)}{x - t} d\mu_\nu(t) \quad (n \geq 0). \tag{2.2.34}$$

Remark 2.2.4 The polynomials $\pi_n^{[1]}(x)$ play an important role in the theory of Padé approximation.

There is a well-known Wronskian-type relation between $\pi_n^{[\nu]}(x)$ and $\pi_n^{[\nu+1]}(x)$ in the form [60, p. 86]

$$D_n^{[\nu]} = \pi_n^{[\nu]}(x)\pi_n^{[\nu+1]}(x) - \pi_{n+1}^{[\nu]}(x)\pi_{n-1}^{[\nu+1]}(x) = \beta_{\nu+1}\beta_{\nu+2}\cdots\beta_{\nu+n} > 0.$$

We can prove this very easily, because of the relation $D_n^{[\nu]} = \beta_{\nu+n} D_{n-1}^{[\nu]}$.

Let $x_{n,k}^{[\nu]}$ ($k = 1, \ldots, n$) be the zeros of $\pi_n^{[\nu]}(x)$ ordered as an increasing sequence. Then the zeros of $\pi_{n+1}^{[\nu]}(x)$ and $\pi_n^{[\nu+1]}(x)$ are interlacing in the same manner as those of $\pi_{n+1}(x)$ and $\pi_n(x)$ (see Theorem 2.2.6). Namely, we have:

Theorem 2.2.19 *The zeros of $\pi_n^{[\nu+1]}(x)$ and $\pi_{n+1}^{[\nu]}(x)$ interlace, i.e.,*

$$x_{n+1,k}^{[\nu]} < x_{n,k}^{[\nu+1]} < x_{n+1,k+1}^{[\nu]} \quad (k = 1, \ldots, n;\ n \in \mathbb{N}).$$

Also, there is a three-term recurrence relation between the polynomials $\pi_n^{[\nu]}(x)$ with different values of ν (cf. [500]):

$$\pi_{n+1}^{[\nu]}(x) = (x - \alpha_\nu)\pi_n^{[\nu+1]}(x) - \beta_{\nu+1}\pi_{n-1}^{[\nu+2]}(x). \tag{2.2.35}$$

For $\nu = 0$, (2.2.35) can be written in the form

$$\pi_{n+1}(x) = \pi_n^{[1]}(x)\pi_1(x) - \beta_1\pi_{n-1}^{[2]}(x)\pi_0(x).$$

In general, the relation

$$\pi_{n+\nu+1}(x) = \pi_n^{[\nu+1]}(x)\pi_{\nu+1}(x) - \beta_{\nu+1}\pi_{n-1}^{[\nu+2]}(x)\pi_\nu(x) \tag{2.2.36}$$

holds, which can be proved by induction arguments.

Remark 2.2.5 Recently, Skrzipek [441] has generalized this concept of associating for arbitrary polynomials v_n ($v_n \in \mathcal{P}_n$). Especially, if v_n is expanded in terms of π_k, $k = 0, 1, \ldots, n$, their associated polynomials are the Clenshaw polynomials which are used in numerical mathematics.

2.2.3.2 Stieltjes Transform of the Measure and Christoffel Numbers

Let

$$F(z) := \int_{\mathbb{R}} \frac{d\mu(t)}{z - t}, \qquad z \in \mathbb{C} \setminus \mathrm{supp}\,(d\mu). \tag{2.2.37}$$

It is clear that $F(\infty) = 0$ and F is analytic in the whole complex plane excluding the support of the measure.[2] This function has a formal expansion in descending powers of z,

$$F(z) \sim \frac{\mu_0}{z} + \frac{\mu_1}{z^2} + \frac{\mu_2}{z^3} + \cdots, \tag{2.2.38}$$

where μ_k are the moments of the measure $d\mu(t)$.

The function $F(z)$, known as the *Stieltjes transform* of the measure $d\mu(t)$, also has an *associated continued fraction*

$$F(z) \sim \cfrac{\beta_0}{z - \alpha_0 - \cfrac{\beta_1}{z - \alpha_1 - \cfrac{\beta_2}{z - \alpha_2 - \cdots}}} = \frac{\beta_0}{z - \alpha_0 -} \frac{\beta_1}{z - \alpha_1 -} \cdots, \tag{2.2.39}$$

where α_n, β_n are the same coefficients that appear in the three-term recurrence relation (2.2.28). It is easy to see that the n-th *convergent* of this continued fraction is just equal to $\sigma_n(z)/\pi_n(z)$, i.e.,

$$\frac{\beta_0}{z - \alpha_0 -} \frac{\beta_1}{z - \alpha_1 -} \cdots \frac{\beta_{n-1}}{z - \alpha_{n-1} -} = \frac{\sigma_n(z)}{\pi_n(z)}. \tag{2.2.40}$$

Notice that the rational function (2.2.40) has only simple poles at the points $z = x_{n,k}$ ($k = 1, \ldots, n$), which are zeros of the polynomial $\pi_n(x)$. By $\lambda_{n,k}$ we denote the corresponding residues, i.e.,

$$\lambda_{n,k} = \operatorname*{Res}_{z = x_{n,k}} \frac{\sigma_n(z)}{\pi_n(z)} = \lim_{z \to x_{n,k}} (z - x_{n,k}) \frac{\sigma_n(z)}{\pi_n(z)} = \frac{\sigma_n(x_{n,k})}{\pi_n'(x_{n,k})}.$$

According to (2.2.29), we have

$$\lambda_{n,k} = \frac{1}{\pi_n'(x_{n,k})} \int_{\mathbb{R}} \frac{\pi_n(t)}{t - x_{n,k}} d\mu(t), \tag{2.2.41}$$

---

[2] When the measure is supported on $\mathbb{R}$, the function $F(z)$ is analytic separately in the upper and lower half-plane, with different branches in general.

2.2 Orthogonal Polynomials on the Real Line

so that the fractional expansion (2.2.40) gets the following form

$$\frac{\sigma_n(x)}{\pi_n(x)} = \beta_0 \frac{\pi_{n-1}^{[1]}(x)}{\pi_n(x)} = \sum_{k=1}^{n} \frac{\lambda_{n,k}}{x - x_{n,k}}. \qquad (2.2.42)$$

The coefficients $\lambda_{n,k}$ play an important role in numerical integration. They appear in the so-called *Gauss-Christoffel quadrature formulae* as the weight coefficients (*Cotes numbers*). Namely, such a quadrature formula

$$I(f) = \int_{\mathbb{R}} f(x)\,d\mu(x) = \sum_{k=1}^{n} \lambda_{n,k} f(x_{n,k}) + R_n(f), \qquad (2.2.43)$$

provides an approximation to the integral $I(f)$ with the error $R_n(f)$, which is equal to zero for all algebraic polynomials of degree at most $2n - 1$. As we can see, the zeros of $\pi_n(x)$ are the points (nodes) in the quadrature formula (2.2.43). Such quadratures will be considered in Sect. 5.1.

The integral (2.2.41) can be expressed in terms of the Christoffel function $\lambda_n(d\mu; x) = 1/K_{n-1}(x,x)$, given by (2.1.30).

Starting from Theorem 2.2.3 and the connection between orthonormal and corresponding monic polynomials $(p_n(x) = \gamma_n \pi_n(x))$, from the Christoffel-Darboux identity (2.2.7) for $t = x_{n,k}$, we get

$$K_{n-1}(x, x_{n,k}) = \gamma_{n-1}^2 \frac{\pi_n(x)\pi_{n-1}(x_{n,k})}{x - x_{n,k}},$$

whence, by integration with respect to the measure $d\mu(x)$,

$$\int_{\mathbb{R}} K_{n-1}(x, x_{n,k})\,d\mu(x) = \gamma_{n-1}^2 \pi_{n-1}(x_{n,k})\left[\pi_n'(x_{n,k})\lambda_{n,k}\right].$$

Here, because of the orthogonality, the left hand side is equal to 1. On the other hand, we obtain from (2.2.8) for $x = x_{n,k}$

$$K_{n-1}(x_{n,k}, x_{n,k}) = \gamma_{n-1}^2 \pi_n'(x_{n,k})\pi_{n-1}(x_{n,k}).$$

Combining these equalities we find that $1 = K_{n-1}(x_{n,k}, x_{n,k})\lambda_{n,k}$, i.e.,

$$\lambda_{n,k} = \lambda_n(d\mu; x_{n,k}) \quad (k = 1, \ldots, n). \qquad (2.2.44)$$

Thus, the coefficients $\lambda_{n,k} = \lambda_{n,k}(d\mu)$ are the values of the Christoffel function at the zeros of the orthogonal polynomial $\pi_n(d\mu; x)$, and therefore, they are called the *Christoffel numbers* or the *Cotes-Christoffel coefficients*. Notice that the Christoffel numbers are always positive, i.e., $\lambda_{n,k} > 0$ for each $k = 1, \ldots, n$. These numbers also play the role of discrete weights in the orthogonality property

$$\sum_{k=1}^{n} \lambda_{n,k} \pi_\nu(x_{n,k}) \pi_\mu(x_{n,k}) = \|\pi_\nu\|^2 \delta_{\nu\mu}, \quad 0 \le \nu, \mu \le n - 1, \qquad (2.2.45)$$

where $\|\pi_\nu\|^2 = \beta_0\beta_1\cdots\beta_\nu$. This property is referred to as the discrete orthogonality of the polynomials $\pi_\nu(x)$ at the zeros $x_{n,k}$ of the polynomial $\pi_n(x)$. The orthogonality relation (2.2.45) can be verified by using the Gauss-Christoffel quadrature formula (2.2.43).

2.2.3.3 Markov's Moment Problem

The *moment problem* consists in determining the distribution function μ from the sequence of its moments $\mu_k = \int_\mathbb{R} x^k \, d\mu(x)$ $(k = 0, 1, \ldots)$.

Definition 2.2.4 For a given measure $d\mu$, the moment problem is said to be *determined* if the measure $d\mu$ is uniquely determined by the moments μ_k $(k = 0, 1, \ldots)$; otherwise, the moment problem is *indeterminate*.

One important question is the convergence of the continued fraction in (2.2.39) to the integral $F(z)$, given by (2.2.37), i.e.,

$$\lim_{n \to +\infty} \frac{\sigma_n(z)}{\pi_n(z)} = F(z), \qquad z \notin \mathrm{supp}\,(d\mu). \tag{2.2.46}$$

In the case when $[a, b] = \mathrm{supp}\,(d\mu)$ is a finite interval, Markov [285, p. 89] proved that (2.2.46) holds for any $z \notin [a, b]$. The distribution function μ can be recovered from F by the inversion formula of Stieltjes (see [13, p. 259])

$$\mu(c) - \mu(d) = -\frac{1}{\pi} \lim_{y \to 0+} \int_c^d \mathrm{Im}[F(x + iy)] \, dx,$$

where F is defined in (2.2.37). At the points of discontinuity, μ must be redefined so that

$$\mu(x) = \frac{1}{2}[\mu(x+0) + \mu(x-0)].$$

In the case of unbounded intervals, i.e., when $[a, b]$ is a half-infinite interval (e.g. $[0, +\infty]$) or $[a, b] = [-\infty, +\infty]$, then (2.2.46) holds, whenever the moment problem[3] for the moment sequence μ_k $(k = 0, 1, 2, \ldots)$ is determined. A nice survey of the history of Markov's theorem, a proof in the determinate case, as well as a proof of a version of this theorem in the indeterminate case, were given by Berg [31]. Also, the results were applied to the shifted moment problem.

Remark 2.2.6 In the general case, the measures $d\mu$ and $d\mu_1$ (for associated polynomials) can be studied conveniently through their Stieltjes transforms (2.2.37) and

$$F_1(z) = \int_\mathbb{R} \frac{d\mu_1(t)}{z - t}, \qquad z \in \mathbb{C} \setminus \mathrm{supp}\,(d\mu_1),$$

---

[3] Stieltjes or Hamburger moment problem, respectively.

2.2 Orthogonal Polynomials on the Real Line

respectively, which are analytic in their domains of definitions. A classical result in the theory of continued fractions (cf. Shohat and Sherman [434], Sherman [430], Berg [31], van Doorn [489]) gives a connection between these transforms

$$F(z) = \frac{\beta_0}{z - \alpha_0 - F_1(z)}. \tag{2.2.47}$$

Recently, van Doorn [489] (see also [490]) considered the question of the determination of the measure $d\mu$, if the orthogonalizing measure for the associated polynomials is known. Namely, knowing $F_1(z)$ he uses (2.2.47) to find $F(z)$, and then applies the Stieltjes inversion formula to recover the distribution function μ.

2.2.4 Functions of the Second Kind and Stieltjes Polynomials

Let $\{\pi_n(x)\}_{n \in \mathbb{N}_0}$ be a system of monic orthogonal polynomials with respect to the measure $d\mu(x)$ on $\mathbb{R}$. The functions

$$f_n(z) = \int_\mathbb{R} \frac{\pi_n(t)}{z-t} d\mu(t) \qquad (z \in \mathbb{C} \setminus \text{supp}(d\mu),\ n \geq 0) \tag{2.2.48}$$

are known as the *functions of the second kind*. A straightforward analysis shows that these functions satisfy the three-term recurrence relation (2.2.4) with initial conditions

$$f_{-1}(z) = 1, \quad f_0(z) = \int_\mathbb{R} \frac{d\mu(t)}{z-t}. \tag{2.2.49}$$

Notice that f_0 is exactly the Stieltjes transform of the measure $d\mu(t)$ (see (2.2.37)). Using the orthogonality of $\{\pi_n\}$, it is easy to see that

$$\pi_n(z) f_n(z) = \int_\mathbb{R} \frac{\pi_n(z) - \pi_n(t)}{z-t} \pi_n(t) d\mu(t) + \int_\mathbb{R} \frac{\pi_n(t)^2}{z-t} d\mu(t) = \int_\mathbb{R} \frac{\pi_n(t)^2}{z-t} d\mu(t).$$

This means that $\pi_n(z) f_n(z)$ is the Stieltjes transform of the measure $\pi_n(t)^2 d\mu(t)$ and that this transform of the measure has no zeros outside the convex hull of its support.

For $|z|$ sufficiently large, a formal expression of $f_n(z)$ yields

$$f_n(z) \sim \int_\mathbb{R} \pi_n(t) \left(\sum_{k=0}^{+\infty} \frac{t^k}{z^{k+1}} \right) d\mu(t) = \sum_{k=0}^{+\infty} \frac{c_k}{z^{k+1}},$$

where $c_k = \int_\mathbb{R} \pi_n(t) t^k d\mu(t)$. Since $c_k = 0$, for $k < n$, because of the orthogonality, we get

$$f_n(z) \sim \frac{c_n}{z^{n+1}} + \frac{c_{n+1}}{z^{n+2}} + \cdots, \tag{2.2.50}$$

i.e., $f_n(z) = O(z^{-n-1})$ as $|z| \to +\infty$. Notice that (2.2.50), for $n = 0$, reduces to (2.2.38). Some characteristic properties of orthogonal polynomials in terms of functions of the second kind were investigated by Grinshpun [192].

Theorem 2.2.20 *Suppose that the moment problem for the moment sequence μ_k ($k = 0, 1, 2, \ldots$) is determined. Then, for every $\nu \in \mathbb{N}$,*

$$\lim_{n \to +\infty} \frac{\pi_{n-\nu}^{[\nu]}(z)}{\pi_n(z)} = \frac{f_{\nu-1}(z)}{\beta_0 \beta_1 \cdots \beta_{\nu-1}} \tag{2.2.51}$$

holds uniformly on compact subsets of $\mathbb{C} \setminus \mathrm{Co}\,(\mathrm{supp}\,(\mu))$, where the associated polynomials $\pi_n^{[\nu]}(z)$ are defined by (2.2.34).

Proof We use induction with respect to ν. For $\nu = 1$, (2.2.51) reduces to Markov's result:

$$\lim_{n \to +\infty} \frac{\pi_{n-1}^{[1]}(z)}{\pi_n(z)} = \frac{f_0(z)}{\beta_0},$$

uniformly on every compact subset of $\mathbb{C} \setminus \mathrm{Co}\,(\mathrm{supp}\,(\mu))$.

Suppose now that the result (2.2.51) holds up to ν. Then, using (2.2.35) with $x := z$, $\nu := \nu - 1$ and $n := n - \nu$, i.e.,

$$\pi_{n-\nu-1}^{[\nu+1]}(z) = \frac{1}{\beta_\nu} \left[(z - \alpha_{\nu-1}) \pi_{n-\nu}^{[\nu]}(z) - \pi_{n-\nu+1}^{[\nu-1]}(z) \right],$$

dividing each term in this equation by $\pi_n(z)$, and using the induction hypothesis, we obtain

$$\lim_{n \to +\infty} \frac{\pi_{n-\nu-1}^{[\nu+1]}(z)}{\pi_n(z)} = \frac{1}{\beta_\nu} \left[(z - \alpha_{\nu-1}) \frac{f_{\nu-1}(z)}{\beta_0 \beta_1 \cdots \beta_{\nu-1}} - \frac{f_{\nu-2}(z)}{\beta_0 \beta_1 \cdots \beta_{\nu-2}} \right]$$

$$= \frac{f_\nu(z)}{\beta_0 \beta_1 \cdots \beta_\nu}.$$

uniformly on compact subsets of $\mathbb{C} \setminus \mathrm{Co}\,(\mathrm{supp}\,(\mu))$. □

Remark 2.2.7 Van Assche [484] proved that the limit (2.2.51) holds uniformly on compact subsets of $\mathbb{C} \setminus (X_1 \cup X_2)$, where the following sets are defined: $Z_N = \{x_{n,\nu} \mid 1 \le \nu \le n,\ n \ge N\}$, $X_1 = Z_1' = \{\text{accumulation points of } Z_1\}$, $X_2 = \{x \in Z_1 \mid \pi_n(x) = 0 \text{ for infinitely many } n\}$. Notice that

$$\mathrm{supp}\,(d\mu) \subset X_1 \cup X_2 \subset \mathrm{Co}\,(\mathrm{supp}\,(d\mu)).$$

Theorem 2.2.20 is very useful in determining the distribution function μ_ν for the associated polynomials $\{\pi_n^{[\nu]}\}$ when the moment problem for μ is determined (cf.

2.2 Orthogonal Polynomials on the Real Line

[484]). Indeed, for $z \in \mathbb{C} \setminus \mathbb{R}$ and $\int_{\mathbb{R}} d\mu_\nu(t) = \beta_\nu$, according to

$$\lim_{n \to +\infty} \frac{\pi_{n-1}^{[\nu+1]}(z)}{\pi_n^{[\nu]}(z)} = \frac{1}{\beta_\nu} \int_{\mathbb{R}} \frac{d\mu_\nu(t)}{z-t}, \quad \frac{\pi_{n-1}^{[\nu+1]}(z)}{\pi_n^{[\nu]}(z)} = \frac{\pi_{n-1}^{[\nu+1]}(z)}{\pi_{n+\nu}(z)} \cdot \frac{\pi_{n+\nu}(z)}{\pi_n^{[\nu]}(z)},$$

and (2.2.51), we get (see [484])

$$\int_{\mathbb{R}} \frac{d\mu_\nu(t)}{z-t} = \frac{f_\nu(z)}{f_{\nu-1}(z)}, \quad z \in \mathbb{C} \setminus \mathbb{R}. \tag{2.2.52}$$

For $\nu = 0$, (2.2.52) reduces to (2.2.49). Formula (2.2.52) may be used to recover $d\mu_\nu$ by the Stieltjes inversion formula.

Using (2.2.29) and (2.2.37) we have $f_n(z) = \pi_n(z) F(z) - \sigma_n(z)$. According to (2.2.50) we conclude that

$$F(z) - \frac{\sigma_n(z)}{\pi_n(z)} = \frac{f_n(z)}{\pi_n(z)} = O(z^{-2n-1}) \quad \text{as } z \to \infty,$$

i.e., for each $n \in \mathbb{N}$, the expansions in descending powers of z of the functions $F(z)$ and $\sigma_n(z)/\pi_n(z)$, given by (2.2.38) and (2.2.40), respectively, agree up to and including the term with z^{-2n}.

Since

$$\frac{1}{f_n(z)} = z^{n+1} \left(d_n + \frac{d_{n+1}}{z} + \cdots \right) = d_n E_{n+1}(z) + \frac{e_1}{z} + \frac{e_2}{z^2} + \cdots, \tag{2.2.53}$$

where $E_{n+1}(z) \in \mathcal{P}_{n+1}$ and d_k and e_k are some coefficients. Notice that $d_n = 1/c_n = 1/\|\pi_n\|^2$. According to (2.2.50), for some appropriate constants $a_1, a_2, \ldots$, we have

$$f_n(z) E_{n+1}(z) = \|\pi_n\|^2 + \frac{a_1}{z^{n+2}} + \frac{a_2}{z^{n+3}} + \cdots.$$

The (monic) polynomials $E_{n+1}(z)$ are known as the *Stieltjes polynomials*. In 1894, in one of his letters to Hermite, Stieltjes introduced and characterized these polynomials for the constant weight $w(x) = 1$ supported on $[-1, 1]$. In fact, he derived the following representation

$$E_{n+1}(x) = \frac{\|\pi_n\|^2}{2\pi i} \oint_C \frac{dz}{(z-x) f_n(z)}, \tag{2.2.54}$$

where x is an arbitrary point of the complex plane, while the contour integration is made along a circumference C, encircling x, of radius sufficiently large so that the expansions involved are convergent on C. To obtain (2.2.54), Stieltjes multiplies (2.2.53) by $(z-x)^{-1} = z^{-1} + xz^{-2} + x^2 z^{-3} + \cdots$, and computes the residue.

Since

$$\int_{\mathbb{R}} (z^k - x^k) \frac{\pi_n(x)}{z - x} d\mu(x) = \sum_{\nu=1}^{k} z^{k-\nu} \int_{\mathbb{R}} x^{\nu-1} \pi_n(x) d\mu(x) = 0 \quad (k \le n),$$

the following equalities

$$z^k f_n(z) = \int_{\mathbb{R}} x^k \frac{\pi_n(x)}{z-x} d\mu(x), \quad k = 0, 1, \ldots, n, \qquad (2.2.55)$$

hold. Using this fact and (2.2.54), we obtain the remarkable property

$$\int_{\mathbb{R}} x^k E_{n+1}(x) \pi_n(x) d\mu(x) = 0, \quad k = 0, 1, \ldots, n, \qquad (2.2.56)$$

i.e., the polynomial $E_{n+1}(x)$ must be orthogonal to $\mathcal{P}_n$ with respect to the oscillatory measure $d\hat{\mu}(x) = \pi_n(x) d\mu(x)$. The orthogonality conditions (2.2.56) imply the expression

$$E_{n+1}(x) \pi_n(x) = c_0 \pi_{n+1}(x) + c_1 \pi_{n+2}(x) + \cdots + c_n \pi_{2n+1}(x) \quad (c_n = 1),$$

which suggests a method of constructing $E_{n+1}(x)$, namely, determining the constants $c_0, c_1, \ldots, c_{n-1}$, so that the polynomial on the right-hand side vanishes at the zeros of $\pi_n(x)$.

A nice survey with historical remarks, several properties of Stieltjes polynomials and applications in quadrature rules was given by Monegato [368] (see also [107, 108, 366, 367, 369, 389, 391], etc.)

Recently, Peherstorfer and Petras [393] proved the following representation of the Stieltjes polynomials.

Theorem 2.2.21 *The Stieltjes polynomial $E_{n+1}(x)$ is given by*

$$E_{n+1}(x) = \pi_{n+1}(x) - \int_{\mathbb{R}} \frac{\pi_n(x) - \pi_n(t)}{x - t} d\mu_{n+1}(t), \qquad (2.2.57)$$

where $d\mu_{n+1}(t)$ is the measure of the associated polynomials $\pi_k^{[n+1]}$ of order $n+1$, normalized by $\int_{\mathbb{R}} d\mu_{n+1}(x) = \beta_{n+1}$.

Proof By definition, $E_{n+1}(x) \pi_n(x)$ is orthogonal to $\mathcal{P}_n$ with respect to the measure $d\mu$ and hence

$$E_{n+1}(x) \pi_n(x) = \sum_{\nu=0}^{n} c_\nu \pi_{n+\nu+1}(x),$$

where $c_n = 1$ and $c_\nu \in \mathbb{R}$. Using (2.2.36), with n and ν interchanged, i.e.,

$$\pi_{n+\nu+1}(x) = \pi_\nu^{[n+1]}(x) \pi_{n+1}(x) - \beta_{n+1} \pi_{\nu-1}^{[n+2]}(x) \pi_n(x),$$

we get

$$E_{n+1}(x) \pi_n(x) = \left(\sum_{\nu=0}^{n} c_\nu \pi_\nu^{[n+1]}(x) \right) \pi_{n+1}(x) - \beta_{n+1} \left(\sum_{\nu=0}^{n} c_\nu \pi_{\nu-1}^{[n+2]}(x) \right) \pi_n(x). \qquad (2.2.58)$$

Considering (2.2.58) at the zeros of $\pi_n(x)$, we can conclude that

$$\pi_n(x) = \sum_{\nu=0}^{n} c_\nu \pi_\nu^{[n+1]}(x), \qquad (2.2.59)$$

which implies again by (2.2.58) that

$$E_{n+1}(x) = \pi_{n+1}(x) - \beta_{n+1} \sum_{\nu=0}^{n} c_\nu \pi_{\nu-1}^{[n+2]}(x).$$

On the other hand, we have according to (2.2.34) and (2.2.59)

$$\beta_{n+1} \sum_{\nu=0}^{n} c_\nu \pi_{\nu-1}^{[n+2]}(x) = \sum_{\nu=0}^{n} c_\nu \int_{\mathbb{R}} \frac{\pi_\nu^{[n+1]}(x) - \pi_\nu^{[n+1]}(t)}{x-t} d\mu_{n+1}(t)$$

$$= \int_{\mathbb{R}} \frac{\pi_n(x) - \pi_n(t)}{x-t} d\mu_{n+1}(t). \qquad \square$$

2.3 Classical Orthogonal Polynomials

2.3.1 Definition of the Classical Orthogonal Polynomials

A survey on characterization theorems for orthogonal polynomials on the real line was given by Al-Salam [10]. The most important orthogonal polynomials on the real line are the so-called *very classical orthogonal polynomials* (cf. Van Assche [485], Nikiforov and Uvarov [381], Suetin [458]). An extension of the very classical orthogonal polynomials using the difference operators and q-difference operators is known nowadays as the classical orthogonal polynomials (see Andrews and Askey [12], Andrews, Askey, and Roy [13], Askey and Wilson [20], Atakishiyev and Suslov [22]). Such a much larger class of orthogonal polynomials can be arranged in a table, which is known as the Askey table and its q-extension (cf. Koekoek and Swarttouw [232]).

In this subsection we consider only *very* classical orthogonal polynomials. In the sequel we will omit the term "very" and call such polynomials the *classical orthogonal polynomials*. They are distinguished by several particular properties.

Let the inner product be given by

$$(f,g)_w = \int_a^b f(x)g(x)w(x)\,dx. \qquad (2.3.1)$$

Since every interval (a,b) can be transformed by a linear transformation to one of the following intervals: $(-1,1)$, $(0,+\infty)$, $(-\infty,+\infty)$, we will restrict our considerations (without loss of generality) to these intervals only.

Definition 2.3.1 The orthogonal polynomials $\{Q_n(x)\}$ on (a,b) with respect to the inner product (2.3.1) are called *classical orthogonal polynomials* if their weight functions $x \mapsto w(x)$ satisfy the differential equation

$$\frac{d}{dx}(A(x)w(x)) = B(x)w(x), \qquad (2.3.2)$$

where

$$A(x) = \begin{cases} 1 - x^2, & \text{if } (a,b) = (-1,1), \\ x, & \text{if } (a,b) = (0, +\infty), \\ 1, & \text{if } (a,b) = (-\infty, +\infty), \end{cases}$$

and $B(x)$ is a linear polynomial. For such classical weights we will write $w \in CW$.

We note that if $w \in CW$, then $w \in C^1(a,b)$, and also the following property holds.

Theorem 2.3.1 *If $w \in CW$ then we have for each $m = 0, 1, \ldots$*

$$\lim_{x \to a+} x^m A(x) w(x) = 0 \quad \text{and} \quad \lim_{x \to b-} x^m A(x) w(x) = 0. \qquad (2.3.3)$$

According to the definition above, this class of orthogonal polynomials $\{Q_n(x)\}$ on (a,b) can be classified as

1° the *Jacobi polynomials* $P_n^{(\alpha,\beta)}(x)$ ($\alpha, \beta > -1$) on $(-1, 1)$;
2° the *generalized Laguerre polynomials* $L_n^\alpha(x)$ ($\alpha > -1$) on $(0, +\infty)$;
3° the *Hermite polynomials* $H_n(x)$ on $(-\infty, +\infty)$.

Their weight functions and the corresponding polynomials $A(x)$ and $B(x)$ are given in Table 2.3.1.

Special cases of the Jacobi polynomials are:

- the *Gegenbauer polynomials* $C_n^\lambda(x)$ ($\alpha = \beta = \lambda - 1/2$);
- the *Legendre polynomials* $P_n(x)$ ($\alpha = \beta = 0$);
- the *Chebyshev polynomials of the first kind* $T_n(x)$ ($\alpha = \beta = -1/2$);
- the *Chebyshev polynomials of the second kind* $U_n(x)$ ($\alpha = \beta = 1/2$);
- the *Chebyshev polynomials of the third kind* $V_n(x)$ ($\alpha = -\beta = -1/2$);
- the *Chebyshev polynomials of the fourth kind* $W_n(x)$ ($\alpha = -\beta = 1/2$).

Table 2.3.1 Classification of the classical orthogonal polynomials

(a,b)	$w(x)$	$A(x)$	$B(x)$	λ_n
$(-1, 1)$	$(1-x)^\alpha (1+x)^\beta$	$1-x^2$	$\beta - \alpha - (\alpha + \beta + 2)x$	$n(n + \alpha + \beta + 1)$
$(0, +\infty)$	$x^\alpha e^{-x}$	x	$\alpha + 1 - x$	n
$(-\infty, +\infty)$	e^{-x^2}	1	$-2x$	$2n$

2.3 Classical Orthogonal Polynomials

If $\alpha = 0$, the generalized Laguerre polynomials reduce to the *standard Laguerre polynomials* $L_n(x)$.

The Chebyshev polynomials of the first and second kind were already introduced and studied in Sects. 1.1.2 and 1.1.4. Putting $x = \cos\theta$, $-1 \le x \le 1$, these polynomials can be expressed in the form (cf. Sect. 1.1.2)

$$T_n(x) = T_n(\cos\theta) = \cos n\theta \quad \text{and} \quad U_n(x) = U_n(\cos\theta) = \frac{\sin(n+1)\theta}{\sin\theta},$$

respectively. Similarly, for the Chebyshev polynomials of the third and fourth kind the following expressions

$$V_n(\cos\theta) = \frac{\cos(n+1/2)\theta}{\cos\theta/2} \quad \text{and} \quad W_n(\cos\theta) = \frac{\sin(n+1/2)\theta}{\sin\theta/2}$$

hold. Notice that $W_n(-x) = (-1)^n V_n(x)$.

There are many characterizations of the classical orthogonal polynomials.

Similarly as in the case of the well-known inequalities of Landau type [250] and Kolmogorov type [233] for continuously differentiable functions, as well as their generalizations (see, for example, [101, 191, 213, 233, 325, 427], and [454]), it is possible to consider such a kind of inequalities for algebraic polynomials of fixed degree (cf. Varma [493], Bojanov and Varma [46], Alves and Dimitrov [11], Agarwal and Milovanović [3, 4]).

The following characterization of the classical orthogonal polynomials was given by Agarwal and Milovanović [3]:

Theorem 2.3.2 *For all $P(x) \in \mathcal{P}_n$ the inequality*

$$(2\lambda_n + B'(0))\|\sqrt{A}P'\|^2 \le \|AP''\|^2 + \lambda_n^2 \|P\|^2 \tag{2.3.4}$$

holds, with equality if only if $P(x) = cQ_n(x)$, where $Q_n(x)$ is the classical orthogonal polynomial of degree n which is orthogonal to all polynomials of degree $\le n-1$ with respect to the weight function $w(x)$ on (a, b), and c is an arbitrary real constant. The λ_n, $A(x)$, and $B(x)$ are given in Table 2.3.1.

We mention some special cases.

First, for $w(x) = e^{-x^2}$ on $(-\infty, +\infty)$, the inequality (2.3.4) reduces to Varma's inequality

$$\|P'\|^2 \le \frac{1}{2(2n-1)}\|P''\|^2 + \frac{2n^2}{2n-1}\|P\|^2 \quad (P(x) \in \mathcal{P}_n),$$

which reduces to an equality if and only if $P(x) = cH_n(x)$, where $H_n(x)$ is the Hermite polynomial of degree n and c is an arbitrary real constant.

In the generalized Laguerre case, the inequality (2.3.4) becomes

$$\|\sqrt{x}P'\|^2 \le \frac{n^2}{2n-1}\|P\|^2 + \frac{1}{2n-1}\|xP''\|^2,$$

where $w(x) = x^\alpha e^{-x}$ ($\alpha > -1$) on $(0, +\infty)$.

In the Jacobi case we get the inequality

$$((2n-1)(\alpha+\beta) + 2(n^2+n-1))\|\sqrt{1-x^2}P'\|^2$$
$$\leq n^2(n+\alpha+\beta+1)^2\|P\|^2 + \|(1-x^2)P''\|^2,$$

where $w(x) = (1-x)^\alpha(1+x)^\beta$ ($\alpha, \beta > -1$) on $(-1, 1)$.

Weighted polynomial inequalities in the L^2-norm of Markov-Bernstein type, as well as the corresponding connections with the classical orthogonal polynomials, were considered in [198–201].

A characterization of the classical orthogonal polynomials based on the concept of "reversed" continued fraction of Stieltjes type was proposed by Dette and Studden [94]. Using a concept of the dual orthogonal polynomials introduced by de Boor and Saff [84], Vinet and Zhedanov [500] studied their properties and also presented a characterization of the classical and semi-classical orthogonal polynomials (see Sect. 2.4.1).

In the sequel we give the basic common properties of the classical orthogonal polynomials (cf. [4, 334, 359]).

2.3.2 General Properties of the Classical Orthogonal Polynomials

Using the notations of the previous section, we have:

Theorem 2.3.3 *The derivatives of the classical orthogonal polynomials $\{Q_n\}_{n\in\mathbb{N}_0}$ also form a sequence of the classical orthogonal polynomials.*

Proof Put

$$I_{m,n} = \int_a^b x^{m-1} B(x) Q_n(x) w(x)\, dx = \left(x^{m-1}B(x), Q_n\right)_w \quad (m \in \mathbb{N},\, n \in \mathbb{N}_0).$$

For each $n > m$ ($= \deg(x^{m-1}B(x))$), because of the orthogonality, we have $I_{m,n} = 0$. On the other hand, using (2.3.2), (2.3.3) and integration by parts, we obtain

$$I_{m,n} = \int_a^b x^{m-1} Q_n(x) \frac{d}{dx}(A(x)w(x))\, dx$$
$$= -(m-1)\left(x^{m-2}A(x), Q_n\right)_w - \left(x^{m-1}A(x), Q'_n\right)_w.$$

Since $\left(x^{m-2}A(x), Q_n\right)_w = 0$ ($m < n$), we conclude that

$$\left(x^{m-1}A(x), Q'_n\right)_w = \left(x^{m-1}, Q'_n\right)_{Aw} = 0,$$

2.3 Classical Orthogonal Polynomials

i.e., the sequence of polynomials $\{Q'_n\}_{n\in\mathbb{N}}$ is orthogonal with respect to the weight function $x \mapsto w_1(x) = A(x)w(x)$. If $w_1 \in CW$, these orthogonal polynomials are classical. Indeed,

$$\frac{d}{dx}(A(x)w_1(x)) = A'(x)w_1(x) + A(x)\frac{d}{dx}(A(x)w(x))$$
$$= (A'(x) + B(x))A(x)w(x) = B_1(x)w_1(x),$$

where $B_1(x) = A'(x) + B(x)$ is a linear polynomial. □

Applying induction we can prove a more general result:

Theorem 2.3.4 *The sequence $\{Q_n^{(m)}\}_{n=m,m+1,\ldots}$ is a classical orthogonal polynomial sequence on (a,b) with respect to the weight function $x \mapsto w_m(x) = A(x)^m w(x)$. The differential equation for this weight is*

$$(A(x)w_m(x))' = B_m(x)w_m(x),$$

where $B_m(x) = m A'(x) + B(x)$.

Theorem 2.3.5 *The classical orthogonal polynomial $Q_n(x)$ is a particular solution of the second order linear differential equation*

$$L[y] = A(x)y'' + B(x)y' + \lambda_n y = 0 \tag{2.3.5}$$

of hypergeometric type, where

$$\lambda_n = -n\left(\frac{1}{2}(n-1)A''(0) + B'(0)\right). \tag{2.3.6}$$

Proof Let $m < n$. Since $I_{m,n} = (Q'_n, x^{m-1})_{Aw} = 0$ by Theorem 2.3.3, integration by parts yields

$$I_{m,n} = -\frac{1}{m}\int_a^b \frac{d}{dx}(A(x)Q'_n(x)w(x))x^m\,dx = -\frac{1}{m}(\tilde{Q}_n, x^m)_w,$$

where we put $\tilde{Q}_n(x) = A(x)Q''_n(x) + B(x)Q'_n(x)$. This means that the polynomial $\tilde{Q}_n(x)$ is orthogonal to $\mathcal{P}_{n-1}$ with respect to the inner product $(\cdot,\cdot)_w$.

Thus, $\tilde{Q}_n(x)$ must be equal to $Q_n(x)$ up to a multiplicative constant, i.e.,

$$A(x)Q''_n(x) + B(x)Q'_n(x) + \lambda_n Q_n(x) = 0.$$

Comparing coefficients we get λ_n in the form (2.3.6). □

Equation (2.3.5) can be written in the Sturm-Liouville form

$$\frac{d}{dx}\left(A(x)w(x)\frac{dy}{dx}\right) + \lambda_n w(x)y = 0. \tag{2.3.7}$$

The coefficients λ_n are also displayed in Table 2.3.1.

Similarly, the m-th derivative of Q_n satisfies the differential equation

$$\frac{d}{dx}\left(A(x)w_m(x)\frac{dy}{dx}\right) + \lambda_{n,m}w_m(x)y = 0, \qquad (2.3.8)$$

where $\lambda_{n,m} = -(n-m)(\frac{1}{2}(n+m-1)A''(0) + B'(0))$. We note that this expression for $\lambda_{n,m}$ reduces to (2.3.6) for $m = 0$, i.e., $\lambda_{n,0} = \lambda_n$.

Remark 2.3.1 The characterization of the classical orthogonal polynomials by differential equation (2.3.5), i.e. (2.3.7), was proved by Lesky [255], and conjectured by Aczél [2] (see also Bochner [43]). Such a differential equation appears in many mathematical models in atomic physics, electrodynamics and acoustics. As an example we mention the well-known Schrödinger equation.

The classical orthogonal polynomials possess a Rodrigues' type formula (cf. Bateman and Erdélyi [29], Tricomi [479], and Suetin [458]), which can be derived by successively applying (2.3.8) n times.

Theorem 2.3.6 *The classical orthogonal polynomial $Q_n(x)$ can be expressed in the form*

$$Q_n(x) = \frac{C_n}{w(x)} \cdot \frac{d^n}{dx^n}\left(A(x)^n w(x)\right), \qquad (2.3.9)$$

where C_n are non-zero constants.

Using the Cauchy formula for the n-th derivative of a regular function, (2.3.9) can be represented in the following integral form

$$Q_n(x) = \frac{C_n}{w(x)} \cdot \frac{n!}{2\pi i} \oint_\Gamma \frac{A(z)^n w(z)}{(z-x)^{n+1}}\, dz, \qquad (2.3.10)$$

where Γ is a closed contour such that $x \in \text{int}\,\Gamma$.

The constants C_n in (2.3.9) and (2.3.10) can be chosen in different ways (for example, Q_n to be monic, orthonormal, etc.). A historical reason leads to

$$C_n = \begin{cases} \dfrac{(-1)^n}{2^n n!} & \text{for } P_n^{(\alpha,\beta)}(x), \\ \dfrac{1}{n!} & \text{for } L_n^s(x), \\ (-1)^n & \text{for } H_n(x). \end{cases}$$

In addition, for the Gegenbauer and the Chebyshev polynomials we have

$$C_n^\lambda(x) = \frac{(2\lambda)_n}{\left(\lambda+\frac{1}{2}\right)_n} P_n^{(\alpha,\alpha)}(x) \quad (\alpha = \lambda - 1/2), \qquad (2.3.11)$$

2.3 Classical Orthogonal Polynomials

$$T_n(x) = \frac{n!}{\left(\frac{1}{2}\right)_n} P_n^{(-1/2,-1/2)}(x),$$

$$U_n(x) = \frac{(n+1)!}{\left(\frac{3}{2}\right)_n} P_n^{(1/2,1/2)}(x),$$

where $(s)_n$ is the standard notation for Pochhammer's symbol

$$(s)_n = s(s+1)\cdots(s+n-1) = \frac{\Gamma(s+n)}{\Gamma(s)} \quad (\Gamma \text{ is the gamma function}).$$

For such defined polynomials

$$Q_n(x) = k_n\left(x^n + r_n x^{n-1} + \cdots\right), \tag{2.3.12}$$

using (2.3.9) and integration by parts, we can get the following formula for the norm of polynomials (cf. [328, p. 126])

$$\|Q_n\|^2 = (Q_n, Q_n)_w = k_n C_n (-1)^n n! \int_a^b A(x)^n w(x)\,dx. \tag{2.3.13}$$

Also, using the Rodrigues formula (2.3.9), for the leading coefficient k_n in $Q_n(x)$, as well as for the coefficient r_n, we get (cf. [328, p. XXX])

$$k_n = C_n \prod_{\nu=1}^n \left(B'(0) + \frac{1}{2}(2n - \nu - 1)A''(0)\right), \tag{2.3.14}$$

$$r_n = n\,\frac{B(0) + (n-1)A'(0)}{B'(0) + (n-1)A''(0)}. \tag{2.3.15}$$

By $\widehat{Q}_n(x)$ we denote the corresponding monic classical orthogonal polynomials, i.e., $\widehat{Q}_n(x) = k_n^{-1} Q_n(x)$. According to the recurrence relation (2.2.4) for monic polynomials with the recursion coefficients α_n and β_n, we conclude that the polynomials $\{Q_n(x)\}$ satisfy the recurrence relation

$$Q_{n+1}(x) = \frac{k_{n+1}}{k_n}(x - \alpha_n) Q_n(x) - \frac{k_{n+1}}{k_{n-1}} \beta_n Q_{n-1}(x) \quad (n \geq 0), \tag{2.3.16}$$

where the leading coefficients k_n are given in (2.3.14).

In the case of the classical orthogonal polynomials one can also express $Q_n'(x)$ in terms of $Q_n(x)$ and $Q_{n-1}(x)$. Namely,

$$A(x) Q_n'(x) = (e_n x + f_n) Q_n(x) + g_n Q_{n-1}(x), \tag{2.3.17}$$

where

$$e_n = \frac{1}{2} n A''(0), \quad f_n = n A'(0) - \frac{1}{2} r_n A''(0),$$

$$g_n = -\frac{k_n \beta_n}{k_{n-1}} \left[B'(0) + \left(n - \frac{1}{2}\right) A''(0) \right].$$

According to the three-term recurrence relation (2.3.16), (2.3.17) can be also expressed in the form

$$A(x) Q'_n(x) = u_n Q_{n+1}(x) + v_n Q_n(x) + w_n Q_{n-1}(x), \tag{2.3.18}$$

with the corresponding coefficients u_n, v_n, w_n. Such a kind of relation can be taken as a characterization of the classical orthogonal polynomials, supposing that $A(x)$ is a polynomial of degree not exceeding 2 (cf. [287] and [500]).

The classical polynomial $Q_n(x)$ can be expressed in terms of $Q'_{n-1}(x)$, $Q'_n(x)$, and $Q'_{n+1}(x)$ in the following way

$$\omega_n Q_n(x) = \xi_n Q'_{n-1}(x) + \eta_n Q'_n(x) + \zeta_n Q'_{n+1}(x), \tag{2.3.19}$$

where

$$\omega_n = (n+1) m_n, \quad m_n = B'(0) + \frac{1}{2}(n-2) A''(0), \quad \xi_n = -\frac{(n+1) k_n \beta_n}{2 k_{n-1}} A''(0),$$

$$\zeta_n = \frac{k_n m_n}{k_{n+1}}, \quad \eta_n = B(0) + (n-1) A'(0) - \frac{1}{2} r_n A''(0) - (r_{n+1} - r_n) m_n.$$

These coefficients will be obtained later for some particular cases of the classical orthogonal polynomials.

2.3.3 Generating Function

The classical orthogonal polynomials can be considered as the coefficients in the Taylor series of certain analytic functions.

Definition 2.3.2 We call a function $(x, t) \mapsto \Phi(x, t)$ a *generating function* of the system of polynomials $\{Q_k\}_{k \in \mathbb{N}_0}$ if, for sufficiently small t,

$$\Phi(x, t) = \sum_{k=0}^{+\infty} \frac{\widetilde{Q}_k(x)}{k!} t^k,$$

where $\widetilde{Q}_k(x) = Q_k(x) / C_k$ and C_k is the normalized constant which appears in the Rodrigues formula (2.3.9).

According to (2.3.10), i.e.,

$$\frac{\widetilde{Q}_k(x)}{k!} = \frac{1}{w(x)} \cdot \frac{1}{2\pi i} \oint_\Gamma \frac{A(z)^k w(z)}{(z-x)^{k+1}} dz,$$

2.3 Classical Orthogonal Polynomials

where Γ is a closed contour such that $x \in \text{int}\,\Gamma$, we have

$$\Phi(x,t) = \frac{1}{2\pi i}\cdot\frac{1}{w(x)}\oint_\Gamma \frac{w(z)}{z-x}\left(\sum_{k=0}^{+\infty}\left(\frac{A(z)t}{z-x}\right)^k\right)dz.$$

We have, since $\left|\dfrac{A(z)t}{z-x}\right| < 1$ for sufficiently small t,

$$\Phi(x,t) = \frac{1}{2\pi i}\cdot\frac{1}{w(x)}\oint_\Gamma \frac{w(z)}{z-x-A(z)t}\,dt.$$

If $t \to 0$, we conclude that the equation

$$z - x - A(z)t = 0 \qquad (2.3.20)$$

has a root $z \to x$, and the second root, if it exists, tends to ∞. Thus, for a sufficiently small t we may assume that the contour Γ encloses only one root $z = g(x,t)$, which means that the integrand has only one simple pole $z = g(x,t)$ inside the contour Γ. Then

$$\Phi(x,t) = \frac{1}{w(x)}\,\underset{z=g(x,t)}{\text{Res}}\left\{\frac{w(z)}{z-x-A(z)t}\right\} = \frac{1}{w(x)}\cdot\frac{w(z)}{1-A'(z)t}\bigg|_{z=g(x,t)}.$$

where $z = g(x,t)$ is the root of (2.3.20), close to the point $z = x$ for sufficiently small t.

Example 2.3.1 In the case of the Legendre polynomials we have $w(x) = 1$ and $A(x) = 1 - x^2$. Then, (2.3.20) reduces to $z - x - (1-z^2)t = 0$, and we have

$$g(x,t) = \frac{-1+\sqrt{1+4t(t-x)}}{2t},$$

and

$$\Phi(x,t) = \sum_{k=0}^{+\infty}\frac{P_k(x)}{C_k k!}t^k = \frac{1}{1+2zt}\bigg|_{z=g(x,t)} = \frac{1}{\sqrt{1+4tx+4t^2}}.$$

Since $C_k = (-1)^k/(2^k k!)$, we have

$$\frac{1}{\sqrt{1+4tx+4t^2}} = \sum_{k=0}^{+\infty} P_k(x)(-2t)^k,$$

i.e.,

$$\frac{1}{\sqrt{1-2tx+t^2}} = \sum_{k=0}^{+\infty} P_k(x)t^k. \qquad (2.3.21)$$

Similarly, we can get the generating function for the Jacobi polynomials,

$$\Phi(x,t) = \frac{2^{\alpha+\beta}}{R(1-t+R)^\alpha (1+t+R)^\beta} = \sum_{k=0}^{+\infty} P_k^{(\alpha,\beta)}(x) t^k, \qquad (2.3.22)$$

where $R = \sqrt{1-2tx+t^2}$. Taking $\alpha = \beta = \lambda - 1/2$ and using (2.3.11), the generating function (2.3.22) becomes

$$\frac{2^{\lambda-1/2}}{R(1-xt+R)^{\lambda-1/2}} = \sum_{k=0}^{+\infty} \frac{\left(\lambda+\frac{1}{2}\right)_k}{(2\lambda)_k} C_k^\lambda(x) t^k.$$

On the other hand, for the Gegenbauer polynomials, there is another, simpler generating function

$$(1-2tx+t^2)^{-\lambda} = \sum_{k=0}^{+\infty} C_k^\lambda(x) t^k. \qquad (2.3.23)$$

Notice that for $\lambda = 1/2$ both those generating functions reduce to (2.3.21).

Example 2.3.2 For the generalized Laguerre polynomials we have $w(x) = x^\alpha e^{-x}$ ($\alpha > -1$) and $A(x) = x$. From $z - x - zt = 0$ it follows that $g(x,t) = x/(1-t)$, and then

$$\Phi(x,t) = \frac{1}{x^\alpha e^{-x}} \left(\frac{x}{1-t}\right)^\alpha e^{-x/(1-t)} \cdot \frac{1}{1-t} = (1-t)^{-(\alpha+1)} e^{-xt/(1-t)}.$$

Thus,

$$(1-t)^{-(\alpha+1)} e^{-xt/(1-t)} = \sum_{k=0}^{+\infty} L_k^\alpha(x) t^k.$$

Example 2.3.3 For the Hermite polynomials we have $w(x) = e^{-x^2}$ and $A(x) = 1$. Equation (2.3.20), in this case, becomes $z - x - t = 0$, with only one root $g(x,t) = x+t$. According to the previous considerations, we get

$$\Phi(x,t) = e^{-(x+t)^2 + x^2} = e^{-2xt - t^2}.$$

Thus, we have

$$e^{-2xt-t^2} = \sum_{k=0}^{+\infty} \frac{H_k(x)}{(-1)^k k!} t^k,$$

i.e.,

$$e^{2xt-t^2} = \sum_{k=0}^{+\infty} \frac{H_k(x)}{k!} t^k. \qquad (2.3.24)$$

Putting $x/\sqrt{\lambda}$ and $t/\sqrt{\lambda}$ ($\lambda > 0$) instead of x and t, respectively, in (2.3.23) and observing that

$$\lim_{\lambda \to +\infty} \left(1 - 2\frac{xt}{\lambda} + \frac{t^2}{\lambda}\right)^{-\lambda} = e^{2xt-t^2},$$

according to (2.3.24), we conclude

$$\lim_{\lambda \to +\infty} \lambda^{-k/2} C_k^\lambda(x/\sqrt{\lambda}) = \frac{H_k(x)}{k!}.$$

This means that the Hermite polynomials are limits of the Gegenbauer polynomials. Also, it can be proved (cf. [13, p. 306]) that

$$\lim_{\beta \to +\infty} P_k^{(\alpha,\beta)}(1 - 2x/\beta) = L_k^\alpha(x).$$

2.3.4 Jacobi Polynomials

Using the notations of the previous sections, for the Jacobi polynomials $P_n^{(\alpha,\beta)}(x)$ that are orthogonal on $(a, b) = (-1, 1)$ with respect to the weight function

$$w(x) = v^{\alpha,\beta}(x) = (1-x)^\alpha (1+x)^\beta \qquad (\alpha, \beta > -1), \tag{2.3.25}$$

we have $A(x) = 1 - x^2$ and $B(x) = \beta - \alpha - (\alpha + \beta + 2)x$.

The differential equation (2.3.5) becomes

$$(1-x^2)y'' + [\beta - \alpha - (\alpha+\beta+2)x]y' + n(n+\alpha+\beta+1)y = 0. \tag{2.3.26}$$

The Rodrigues' formula (2.3.9), in this case, takes the form

$$P_n^{(\alpha,\beta)}(x) = \frac{(-1)^n}{2^n n!} (1-x)^{-\alpha}(1+x)^{-\beta} \frac{d^n}{dx^n}\left((1-x)^{n+\alpha}(1+x)^{n+\beta}\right),$$

from which the following explicit expression follows

$$P_n^{(\alpha,\beta)}(x) = \frac{1}{2^n} \sum_{\nu=0}^n \binom{n+\alpha}{\nu}\binom{n+\beta}{n-\nu}(x-1)^{n-\nu}(x+1)^\nu.$$

Notice that

$$P_n^{(\alpha,\beta)}(-x) = (-1)^n P_n^{(\beta,\alpha)}(x) \tag{2.3.27}$$

and

$$P_n^{(\alpha,\beta)}(1) = \binom{n+\alpha}{n} = \frac{(\alpha+1)_n}{n!}.$$

According to (2.3.13), (2.3.14), and (2.3.15) we find the norm, the leading coefficient k_n and r_n in the expansion $P_n^{(\alpha,\beta)}(x) = k_n(x^n + r_n x^{n-1} + \cdots)$:

$$\|P_n^{(\alpha,\beta)}\|^2 = \frac{2^{\alpha+\beta+1}\Gamma(n+\alpha+1)\Gamma(n+\beta+1)}{n!(2n+\alpha+\beta+1)\Gamma(n+\alpha+\beta+1)},$$

$$k_n = \frac{(n+\alpha+\beta+1)_n}{2^n n!}, \qquad r_n = \frac{n(\alpha-\beta)}{2n+\alpha+\beta}.$$

Using the asymptotic formula $\Gamma(n+\alpha)/\Gamma(n) = n^\alpha[1+O(1/n)]$ $(n \to +\infty)$ for a fixed α (cf. [386, p. 15]), we conclude that

$$\|P_n^{(\alpha,\beta)}\|^2 = O(1/n).$$

By changing the variable $x = 1 - 2t$, the differential equation (2.3.26) can be transformed to the Gauss hypergeometric equation, and then the Jacobi polynomial of degree n is a terminating hypergeometric series, i.e.,

$$P_n^{(\alpha,\beta)}(x) = \binom{n+\alpha}{n} {}_2F_1\left(-n, n+\alpha+\beta+1; \alpha+1; \frac{1-x}{2}\right)$$

$$= \binom{n+\alpha}{n} \sum_{\nu=0}^{n} \frac{(-n)_\nu (n+\alpha+\beta+1)_\nu}{(\alpha+1)_\nu \nu!} \left(\frac{1-x}{2}\right)^\nu.$$

Using one of several relations for the hypergeometric functions (see Andrews, Askey, and Roy [13, pp. 124–186 & p. 248]) we get the three-term recurrence relation for the Jacobi polynomials

$$2(n+1)(n+\alpha+\beta+1)(2n+\alpha+\beta) P_{n+1}^{(\alpha,\beta)}(x)$$
$$= (2n+\alpha+\beta+1)[(2n+\alpha+\beta)(2n+\alpha+\beta+2)x - \alpha^2 - \beta^2] P_n^{(\alpha,\beta)}(x)$$
$$- 2(n+\alpha)(n+\beta)(2n+\alpha+\beta+2) P_{n-1}^{(\alpha,\beta)}(x).$$

Here, $P_0^{(\alpha,\beta)}(x) = 1$ and $P_1^{(\alpha,\beta)}(x) = \frac{1}{2}(\alpha+\beta+2)x + \frac{1}{2}(\alpha-\beta)$.

The coefficients α_n and β_n in the three-term recurrence relation for the monic Jacobi polynomials $\hat{P}_n^{(\alpha,\beta)}(x) = 2^n n!/((n+\alpha+\beta+1)_n) P_n^{(\alpha,\beta)}(x)$ (cf. (2.2.4)) are

$$\alpha_n = \frac{\beta^2 - \alpha^2}{(2n+\alpha+\beta)(2n+\alpha+\beta+2)} \qquad (n \geq 0), \qquad (2.3.28)$$

$$\beta_n = \frac{4n(n+\alpha)(n+\beta)(n+\alpha+\beta)}{(2n+\alpha+\beta)^2((2n+\alpha+\beta)^2 - 1)} \qquad (n \geq 1). \qquad (2.3.29)$$

The coefficient β_0 can be defined as

$$\beta_0 = \mu_0 = \int_{-1}^{1} v^{\alpha,\beta}(x)\,dx = \frac{2^{\alpha+\beta+1}\Gamma(\alpha+1)\Gamma(\beta+1)}{\Gamma(\alpha+\beta+2)}.$$

2.3 Classical Orthogonal Polynomials

It is easy to prove that the following asymptotic relations hold for the coefficients in (2.3.28) and (2.3.29)

$$\alpha_n = \frac{\beta^2 - \alpha^2}{4n^2} + O(n^{-3}), \quad \beta_n = \frac{1}{4} + \frac{1 - 2(\alpha^2 + \beta^2)}{16(n+1)^2} + O(n^{-3}).$$

The coefficients b_n and a_n in the corresponding recurrence relation for the orthonormal Jacobi polynomials (see relation (2.2.3))

$$p_n^{(\alpha,\beta)}(x) = \frac{P_n^{(\alpha,\beta)}(x)}{\|P_n^{(\alpha,\beta)}\|} = \gamma_n(v^{\alpha,\beta}) x^n + \cdots$$

are given by $b_n = \sqrt{\beta_n}$ and $a_n = \alpha_n$, and their leading coefficients by

$$\gamma_n(v^{\alpha,\beta}) = \sqrt{\frac{2n+\alpha+\beta+1}{2^{2n+\alpha+\beta+1} n!}} \cdot \frac{\Gamma(2n+\alpha+\beta+1)}{\sqrt{\Gamma(n+\alpha+1)\Gamma(n+\beta+1)\Gamma(n+\alpha+\beta+1)}}.$$

Note that

$$a_n = O\left(\frac{1}{n^2}\right) \quad \text{and} \quad b_n = \frac{1}{2} + O\left(\frac{1}{n^2}\right), \quad \text{as } n \to +\infty. \tag{2.3.30}$$

We also list the coefficients in the relation (2.3.17),

$$e_n = -n, \quad f_n = \frac{n(\alpha - \beta)}{2n + \alpha + \beta}, \quad g_n = \frac{2(n+\alpha)(n+\beta)}{2n + \alpha + \beta},$$

as well as in the relation (2.3.19),

$$\xi_n = \frac{2(n+1)(n+\alpha)(n+\beta)}{(2n+\alpha+\beta)(2n+\alpha+\beta+1)}, \quad \eta_n = \frac{2(n+1)(n+\alpha+\beta)(\beta-\alpha)}{(2n+\alpha+\beta)(2n+\alpha+\beta+2)},$$

$$\zeta_n = -\frac{2(n+1)(n+\alpha+\beta)(n+\alpha+\beta+1)}{(2n+\alpha+\beta+1)(2n+\alpha+\beta+2)}, \quad \omega_n = -(n+1)(n+\alpha+\beta).$$

According to Theorem 2.3.3 we have

$$\frac{d}{dx} P_n^{(\alpha,\beta)}(x) = \frac{1}{2}(n+\alpha+\beta+1) P_{n-1}^{(\alpha+1,\beta+1)}(x).$$

2.3.4.1 Special Cases

In the symmetric case when $\alpha = \beta = \lambda - 1/2$ ($\lambda > -1/2$), the corresponding polynomials are known as the Gegenbauer or *ultraspherical* polynomials (see (2.3.11))

$$C_n^\lambda(x) = \frac{(2\lambda)_n}{\left(\lambda + \frac{1}{2}\right)_n} P_n^{(\lambda-1/2, \lambda-1/2)}(x). \tag{2.3.31}$$

In the limit case, when $\lambda \to 0$, we have

$$\lim_{\lambda \to 0} \frac{C_n^\lambda(x)}{\lambda} = \frac{2}{n} T_n(x) \quad (n \in \mathbb{N}),$$

where T_n is the Chebyshev polynomial of the first kind.

Since the weight $w(x) = v^{\lambda-1/2, \lambda-1/2}(x) = (1-x^2)^{\lambda-1/2}$ is an even function, we have $C_n^\lambda(-x) = (-1)^n C_n^\lambda(x)$. Also,

$$C_n^\lambda(1) = \binom{n+2\lambda-1}{n} = \frac{(2\lambda)_n}{n!}.$$

The Gegenbauer polynomials, which have the following explicit expansion

$$C_n^\lambda(x) = \sum_{v=0}^{[n/2]} \frac{(-1)^v (\lambda)_{n-v}}{v!(n-2v)!} (2x)^{n-2v},$$

can also be represented by the Gauss hypergeometric function in the form

$$C_{2k}^\lambda(x) = (-1)^k \frac{(\lambda)_k}{k!} {}_2F_1\left(-k, k+\lambda; \frac{1}{2}; x^2\right)$$

$$C_{2k+1}^\lambda(x) = (-1)^k \frac{(\lambda)_{k+1}}{k!} 2x \, {}_2F_1\left(-k, k+\lambda+1; \frac{3}{2}; x^2\right).$$

These polynomials can be expressed in terms of the Jacobi polynomials,

$$C_{2k}^\lambda(x) = \frac{(\lambda)_k}{\left(\frac{1}{2}\right)_k} P_k^{(\lambda-1/2, -1/2)}(2x^2-1),$$

$$C_{2k+1}^\lambda(x) = \frac{(\lambda)_{k+1}}{\left(\frac{1}{2}\right)_{k+1}} x P_k^{(\lambda-1/2, 1/2)}(2x^2-1).$$

According to the last formulas, we get the following relations for the Chebyshev polynomials,

$$T_{2k+1}(x) = \frac{k!}{\left(\frac{1}{2}\right)_k} x P_k^{(-1/2, 1/2)}(2x^2-1)$$

and

$$U_{2k}(x) = \frac{k!}{\left(\frac{1}{2}\right)_k} P_k^{(1/2, -1/2)}(2x^2-1).$$

In the simplest case $\lambda = 1/2$, the Gegenbauer polynomials reduce to the well-known Legendre polynomials $P_n(x)$. We mention here the interesting formula

$$P_n(x) = \frac{1}{\pi} \int_0^\pi \left(x + \sqrt{x^2-1} \cos\varphi\right)^n d\varphi, \qquad (2.3.32)$$

2.3 Classical Orthogonal Polynomials

which is known as the *Laplace integral formula* for the Legendre polynomials. Notice that $P_n(\pm 1) = (\pm 1)^n$.

Changing the variables, $u = x + \sqrt{x^2 - 1} \cos\varphi$, $x = \cos\theta$, in (2.3.32), and then putting $u = e^{i\varphi}$, we obtain the integral representation

$$P_n(\cos\theta) = \frac{\sqrt{2}}{\pi} \int_0^\theta \frac{\cos(n + \frac{1}{2})\varphi}{\sqrt{\cos\varphi - \cos\theta}} d\varphi, \quad 0 < \theta < \pi,$$

which is known as *Dirichlet-Mehler formula*. A general formula for the Jacobi polynomials can be found in [13, pp. 313–316]).

An interesting connection between the Chebyshev polynomials of the first and second kind can be done by the Hilbert transform. Namely,

$$T_k(x) = -\frac{1}{\pi} \text{P.V.} \int_{-1}^1 \frac{U_{k-1}(t)}{t - x} \cdot \frac{dt}{\sqrt{1 - t^2}}$$

and

$$U_{k-1}(x) = \frac{1}{\pi} \text{P.V.} \int_{-1}^1 \frac{T_k(t)}{t - x} \cdot \frac{dt}{\sqrt{1 - t^2}}.$$

2.3.4.2 Zeros

Let $x_{n,k}$ ($k = 1, \ldots, n$) be the zeros of the Jacobi polynomial $P_n^{(\alpha,\beta)}(x)$ ordered as an increasing sequence, i.e.,

$$-1 < x_{n,1} < x_{n,2} < \cdots < x_{n,n-1} < x_{n,n} < 1,$$

and let $x_{n,k} = \cos\theta_{n,k}$, with $\theta_{n,0} = \pi$, $\theta_{n,n+1} = 0$.

A characterization of $P_n^{(\alpha,\beta)}(\cos\theta)$ for $\theta = O(1/n)$ can be done by the formula of Mehler-Heine type (cf. Szegő [470, p. 167 and p. 192])

$$\lim_{n \to +\infty} n^{-\alpha} P_n^{(\alpha,\beta)}\left(\cos\frac{z}{n}\right) = \lim_{n \to +\infty} n^{-\alpha} P_n^{(\alpha,\beta)}\left(1 - \frac{z^2}{2n^2}\right) = (2/z)^\alpha J_\alpha(z),$$

where $J_\alpha(z)$ is the Bessel function of order α. This formula holds uniformly in every bounded domain of the complex z-plane.

Theorem 2.3.7 *For a fixed k, we have*

$$\lim_{n \to +\infty} n\theta_{n,n-k+1} = j_k, \quad (2.3.33)$$

where j_k is the k-th positive zero of $J_\alpha(z)$.

Putting $N = n + (\alpha + \beta + 1)/2$, Vértesi [496] proved that

$$\theta_{n,n-k+1} = \frac{2k + \alpha - \frac{1}{2}}{2N} + \varrho_{n,k}, \quad |\varrho_{n,k}| \leq \frac{c}{kn}, \quad 1 \leq k \leq (1 - \varepsilon)n. \quad (2.3.34)$$

Combining (2.3.33), (2.3.27), and (2.3.34), an important property of the Jacobi zeros can be given in the form (cf. [465, pp. 282–283])

$$\theta_{n,k} - \theta_{n,k+1} \sim \frac{1}{n}, \quad 0 \le k \le n. \tag{2.3.35}$$

Remark 2.3.2 In the case of the Chebyshev polynomials of the first kind (see (1.1.19)), the distance $\theta_{n,k} - \theta_{n,k+1}$ is exactly π/n for each $1 < k < n$, and $\pi/(2n)$ when $k = 0$ and $k = n$.

2.3.4.3 Inequalities and Asymptotics

According to Theorem 2.2.10, the following simple inequality for Legendre polynomials

$$|P_n(x)| \le 1 \quad (-1 \le x \le 1)$$

holds. We can also obtain this inequality by taking $-1 \le x \le 1$ in (2.3.32). Namely,

$$|P_n(x)| \le \frac{1}{\pi} \int_0^\pi |x + i\sqrt{1-x^2}\cos\theta|^n \, d\theta$$

$$= \frac{1}{\pi} \int_0^\pi \sqrt{(\cos^2\theta + x^2\sin^2\theta)^n} \, d\theta \le \frac{1}{\pi} \int_0^\pi d\theta = 1.$$

In the general case for the Jacobi polynomials we have

$$\max_{-1 \le x \le 1} |P_n^{(\alpha,\beta)}(x)| = \max |P_n^{(\alpha,\beta)}(\pm 1)| = \binom{n+q}{n},$$

where

$$q = \max(\alpha, \beta) \ge -\frac{1}{2}, \quad \alpha, \beta > -1.$$

If $-1 < \alpha, \beta < -1/2$ and $x_0 = (\beta - \alpha)/(\alpha + \beta + 1)$, then

$$\max_{-1 \le x \le 1} |P_n^{(\alpha,\beta)}(x)| = |P_n^{(\alpha,\beta)}(x')|,$$

where x' is one of the two maximum points closest to x_0 (see Szegő [470, p. 168]). This maximum has the order $1/\sqrt{n}$, when $n \to +\infty$.

Bernstein proved that the Legendre polynomials $P_n(x)$ satisfy the inequality

$$\sqrt{\sin\theta}\, |P_n(\cos\theta)| < \sqrt{2/n\pi}, \quad 0 \le \theta \le \pi.$$

The constant $\sqrt{2/\pi}$ is the least possible. Antonov and Holševnikov [15] (see also Lorch [266]) improved the inequality to $\sqrt{\sin\theta}\,|P_n(\cos\theta)| < \sqrt{2/\pi}(n+\frac{1}{2})^{-1/2}$, and Lorch [267] generalized the result to ultraspherical polynomials by proving that

$$(\sin\theta)^\lambda |C_n^\lambda(\cos\theta)| < 2^{1-\lambda}[\Gamma(\lambda)]^{-1}(n+\lambda)^{\lambda-1},$$

2.3 Classical Orthogonal Polynomials

for $0 < \lambda < 1$, $0 \leq \theta \leq \pi$. In 1994 Chow, Gatteschi, and Wong [61], using Gasper's Mehler-type integral for the Jacobi polynomials and estimating a contour integral, proved an inequality generalizing the ultraspherical inequality to the Jacobi polynomials,

$$|P_n^{(\alpha,\beta)}(\cos\theta)| \leq k(\theta)\Gamma(q+1)\binom{n+q}{n}N^{-q-1/2},$$

where $q = \max(\alpha, \beta)$, $N = n + \frac{1}{2}(\alpha + \beta + 1)$ and

$$k(\theta) = \pi^{-1/2}\left(\sin\frac{\theta}{2}\right)^{-\alpha-1/2}\left(\cos\frac{\theta}{2}\right)^{-\beta-1/2}. \tag{2.3.36}$$

In [470, p. 196] we can also find an important formula on this subject due to Darboux

$$P_n^{(\alpha,\beta)}(\cos\theta) = n^{-1/2}k(\theta)\cos(N\theta + \gamma) + O(n^{-3/2}), \tag{2.3.37}$$

where $\gamma = -(\alpha + 1/2)\pi/2$, $0 < \theta < \pi$. The bound for the error term holds uniformly in the interval $[\varepsilon, \pi - \varepsilon]$. Furthermore,

$$P_n^{(\alpha,\beta)}(\cos\theta) = n^{-1/2}k(\theta)\bigl[\cos(N\theta + \gamma) + (n\sin\theta)^{-1}O(1)\bigr], \tag{2.3.38}$$

is a more precise formula than (2.3.37), for $\alpha, \beta > -1$ and $c/n \leq \theta \leq \pi - c/n$, where c is a fixed positive number and the constant in $O(1)$ is independent of n. Formula (2.3.38) was proved by Szegő (cf. [470, pp. 197–198]).

Also, we give the corresponding formula for the first derivative of $P_n^{(\alpha,\beta)}(\cos\theta)$,

$$\frac{d}{d\theta}[(\cos\theta)] = n^{1/2}k(\theta)\bigl[-\sin(N\theta + \gamma) + (n\sin\theta)^{-1}O(1)\bigr], \tag{2.3.39}$$

which holds under the same conditions as (2.3.38). According to (2.3.36) note that $k'(\theta) = k(\theta)(\sin\theta)^{-1}O(1)$. For a proof of (2.3.39) see [470, pp. 236–237].

Applying Darboux's method one can obtain the following asymptotic formula (cf. [383, pp. 154–155])

$$P_n^{(\alpha,\beta)}(z) = \frac{1}{\sqrt{\pi n}}C(z)\varrho(z)^n\bigl[1 + O(n^{-1/2})\bigr], \quad n \to +\infty,$$

which holds uniformly with respect to z on compact subsets of $\mathbb{C} \setminus [-1, 1]$, where $\varrho(z) = z + \sqrt{z^2 - 1}$ ($|\varrho(z)| > 1$),

$$C(z) = \left(1 + \sqrt{\frac{z+1}{z-1}}\right)^{\alpha}\left(1 + \sqrt{\frac{z-1}{z+1}}\right)^{\beta}\sqrt{\frac{z + \sqrt{z^2-1}}{2\sqrt{z^2-1}}}, \tag{2.3.40}$$

and the branches of the multi-valued functions in (2.3.40) are chosen so that $C(\infty) = 2^{\alpha+\beta}$. For some recent and more general asymptotic relations see Sect. 2.4.3.

An important bound for orthonormal Jacobi polynomials can be given in the form ([376])

$$|p_n^{(\alpha,\beta)}(x)| \leq C\left(\sqrt{1-x} + \frac{1}{n}\right)^{-\alpha-1/2}\left(\sqrt{1+x} + \frac{1}{n}\right)^{-\beta-1/2}, \quad |x| \leq 1, \quad (2.3.41)$$

where $C \neq C(n, x)$.

For such polynomials, Nevai, Erdélyi, and Magnus [380] proved the following result:

Theorem 2.3.8 *For all Jacobi weight functions $v^{\alpha,\beta}(x) = (1-x)^\alpha(1+x)^\beta$ with $\alpha \geq -1/2$ and $\beta \geq -1/2$, the inequalities*

$$\max_{x \in [-1,1]} \frac{[p_n^{(\alpha,\beta)}(x)]^2}{\sum_{k=0}^n [p_k^{(\alpha,\beta)}(x)]^2} \leq \frac{4\left(2 + \sqrt{\alpha^2 + \beta^2}\right)}{2n + \alpha + \beta + 2}$$

and

$$\max_{x \in [-1,1]} (1-x)^{\alpha+1/2}(1+x)^{\beta+1/2}[p_n^{(\alpha,\beta)}(x)]^2 \leq \frac{2e\left(2 + \sqrt{\alpha^2 + \beta^2}\right)}{\pi} \quad (2.3.42)$$

hold for each $n \in \mathbb{N}_0$.

According to certain numerical computation, they conjectured that the maximum on the left hand side in (2.3.42) is $O((\alpha^2 + \beta^2)^{1/4})$. Recently, Krasikov [239] (see also [238]) has confirmed this conjecture in the ultraspherical case $\alpha = \beta \geq (1 + \sqrt{2})/4$, even in a stronger form by giving very explicit upper bounds. Taking

$$\delta = \sqrt{1 - \frac{4\alpha^2 - 1}{(2k + 2\alpha + 1)^2 - 4}},$$

Krasikov [239] also showed that

$$\sqrt{\delta^2 - x^2}(1 - x^2)^\alpha [p_{2k}^{(\alpha,\alpha)}(x)]^2 < \frac{2}{\pi}\left(1 + \frac{1}{8(2k + \alpha)^2}\right),$$

where the interval $(-\delta, \delta)$ contains all the zeros of $p_{2k}^{(\alpha,\alpha)}(x)$. For polynomials of odd degree he obtained slightly weaker bounds.

Finally, we mention here an interesting simple inequality for the Chebyshev polynomials

$$T_n(xy) \leq T_n(x)T_n(y), \quad x, y \geq 1,$$

which can be verified by using the extremal property of the Chebyshev polynomials given by Theorem 1.1.8 in Sect. 1.1.4. This inequality also follows from

$$\frac{d^2}{du^2} \log T_n(e^u) \leq 0, \quad u \geq 0.$$

2.3 Classical Orthogonal Polynomials

Various proofs for these inequalities, as well as several generalizations for other classes of polynomials, were given by Askey, Gasper, and Harris [21].

2.3.4.4 Christoffel Function and Christoffel Numbers

In order to find the Christoffel function for the Jacobi weight we need (2.2.8), i.e.,

$$K_{n-1}^{(\alpha,\beta)}(x,x) = \sum_{k=0}^{n-1} p_k(v^{\alpha,\beta};x)^2$$

$$= \sqrt{\beta_n}\left(p'_n(v^{\alpha,\beta};x)p_{n-1}(v^{\alpha,\beta};x) - p'_{n-1}(v^{\alpha,\beta};x)p_n(v^{\alpha,\beta};x)\right),$$

where $p_k(v^{\alpha,\beta};x) = P_k^{(\alpha,\beta)}(x)/\|P_k^{(\alpha,\beta)}\|$ and β_n is given by (2.3.29). Since

$$G_n^{(\alpha,\beta)} = \frac{\sqrt{\beta_n}}{\|P_n^{(\alpha,\beta)}\|\|P_{n-1}^{(\alpha,\beta)}\|} = \frac{2^{-(\alpha+\beta)}n!\Gamma(n+\alpha+\beta+1)}{(2n+\alpha+\beta)\Gamma(n+\alpha)\Gamma(n+\beta)},$$

we have

$$K_{n-1}^{(\alpha,\beta)}(x,x) = G_n^{(\alpha,\beta)}\left(P_{n-1}^{(\alpha,\beta)}(x)\frac{d}{dx}P_n^{(\alpha,\beta)}(x) - P_n^{(\alpha,\beta)}(x)\frac{d}{dx}P_{n-1}^{(\alpha,\beta)}(x)\right),$$

so that, according to (2.1.30), the corresponding Christoffel function becomes

$$\lambda_n^{(\alpha,\beta)}(x) = \lambda_n(v^{\alpha,\beta};x) = \left[K_{n-1}^{(\alpha,\beta)}(x,x)\right]^{-1}. \tag{2.3.43}$$

Putting $x = x_{n,\nu}$ (a zero of $P_n^{(\alpha,\beta)}(x)$) in (2.3.43) we get the Christoffel numbers (or the Cotes-Christoffel coefficients), $\lambda_{n,\nu}^{(\alpha,\beta)} = \lambda_n^{(\alpha,\beta)}(x_{n,\nu})$. Using the relation (2.3.17) for the Jacobi polynomials $P_n^{(\alpha,\beta)}(x)$ at the point $x = x_{n,\nu}$, as well as the expression for $G_n^{(\alpha,\beta)}$, we get

$$\lambda_{n,\nu}^{(\alpha,\beta)} = \frac{\Gamma(n+\alpha+1)\Gamma(n+\beta+1)}{n!\Gamma(n+\alpha+\beta+1)} \cdot \frac{2^{\alpha+\beta+1}}{\left(1-x_{n,\nu}^2\right)\left[\frac{d}{dx}P_n^{(\alpha,\beta)}(x_{n,\nu})\right]^2}. \tag{2.3.44}$$

In the Chebyshev case of the first kind ($\alpha = \beta = -1/2$) and of the second kind ($\alpha = \beta = 1/2$), (2.3.44) reduces to

$$\lambda_{n,\nu}^{(-1/2,-1/2)} = \frac{\pi}{n}, \quad \nu = 1, \ldots, n, \tag{2.3.45}$$

and

$$\lambda_{n,\nu}^{(1/2,1/2)} = \frac{\pi}{n+1}\sin^2\frac{\nu\pi}{n+1}, \quad \nu = 1, \ldots, n, \tag{2.3.46}$$

respectively.

Taking $x_{n,v} = \cos\theta_v$ and using the asymptotic formula (2.3.39) we can get an asymptotic estimate of the Christoffel numbers (2.3.44) in the form (cf. Szegő [470, p. 253])

$$\lambda_{n,v}^{(\alpha,\beta)} \cong \frac{2^{\alpha+\beta+1}}{nk(\theta_v)^2} = \frac{2^{\alpha+\beta+1}\pi}{n}\left(\sin\frac{\theta_v}{2}\right)^{2\alpha+1}\left(\cos\frac{\theta_v}{2}\right)^{2\beta+1}. \quad (2.3.47)$$

For $\alpha = \beta = -1/2$ the symbol $\cong$ can be replaced by $=$, according to (2.3.45). Also, the same is true in the Chebyshev case of the second kind if we replace n by $n+1$ (see (2.3.46)).

2.3.5 Generalized Laguerre Polynomials

For the generalized Laguerre polynomials $L_n^\alpha(x)$, which are orthogonal on $(a,b) = (0,+\infty)$ with respect to the weight function $w(x) = w_\alpha(x) = x^\alpha e^{-x}$ ($\alpha > -1$), we have $A(x) = x$ and $B(x) = \alpha + 1 - x$.

The differential equation with a particular polynomial solution $y = L_n^\alpha(x)$ has the form

$$xy'' + (1 - \alpha + x)y' + ny = 0.$$

The Rodrigues type formula for the generalized Laguerre polynomials and their explicit representation are

$$L_n^\alpha(x) = \frac{x^{-\alpha}e^x}{n!} \cdot \frac{d^n}{dx^n}\left(x^{n+\alpha}e^{-x}\right)$$

and

$$L_n^\alpha(x) = \sum_{v=0}^n \binom{n+\alpha}{n-v}\frac{(-x)^v}{v!},$$

respectively. Notice that $L_n^\alpha(0) = \binom{n+\alpha}{n}$.

These polynomials satisfy the following recurrence relation

$$(n+1)L_{n+1}^\alpha(x) = (2n + \alpha + 1 - x)L_n^\alpha(x) - (n+\alpha)L_{n-1}^\alpha(x), \quad (2.3.48)$$

with $L_0^\alpha(x) = 1$ and $L_1^\alpha(x) = \alpha + 1 - x$.

The norm, the leading coefficient k_n and the coefficient r_n in the expansion $L_n^\alpha(x) = k_n(x^n + r_n x^{n-1} + \cdots)$ are

$$\|L_n^\alpha\|^2 = \frac{\Gamma(n+\alpha+1)}{n!}, \quad k_n = \frac{(-1)^n}{n!}, \quad r_n = -n(n+\alpha).$$

According to Theorem 2.3.3, (2.3.17) and (2.3.19), we get the following relations:

$$\frac{d}{dx}L_n^\alpha(x) = -L_{n-1}^{\alpha+1}(x), \quad x\frac{d}{dx}L_n^\alpha(x) = nL_n^\alpha(x) - (n+\alpha)L_{n-1}^\alpha(x),$$

2.3 Classical Orthogonal Polynomials

$$L_n^\alpha(x) = L_n^{\alpha+1}(x) - L_{n-1}^{\alpha+1}(x).$$

An interesting integral representation of the Laguerre polynomials can be given in terms of the Bessel functions (cf. Szegő [470, p. 103])

$$x^{\alpha/2}e^{-x}L_n^\alpha(x) = \frac{1}{n!}\int_0^{+\infty} e^{-t}t^{n+\alpha/2}J_\alpha(2\sqrt{xt})dt, \quad \alpha > -1.$$

Putting $\hat{L}_n^\alpha(x) = (-1)^n n! L_n^\alpha(x)$ in (2.3.48), we get the three-term recurrence relation for the monic generalized Laguerre polynomials

$$\hat{L}_{n+1}^\alpha(x) = [x - (2n+\alpha+1)]\hat{L}_n^\alpha(x) - n(n+\alpha)\hat{L}_{n-1}^\alpha(x).$$

Thus, the recursion coefficients are: $\alpha_n = 2n + \alpha + 1$ ($n \geq 0$) and $\beta_n = n(n+\alpha)$ ($n \geq 1$), with $\beta_0 = \mu_0 = \int_0^{+\infty} x^{\alpha+1}e^{-x}\,dx = \Gamma(\alpha+1)$.

The corresponding coefficients in the relation for the orthonormal polynomials $p_n(w_\alpha; x) = (-1)^n L_n^\alpha(x)/\|L_n^\alpha\|$ are $b_n = \sqrt{n(n+\alpha)}$, $a_n = 2n + \alpha + 1$, and their leading coefficients $\gamma_n = \gamma_n(w_\alpha) = [n!\Gamma(n+\alpha+1)]^{-1/2}$.

2.3.5.1 Zeros

Let $x_k = x_{n,k}$ ($k = 1,\ldots,n$) be the zeros of the generalized Laguerre polynomial $L_n^\alpha(x)$ ordered in an increasing sequence. Then (cf. Szegő [470, p. 127])

$$x_k > \frac{(j_k/2)^2}{n + (\alpha+1)/2}, \quad \nu = 1,\ldots,n,$$

where j_k is the k-th positive zero of $J_\alpha(z)$. For a fixed k, we have $\lim_{n\to+\infty} nx_{n,k} = (j_k/2)^2$. A slightly better bound for the largest zero can be proved (see [470, p. 128])

$$x_n < 2n + \alpha + 1 + \left[(2n+\alpha+1)^2 + \frac{1}{4} - \alpha^2\right]^{1/2} \cong 4n.$$

Also, we have

$$\frac{C}{n} \leq x_1 < x_2 < \cdots < x_n < 4n + 2\alpha + 2 - C\sqrt[3]{4n}, \tag{2.3.49}$$

and

$$x_k = x_{n,k} = C_{n,k}\frac{(k+1)^2}{n}, \quad k = 1,\ldots,n, \tag{2.3.50}$$

where $(3\pi/16)^2 < C_{n,k} < 4$ (cf. [470, p. 129]).

Freud [135, 136] (see also Nevai [374] and Joó [224]) proved that

$$\Delta x_k = x_{k+1} - x_k \sim \sqrt{\frac{x_k}{4n - x_k}}, \quad k = 1,\ldots, n-1.$$

Defining

$$\varphi_n(x) := \sqrt{\frac{x+1/n}{|4n-x|+(4n)^{1/3}}} \qquad (2.3.51)$$

we can see that also

$$\Delta x_k \sim \varphi_n(x_k), \quad k=1,\ldots,n-1. \qquad (2.3.52)$$

According to (2.3.50) we deduce that, for $x_k \leq x \leq x_{k+1}, k=1,\ldots,n-1$,

$$\varphi_n(x_k) \sim \varphi_n(x) \sim \varphi_n(x_{k+1})$$

uniformly in k and n.

As in the case of the Jacobi polynomials, a similar formula holds for the generalized Laguerre polynomials. Namely, for an arbitrary real α and an arbitrary complex z we have (cf. Szegő [470, p. 169])

$$\lim_{n \to +\infty} n^{-\alpha} L_n^\alpha(z/n) = z^{-\alpha/2} J_\alpha\left(2z^{1/2}\right),$$

where $J_\alpha(z)$ is the Bessel function of order α. This formula holds uniformly if z is bounded.

2.3.5.2 Inequalities

Classical estimates for the generalized Laguerre polynomials, for each $n \in \mathbb{N}_0$ and $x \geq 0$, like

$$|L_n^\alpha(x)| \leq \frac{(\alpha+1)_n}{n!} e^{x/2} \qquad (\alpha \geq 0), \qquad (2.3.53)$$

$$|L_n^\alpha(x)| \leq \left(2 - \frac{(\alpha+1)_n}{n!}\right) e^{x/2} \qquad (-1 < \alpha \leq 0), \qquad (2.3.54)$$

were established by Szegő (cf. Abramowitz and Stegun [1, p. 786]). The estimate (2.3.54) has been improved by Rooney [420] in the following way

$$|L_n^\alpha(x)| \leq 2^{-\alpha} q_n e^{x/2} \qquad (\alpha \leq -1/2),$$

$$|L_n^\alpha(x)| \leq \sqrt{2} \, \frac{q_n(\alpha+1)_n}{\left(\frac{1}{2}\right)_n} e^{x/2} \qquad (\alpha \geq -1/2),$$

where $q_n = 2^{-n-1/2}\sqrt{(2n)!}/n!$ and $q_n \sim 1/\sqrt[4]{4\pi n}$, when $n \to +\infty$.

Using the representation

$$L_n^\alpha(x) = \frac{(-1)^n}{(2\alpha+1)_n \Gamma(\alpha+1)} \int_0^{+\infty} (t+x)^n C_n^{\alpha+1/2}\left(\frac{x-t}{x+t}\right) t^\alpha e^{-t} \, dt,$$

2.3 Classical Orthogonal Polynomials

which holds for $\alpha > -1/2$ and $x \geq 0$, Lewandowski and Szynal [263] improved (2.3.54) in the form

$$|L_n^\alpha(x)| \leq \frac{(\alpha+1)_n}{n!} \sigma_n^{(\alpha)}(\exp x) \qquad (\alpha \geq -1/2,\ x \geq 0), \qquad (2.3.55)$$

where σ_n^α, $\alpha > -1$, denotes the Cesàro mean of the formal series $\sum_{k=0}^{+\infty} c_k$, defined by

$$\sigma_n^\alpha\left(\sum_{k=0}^{+\infty} c_k\right) = \frac{n!}{(\alpha+1)_n} \sum_{k=0}^{n} \frac{(\alpha+1)_{n-k}}{(n-k)!} c_k.$$

For example, for $\alpha = 0$ the last estimate reduces to

$$|L_n(x)| \leq \sigma_n^{(0)}(\exp x) = 1 + \frac{x}{1!} + \frac{x^2}{2!} + \cdots + \frac{x^n}{n!} \qquad (x \geq 0).$$

Recently, the bound (2.3.55) has been improved by Michalska and Szynal [323] in the form

$$|L_n^\alpha(x)| \leq \frac{(\alpha+1)_n}{n!}\left\{\sigma_n^{(\alpha)}(\exp x) - A_n(\alpha)\frac{4x}{n+\alpha}\sigma_{n-2}^{(\alpha+1)}(\exp x)\right\},$$

where

$$A_n(\alpha) = 1 - \frac{\Gamma(\alpha+1)\Gamma((n+1)/2)}{\sqrt{\pi}\,\Gamma(n/2+\alpha+1)}$$

and $\alpha \geq -1/2$, $x \geq 0$, $n \geq 2$.

An important estimate for the orthonormal generalized Laguerre polynomials $p_n(w_\alpha;x)\ (=(-1)^n L_n^\alpha(x)/\|L_n^\alpha\|)$ is given in [19]:

$$\sqrt{w_\alpha(x)}\,|p_n(w_\alpha;x)| \qquad (2.3.56)$$

$$\leq C \begin{cases} n^{-1/4}(\nu - x)^{-1/4}, & \text{if } \delta n \leq x \leq \nu - \nu^{1/3}, \\ n^{-1/3}, & \text{if } \nu - \nu^{1/3} \leq x \leq \nu + \nu^{1/3}, \\ n^{-1/4}(x - \nu)^{-1/4}\exp(-\eta(x-\nu)^{3/2}\nu^{-1/2}), & \text{if } \nu + \nu^{1/3} < x < (1+\lambda)\nu, \\ \exp(-\xi x), & \text{if } x \geq (1+\lambda)\nu, \end{cases}$$

where $\nu = 4n + 2\alpha + 2$, $0 < \eta < 2/3$, $0 < \xi < 1/2$, and δ and λ are sufficiently small but fixed positive numbers.

If for any $x \in [0, 4n]$, we denote by $x_d = x_{d(x)}$ a zero of $p_n(w_\alpha)$ closest to x, i.e., $|x - x_d| = \min_{1 \leq k \leq n} |x - x_k|$, then following [301], we have

$$w_\alpha(x)p_n(w_\alpha;x)^2\left(x+\frac{1}{n}\right)^{\alpha+1/2}\sqrt{|4n-x|+(4n)^{1/3}} \sim \left(\frac{x-x_d}{x_d - x_{d\pm 1}}\right)^2,$$

as well as

$$\sqrt{w_\alpha(x)}\,|p_n(w_\alpha;x)| \leq \frac{C}{\sqrt[4]{x(|4n-x|+(4n)^{1/3})}}.$$

2.3.5.3 Christoffel Function and Christoffel Numbers

Since

$$G_n^{(\alpha)} = \frac{\sqrt{\beta_n}}{\|L_n^\alpha\|\|L_{n-1}^\alpha\|} = \frac{n!}{\Gamma(n+\alpha)},$$

as in the Jacobi case, we find

$$K_{n-1}^{(\alpha)}(x,x) = G_n^{(\alpha)}\left(-L_{n-1}^\alpha(x)\frac{d}{dx}L_n^\alpha(x) + L_n^\alpha(x)\frac{d}{dx}L_{n-1}^\alpha(x)\right),$$

so that, according to (2.1.30), the corresponding Christoffel function becomes

$$\lambda_n^{(\alpha)}(x) = \lambda_n(w_\alpha;x) = \left[K_{n-1}^{(\alpha)}(x,x)\right]^{-1}. \tag{2.3.57}$$

Putting $x = x_{n,\nu}$ (a zero of $L_n^\alpha(x)$) in (2.3.57) we get the Christoffel numbers, $\lambda_{n,\nu}^{(\alpha)} = \lambda_n^{(\alpha)}(x_{n,\nu})$. Using the relation (2.3.17) for the generalized Laguerre polynomials $L_n^\alpha(x)$ at the point $x = x_{n,\nu}$, as well as the expression for $G_n^{(\alpha)}$, we get

$$\lambda_{n,\nu}^{(\alpha)} = \lambda_{n,\nu}(w_\alpha) = \frac{\Gamma(n+\alpha+1)}{n!} \cdot \frac{1}{\left[\frac{d}{dx}L_n^\alpha(x_{n,\nu})\right]^2}. \tag{2.3.58}$$

The following result was proved in [301]:

Theorem 2.3.9 *Let $w_\alpha(x) = x^\alpha e^{-x}$, $\alpha > -1$, and $0 \leq x \leq 4n$. Then, there exists a positive constant $C \neq C(n,x)$ such that*

$$\frac{1}{C}\varphi_n(x) \leq \frac{\lambda_n(w_\alpha;x)}{\left(x+\frac{1}{n}\right)^\alpha e^{-x}} \leq C\varphi_n(x), \tag{2.3.59}$$

where $\varphi_n(x)$ is defined in (2.3.51).

The estimate for the Christoffel numbers (2.3.58) can be obtained from (2.3.59), putting $x = x_{n,\nu}$,

$$\lambda_{n,\nu}(w_\alpha) \sim w_\alpha(x_\nu)\sqrt{\frac{x_\nu}{4n-x_\nu}} \sim w_\alpha(x_\nu)\Delta x_\nu. \tag{2.3.60}$$

2.3.6 Hermite Polynomials

Here we have $(a,b) = (-\infty, +\infty)$, $w(x) = e^{-x^2}$, $A(x) = 1$, $B(x) = -2x$.

The corresponding differential equation $y'' - 2xy' + 2ny = 0$ gives an explicit polynomial solution (Hermite polynomial):

$$H_n(x) = n! \sum_{\nu=0}^{[n/2]} \frac{(-1)^\nu}{\nu!(n-2\nu)!} (2x)^{n-2\nu},$$

which can also be expressed by a Rodrigues type formula

$$H_n(x) = (-1)^n e^{x^2} \frac{d^n}{dx^n}\left(e^{-x^2}\right).$$

The Hermite polynomials satisfy the following relations

$$H_{n+1}(x) = 2x H_n(x) - 2n H_{n-1}(x), \qquad H_0(x) = 1, \ H_1(x) = 2x,$$

and $H_n'(x) = 2n H_{n-1}(x)$. We list a few first Hermite polynomials $H_n(x)$:

$$H_0(x) = 1, \quad H_1(x) = 2x, \quad H_2(x) = 4x^2 - 2, \quad H_3(x) = 8x^3 - 12x,$$
$$H_4(x) = 16x^4 - 48x^2 + 12, \quad H_5(x) = 32x^5 - 160x^3 + 120x,$$
$$H_6(x) = 64x^6 - 480x^4 + 720x^2 - 120.$$

We now mention some useful properties of Hermite polynomials:

$$H_n(-x) = (-1)^n H_n(x), \quad H_{2n}(0) = (-1)^n \frac{(2n)!}{n!}, \quad H_{2n+1}(0) = 0;$$

$$k_n = 2^n, \quad r_n = 0, \quad \|H_n\|^2 = 2^n n! \sqrt{\pi},$$

as well as some integrals of the Hermite polynomials

$$\int_0^x e^{-t^2} H_n(t)\, dt = H_{n-1}(0) - e^{-x^2} H_{n-1}(x),$$

$$\int_{-\infty}^{+\infty} e^{-t^2} H_{2n}(xt)\, dt = \sqrt{\pi}\, \frac{(2n)!}{n!} \left(x^2 - 1\right)^n,$$

$$\int_{-\infty}^{+\infty} t e^{-t^2} H_{2n+1}(xt)\, dt = \sqrt{\pi}\, \frac{(2n+1)!}{n!} x \left(x^2 - 1\right)^n,$$

$$\int_{-\infty}^{+\infty} e^{-t^2} t^n H_n(xt)\, dt = \sqrt{\pi}\, n!\, P_n(x),$$

where P_n is the Legendre polynomial of degree n.

According to Theorem 2.2.12, a connection between the Hermite and generalized Laguerre polynomials can be given in the form

$$H_{2n}(x) = (-1)^n 2^{2n} L_n^{-1/2}(x^2), \quad H_{2n+1}(x) = (-1)^n 2^{2n+1} x L_n^{1/2}(x^2).$$

The coefficients in the three-term recurrence relation for the monic Hermite polynomials $\hat{H}_n(x) = 2^{-n} H_n(x)$ are $\alpha_n = 0$, $\beta_n = n/2$, with $\beta_0 = \mu_0 = \sqrt{\pi}$.

The coefficients in the corresponding relation for the orthonormal Hermite polynomials $h_n(x) = H_n(x)/\|H_n\| = \gamma_n x^n + \cdots$ are given by $b_n = \sqrt{n/2}$, $a_n = 0$, and their leading coefficients by $\gamma_n = 2^{n/2}/\sqrt{n!\sqrt{\pi}}$.

Remark 2.3.3 Taking $0 < A < 1$ and

$$c_n(A) = \frac{\pi\sqrt{A}}{1-A}\left(2\frac{1+A}{1-A}\right)^n n!,$$

and defining the polynomials $h_n^A(z) := (c_n(A))^{-1/2} H_n(z)$, we can prove the following orthogonality relation over the complex plane (cf. [466, 491])

$$\int_{\mathbb{C}} h_n^A(z)\overline{h_m^A(z)} \exp\left[-(1-A)x^2 - \left(\frac{1}{A}-1\right)y^2\right] dx\,dy = \delta_{n,m}, \quad z = x+iy.$$

The Hermite polynomials can be considered as a special case of the Sonin-Markov polynomials or Freud polynomials (see Sects. 2.4.4 and 2.4.5).

2.4 Nonclassical Orthogonal Polynomials

2.4.1 Semi-classical Orthogonal Polynomials

There are several classes of orthogonal polynomials which are in a certain sense close to the classical orthogonal polynomials. For example, when the weight $W(x)$ is the product of a classical weight $w(x)$ and a polynomial, Ronveaux [417] found the second-order differential equation for the corresponding orthogonal polynomials. Ronveaux and Thiry [419] developed a REDUCE package giving such differential equations. The following cases have been studied by Ronveaux and Marcellán [418]:

1° *Rational Case.* $W(x) = R(x)w(x)$, where R is a rational function with poles and zeros outside the support of w;
2° δ *Dirac distribution.*

$$W(x) = w(x) + \sum_{k=1}^{m} w_k \delta(x-x_k),$$

where the positive mass w_k is located at x_k (x_k outside or inside the support of w).

2.4 Nonclassical Orthogonal Polynomials

In both cases, the orthogonal polynomials are *semi-classical* (see Maroni [286]). A nice survey on orthogonal polynomials and spectral theory was given by Everitt and Littlejohn [121]. A continuation of this survey has recently been presented at the *Fifth International Symposium on Orthogonal Polynomials, Special Functions and their Applications* (Patras, 1999) by Everitt, Kwon, Littlejohn, and Wellman [122].

The *semi-classical* (monic) orthogonal polynomials $\{\pi_n(x)\}$ can be defined by the relation

$$A(x)\pi_n'(x) = \sum_{\nu=1}^{r+1} \xi_{n,\nu} \pi_{n+m-\nu}(x), \tag{2.4.1}$$

where $A(x)$ is a polynomial of exactly mth degree, r is a fixed nonnegative integer (not depending on n) and $\xi_{n,\nu}$ are some coefficients. In fact, this definition is one of the possibly different (but equivalent) ways to define semi-classical orthogonal polynomials (for details see [287]).

According to (2.4.1) and (2.3.18), we can see that the classical orthogonal polynomials are a special case of semi-classical polynomials (for $m = r = 2$).

A survey on semi-classical orthogonal polynomials was given by Maroni [288].

2.4.2 Generalized Gegenbauer Polynomials

Let $w(x) = |x|^\gamma (1-x^2)^\alpha$, $\gamma, \alpha > -1$, on $[-1, 1]$. The (monic) generalized Gegenbauer polynomials $W_k^{(\alpha,\beta)}(x)$, $\beta = (\gamma-1)/2$, were introduced by Laščenov [251] (see, also, Chihara [60, pp. 155–156]). Their natural generalization are the generalized Jacobi polynomials (see Sect. 2.4.3).

The generalized Gegenbauer polynomials can be expressed in terms of the Jacobi polynomials,

$$W_{2k}^{(\alpha,\beta)}(x) = \frac{k!}{(k+\alpha+\beta+1)_k} P_k^{(\alpha,\beta)}(2x^2-1),$$

$$W_{2k+1}^{(\alpha,\beta)}(x) = \frac{k!}{(k+\alpha+\beta+2)_k} x P_k^{(\alpha,\beta+1)}(2x^2-1).$$

Notice that $W_{2k+1}^{(\alpha,\beta)}(x) = x W_{2k}^{(\alpha,\beta+1)}(x)$. Their three-term recurrence relation is

$$W_{k+1}^{(\alpha,\beta)}(x) = x W_k^{(\alpha,\beta)}(x) - \beta_k W_{k-1}^{(\alpha,\beta)}(x), \quad k = 0, 1, \ldots,$$

$$W_{-1}^{(\alpha,\beta)}(x) = 0, \quad W_0^{(\alpha,\beta)}(x) = 1,$$

where

$$\beta_{2k} = \frac{k(k+\alpha)}{(2k+\alpha+\beta)(2k+\alpha+\beta+1)}, \quad \beta_{2k-1} = \frac{(k+\beta)(k+\alpha+\beta)}{(2k+\alpha+\beta-1)(2k+\alpha+\beta)},$$

for $k = 1, 2, \ldots$, except when $\alpha + \beta = -1$; then $\beta_1 = (\beta+1)/(\alpha+\beta+2)$.

Remark 2.4.1 Some applications of these polynomials in numerical quadratures and least square approximation with constraint were given in [236] and [351], respectively.

A more general case with

$$w(x) = \begin{cases} |x+c|^\gamma (x^2 - \xi^2)^\beta (1 - x^2)^\alpha, & x \in (-1, -\xi) \cup (\xi, 1), \\ 0, & \text{otherwise,} \end{cases}$$

where $0 < \xi < 1$, $\alpha, \beta > 0$, and $\gamma \in \mathbb{R}$, was studied by Barkov [26]. Here, the measure is supported on two disjoint intervals $[-1, -\xi]$ and $[\xi, 1]$. The special (symmetric) case $c = 0$, $\gamma = 1$, $\alpha = \beta = -1/2$, $\xi = (1-\varrho)/(1+\varrho)$, $0 < \varrho < 1$, arises in the study of the diatomic linear chain (cf. [504] and [166, p. 5]). For such a symmetric case Gautschi [151] obtained the coefficients in the corresponding three-term recurrence relation in an explicit form.

2.4.3 Generalized Jacobi Polynomials

Let $v^{\alpha,\beta}(x) = (1-x)^\alpha (1+x)^\beta$. We consider the generalized Jacobi weight function

$$w(x) = v^{\alpha,\beta}(x) \prod_{\nu=1}^{r} |x - t_\nu|^{\gamma_\nu}, \qquad \alpha, \beta, \gamma_1, \ldots, \gamma_r > -1, \qquad (2.4.2)$$

where $-1 < t_1 < \cdots < t_r < 1$ and put

$$w_n(x) = \left(\sqrt{1+x} + \frac{1}{n}\right)^{2\beta} \prod_{\nu=1}^{r} \left(|x - t_\nu| + \frac{1}{n}\right)^{\gamma_\nu} \left(\sqrt{1-x} + \frac{1}{n}\right)^{2\alpha}. \qquad (2.4.3)$$

The corresponding orthonormal polynomials will be denoted by $p_n(w; x) = \gamma_n(w) x^n + \cdots$, $\gamma_n(w) \sim 2^n$, and their zeros by $x_k = x_{n,k} = \cos \theta_{n,k}$ ($k = 1, \ldots, n$), where

$$-1 < x_{n,1} < x_{n,2} < \cdots < x_{n,n} < 1.$$

Putting $\theta_{n,0} = \pi$ and $\theta_{n,n+1} = 0$, it can be proved the "arc sine distribution" of zeros, i.e.,

$$\theta_{n,k} - \theta_{n,k+1} \sim \frac{1}{n} \qquad (0 \leq k \leq n) \qquad (2.4.4)$$

(see Nevai [375, Theorems 20 & 21]).

Let x_d ($= x_{n,d}$) be a zero closest to x, i.e.,

$$|x - x_d| = \min_{1 \leq k \leq n} |x - x_{n,k}|,$$

2.4 Nonclassical Orthogonal Polynomials

and let $\ell_{n,k}(x)$ be the corresponding fundamental Lagrange polynomial (see (1.3.5)). Then, it can be proven that $\ell_{n,d}(x)^2 \sim 1$ (cf [375, Theorem 33]).

For the Christoffel function $\lambda_n(w;x) = \left(\sum_{k=0}^{n-1} |p_k(w;x)|^2\right)^{-1}$, in this case, we have (cf. Nevai [375, Theorem 28])

$$\lambda_n(w;x) \sim \left(\frac{\sqrt{1-x^2}}{n} + \frac{1}{n^2}\right) w_n(x). \tag{2.4.5}$$

Using (2.4.4) and (2.4.5), we can prove that for the Christoffel numbers the following relation

$$\lambda_{n,k}(w) = \lambda_n(w;x_{n,k}) \sim \frac{\sqrt{1-x_{n,k}^2}}{n} v^{\alpha,\beta}(x_{n,k}) \prod_{v=1}^{r}\left(|x_{n,k}-t_v|+\frac{1}{n}\right)^{\gamma_v} \tag{2.4.6}$$

holds.

An interesting inequality for the orthonormal polynomials $p_n(w;x)$,

$$|p_n(w;x)| \leq \frac{C}{\sqrt{n\lambda_n(w;x)}} \qquad (-1 \leq x \leq 1), \tag{2.4.7}$$

was proved by Badkov [25] (see also Nevai [375, Lemma 29]). Also, the following relation can be found in Nevai [375, Theorem 31]

$$w_n(x_{n,k})\left[p_{n-1}(w;x_{n,k})\right]^2 \sim \sqrt{1-x_{n,k}^2}. \tag{2.4.8}$$

Remark 2.4.2 For the Jacobi polynomials the estimate (2.4.7) reduces to (2.3.41).

We also mention an interesting L^p-inequality, which holds for all $p \geq 1$.

Theorem 2.4.1 *For each fixed $a > 0$ we define*

$$A_n = \left(-1+\frac{a}{n^2}, 1-\frac{a}{n^2}\right) \bigcup_{v=1}^{r}\left[t_k - \frac{a}{n}, t_k + \frac{a}{n}\right].$$

Let $1 \leq p \leq +\infty$. Then, for each $P \in \mathcal{P}_n$, the following inequality

$$\|wP\|_p \leq C\|wP\|_{L^p(A_n)}, \qquad C = C(a) \neq C(P),$$

holds.

All previous results also hold for the generalized Ditzian-Totik weight function defined by

$$w(x) = \prod_{v=0}^{r+1} |x-t_v|^{\gamma_v} W_v\left(|x-t_v|^{\delta_v}\right), \tag{2.4.9}$$

where $-1 = t_0 < t_1 < \cdots < t_r < t_{r+1} = 1$, $\gamma_\nu > -1$, $\nu = 0, 1, \ldots, r+1$, $\delta_0 = \delta_{r+1} = 1/2$, and $\delta_\nu = 1$ for $\nu = 1, \ldots, r$. The function W_ν is either equal to 1 or is a concave modulus of continuity of the first order (i.e., W_ν is a semi-additive, non-negative, continuous and non-decreasing function on $[0, 1]$, with $W_\nu((a+b)/2) \geq (W_\nu(a) + W_\nu(b))/2$ for $a, b \in [0, 1]$). Here, instead of (2.4.3), we should put

$$w_n(x) = \left(\sqrt{1+x} + \frac{1}{n}\right)^{2\gamma_0} W_0\left(\sqrt{1+x} + \frac{1}{n}\right) \prod_{\nu=1}^{r} \left(|x - t_\nu| + \frac{1}{n}\right)^{\gamma_k} W_\nu\left(|x - t_\nu| + \frac{1}{n}\right)$$

$$\times \left(\sqrt{1-x} + \frac{1}{n}\right)^{2\gamma_{r+1}} W_{r+1}\left(\sqrt{1-x} + \frac{1}{n}\right).$$

An example is

$$w(x) = \prod_{\nu=0}^{r+1} |x - t_\nu|^{\gamma_\nu} \log^{\beta_\nu} \frac{e}{|x - t_\nu|}, \quad \beta_\nu \in \mathbb{R}.$$

Several authors considered this type of weights (cf. [25, 290, 306, 314]).

Recently, Vanlessen [492] has studied asymptotic properties of the recurrence coefficients of orthonormal polynomials associated with the generalized Jacobi weight function (2.4.2), introducing an additional real analytic factor h, strictly positive on $[-1, 1]$, i.e.,

$$w(x) = v^{\alpha,\beta}(x) h(x) \prod_{\nu=1}^{r} |x - t_\nu|^{\gamma_\nu}, \quad \alpha, \beta, \gamma_r > -1, \; \gamma_r \neq 0. \quad (2.4.10)$$

The recurrence coefficients in (2.2.3), i.e.,

$$x p_n(x) = b_{n+1} p_{n+1}(x) + a_n p_n(x) + b_n p_{n-1}(x), \quad p_{-1}(x) = 0, \quad (2.4.11)$$

can be written in terms of the solution of the corresponding Riemann–Hilbert (RH) problem for orthogonal polynomials (see Sect. 2.2.2). Using the steepest descent method of Deift and Zhou [87], Vanlessen [492] analyzed the RH problem, and obtained complete asymptotic expansions of b_n and a_n.

Theorem 2.4.2 *The recurrence coefficients b_n and a_n in (2.4.11) for orthonormal polynomials associated to the generalized Jacobi weight (2.4.10) on $[-1, 1]$ have a complete asymptotic expansion of the form*

$$a_n \sim \sum_{k=1}^{+\infty} \frac{A_k(n)}{n^k}, \quad b_n \sim \frac{1}{2} + \sum_{k=1}^{+\infty} \frac{B_k(n)}{n^k},$$

2.4 Nonclassical Orthogonal Polynomials

as $n \to +\infty$. The coefficients $A_k(n)$ and $B_k(n)$ are explicitly computable for every k, and the coefficients with the $1/n$ term in the expansions are given by

$$A_1(n) = -\frac{1}{2} \sum_{v=1}^{r} \gamma_v \sqrt{1-t_v^2} \cos[(2n+1) \arccos t_v - \Phi_v],$$

$$B_1(n) = -\frac{1}{4} \sum_{v=1}^{r} \gamma_v \sqrt{1-t_v^2} \cos[2n \arccos t_v - \Phi_v],$$

where

$$\Phi_v = \left(\alpha + \frac{1}{2} \gamma_v + \sum_{k=v+1}^{r} \gamma_v \right) \pi - \left(\alpha + \beta + \sum_{k=1}^{r} \gamma_v \right) \arccos t_v$$

$$- \frac{\sqrt{1-t_v^2}}{\pi} \text{P.V.} \int_{-1}^{1} \frac{\log h(t)}{\sqrt{1-t^2}} \frac{dt}{t-t_v}. \qquad (2.4.12)$$

Note that $2na_n \sim 2A_1(n) + \cdots$ and $2n(2b_n - 1) \sim 4B_1(n) + \cdots$ are oscillatory and behave asymptotically like a superposition of r wave functions of the form $R_v \cos(n\omega_v + \varphi_v)$ with amplitudes $|\gamma_v|\sqrt{1-t_v^2}$, frequencies $\omega_v = 2 \arccos t_v$, and phase shifts φ_v which are different for $2A_1(n)$ and $4B_1(n)$. The amplitude R_v depends on the location and the strength of the singularity t_v, while the frequency ω_v depends only on the location of the singularity t_v. As we can see from (2.4.12), the strengths of the other singularities has an influence on the phase shift φ_v.

Remark 2.4.3 The RH approach also gives strong asymptotics of the orthonormal polynomials near the algebraic singularities in terms of the Bessel functions (see [492]).

It is important to mention that if we have no singularities in the weight (2.4.10), i.e., if $\gamma_1 = \cdots = \gamma_r = 0$, all the amplitudes in the wave functions vanish. This means that the terms of order $1/n$ in the expansions of the recurrence coefficients vanish, which is in accordance with the case of the pure Jacobi weight (see (2.3.30)), as well as with the case of the modified Jacobi weight $w(x) = v^{\alpha,\beta}(x) h(x)$, where h is a real analytic and strictly positive function on $[-1, 1]$ (see [242]). Note that these modified Jacobi weights satisfy the Szegő condition (2.2.17), i.e., $w \in \mathcal{S}$ (see Definition 2.2.1). For such modified Jacobi weights, full asymptotic expansions for the monic and orthonormal polynomials outside the interval $[-1, 1]$, for the recurrence coefficients and for the leading coefficients $\gamma_n(w)$ of the orthonormal polynomials $p_n(x; w) = \gamma_n(w) x^n + \cdots$ were obtained in [242].

If $\varrho: \mathbb{C} \setminus [-1, 1] \to \mathbb{C}$ is defined as before in (2.2.20) by $\varrho(z) = z + \sqrt{z^2 - 1}$ and a non-zero analytic function $D: \mathbb{C} \setminus [-1, 1] \to \mathbb{C}$ is introduced by

$$D(z) = \exp \left(\frac{\varrho(z) - z}{2\pi} \int_{-1}^{1} \frac{\log w(x)}{\sqrt{1-x^2}} \frac{dx}{z-x} \right),$$

then for $n \to +\infty$, we have (see [242])

$$2^{-n}\gamma_n(w) \sim \frac{1}{\sqrt{\pi D_\infty}}\left\{1 + \frac{\Gamma_1}{n} + \frac{\Gamma_2}{n^2} + \cdots\right\}$$

and

$$\frac{p_n(w;z)}{\varrho(z)^n} \sim \frac{\varrho(z)^{1/2}}{\sqrt{2\pi}(z^2-1)^{1/4}D(z)}\left\{1 + \frac{P_1(z)}{n} + \frac{P_2(z)}{n^2} + \cdots\right\}, \quad z \in \mathbb{C}\setminus[-1,1],$$

where

$$D_\infty = \lim_{z\to\infty} D(z) = \exp\left(\frac{1}{2\pi}\int_{-1}^1 \frac{\log w(x)}{\sqrt{1-x^2}}dx\right),$$

and the coefficients Γ_k and the functions $P_k(z)$ are explicitly computable. For example,

$$\Gamma_1 = -\frac{4\alpha^2-1}{16} - \frac{4\beta^2-1}{16}, \quad P_1(z) = -\frac{4\alpha^2-1}{16}\frac{\varrho(z)+1}{\varrho(z)-1} - \frac{4\beta^2-1}{16}\frac{\varrho(z)-1}{\varrho(z)+1}.$$

The functions $P_k(z)$ are analytic on $\mathbb{C}\setminus[-1,1]$.

2.4.4 Sonin-Markov Orthogonal Polynomials

Let $w(x) = w^\beta(x) = |x|^\beta e^{-x^2}$ ($\beta > -1$) and let $\{p_n(w^\beta)\}$ denote the corresponding system of orthonormal polynomials on $\mathbb{R}$ with positive leading coefficients. These polynomials are known as the *Sonin-Markov* or *generalized Hermite polynomials*.

Putting $\alpha = (\beta-1)/2$ and

$$c_n = (-1)^n\sqrt{\frac{n!}{\Gamma(n+1+\alpha)}}, \quad d_n = (-1)^n\sqrt{\frac{n!}{\Gamma(n+2+\alpha)}},$$

according to Theorem 2.2.12, the Sonin-Markov polynomials can be expressed in terms of the generalized Laguerre polynomials in the form (cf. Kis [230])

$$p_{2n}(w^\beta;x) = c_n L_n^\alpha(x^2), \quad p_{2n+1}(w^\beta;x) = d_n x L_n^{\alpha+1}(x^2).$$

Let $x_k = x_{n,k}$ denote the zeros of $p_n(w^\beta;x)$. Then the following relations (see [303])

$$-\sqrt{2n} + \frac{C}{n^{1/6}} < x_1 < \cdots < x_n < \sqrt{2n} - \frac{C}{n^{1/6}}, \quad C \neq C(n),$$

2.4 Nonclassical Orthogonal Polynomials

hold. Furthermore, setting

$$\varphi_n(x) := \frac{1}{\sqrt{2n - x^2 + (2n)^{1/3}}},$$

we have, for $k = 0, 1, \ldots, n$,

$$\Delta x_k := x_{k+1} - x_k \sim \varphi_n(x_k) \sim \frac{1}{\sqrt{2n - x_k^2}}, \qquad -x_0 = x_{n+1} = \sqrt{2n},$$

as well as

$$|x_k| \sim \frac{k}{\sqrt{n}}, \qquad k = [n/2], \ldots, n.$$

Regarding the Christoffel function $\lambda_n(w^\beta; x) = \left(\sum_{k=0}^{n-1} |p_k(w^\beta; x)|^2 \right)^{-1}$ and the Christoffel numbers $\lambda_{n,k}(w^\beta) = \lambda_n(w^\beta; x_k)$, $k = 1, \ldots, n$, Mastroianni and Occorsio [303] proved the following results:

Theorem 2.4.3 *We have*

$$\lambda_n(w^\beta; x) \sim \left(|x| + \frac{1}{\sqrt{n}} \right)^\beta \varphi_n(x) e^{-x^2}, \qquad |x| \leq \sqrt{2n}, \qquad (2.4.13)$$

and

$$\lambda_{n,k}(w^\beta) \sim w^\beta(x_k) \Delta x_k, \qquad k = 1, \ldots, n. \qquad (2.4.14)$$

Theorem 2.4.4 *Let $|x| \leq \sqrt{2n}$ and let x_d be a zero of $p_n(w^\beta; x)$ closest to x, i.e.,*

$$|x - x_d| = \min_{1 \leq k \leq n} |x - x_k|.$$

Then,

$$[p_n(w^\beta; x)]^2 e^{-x^2} \left(|x| + \frac{1}{\sqrt{n}} \right)^\beta \sqrt{2n - x^2 + (2n)^{1/3}} \sim \left(\frac{x - x_d}{x_d - x_{d \pm 1}} \right)^2. \tag{2.4.15}$$

By (2.4.15) we get

$$|p_n(w^\beta; x)| \sqrt{w^\beta(x)} \sqrt[4]{2n - x^2 + (2n)^{1/3}} \leq C, \qquad |x| \leq \sqrt{2n},$$

for some positive constant $C \neq C(n, x, d)$.

2.4.5 Freud Orthogonal Polynomials

As we mentioned in Sect. 2.2.2, Géza Freud was the first who started in the 1960's with an investigation of polynomials orthogonal with respect to certain general weights on the real line. In recent years, a significant progress in the theory of orthogonal polynomials for weights on $\mathbb{R}$ has been made so that these polynomials for certain classes of weights can be treated as ones for weights on finite intervals (cf. [258, 316, 378]). Especially, the book by Levin and Lubinsky [258] contains very recent results and detailed proofs, using the latest available tools and techniques, such as weighted (logarithmic) potential theory (see Saff and Totik [423]).

In this section we consider the *Freud weights*

$$w(x) = W(x)^2 := e^{-2Q(x)}, \tag{2.4.16}$$

where $Q \colon \mathbb{R} \to \mathbb{R}$ is even, convex and of smooth polynomial growth at infinity. A typical example of such weights is

$$w(x) = W_\alpha(x)^2 := \exp\left(-|x|^\alpha\right), \quad \alpha \geq 1. \tag{2.4.17}$$

We denote the orthonormal polynomials for W^2 by

$$p_n(x) = p_n(W^2; x) = \gamma_n x^n + \text{lower degree terms}, \quad \gamma_n = \gamma_n(W^2) > 0, \tag{2.4.18}$$

so that

$$\int_\mathbb{R} p_n(W^2; x) p_m(W^2; x) W(x)^2 \, dx = \delta_{nm}.$$

According to Theorem 2.2.1 the three-term recurrence relation for the system of orthonormal polynomials (2.4.18) has the form

$$x p_n(x) = b_{n+1} p_{n+1}(x) + b_n p_{n-1}(x) \quad (n \geq 0), \tag{2.4.19}$$

where $p_{-1}(x) = 0$ and the coefficient $b_n = b_n(W^2)$ is given by

$$b_n = \int_\mathbb{R} x p_{n-1}(x) p_n(x) W(x)^2 \, dx = \frac{\gamma_{n-1}}{\gamma_n}.$$

Let the zeros of $p_n(W^2; x)$ be indexed in decreasing size, as

$$-\infty < x_{n,n} < x_{n,n-1} < \cdots < x_{n,2} < x_{n,1} < +\infty.$$

2.4.5.1 Mhaskar-Rakhmanov-Saff Number

An important parameter associated with the weight W^2 is the so-called *Mhaskar-Rakhmanov-Saff number* M_n, the positive root of the equation

$$n = \frac{2}{\pi} \int_0^1 \frac{M_n t Q'(M_n t)}{\sqrt{1-t^2}} \, dt. \tag{2.4.20}$$

2.4 Nonclassical Orthogonal Polynomials

It was independently defined by Rakhmanov [405] and Mhaskar and Saff [318, 319].

For example, for the weight given by (2.4.17), we get

$$M_n = C(\alpha) n^{1/\alpha}, \quad C(\alpha) = \left(\frac{2^{\alpha-1} \Gamma(\alpha/2)^2}{\Gamma(\alpha)} \right)^{1/\alpha}. \quad (2.4.21)$$

In the standard Hermite case ($\alpha = 2$), this number becomes $M_n = \sqrt{2n}$.

It turns out that $p_n(W^2; x)$ behaves on the interval $[-M_n, M_n]$ much like an orthonormal polynomial for a weight from Szegő's class on $[-1, 1]$ (see Definition 2.2.1). The zeros of $p_n(W^2; x)$ lie inside, or close to, $[-M_n, M_n]$ and have a specific asymptotic distribution there. An important identity for an arbitrary polynomial $P \in \mathcal{P}_n$ in the uniform norm,

$$\|PW\|_{\mathbb{R}} = \|PW\|_{[-M_n, M_n]}, \quad (2.4.22)$$

was established by Mhaskar and Saff [319], proving also that M_n is asymptotically the smallest number for which the identity (2.4.22) holds. A similar investigation in the L_p-norm was made in [320].

2.4.5.2 Basic Properties of Freud Polynomials

Levin and Lubinsky [257] (see also [258]) studied in detail the sequence of orthonormal polynomials $p_n(W^2; x)$ on $\mathbb{R}$, where $Q: \mathbb{R} \to \mathbb{R}$ is even and continuous in $\mathbb{R}$, Q'' is continuous in $(0, +\infty)$ and $Q'(x) > 0$ in $(0, +\infty)$. Furthermore, for some constants A and B, Q satisfies the following condition

$$1 < A \leq \frac{1}{Q'(x)} \left\{ \frac{d}{dx}(xQ'(x)) \right\} \leq B, \quad x \in (0, +\infty). \quad (2.4.23)$$

They obtained several interesting properties of such polynomials. Recently, Kasuga and Sakai [226] have considered the generalized Freud-type weight $W_r(x)^2 = |x|^{2r} e^{-2Q(x)}$, where $r > -1/2$. Their results are similar to those for the Freud weight (2.4.16), obtained by Levin and Lubinsky [257].

We mention here some important properties of the (generalized) Freud polynomials when Q satisfies the previous conditions including (2.4.23) (for details see [257, 258] and [226]).

Theorem 2.4.5 *We assume $pr + 1 > 0$ if $0 < p < +\infty$, and $r \geq 0$ if $p = +\infty$. Let $K > 0$. Then, we have for every $P \in \mathcal{P}_n$*

$$\|PW_r\|_{L^p(\mathbb{R})} \leq C \|PW_r\|_{L^p(|x| \leq M_n(1 - Kn^{-2/3}))}. \quad (2.4.24)$$

This theorem was proved by Kasuga and Sakai [226] and it is an improvement of Bauldry's result [30, Theorem 3.1]. The case $r = 0$ is given by Levin and Lubinsky [257, Theorem 1.8]. The inequality (2.4.24) is called the infinite-finite range inequality. In general, such inequalities show that the norm of weighted polynomials on $\mathbb{R}$ lives on a smaller interval $[-M_n, M_n]$ (or $[M_{-n}, M_n]$ in nonsymmetric cases), where the endpoints are the Mhaskar-Rakhmanov-Saff numbers.

Theorem 2.4.6 *Let $x_j = x_{n,j}$, $j = 1, \ldots, n$, be the zeros of $p_n(W^2; x)$ and M_n be the Mhaskar-Rakhmanov-Saff number defined by (2.4.20).*

1° *Then*
$$-M_n < x_n < \cdots < x_2 < x_1 < M_n;$$

2° *There is a certain positive constant C such that*
$$1 - \frac{x_1}{M_n} \leq Cn^{-2/3};$$

3° *For $j = 2, \ldots, n$,*
$$\Delta x_j := x_{j-1} - x_j \sim \frac{M_n}{n}\left(\max\left\{n^{-2/3}, 1 - \frac{|x_j|}{M_n}\right\}\right)^{-1/2}.$$

Theorem 2.4.7 *Let $J_n := \{x \in \mathbb{R} \mid |x| \leq M_n(1 + Ln^{-2/3})\}$ for a given fixed $L > 0$. Then, for the Christoffel function $\lambda_n(W^2; x)$ we have uniformly for $n \geq 1$ and $x \in J_n$,*
$$\frac{\lambda_n(W^2; x)}{W^2(x)} \sim \frac{M_n}{n}\left(\max\left\{n^{-2/3}, 1 - \frac{|x|}{M_n}\right\}\right)^{-1/2}.$$

Moreover, for all $x \in \mathbb{R}$, and $n \geq 1$,
$$\frac{\lambda_n(W^2; x)}{W^2(x)} \geq C\frac{M_n}{n}\left(\max\left\{n^{-2/3}, 1 - \frac{|x|}{M_n}\right\}\right)^{-1/2},$$

for some $C > 0$.

Theorem 2.4.8 *We have uniformly for $n \geq 1$ and $1 \leq j \leq n$,*
$$\frac{M_n}{n}|p'_n(W^2; x_j)|W(x_j) \sim |p_{n-1}(W^2; x_j)|W(x_j)$$
$$\sim M_n^{-1/2}\left(\max\left\{n^{-2/3}, 1 - \frac{|x_j|}{M_n}\right\}\right)^{1/4}.$$

Theorem 2.4.9 *We have*
$$\sup_{x \in \mathbb{R}} |p_n(W^2; x)|W(x)\left|1 - \frac{|x|}{M_n}\right|^{1/4} \sim M_n^{-1/2}$$

and
$$\sup_{x \in \mathbb{R}} |p_n(W^2; x)|W(x) \sim n^{1/6}M_n^{-1/2}.$$

2.4 Nonclassical Orthogonal Polynomials

For the special weights $\exp(-|x|^\alpha)$, $\alpha > 1$, the last result becomes a fairly complete resolution of a problem of Nevai proposed in 1976.

In 1995 Criscuolo, Della Vecchia, Lubinsky and Mastroianni [67] investigated the functions of the second kind, defined by $f_n(x) = \int_{\mathbb{R}} (p_n(t) W^2(t)/(x-t))\,dt$. This integral is well defined when x is in the complex plane, away from the real line. In the case when x is real, the Cauchy principal value is taken so that f_n becomes a Hilbert transform of $p_n W^2$. The authors obtained bounds of these functions of the second kind in the L_∞ and L_p norms.

Remark 2.4.4 Very recently, Levin and Lubinsky [259] have considered polynomials orthogonal with respect to the weight function $x^{2\rho} e^{-2Q(x)}$ on $[0,d)$, where $d \leq +\infty$, $\rho > -1/2$, and Q satisfies some specific conditions. It is a complete generalization of the Laguerre polynomials (see Sect. 2.3.5).

2.4.5.3 Strong Asymptotics

As in the case of orthogonal polynomials on finite intervals, the Riemann-Hilbert approach in getting strong asymptotics has also been applied to orthogonal polynomials on $\mathbb{R}$. The case with the weight (2.4.17) can be reduced to a problem with the weight function

$$w_\alpha(x) := \exp\left(-\kappa_\alpha |x|^\alpha\right), \quad \text{with} \quad \kappa_\alpha = C(\alpha)^\alpha = \frac{2^{\alpha-1} \Gamma(\alpha/2)^2}{\Gamma(\alpha)}. \qquad (2.4.25)$$

As we can see, the constant κ_α is chosen so that the largest zero of $p_n(w_\alpha; x)$ ($= \gamma_n x^n + \cdots$) satisfies $x_{n,1} n^{-1/\alpha} \to 1$ as $n \to +\infty$ (cf. [318] and [92]). In this case asymptotics of the leading coefficient γ_n, the recurrence coefficient b_n ($a_n = 0$), and zeros of $p_n(w_\alpha; x)$ were derived by Kriecherbauer and McLaughlin in [240] (see also [92] and [317]):

Theorem 2.4.10 *Let $\alpha > 0$ and $\{p_n(w_\alpha; x)\}$ ($p_n(w_\alpha; x) = \gamma_n x^n + \cdots$) be the system of polynomials orthonormal on $\mathbb{R}$ with respect to $w_\alpha(x)$ defined in (2.4.25). Let*

$$c_\alpha := \frac{2}{\pi}(-1)^{n+1} \Gamma(\alpha+1) \left(\frac{\alpha-1}{2\alpha}\right)^{\alpha+1} \left(\sum_{j=0}^{+\infty}(2j+1-\alpha)^{-1} \prod_{\ell=1}^{j} \frac{2\ell-1}{2\ell}\right) \cos\frac{\pi\alpha}{2}.$$

$1°$ *For the leading coefficient γ_n, we have*

$$\gamma_n \sqrt{\pi} n^{(2n+1)/(2\alpha)} e^{-n/\alpha} 2^{-n} = 1 + \left(\frac{\alpha-4}{24\alpha}\right)\frac{1}{n} + r_\alpha^\gamma(n), \qquad (2.4.26)$$

where

$$r_\alpha^\gamma(n) = \begin{cases} O(n^{-2}) & \text{if } 0 < \alpha \le 1/2 \text{ or } \alpha \ge 2, \\ O(n^{-1/\alpha}) & \text{if } 1/2 < \alpha < 1, \\ \dfrac{(-1)^{n+1}}{4n(\log n)^2}(1+o(1)) & \text{if } \alpha = 1, \\ c_\alpha n^{-\alpha} + O(n^{-(2\alpha-1)}) + O(n^{-2}) & \text{if } 1 < \alpha < 2; \end{cases}$$

2° *For the recurrence coefficient b_n, we have*

$$\frac{b_n}{n^{1/\alpha}} = \frac{1}{2} + \begin{cases} O(n^{-2}) & \text{if } 0 < \alpha \le 1/2 \text{ or } \alpha \ge 2, \\ O(n^{-1/\alpha}) & \text{if } 1/2 < \alpha < 1, \\ \dfrac{(-1)^{n+1}}{4n(\log n)^2}(1+o(1)) & \text{if } \alpha = 1, \\ -c_\alpha n^{-\alpha} + O(n^{-(2\alpha-1)}) + O(n^{-2}) & \text{if } 1 < \alpha < 2; \end{cases}$$

3° *For $\alpha \ge 1$, the zeros $x_{n,k}$, $k = 1, \ldots, n$, of $p_n(w_\alpha; x)$ satisfy*

$$\frac{x_{n,k}}{n^{1/\alpha}} = 1 - (2\alpha^2)^{-1/3}\frac{\iota_k}{n^{2/3}} + O(n^{-1}), \qquad (2.4.27)$$

where $-\iota_k$ is the k-th zero of the Airy function[4] *Ai.*

Similar investigations for $w(x) = e^{-Q(x)}$ on the real line, where

$$Q(x) = \sum_{k=0}^{2m} q_k x^k, \quad q_{2m} > 0, \ m > 0,$$

and for varying weights $w(x) = w_n(x) = e^{-nV(x)}$, where $V: \mathbb{R} \to \mathbb{R}$ is an arbitrary real analytic function satisfying $\lim_{|x| \to +\infty} V(x)/\log(1+x^2) = +\infty$, were given in [91] and [88, 90], respectively.

Statement 2° of Theorem 2.4.10 contains the well-known Freud conjecture, $b_n n^{-1/\alpha} \to 1/2$ as $n \to +\infty$. This limit was first established by Magnus [280] for even integers α, and by Lubinsky, Mhaskar, and Saff [276] for an arbitrary $\alpha > 0$. Moreover, these authors proved that $b_n/M_n \to 1/2$ (and $a_n/M_n \to 0$ in nonsymmetric cases) for the weights $W(x) = g(x)\exp(-Q(x))$ on $\mathbb{R}$, where $g(x)$ is a 'generalized Jacobi factor',[5] $Q(x)$ satisfies various restrictions, and M_n is the corresponding Mhaskar-Rakhmanov-Saff number. Several applications of Freud's conjecture were discussed by Nevai [378].

Also, asymptotics for the leading coefficients γ_n were obtained by Lubinsky and Saff [275] and by Rakhmanov [406] (see also Totik [476]).

---

[4]The Airy function is defined as in [1, 10.4]. Note, that the function Ai is uniquely determined as the solution of $(d^2/dz^2)Ai(z) = zAi(z)$, satisfying $\lim_{x \to \infty} Ai(x)\sqrt{4\pi}\,x^{1/4}\exp(2x^{3/2}/3) = 1$.
[5]$g(x) := \prod_{j=1}^N |x - z_j|^{\Delta_j}$, $N \ge 1$; $z_1, \ldots, z_N$ are distinct complex numbers, $\Delta_1, \ldots, \Delta_N \in \mathbb{R}$, and, for each real z_j, the corresponding $\Delta_j > -1/2$.

2.4.6 Orthogonal Polynomials with Respect to Abel, Lindelöf, and Logistic Weights

There are three interesting weight functions w_ν, $\nu = 1, 2, 3$, on $\mathbb{R}$ for which we explicitly know the recursion coefficients α_n^ν and β_n^ν for the corresponding (monic) orthogonal polynomials $\pi_n^\nu(\cdot) = \pi_n(w_\nu; \cdot)$,

$$\pi_{n+1}^\nu(x) = (x - \alpha_n^\nu)\pi_n^\nu(x) - \beta_n^\nu \pi_{n-1}^\nu(x), \qquad n = 0, 1, 2, \ldots, \tag{2.4.28}$$

where $\pi_{-1}^\nu(x) = 0$ and $\pi_0^\nu(x) = 1$. These weights are known as Abel, Lindelöf, and logistic weights, and they are defined on $\mathbb{R}$ by

$$w_1(x) = \frac{x}{e^{\pi x} - e^{-\pi x}}, \quad w_2(x) = \frac{1}{e^{\pi x} + e^{-\pi x}}, \quad w_3(x) = \frac{e^{-x}}{(1+e^{-x})^2},$$

respectively. Since the weight functions are even, it is clear that $\alpha_n^\nu = 0$. The corresponding coefficients β_n^ν are

$$\beta_n^1 = \frac{n(n+1)}{4}, \qquad \beta_n^2 = \frac{n^2}{4}, \qquad \beta_n^3 = \frac{n^4 \pi^2}{4n^2 - 1}.$$

Notice that $\beta_n^\nu = O(n^2)$, when $n \to +\infty$. Some additional information on these polynomials can be found in [69–71] and [363].

2.4.7 Strong Non-classical Orthogonal Polynomials

A system of orthogonal polynomials for which the recursion coefficients are not known explicitly will be said to be *strong non-classical* orthogonal polynomials. In such cases there are a few known approaches to compute the first n coefficients α_k, β_k, $k = 0, 1, \ldots, n-1$. These then allow us to compute all orthogonal polynomials of degree $\leq n$ by a straightforward application of the three-term recurrence relation (2.2.4).

One of the approaches for the numerical construction of the monic orthogonal polynomials $\{\pi_k(x)\}$ is the *method of moments*, or more precisely, the *modified Chebyshev algorithm*.

The second method makes use of the explicit representations

$$\alpha_k = \frac{(x\pi_k, \pi_k)}{(\pi_k, \pi_k)} \quad (k \geq 0), \quad \beta_0 = (\pi_0, \pi_0), \quad \beta_k = \frac{(\pi_k, \pi_k)}{(\pi_{k-1}, \pi_{k-1})} \quad (k \geq 1),$$

in terms of the inner product $(.\,,.)$. The method is known as the *Stieltjes procedure*. Using a discretization of the inner product by some appropriate quadrature

$$(f, g) \approx (f, g)_N = \sum_{k=1}^{N} w_k f(x_k) g(x_k), \quad w_k > 0,$$

we get a very efficient method. In the sequel we refer to this method as the *discretized Stieltjes-Gautschi procedure*. An alternative approach is the *Lanczos algorithm*.

In the next section we briefly describe only the modified Chebyshev algorithm and discretized Stieltjes-Gautschi procedure.

2.4.8 Numerical Construction of Orthogonal Polynomials

2.4.8.1 Modified Chebyshev Algorithm

Let $\{\pi_k(x)\}_{k \in \mathbb{N}_0}$ be a system of the monic polynomials orthogonal with respect to the measure $d\mu(x)$ on the real line and let $\mu_k = \int_{\mathbb{R}} x^k \, d\mu(x)$, $k \in \mathbb{N}_0$, be the corresponding moments. The first $2n$ moments $\mu_0, \mu_1, \ldots, \mu_{2n-1}$ uniquely determine the first n recurrence coefficients $\alpha_k(d\mu)$ and $\beta_k(d\mu)$, $k = 0, 1, \ldots, n-1$, in (2.2.4), i.e.,

$$\pi_{k+1}(x) = (x - \alpha_k)\pi_k(x) - \beta_k \pi_{k-1}(x), \qquad k = 0, 1, 2, \ldots, \qquad (2.4.29)$$

where $\pi_{-1}(x) = 0$ and $\pi_0(x) = 1$. However, the corresponding moment map $\mathbb{R}^{2n} \to \mathbb{R}^{2n}$, defined by (see Remark 2.2.1)

$$[\mu_0 \ \mu_1 \ \mu_2 \ \ldots \ \mu_{2n-1}]^\mathrm{T} \mapsto [\alpha_0 \ \beta_0 \ \alpha_1 \ \beta_1 \ \ldots \ \alpha_{n-1} \ \beta_{n-1}]^\mathrm{T},$$

is severely ill-conditioned when n is large. Namely, this map is very sensitive with respect to small perturbations in the moment information (the first $2n$ moments). An analysis of such maps in detail can be found in the recent book of Gautschi [166, Chap. 2]. Sometimes, using the so-called *modified moments*

$$m_k = \int_{\mathbb{R}} q_k(x) \, d\mu(x), \qquad k = 0, 1, \ldots, \qquad (2.4.30)$$

where $\{q_k(x)\}_{k \in \mathbb{N}_0}$ ($\deg q_k(x) = k$) is a given system of polynomials chosen to be close in some sense to the desired orthogonal polynomials $\{\pi_k\}_{k \in \mathbb{N}_0}$, the corresponding map

$$[m_0 \ m_1 \ m_2 \ \ldots \ m_{2n-1}]^\mathrm{T} \mapsto [\alpha_0 \ \beta_0 \ \alpha_1 \ \beta_1 \ \ldots \ \alpha_{n-1} \ \beta_{n-1}]^\mathrm{T},$$

can become remarkably well-conditioned, especially for measures supported on a finite interval.

In this section we present an algorithm (see [147, §2.4], and also [166, pp. 76–78]), known as the *modified Chebyshev algorithm*. In fact, it is a generalization from ordinary to modified moments of an algorithm due to Chebyshev.

We suppose that the polynomials q_k are also monic and satisfy a three-term recurrence relation

$$q_{k+1}(x) = (x - a_k)q_k(x) - b_k q_{k-1}(x), \qquad k = 0, 1, 2, \ldots, \qquad (2.4.31)$$

2.4 Nonclassical Orthogonal Polynomials

where $q_{-1}(x) = 0$ and $q_0(x) = 1$, with given coefficients $a_k \in \mathbb{R}$ and $b_k \geq 0$. In the case $a_k = b_k = 0$, (2.4.31) gives the monomials $q_k(x) = x^k$, and m_k reduce to the ordinary moments μ_k ($k \in \mathbb{N}_0$).

Following Gautschi [166, pp. 76–78], we introduce the "mixed moments"

$$\sigma_{k,i} = (\pi_k, q_i) = \int_{\mathbb{R}} \pi_k(x) q_i(x) \, d\mu(x), \quad k, i \geq -1. \tag{2.4.32}$$

Then, $\sigma_{0,i} = m_i$, $\sigma_{-1,i} = 0$ and, because of the orthogonality, $\sigma_{k,i} = 0$ for $k > i$. Also, we take $\sigma_{0,0} = m_0 =: \beta_0$.

According to (2.4.29) we have

$$(\pi_{k+1}, q_i) = (\pi_k, x q_i) - \alpha_k (\pi_k, q_i) - \beta_k (\pi_{k-1}, q_i).$$

Now, using (2.4.31), i.e., $x q_i(x) = q_{i+1}(x) + a_i q_i(x) + b_i q_{i-1}(x)$, we get

$$\sigma_{k+1,i} = \sigma_{k,i+1} - (\alpha_k - a_i)\sigma_{k,i} - \beta_k \sigma_{k-1,i} + b_i \sigma_{k,i-1}. \tag{2.4.33}$$

Finally, putting $i := k - 1$ and $i := k$, we obtain from (2.4.33)

$$0 = \sigma_{k,k} - \beta_k \sigma_{k-1,k-1} \quad \text{and} \quad 0 = \sigma_{k,k+1} - (\alpha_k - a_k)\sigma_{k,k} - \beta_k \sigma_{k-1,k},$$

respectively, i.e.,

$$\alpha_k = a_k + \frac{\sigma_{k,k+1}}{\sigma_{k,k}} - \frac{\sigma_{k-1,k}}{\sigma_{k-1,k-1}}, \quad \beta_k = \frac{\sigma_{k,k}}{\sigma_{k-1,k-1}} \quad (k \geq 1).$$

For $k = 0$, we have

$$\alpha_0 = a_0 + \frac{m_1}{m_0}, \quad \beta_0 = m_0.$$

According to the previous considerations we can state the following algorithm:

1° *Initialization:* $\alpha_0 = a_0 + m_1/m_0$, $\beta_0 = m_0$, and

$$\sigma_{0,i} = m_i \ (0 \leq i \leq 2n - 1), \quad \sigma_{-1,i} = 0 \ (1 \leq i \leq 2n - 2);$$

2° *Continuation (if $n > 1$): for $k = 1, 2, \ldots, n - 1$ do*

$$\sigma_{k,i} = \sigma_{k-1,i+1} - (\alpha_{k-1} - a_i)\sigma_{k-1,i} - \beta_{k-1}\sigma_{k-2,i} + b_i \sigma_{k-1,i-1}$$

$$(i = k, k+1, \ldots, 2n - k - 1),$$

$$\alpha_k = a_k + \frac{\sigma_{k,k+1}}{\sigma_{k,k}} - \frac{\sigma_{k-1,k}}{\sigma_{k-1,k-1}}, \quad \beta_k = \frac{\sigma_{k,k}}{\sigma_{k-1,k-1}}.$$

Thus, this algorithm requires as input the $2n$ modified moments m_i, $i = 0, 1, \ldots, 2n - 1$, given by (2.4.30), and the coefficients a_k and b_k, $k = 0, 1, \ldots, 2n - 2$, in the recurrence relation (2.4.30). Then, it produces the desired coefficients α_k and β_k, $k = 0, 1, \ldots, n - 1$.

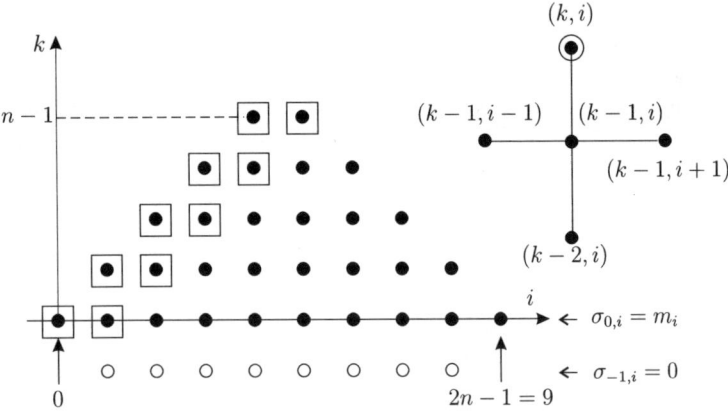

Fig. 2.4.1 The scheme of the modified Chebyshev algorithm

Figure 2.4.1 displays the trapezoidal array of the mixed moments (2.4.32) and the computing stencil showing that the circled entry is computed in terms of the four entries below. The entries in the boxes are those used to compute the coefficients α_k and β_k. Several illustrations and examples of this algorithm can be found in the book [166, Chap. 2].

A useful collection of modified moments, including recurrence relations for their calculation, was given by Piessens and Branders [398].

2.4.8.2 Discretized Stieltjes-Gautschi Procedure

First we introduce a *discrete measure* $d\mu_N$ as

$$d\mu_N(x) = \sum_{k=1}^{N} w_k \delta(x - x_k)\,dx, \quad x_1 < x_2 < \cdots < x_N, \tag{2.4.34}$$

where δ is the Dirac delta function, and usually $w_k > 0$. The support of $d\mu_N$ consists of its N support points $x_1, x_2, \ldots, x_N$. In general, x_k and w_k depend on N.

Let the measure $d\mu(x)$ be given as before and let

$$(p, q) = (p, q)_{d\mu} = \int_{\mathbb{R}} p(x)q(x)\,d\mu(x) \quad (p, q \in \mathcal{P}).$$

The basic idea of discretization methods consists in the approximation of the given measure $d\mu(x)$ on $\mathbb{R}$ by a discrete N-point measure (2.4.34) and the computation of the recursion coefficients $\alpha_{k,N} = \alpha_{k,N}(d\mu_N)$ and $\beta_{k,N} = \beta_{k,N}(d\mu_N)$ in the corresponding three-term recurrence relation for monic polynomials orthogonal with respect to this discrete measure,

$$\pi_{k+1,N}(x) = (x - \alpha_{k,N})\pi_{k,N}(x) - \beta_{k,N}\pi_{k-1,N}(x), \quad 0 \le k \le N-1, \tag{2.4.35}$$

2.4 Nonclassical Orthogonal Polynomials

with $\pi_{0,N}(x) = 1$ and $\pi_{-1,N}(x) = 0$. By definition, $\beta_{0,N} = (\pi_{0,N}, \pi_{0,N})_{d\mu_N} = (1,1)_{d\mu_N}$. Of course, for these coefficients Darboux's formulae

$$\alpha_{k,N} = \frac{(x\pi_{k,N}, \pi_{k,N})_{d\mu_N}}{(\pi_{k,N}, \pi_{k,N})_{d\mu_N}} \qquad (0 \le k \le N-1), \tag{2.4.36}$$

$$\beta_{k,N} = \frac{(\pi_{k,N}, \pi_{k,N})_{d\mu_N}}{(\pi_{k-1,N}, \pi_{k-1,N})_{d\mu_N}} \qquad (1 \le k \le N-1) \tag{2.4.37}$$

hold.

Roughly speaking, if the inner product

$$(p,q)_{d\mu_N} := \int_{\mathbb{R}} p(x)q(x)\, d\mu_N(x) = \sum_{k=1}^{N} w_k p(x_k) q(x_k) \quad (p,q \in \mathcal{P}) \tag{2.4.38}$$

converges to $(p,q)_{d\mu}$ as $N \to +\infty$, the approximation $d\mu(x) \approx d\mu_N(x)$ is "well-stated" and it is reasonable to expect that the discrete orthogonal polynomials $\pi_{k,N}(x)$ tend to $\pi_k(x)$ as $N \to +\infty$ (cf. [166, Theorem 2.32] for positive measures on $[-1, 1]$).

Beside this convergence problem, two additional problems appear: (a) how to choose an appropriate discrete measure (2.4.34), (i.e., how to discretize the original measure $d\mu(x)$); (b) how to compute the recursion coefficients $\alpha_{k,N} = \alpha_{k,N}(d\mu_N)$ and $\beta_{k,N} = \beta_{k,N}(d\mu_N)$, given by (2.4.36) and (2.4.37), respectively.

An elegant solution of the second problem can be given by an old Stieltjes' idea from 1884, which is based on a combination of Darboux's formulae with the basic three-term recurrence relation. Thus, we apply Darboux's formulae, with inner products in discretized form (2.4.38), in tandem with the basic linear relation (2.4.35).

Since $\pi_{0,N}(x) = 1$, we can compute $\alpha_{0,N}$ from (2.4.36) for $k = 0$, and $\beta_{0,N} = w_1 + \cdots + w_N$. Having obtained $\alpha_{0,N}$, we then use (2.4.35) with $k = 0$ to compute $\pi_{1,N}(x)$ for $x = x_k$, $k = 1, \ldots, N$. In such a way we completed all values needed to reapply Darboux's formulae (2.4.36) and (2.4.37), with $k = 1$. Now, with $\alpha_{1,N}$ and $\beta_{1,N}$, using (2.4.35) for $k = 1$, we calculate $\pi_{2,N}(x)$ for $x = x_k$, $k = 1, \ldots, N$. Thus, in this way, alternating between Darboux's formulae and the three-term recurrence relation (2.4.35), we can determine all desired coefficients $\alpha_{k,N}, \beta_{k,N}, k \le n-1$.

This Stieltjes-Gautschi procedure is quite effective, at least when $n \ll N$.

Remark 2.4.5 Stieltjes-Gautschi procedure was useful applied in numerical construction of polynomials orthogonal on the radial rays in the complex plane [341], as well as for computing recursive coefficients in the so-called multiple orthogonal polynomials [356].

A good way of discretizing the original measure $d\mu(x)$ (the problem (a) mentioned before) can be obtained by applying suitable quadrature formulae. We mention here only one a relatively general approach in the case $d\mu(x) = w(x)dx$, where w is continuous on $(-1, 1)$, with possible integrable singularities at ± 1 (see Gautschi [166, § 2.2.2]). Namely, we use an approximation of the form

$$\int_{-1}^{1} f(x)w(x)\,dx \approx \sum_{k=1}^{N} A_k f(x_k) w(x_k) \tag{2.4.39}$$

in order to get the discrete inner product (2.4.38)

$$(p,q)_{d\mu_N} = \sum_{k=1}^{N} A_k w(x_k) p(x_k) q(x_k). \tag{2.4.40}$$

Thus, in this case[6] the weights in (2.4.34) (resp. (2.4.38)) are given by $w_k = A_k w(x_k)$, $k = 1, \ldots, N$. As we can see, this approach is realized practically by the quadrature formula $\int_{-1}^{1} g(x)\,dx \approx \sum_{k=1}^{N} A_k g(x_k)$. As a good choice is the well-known *Fejér quadrature formula* with the Chebyshev nodes (see (1.1.18))

$$x_k = x_k^F = \cos\theta_k^F, \quad \theta_k^F = \frac{(2k-1)\pi}{2N}, \quad k = 1, \ldots, N, \tag{2.4.41}$$

which is exact for all algebraic polynomials of degree at most $N - 1$. The corresponding weight coefficients are explicitly known,

$$A_k = A_k^F = \frac{2}{N}\left(1 - 2\sum_{\nu=1}^{[N/2]} \frac{\cos 2\nu\theta_k^F}{4\nu^2 - 1}\right), \quad k = 1, \ldots, N. \tag{2.4.42}$$

Fejér [126] proved that $A_k^F > 0$ for all k, which is in this case sufficient for convergence $\sum_{k=1}^{N} A_k^F g(x_k^F)$ to $\int_{-1}^{1} g(x)\,dx$, $N \to +\infty$, for all continuous functions g (see Sect. 5.1). A similar quadrature with the Chebyshev nodes of the second kind was also proved by Fejér [126].

Remark 2.4.6 If $d\mu(x) = w(x)dx$ is supported on an arbitrary finite interval $[a, b]$, then, using the following linear transformation $x = \varphi(y) = [a + b + (b - a)y]/2$, the problem reduces to previous one, because

$$\int_{a}^{b} f(x)w(x)\,dx = \int_{-1}^{1} f(\varphi(y))w(\varphi(y))\varphi'(y)\,dy.$$

In the case $b = +\infty$, we can apply a bilinear transformation

$$x = \varphi(y) = a + (1 + y)/(1 - y).$$

---

[6] Also, some weighted quadratures (see Sect. 5.1) can be used for approximating the integral in (2.4.39).

2.4 Nonclassical Orthogonal Polynomials

Remark 2.4.7 In order to accelerate the convergence we can use a suitable partition of the interval $[a,b] = \bigcup_{i=1}^{m}[a_i, b_i]$, with

$$a = a_1 < b_1 = a_2 < b_2 = a_3 < \cdots < b_{m-1} = a_m < b_m = b,$$

and then apply the Fejér quadrature to each of subintervals $[a_i, b_i]$ (cf. Gautschi and Milovanović [169]).

In the sequel we mention some non-classical measures $d\mu(x) = w(x)\,dx$ for which the recursion coefficients $\alpha_k(d\mu)$, $\beta_k(d\mu)$, $k = 0, 1, \ldots, n-1$, have been tabulated in the literature and used in the construction of Gaussian quadratures.

1° *One-sided Hermite weight* $w(x) = \exp(-x^2)$ on $[0, c]$, $0 < c \le +\infty$. This distribution $w(x)\,dx$ is known as the *Maxwell (velocity) distribution*. The cases $c = 1, n = 10$ and $c = +\infty, n = 15$ were considered by Steen, Byrne and Gelbard [455] (see also Gautschi [155]).

2° *Logarithmic weight* $w(x) = x^\alpha \log(1/x)$, $\mu > -1$ on $(0, 1)$. Piessens and Branders [399] considered cases when $\alpha = 0, \pm 1/2, \pm 1/3, -1/4, -1/5$ (see also Gautschi [154]).

3° *Airy weight* $w(x) = \exp(-x^3/3)$ on $(0, +\infty)$. The inhomogeneous Airy functions $\mathrm{Hi}(x)$ and $\mathrm{Gi}(x)$, arise in theoretical chemistry (e.g. in harmonic oscillator models for large quantum numbers) and their integral representations (see Lee [253]) are given by

$$\mathrm{Hi}(t) = \frac{1}{\pi}\int_0^{+\infty} w(x)e^{xt}\,dx,$$

$$\mathrm{Gi}(t) = -\frac{1}{\pi}\int_0^{+\infty} w(x)e^{-xt/2}\cos\left(\frac{\sqrt{3}}{2}xt + \frac{2\pi}{3}\right)dx.$$

These functions can effectively be evaluated by the Gaussian quadrature relative to the Airy weight $w(x)$. It needs orthogonal polynomials with respect to this weight. Gautschi [149] computed the recursion coefficients for $n = 15$ with 16 decimal digits after the decimal point (D).

4° *Reciprocal gamma function* $w(x) = 1/\Gamma(x)$ on $(0, +\infty)$. Gautschi [148] determined the recursion coefficients for $n = 40$ with 20 significant decimal digits (S). This function could be useful as a probability density function in reliability theory (see Fransén [131]).

5° *Einstein's* and *Fermi's weight functions* on $(0, +\infty)$,

$$w_1(x) = \varepsilon(x) = \frac{x}{e^x - 1} \quad \text{and} \quad w_2(x) = \varphi(x) = \frac{1}{e^x + 1}.$$

These functions arise in solid state physics. Integrals with respect to the measure $d\mu(x) = \varepsilon(x)^r\,dx$, $r = 1$ and $r = 2$, are widely used in phonon statistics and lattice specific heats and occur also in the study of radiative recombination processes. Similarly, integrals with $\varphi(x)$ are encountered in the dynamics of electrons in metals. For $w_1(x)$, $w_2(x)$, $w_3(x) = \varepsilon(x)^2$ and $w_4(x) = \varphi(x)^2$, Gautschi and Milovanović [169] determined the recursion coefficients α_k and β_k, for $n = 40$ with 25 S, and gave an application of the corresponding Gauss-Christoffel quadratures to the summation of slowly convergent series (see Sects. 5.1.3 and 5.4.1).

6° *The hyperbolic weights* on $(0, +\infty)$,

$$w_1(x) = \frac{1}{\cosh^2 x} \quad \text{and} \quad w_2(x) = \frac{\sinh x}{\cosh^2 x}.$$

The recursion coefficients α_k, β_k, for $n = 40$ with 30 S, were obtained by Milovanović [330–332]. The discretization was based on the Gauss-Laguerre quadrature rule (see Sect. 5.1.3). Such quadratures were used in summation formulas (for details see Sect. 5.4.2). A general approach to summation formulas due to Plana, Lindelöf and Abel was recently given by Dahlquist [69–71].

7° *Modified Bessel functions and Airy function* on $(0, +\infty)$. Gautschi [165] described procedures for the high-precision calculation of the modified Bessel function $K_\nu(x)$, $0 < \nu < 1$, and the Airy function $Ai(x)$, for positive arguments x, as prerequisites for generating Gaussian quadrature rules having these functions as a weight function.

2.5 Weighted Polynomial Approximation

2.5.1 Weighted Functional Spaces, Moduli of Smoothness and K-functionals

In several applications we have to approximate functions f that are not bounded in a finite number of points of the considered interval. Such functions cannot be classified as continuous or L^p-functions, but usually their product with a suitable function u (a *weight function*) succeeds in compensating the irregular parts of f and gives a function fu which is a continuous or L^p-function.

In these cases we study the so-called *weighted approximation*. Several types of weight functions have been introduced in dependence on the kind of irregularity of the functions to be approximated (see [104] for a general approach).

As we know, the most common weight function on the interval $[-1, 1]$ is the Jacobi weight defined as

$$u(x) := (1-x)^\alpha (1+x)^\beta, \qquad \alpha, \beta > -1.$$

Sometimes the notation $v^{\alpha,\beta}(x) := (1-x)^\alpha (1+x)^\beta$ is also used to underline the dependence on the exponents α and β (see Sect. 2.3.4).

For a Jacobi weight u and for $1 \leq p < +\infty$, the space L_u^p is defined by

$$L_u^p := \left\{ f \mid fu \in L^p[-1, 1] \right\}$$

2.5 Weighted Polynomial Approximation

and equipped with the norm

$$\|f\|_{L^p_u} := \|fu\|_p = \left(\int_{-1}^{1} |f(x)|^p u^p(x) dx\right)^{1/p},$$

so that L^p_u is a Banach space.

In the case when $p = +\infty$ and $u = v^{\alpha,\beta}$ has non negative exponents, we set $L^\infty_u \equiv C^0_u$, where C^0_u is defined as

$$C^0_u := \left\{ f \in C(-1,1) \,\Big|\, \lim_{|x| \to 1} |(fu)(x)| = 0 \right\},$$

in the case $\alpha, \beta > 0$, while if $\alpha = 0$ (respectively $\beta = 0$) C^0_u consists of all continuous functions on $(-1, 1]$ (resp. $[-1, 1)$) such that

$$\lim_{x \to -1} (fu)(x) = 0 \quad \left(\text{resp.} \lim_{x \to +1} (fu)(x) = 0\right).$$

Finally, in the case $\alpha = \beta = 0$ (i.e., $u(x) = 1$) we assume $C^0_u = C[-1, 1]$.
The space C^0_u equipped with the norm

$$\|f\|_{C^0_u} := \|fu\|_\infty = \sup_{|x| \le 1} |(fu)(x)|,$$

is a Banach space.

We point out that the limiting conditions required in the definition of C^0_u are necessary for the validity of the Weierstrass theorem in this space, since it is obvious that for each polynomial P, we have

$$\|(f - P)u\|_\infty \ge |(fu)(\pm 1)|,$$

assuming that $u = v^{\alpha,\beta}$ with $\alpha, \beta > 0$ (so that $u(\pm 1) = 0$).

For general $p \in [1, +\infty]$, the well-known subspaces of L^p_u are Sobolev-type spaces, defined as

$$W^r_p(u) := \left\{ f \in L^p_u \,\Big|\, f^{(r-1)} \in AC(-1, 1) \text{ and } \|f^{(r)} \varphi^r u\|_p < +\infty \right\},$$

$r = 1, 2, \ldots,$

with

$$\|f\|_{W^r_p(u)} := \|fu\|_p + \|f^{(r)} \varphi^r u\|_p,$$

$\varphi(x) := \sqrt{1 - x^2}$, and $\alpha, \beta > -1/p$ if $p < +\infty$ and $\alpha, \beta \ge 0$ if $p = +\infty$.

In the L^p_u spaces ($1 \le p \le +\infty$) we can define the *main part modulus of continuity* of order $k \in \mathbb{N}$ as follows (see [98, pp. 59–62])

$$\Omega^k_\varphi(f, t)_{u,p} := \sup_{0 < h \le t} \|(\Delta^k_{h\varphi} f) u\|_{L^p[-1+4k^2h^2, 1-4k^2h^2]}, \quad (2.5.1)$$

where $\varphi(x) := \sqrt{1-x^2}$ and

$$\Delta_{h\varphi}^k f(x) := \sum_{i=0}^{k}(-1)^i \binom{k}{i} f\left(x + \frac{kh}{2}\varphi(x) - ih\varphi(x)\right) \qquad (2.5.2)$$

is the central finite difference of order k and the variable step $h\varphi(x)$.

The *global moduli of smoothness* are given by ([98, Chap. 6])

$$\omega_\varphi^k(f,t)_{u,p} := \Omega_\varphi^k(f,t)_{u,p} + \inf_{P \in \mathcal{P}_{k-1}} \|(f-P)u\|_{L^p[-1,-1+4k^2t^2]}$$

$$+ \inf_{P \in \mathcal{P}_{k-1}} \|(f-P)u\|_{L^p[1-4k^2t^2,1]}, \qquad (2.5.3)$$

where $\mathcal{P}_{k-1}$ denotes the set of all algebraic polynomials of degree at most $k-1$, and, for $\alpha, \beta > 0$,

$$^*\omega_\varphi^k(f,t)_{u,p} := \Omega_\varphi^k(f,t)_{u,p} + \sup_{0<h\leq 2k^2t^2} \|(\overrightarrow{\Delta}_h^k f)u\|_{L^p(-1,-1+4k^2t^2]}$$

$$+ \sup_{0<h\leq 2k^2t^2} \|(\overleftarrow{\Delta}_h^k f)u\|_{L^p[1-4k^2t^2,1)}, \qquad (2.5.4)$$

where $\overrightarrow{\Delta}_h^k$ and $\overleftarrow{\Delta}_h^k$ are the forward and backward differences of order k and step h, given by

$$\overrightarrow{\Delta}_h^k f(x) := \sum_{i=0}^{k}(-1)^i \binom{k}{i} f(x+kh-ih), \qquad (2.5.5)$$

$$\overleftarrow{\Delta}_h^k f(x) := \sum_{i=0}^{k}(-1)^i \binom{k}{i} f(x-ih). \qquad (2.5.6)$$

It can be shown that

$$\omega_\varphi^k(f,t)_{u,p} \sim {}^*\omega_\varphi^k(f,t)_{u,p}. \qquad (2.5.7)$$

Hence in the sequel we assume either (2.5.3) or (2.5.4) as a definition of the φ-modulus $\omega_\varphi^k(f,t)_{u,p}$. Moreover, in the case $k=1$ we set $\omega_\varphi(f,t)_{u,p} := \omega_\varphi^1(f,t)_{u,p}$.

The weighted φ-modulus $\omega_\varphi(f,t)_{u,\infty}$ characterizes the space C_u^0 as follows

$$f \in C_u^0 \iff \lim_{t \to 0} \omega_\varphi(f,t)_{u,\infty} = 0. \qquad (2.5.8)$$

Notice that the obvious inequality (take $u = v^{\alpha,\beta}$ with $\alpha, \beta > 0$)

$$\omega_\varphi(f,t)_{u,\infty} \geq |(fu)(-1)| + |(fu)(1)|$$

2.5 Weighted Polynomial Approximation

shows that the limiting conditions $\lim_{x \to \pm 1} (fu)(x) = 0$ are essential for the equivalence (2.5.8).

Denoting by AC_{loc} the set of all locally absolutely continuous functions on $(-1, 1)$, we define the K-functional of Peetre

$$K_\varphi^k(f, t^k)_{u,p} := \inf_{g^{(k-1)} \in AC_{\text{loc}}} \left\{ \|(f-g)u\|_p + t^k \|g^{(k)} \varphi^k u\|_p \right\} \tag{2.5.9}$$

and the following "main part" of K-functional

$$\widetilde{K}_\varphi^k(f, t^k)_{u,p} := \sup_{0 < h \leq t} \inf_{g^{(k-1)} \in AC_{\text{loc}}} \left\{ \|(f-g)u\|_{L^p(I_{h,k})} + h^k \|g^{(k)} \varphi^k u\|_{L^p(I_{h,k})} \right\}, \tag{2.5.10}$$

where $I_{h,k} := [-1 + 4k^2 h^2,\ 1 - 4k^2 h^2]$.

For such functionals it is possible to state the following interesting result (see [98, p. 60]):

Theorem 2.5.1 *For all $1 \leq p \leq +\infty$ and $f \in L_u^p$ the equivalences*

$$\omega_\varphi^k(f, t)_{u,p} \sim K_\varphi^k(f, t^k)_{u,p} \tag{2.5.11}$$

and

$$\Omega_\varphi^k(f, t)_{u,p} \sim \widetilde{K}_\varphi^k(f, t^k)_{u,p} \tag{2.5.12}$$

hold, where the constants in $\sim$ depend only on $k \in \mathbb{N}$.

The equivalences stated in the previous theorem turn out to be very useful tools in proving the main properties of the moduli of smoothness, as well as for estimating such moduli.

In particular, we point out that $\widetilde{K}_\varphi^k$ is not a K-functional, but, by virtue of (2.5.12), it gives a useful estimate of Ω_φ^k for a wide class of functions. In fact, by (2.5.12), we conclude that

$$\Omega_\varphi^k(f, t)_{u,p} \leq C \sup_{h \leq t} h^k \| f^{(k)} \varphi^k u \|_{L^p(I_{hk})}. \tag{2.5.13}$$

Moreover, using (2.5.12), it is possible to prove that the following properties of $\Omega_\varphi^k(f, t)_{u,p}$ hold (see [98]):

(i) $t_1 \leq t_2 \implies \Omega_\varphi^k(f, t_1)_{u,p} \leq \Omega_\varphi^k(f, t_2)_{u,p}$;
(ii) $\Omega_\varphi^k(f, \lambda t)_{u,p} \leq C \lambda^k \Omega_\varphi^k(f, t)_{u,p}$ $\quad (\lambda > 1)$;
(iii) $\Omega_\varphi^k(f, t)_{u,p} \leq C \Omega_\varphi^{k-1}(f, t)_{u,p}$,
(iv) $\Omega_\varphi^k(f, t)_{u,p} \leq C t^r \Omega_\varphi^{k-r}(f^{(r)}, t)_{u\varphi^r, p}$ $\quad \left(f \in W_r^p(u),\ f^{(r-1)} \in AC_{\text{loc}} \right)$.

The analogous properties for $\omega_\varphi^k(f, t)_{u,p}$ also hold.

By means of the main part modulus $\Omega_\varphi^k(f,t)_{u,p}$ we can define the intermediate spaces of a couple of Sobolev spaces, namely the Besov spaces $B_{r,q}^p(u)$. For all $r > 0$ and $1 \leq p, q \leq +\infty$, such spaces are defined as

$$B_{r,q}^p(u) := \left\{ f \in L_u^p \,\Big|\, \|f\|_{B_{r,q}^p(u)} := \|fu\|_p + \|f\|_{p,q,r,u} < +\infty \right\}, \qquad (2.5.14)$$

where $\|\cdot\|_{p,q,r,u}$ denotes the following semi-norm

$$\|f\|_{u,q,r} := \begin{cases} \left(\displaystyle\int_0^1 \left[\dfrac{\Omega_\varphi^k(f,t)_u}{t^r} \right]^q \dfrac{dt}{t} \right)^{1/q} & \text{if } 1 \leq q < +\infty, \\[2ex] \displaystyle\sup_{t>0} \dfrac{\Omega_\varphi^k(f,t)_u}{t^r} & \text{if } q = +\infty, \end{cases} \qquad (2.5.15)$$

where $k > r$. The Besov spaces $B_{r,q}^p(u)$ were introduced in [99]. They get more and more interest in approximation theory and, as we will see, some useful boundedness properties that are not valid in the whole L_u^p space, can be satisfied if we restrict our considerations to Besov subspaces.

2.5.2 Weighted Best Polynomial Approximation on $[-1, 1]$

Let be $1 \leq p \leq +\infty$ and assume that $u \in L^p$ is a fixed Jacobi weight. Denoting by $\mathcal{P}_n$ the set of all algebraic polynomials of degree at most n, the error of the best polynomial approximation in L_u^p is defined as

$$E_n(f)_{u,p} := \inf_{P \in \mathcal{P}_n} \|(f - P)u\|_p. \qquad (2.5.16)$$

As in the trigonometric case, we can state a Jackson type inequality

$$E_n(f)_{u,p} \leq C \omega_\varphi^k \left(f, \frac{1}{n} \right)_{u,p}, \qquad C = C(k) \neq C(n, f), \quad k < n, \qquad (2.5.17)$$

and a weak inverse Stechkin type inequality

$$\omega_\varphi^k(f,t)_{u,p} \leq C t^k \sum_{0 \leq i \leq 1/t} (1+i)^{k-1} E_i(f)_{u,p}, \qquad C \neq C(t, f). \qquad (2.5.18)$$

The main tool for the proof of (2.5.18) is the following Bernstein inequality

$$\|P'\varphi u\|_p \leq C n \|Pu\|_p, \qquad C \neq C(n),$$

that holds for all $P \in \mathcal{P}_n$, $1 \leq p \leq +\infty$, and $\varphi(x) = \sqrt{1-x^2}$.

2.5 Weighted Polynomial Approximation

Moreover, we have the following weak Jackson estimate

$$E_n(f)_{u,p} \leq C \sum_{i=0}^{+\infty} \Omega_\varphi^k\left(f, \frac{1}{2^i n}\right)_{u,p} \sim \int_0^{1/n} \frac{\Omega_\varphi^k(f,t)_{u,p}}{t} dt, \qquad (2.5.19)$$

where $C \neq C(n,f)$.

We note that in [98] the estimate (2.5.19) was proved instead of (2.5.17). For the sake of completeness we will give a simple proof of (2.5.17). By Lemma 8.2.3 from [98, p. 96], there exists a polynomial sequence $\{P_n\}_{n \in \mathbb{N}_0}$ such that, for every function $f \in AC([-1,1])$ such that $\|f'\|_p < +\infty$, the following inequality

$$\|(f - P_n)u\|_p \leq \frac{C}{n} \|f' u_n \varphi_n\|_p, \quad C \neq C(f,n), \quad 1 \leq p \leq +\infty, \qquad (2.5.20)$$

holds, where $u(x) = (1-x)^\alpha (1+x)^\beta$, $\varphi_n(x) = \sqrt{1-x^2} + n^{-1}$ and $u_n(x) = (\sqrt{1-x} + n^{-1})^{2\alpha} (\sqrt{1+x} + n^{-1})^{2\beta}$.

Now, for every function $f \in W_1^p$, we define

$$f_n := \begin{cases} f & \text{in } A_n = [-1 + n^{-2}, 1 - n^{-2}], \\ f(-1 + n^{-2}) & \text{in } [-1, -1 + n^{-2}], \\ f(1 - n^{-2}) & \text{in } [1 - n^{-2}, 1]. \end{cases}$$

Obviously $f_n \in AC([-1,1])$ and $\|f_n'\|_p < +\infty$. Applying (2.5.20) to f_n we get

$$E_n(f_n)_{u,p} \leq \|(f_n - P_n)u\|_p \leq \frac{C}{n} \|f_n' \varphi_n u_n\|_p = \frac{C}{n} \|f' \varphi u\|_{L^p(A_n)},$$

since in the interval A_n we have $n^{-1} \leq \sqrt{1 - |x|}$. Consequently, we obtain for every function $f \in W_1^p$

$$E_n(f)_{u,p} \leq \|(f - f_n)u\|_p + E_n(f_n)_{u,p} \leq \|(f - f_n)u\|_p + \frac{C}{n} \|f' \varphi u\|_p.$$

Since $\|(f - f_n)u\|_p \leq Cn^{-1} \|f' \varphi u\|_p$, we find

$$E_n(f)_{u,p} \leq \frac{C}{n} \|f' \varphi u\|_p, \quad C \neq C(f,n), \quad 1 \leq p \leq +\infty. \qquad (2.5.21)$$

Now, let Q be an arbitrary polynomial in $\mathcal{P}_n$ and let $q = Q'$. We obtain by (2.5.21)

$$E_n(f)_{u,p} = E_n(f - Q)_{u,p} \leq \frac{C}{n} \|(f' - q)\varphi u\|_p.$$

Taking infimum over $q \in \mathcal{P}_{n-1}$ and recalling that φu is a Jacobi weight too, the following Favard's inequality

$$E_n(f)_{u,p} \leq \frac{C}{n} E_{n-1}(f')_{\varphi u, p}$$

holds. Iterating this inequality, we get the estimate

$$E_n(f)_{u,p} \leq \frac{C}{n^r} \|f^{(k)}\varphi^k u\|_p \qquad (2.5.22)$$

for each function $f \in W_k^p$. In order to prove (2.5.17) for every $f \in L_u^p$ and $g \in W_k^p$, we can write

$$E_n(f)_{u,p} \leq E_n(f-g)_{u,p} + E_n(g)_{u,p} \leq C\left\{\|(f-g)u\|_p + \left(\frac{1}{n}\right)^r \|g^{(k)}\varphi^k u\|_p\right\}.$$

Then, taking infimum over $g \in W_r^p$, recalling (2.5.9) and (2.5.11), we get

$$E_n(f)_{u,p} \leq CK_\varphi\left(f, \left(\frac{1}{n}\right)^r\right)_{u,p} \leq \omega_\varphi^r\left(f, \frac{1}{n}\right)_{u,p}.$$

In (2.5.8) we saw that the space C_u^0 can be characterized by the φ-modulus $\omega_\varphi^k(f,t)_{u,p}$. From (2.5.17) and (2.5.18) we deduce that it can be characterized by the error of the best approximation as well, i.e., we have

$$f \in C_u^0 \iff \lim_{n \to +\infty} E_n(f)_{u,\infty} = 0. \qquad (2.5.23)$$

More generally, by (2.5.17)–(2.5.19), we have for $1 \leq p \leq +\infty$ and $r \in \mathbb{R}^+$ the following equivalences

$$\sup_{n\geq 1} n^r E_n(f)_{u,p} \sim \sup_{t>0} \frac{\omega_\varphi^k(f,t)_{u,p}}{t^r} \sim \sup_{t>0} \frac{\Omega_\varphi^k(f,t)_{u,p}}{t^r}, \qquad (2.5.24)$$

where $n > r > 0$, $k > r$ and the constants in "$\sim$" are independent of f and n. They can be generalized as follows

$$\left(\sum_{n=1}^{+\infty}\left[(n+1)^{r-1/q} E_n(f)_{u,p}\right]^q\right)^{1/q} \sim \left(\int_0^1 \left[\frac{\omega_\varphi^k(f,t)_{u,p}}{t^{r+1/q}}\right]^q dt\right)^{1/q}, \qquad (2.5.25)$$

where $1 \leq q < +\infty$.

Recalling (2.5.15) we can recollect (2.5.24) and (2.5.25) and write for all $r > 0$ and $1 \leq p, q \leq +\infty$,

$$\|f\|_{p,q,r,u} \sim \begin{cases} \left(\sum_{n=1}^{+\infty}\left[(n+1)^{r-1/q} E_n(f)_{u,p}\right]^q\right)^{1/q} & \text{if } 1 \leq q < +\infty, \\ \sup_n (n+1)^r E_n(f)_{u,p} & \text{if } q = +\infty, \end{cases} \qquad (2.5.26)$$

where we point out that the left-hand side term is based on the smoothness properties of f, while the right-hand side terms take into account the rate of convergence

2.5 Weighted Polynomial Approximation

to zero of the best approximation error. Thus, these two concepts are strictly connected, and in analogy with the trigonometric case, the Besov spaces $B_{r,q}^p(u)$ defined in (2.5.14), are characterized by the best approximation in the following sense

$$\|f\|_{B_{r,q}^p(u)} \sim \|fu\|_p + \begin{cases} \left(\sum_{n=1}^{+\infty}\left[(n+1)^{r-1/q}E_n(f)_{u,p}\right]^q\right)^{1/q} & \text{if } 1 \leq q < +\infty, \\ \sup_n (n+1)^r E_n(f)_{u,p} & \text{if } q = +\infty. \end{cases}$$

The special case $B_{s,2}^2(\sqrt{u})$ is interesting. Taking the Jacobi orthonormal system $\{p_k(u)\}_{k\in\mathbb{N}_0}$ and using the corresponding Fourier coefficients $c_k := c_k(f)_u = \int_{-1}^1 p_k(u)fu$, $k = 0, 1, \ldots$, as well as the identity

$$E_n(f)_{\sqrt{u},2}^2 = \sum_{k>n} c_k^2,$$

we get

$$\|f\|_{B_{s,2}^2(\sqrt{u})} \sim \|f\|_{B_{s,2}^2(\sqrt{u})}^* \sim \left(\sum_{k=0}^{+\infty}(k+1)^{2s}c_k^2\right)^{1/2}, \quad s > 0. \qquad (2.5.27)$$

Now, we define the space

$$L_{\sqrt{u},s}^2 := \left\{f \in L_{\sqrt{u}}^2 \,\bigg|\, \sum_{k=0}^{+\infty}(1+k)^{2s}c_k^2 < +\infty\right\}, \quad s \geq 0,$$

equipped with the norm

$$\|f\|_{L_{\sqrt{u},s}^2} := \left(\sum_{k=0}^{+\infty}(1+k)^{2s}c_k^2\right)^{1/2}.$$

Using the Parseval equality, the identity

$$c_k^2(f)_u \sim \frac{1}{k^2}c_{k-1}^2(f')_{u\varphi^2},$$

and the equivalence (2.5.27), we obtain the following expressions for the $L_{\sqrt{u},s}^2$ norm

$$\|f\|_{L_{\sqrt{u},s}^2} \sim \begin{cases} \|f\sqrt{u}\|_2 & s = 0, \\ \|f\sqrt{u}\|_2 + \|f^{(s)}\varphi^s\sqrt{u}\|_2 & s \in \mathbb{N}, \\ \|f\sqrt{u}\|_2 + \left(\int_0^1 \left[\frac{\Omega_\varphi^r(f,t)_{\sqrt{u},2}}{t^{s+1/2}}\right]^2 dt\right)^{1/2} & s \text{ is a positive real.} \end{cases}$$

This completes some results in [40, 62] and [306].

Finally, in several contexts the following equivalence is useful:

$$\omega_\varphi^k(f,t)_{u,p} \sim \inf_{P \in \mathcal{P}_{[1/t]}} \left\{ \|(f-P)u\|_p + t^k \|P^{(k)} \varphi^k u\|_p \right\},$$

where $k \geq 1$ and $1 \leq p \leq +\infty$ (see [98]).

2.5.3 Weighted Approximation on the Semi-axis

On the real semi-axis $[0,+\infty)$ the typical weight function is the classical generalized Laguerre weight

$$w_\alpha(x) := x^\alpha e^{-x}, \qquad \alpha > -1, \quad x \geq 0.$$

Similarly to the case of bounded intervals, for $1 \leq p < +\infty$, the space $L^p_{w_\alpha}$ consists of all functions f such that

$$\|f\|_{L^p_{w_\alpha}} := \|f w_\alpha\|_p = \left(\int_0^{+\infty} |f(x) w_\alpha(x)|^p dx \right)^{1/p} < +\infty.$$

Moreover, for $\alpha \geq 0$ the space $C^0_{w_\alpha}$ is defined as follows

$$C^0_{w_\alpha} := \left\{ f \in C(0,\infty) \;\Big|\; \lim_{x \to 0} |(f w_\alpha)(x)| = 0 = \lim_{x \to +\infty} |(f w_\alpha)(x)| \right\} \quad (\alpha > 0)$$

and

$$C^0_{w_0} := \left\{ f \in C[0,\infty) \;\Big|\; \lim_{x \to +\infty} |f(x) e^{-x}| = 0 \right\} \quad (\alpha = 0).$$

In all cases $\alpha \geq 0$, $C^0_{w_\alpha}$ is equipped with the norm

$$\|f\|_{C^0_{w_\alpha}} := \|f w_\alpha\|_\infty = \sup_{x \geq 0} |(f w_\alpha)(x)|.$$

Moreover, we assume $L^\infty_{w_\alpha} \equiv C^0_{w_\alpha}$.

For smoother functions and $1 \leq p \leq +\infty$, the weighted Sobolev space of order $k \in \mathbb{N}$ is given by

$$W_k^p(w_\alpha) := \left\{ f \in L^p_{w_\alpha} \;\Big|\; f^{(r-1)} \in AC(\mathbb{R}^+) \text{ and } \|f^{(k)} \varphi^k w_\alpha\|_p < +\infty \right\},$$

$$\varphi(x) := \sqrt{x},$$

and equipped with the norm

$$\|f\|_{W_k^p(w_\alpha)} := \|f w_\alpha\|_p + \|f^{(k)} \varphi^k w_\alpha\|_p.$$

2.5.3.1 Weighted K-functionals and Moduli of Smoothness

In analogy with the case of bounded intervals, for a Laguerre weight w_α and for all $1 \leq p \leq +\infty$, we define[7]

1° The Peetre K-functional

$$K_\varphi^k(f, t^k)_{w_\alpha, p} := \inf_{g^{(k-1)} \in AC_{\text{loc}}} \left\{ \|(f-g)w_\alpha\|_p + t^k \|g^{(k)} \varphi^k w_\alpha\|_p \right\};$$

2° The "main part" of K-functional

$$\widetilde{K}_\varphi^k(f, t^k)_{w_\alpha, p} := \sup_{0 < h \leq t} \inf_{g^{(k-1)} \in AC_{\text{loc}}} \left\{ \|(f-g)w_\alpha\|_{L^p(I_{h,k})} \right. $$
$$ \left. + h^k \|g^{(k)} \varphi^k w_\alpha\|_{L^p(I_{h,k})} \right\};$$

3° The main part of the φ-modulus of smoothness

$$\Omega_\varphi^k(f, t)_{w_\alpha, p} := \sup_{0 < h \leq t} \|(\Delta_{h\varphi}^k f) w_\alpha\|_{L^p(I_{h,k})};$$

4° The φ-moduli of smoothness

$$\omega_\varphi^k(f, t)_{w_\alpha, p} := \Omega_\varphi^k(f, t)_{w_\alpha, p} + \inf_{P \in \mathcal{P}_{k-1}} \|(f-P)w_\alpha\|_{L^p[0, 4k^2 t^2]}$$
$$+ \inf_{P \in \mathcal{P}_{k-1}} \|(f-P)w_\alpha\|_{L^p[1/t^2, \infty)},$$

$${}^*\omega_\varphi^k(f, t)_{w_\alpha, p} := \Omega_\varphi^k(f, t)_{w_\alpha, p} + \sup_{0 < h \leq t^2} \left\| (\overrightarrow{\Delta}_h^k f) w_\alpha \right\|_{L^p(0, 4k^2 t^2)}$$
$$+ \inf_{P \in \mathcal{P}_{k-1}} \|(f-P)w_\alpha\|_{L^p[1/t^2, \infty)}, \quad \alpha \geq 0,$$

where the finite differences $\Delta_{h, \varphi}^k$ and $\overrightarrow{\Delta}_h^k f$ are defined as in (2.5.2) and (2.5.5), respectively, but now $\varphi(x) := \sqrt{x}$, $I_{h,k} = [4Ck^2 h^2, C/h^2]$, and C is a fixed positive constant.

Now, an analogue of Theorem 2.5.1 holds ([81]):

Theorem 2.5.2 *For all $1 \leq p \leq +\infty$ and $f \in L_{w_\alpha}^p$, with $w_\alpha(x) = x^\alpha e^{-x}$, we have*

$$\omega_\varphi^k(f, t)_{w_\alpha, p} \sim K_\varphi^k(f, t^k)_{w_\alpha, p} \qquad (2.5.28)$$

---

[7] See the corresponding definitions for a finite interval, which are given in Sect. 2.5.1.

and

$$\Omega_\varphi^k(f,t)_{w_\alpha,p} \sim \widetilde{K}_\varphi^k(f,t^k)_{w_\alpha,p}, \qquad (2.5.29)$$

where the constants in "$\sim$" depend only on $k \in \mathbb{N}$.

Finally, the analogous results of (2.5.8) and (2.5.13) in the case of a Laguerre weight also hold, i.e., the space $C_{w_\alpha}^0$ can be characterized as follows

$$f \in C_{w_\alpha}^0 \iff \lim_{t \to 0} \omega_\varphi(f,t)_{w_\alpha,\infty} = 0, \qquad \alpha \geq 0, \qquad (2.5.30)$$

and the "main part" of the modulus of continuity can be estimated by

$$\Omega_\varphi^k(f,t)_{u,p} \leq C \sup_{h \leq t} h^r \| f^{(k)} \varphi^k u \|_{L^p(I_{h,k})}. \qquad (2.5.31)$$

2.5.3.2 Weighted Best Polynomial Approximation

The error of the best polynomial approximation in $L_{w_\alpha}^p$ ($1 \leq p \leq +\infty$) is defined in a usual way by

$$E_n(f)_{w_\alpha,p} := \inf_{P \in \mathcal{P}_n} \|(f-P)w_\alpha\|_p.$$

In this case, the Jackson and Stechkin type inequalities are (see [81]):

$$E_n(f)_{w_\alpha,p} \leq C \omega_\varphi^k \left(f, \frac{1}{\sqrt{n}} \right)_{w_\alpha,p}, \qquad C \neq C(n,f), \qquad (2.5.32)$$

and

$$\omega_\varphi^k(f,t)_{w_\alpha,p} \leq C t^k \sum_{0 \leq i \leq 1/t^2} (1+i)^{k/2-1} E_i(f)_{w_\alpha,p}, \qquad C \neq C(t,f). \qquad (2.5.33)$$

Furthermore, concerning the main part of the φ-modulus, we have the weak Jackson estimate

$$E_n(f)_{w_\alpha,p} \leq C \int_0^{1/\sqrt{n}} \frac{\Omega_\varphi^k(f,t)_{w_\alpha,p}}{t} \, dt, \qquad C \neq C(n,f). \qquad (2.5.34)$$

From (2.5.32), (2.5.33), and (2.5.30), the space $C_{w_\alpha}^0$, $\alpha \geq 0$, can be characterized by the error of the best approximation

$$f \in C_{w_\alpha}^0 \iff \lim_{n \to \infty} E_n(f)_{w_\alpha,\infty} = 0. \qquad (2.5.35)$$

In particular, for any $k > r > 0$, we have

$$E_n(f)_{w_\alpha,p} = O(n^{-r/2}) \iff \omega_\varphi^k(f,t)_{w_\alpha,p} = O(t^r) \iff \Omega_\varphi^k(f,t)_{w_\alpha,p} = O(t^r).$$

Finally, in the Sobolev space $W_k^p(w_\alpha)$, $k \in \mathbb{N}$, $1 \leq p \leq +\infty$, the following estimate [81]

$$E_n(f)_{w_\alpha,p} \leq \frac{C}{n^{k/2}} \|f^{(k)} \varphi^k w_\alpha\|_p$$

holds, with $\varphi(x) = \sqrt{x}$ and $C \neq C(n, f)$.

In conclusion we recall a useful equivalence from [81],

$$\omega_\varphi^k(f,t)_{w_\alpha,p} \sim \inf_{P \in \mathcal{P}_{[1/t]}} \left\{ \|(f-P)w_\alpha\|_p + t^{k/2} \|P^{(k)} \varphi^k w_\alpha\|_p \right\},$$

where $k \geq 1$ and $1 \leq p \leq +\infty$.

2.5.3.3 Weighted Besov Type Spaces

For all $r > 0$ and $1 \leq p, q \leq +\infty$, the Besov type spaces $B_{r,q}^p(w_\alpha)$, with respect to the generalized Laguerre weight $w_\alpha(x) = x^\alpha e^{-x}$, can be defined by means of the main part φ-modulus $\Omega_\varphi^k(f,t)_{w_\alpha,p}$ similarly to the case of Jacobi weights. Thus, we have

$$B_{r,q}^p(w_\alpha) := \left\{ f \in L_{w_\alpha}^p \,\Big|\, \|f\|_{B_{r,q}^p(w_\alpha)} < +\infty \right\}, \qquad (2.5.36)$$

where the Besov norm is given by

$$\|f\|_{B_{r,q}^p(w_\alpha)} := \|f w_\alpha\|_p + \begin{cases} \left(\int_0^1 \left[\frac{\Omega_\varphi^k(f,t)_{w_\alpha}}{t^{r+1/q}} \right]^q \frac{dt}{t} \right)^{1/q} & \text{if } 1 \leq q < +\infty, \\ \sup_{t>0} \frac{\Omega_\varphi^k(f,t)_{w_\alpha}}{t^r} & \text{if } q = +\infty, \end{cases}$$

for $k > r$. By virtue of (2.5.34) and (2.5.33), it can equivalently be expressed in terms of the best approximation error as follows

$$\|f\|_{B_{r,q}^p(w_\alpha)} \sim \|f w_\alpha\|_p + \left(\sum_{n=1}^{+\infty} \left[(n+1)^{r/2-1/q} E_n(f)_{w_\alpha,p} \right]^q \right)^{1/q}$$

if $1 \leq q < +\infty$, and

$$\|f\|_{B_{r,q}^p(w_\alpha)} \sim \|f w_\alpha\|_p + \sup_n (n+1)^{r/2} E_n(f)_{w_\alpha,p}$$

if $q = +\infty$.

As in the case of the Jacobi weight, the case $p = q = 2$ is of special interest.
Using the Fourier coefficients with respect to the Laguerre orthonormal system $\{p_k(w_\alpha)\}_{k \in \mathbb{N}_0}$, $c_k := c_k(f)_{w_\alpha} = \int_0^{+\infty} p_k(w_\alpha) f w_\alpha$, $k = 0, 1, \ldots$, and the identity

$$E_n(f)_{\sqrt{w_\alpha},2}^2 = \sum_{k>n} c_k^2,$$

we get

$$\|f\|_{B^2_{s,2}(\sqrt{w_\alpha})} \sim \|f\|^*_{B^2_{s,2}(\sqrt{w_\alpha})} \sim \left(\sum_{k=0}^{+\infty}(k+1)^s c_k^2\right)^{1/2}, \quad s>0. \qquad (2.5.37)$$

Now, we define the space

$$L^2_{\sqrt{w_\alpha},s} := \left\{ f \in L^2_{\sqrt{w_\alpha}} \mid \sum_{k=0}^{+\infty}(1+k)^s c_k^2 < +\infty \right\}, \quad s \geq 0,$$

equipped with the norm

$$\|f\|_{L^2_{\sqrt{w_\alpha},s}} := \left(\sum_{k=0}^{+\infty}(1+k)^s c_k^2\right)^{1/2}.$$

Using the Parseval equality, the identity

$$c_k^2(f)_{w_\alpha} = \frac{1}{k} c_{k-1}^2(f')_{w_{\alpha+1}},$$

and (2.5.37), we obtain the following expressions for the $L^2_{\sqrt{w_\alpha},s}$ norm

$$\|f\|_{L^2_{\sqrt{w_\alpha},s}} \sim \begin{cases} \|f\sqrt{w_\alpha}\|_2 & s=0, \\ \|f\sqrt{w_\alpha}\|_2 + \|f^{(s)}\varphi^s\sqrt{w_\alpha}\|_2 & s \in \mathbb{N}, \\ \|f\sqrt{w_\alpha}\|_2 + \left(\displaystyle\int_0^1 \left[\frac{\Omega^r_\varphi(f,t)_{\sqrt{w_\alpha},2}}{t^{s+1/2}}\right]^2 dt\right)^{1/2} & s \text{ is a positive real.} \end{cases}$$

This completes a result in [297].

The previous results can be found in [81] together with the corresponding proofs. Following such proofs, it is easy to see that the previous relations also hold if the weight w_α is replaced with $\widetilde{w}_{\alpha,\beta}(x) = w_\alpha(x)(1+x)^\beta$, $\beta \in \mathbb{R}$.

2.5.4 Weighted Approximation on the Real Line

On the real line a weight function $u(x)$ must satisfy the condition ([5])

$$\int_{-\infty}^{+\infty} \frac{\log u(x)}{1+x^2} dx = -\infty$$

in order that the set of all polynomials is dense in C^0_u. We limit our study only to the Hermite weight

$$w(x) := e^{-x^2}, \quad x \in \mathbb{R}.$$

2.5 Weighted Polynomial Approximation

As usual the space L_w^p, $1 \leq p < +\infty$, is defined as the space of all functions f such that

$$\|f\|_{L_w^p} := \|fw\|_p = \left(\int_{-\infty}^{+\infty} |f(x)w(x)|^p dx \right)^{1/p} < +\infty.$$

Moreover, $C_w^0 \equiv L_w^\infty$ is given by

$$C_w^0 := \left\{ f \in C(\mathbb{R}) \,\Big|\, \lim_{|x| \to +\infty} (fw)(x) = 0 \right\}$$

and equipped with the norm

$$\|f\|_{C_w^0} := \|fw\|_\infty = \sup_{x \in \mathbb{R}} |(fw)(x)|.$$

For all $1 \leq p \leq +\infty$, the weighted Sobolev space of order $k \in \mathbb{N}$ is given by

$$W_k^p(w) := \left\{ f \in L_w^p \,\Big|\, f^{(k-1)} \in AC \text{ and } \|f\|_{W_k^p(w)} := \|fw\|_p + \|f^{(k)}w\|_p < +\infty \right\}.$$

Concerning the Sobolev spaces, defined in the previous subsections, we note that now there is no the additional weight function $\varphi(x)$ since we consider the approximation on $\mathbb{R}$ which is unbounded on the upper and lower part. In this sense, the Hermite case appears simpler and also the corresponding definitions of the modulus of smoothness and the K-functional are simplified.

The modulus of smoothness is defined by (see [98, pp. 180–195])

$$\omega^k(f,t)_{w,p} := \Omega^k(f,t)_{w,p} + \inf_{P \in \mathcal{P}_{k-1}} \|(f-P)w\|_{L^p(-\infty,-1/t]}$$

$$+ \inf_{P \in \mathcal{P}_{k-1}} \|(f-P)w\|_{L^p[1/t,\infty)},$$

where $\Omega^k(f,t)_{w,p}$ is the main part of modulus of smoothness

$$\Omega_\varphi^k(f,t)_{w,p} := \sup_{0 < h \leq t} \|(\Delta_h^k f)w\|_{L^p([-1/h,1/h])}$$

and $\Delta_h^k f$ is the standard central finite difference of order k and step h,

$$\Delta_h^k f(x) := \sum_{i=0}^{k} (-1)^i \binom{k}{i} f\left(x + \frac{kh}{2} - ih \right).$$

The Peetre K-functional is given by

$$K^k(f,t^k)_{w,p} := \inf_{g^{(k-1)} \in AC(\mathbb{R})} \left\{ \|(f-g)w\|_p + t^k \|g^{(k)}w\|_p \right\}.$$

For all $1 \leq p \leq +\infty$ and $f \in L_w^p$, the equivalence

$$\omega^k(f,t)_{w,p} \sim K^k(f,t^k)_{w,p} \tag{2.5.38}$$

holds, with the constants in $\sim$ depending only on $k \in \mathbb{N}$.

For $1 \leq p \leq +\infty$, the error of the best polynomial approximation in L_w^p,

$$E_n(f)_{w,p} := \inf_{P \in \mathcal{P}_n} \|(f-P)w\|_p,$$

is connected to the previous modulus of smoothness $\omega(f,t)_{w,p}$ by the following Jackson and Stechkin type inequalities ([81])

$$E_n(f)_{w,p} \leq C\omega^k\left(f, \frac{1}{\sqrt{n}}\right)_{w,p}, \qquad C \neq C(n,f), \tag{2.5.39}$$

and

$$\omega^k(f,t)_{w,p} \leq Ct^k \sum_{0 \leq i \leq 1/t^2} (1+i)^{k/2-1} E_i(f)_{w,p}, \qquad C \neq C(t,f). \tag{2.5.40}$$

Moreover, the weak Jackson estimate

$$E_n(f)_{w,p} \leq C \int_0^{1/\sqrt{n}} \frac{\Omega^k(f,t)_{w,p}}{t} \, dt, \qquad C \neq C(n,f), \tag{2.5.41}$$

holds.

The space C_w^0 is characterized by the modulus of smoothness

$$f \in C_w^0 \iff \lim_{t \to 0} \omega(f,t)_{w,\infty} = 0, \tag{2.5.42}$$

as well as by the error of the best approximation

$$f \in C_w^0 \iff \lim_{n \to +\infty} E_n(f)_{w,\infty} = 0. \tag{2.5.43}$$

For any $k > r > 0$, we also have

$$E_n(f)_{w,p} = O(n^{-r/2}) \iff \omega^k(f,t)_{w,p} = O(t^r) \iff \Omega^k(f,t)_{w,p} = O(t^r).$$

Moreover, in the Sobolev space $W_k^p(w)$, $k \in \mathbb{N}$, $1 \leq p \leq +\infty$, the error of the best approximation can be estimated in the following form [98, pp. 180–195] (see also [137, 138])

$$E_n(f)_{w,p} \leq \frac{C}{n^{k/2}} \|f^{(k)}w\|_p, \qquad C \neq C(n,f).$$

Finally, for $r > 0$ and $1 \leq p,q \leq +\infty$, the Besov type spaces $B_{r,q}^p(w)$, with respect to the Hermite weight $w(x) = e^{-x^2}$, are defined by means of the main part

2.5 Weighted Polynomial Approximation

modulus $\Omega^k(f,t)_{w,p}$ or equivalently by the best approximation error $E_n(f)_{w,p}$, exactly as in the case of the generalized Laguerre weight, i.e., we have

$$B_{r,q}^p(w) := \left\{ f \in L_w^p \,\Big|\, \|f\|_{B_{r,q}^p(w)} < +\infty \right\}, \qquad (2.5.44)$$

where

$$\|f\|_{B_{r,q}^p(w)} := \|fw\|_p + \begin{cases} \left(\int_0^1 \left[\dfrac{\Omega^k(f,t)_w}{t^r} \right]^q \dfrac{dt}{t} \right)^{1/q} & \text{if } 1 \le q < +\infty, \\[2ex] \sup_{t>0} \dfrac{\Omega^k(f,t)_w}{t^r} & \text{if } q = +\infty, \end{cases}$$

with $k > r$, and

$$\|f\|_{B_{r,q}^p(w)} \sim \|fw\|_p + \begin{cases} \left(\displaystyle\sum_{n=1}^{+\infty} \left[(n+1)^{r-1/q} E_n(f)_{w,p} \right]^q \right)^{1/q} & \text{if } 1 \le q < +\infty, \\[2ex] \sup_n (n+1)^r E_n(f)_{w,p} & \text{if } q = +\infty. \end{cases}$$

In an analogous way to the Laguerre weight we can define the space

$$L^2_{\sqrt{w},s} = \left\{ f \in L^2 \,\Big|\, \sum_{k=0}^{+\infty} (1+k)^s c_k^2 < +\infty \right\},$$

equipped with the norm

$$\|f\|_{L^2_{\sqrt{w},s}} = \left(\sum_{k=0}^{+\infty} (1+k)^s c_k^2 \right)^{1/2},$$

where $c_k := c_k(f)_w = \int_{\mathbb{R}} p_k(w) f w$, $k = 0, 1, \ldots$, are the Fourier coefficients of a function f in the orthonormal Hermite system $\{p_k(w)\}_{k \in \mathbb{N}_0}$. Using the previous argument, we obtain

$$\|f\|_{L^2_{\sqrt{w},s}} \sim \begin{cases} \|f\sqrt{w}\|_2 & s = 0, \\[1ex] \|f\sqrt{w}\|_2 + \|f^{(s)}\sqrt{w}\|_2 & s \in \mathbb{N}, \\[1ex] \|f\sqrt{w}\|_2 + \left(\displaystyle\int_0^1 \left[\dfrac{\Omega^r(f,t)_{\sqrt{w},2}}{t^{s+1/2}} \right]^2 dt \right)^{1/2} & s \text{ is a positive real.} \end{cases}$$

In order to complete this subsection, we note that all relations relative to the Hermite weight $w(x) = e^{-x^2}$ (except the ones concerning the norm in $L^2_{\sqrt{w},s}$) hold true, mutatis mutandis, also for the weight $e^{-|x|^\alpha}$, $\alpha > 1$, and for more general Freud weights $e^{-Q(x)}$, with Q satisfying suitable conditions (for details see [98, p. 180], [97, 99]).

2.5.5 Weighted Polynomial Approximation of Functions Having Isolated Interior Singularities

We first consider the case of piecewise continuous functions in $(-1, 1)$ or continuous functions in $(-1, 1)$ whose derivatives are not continuous in some isolated inner point. Examples of such functions are: $x \mapsto \text{sgn}(x) \log(1 - x^2)$ and $x \mapsto |x| \sqrt[4]{1+x}$. To simplify notation, we consider functions having only one critical point at zero. In other words we consider normed functional spaces whose weight is given by

$$w(x) = v^{\alpha,\beta}(x)|x|^\gamma = (1-x)^\alpha |x|^\gamma (1+x)^\beta, \quad \alpha, \beta, \gamma > -1. \quad (2.5.45)$$

The results we show in this subsection can be found in [82] (see also [24, 308, 309] and [310]).

Denote by L^p, $1 \le p < +\infty$, the set of all measurable functions in $(-1, 1)$ such that

$$\|f\|_p^p := \int_{-1}^{1} |f(x)|^p dx < +\infty.$$

If $fw \in L^p$ then we write $f \in L^p_w$ and assume $\beta, \alpha > -1/p$ and $\gamma \ge 0$. In the case $p = +\infty$ we assume $\beta, \gamma, \alpha \ge 0$ and write $f \in L^\infty_w$ if f is continuous in $(-1,1) \setminus \{0\}$ and $\lim_{|x| \to 1} (fw)(x) = 0 = \lim_{x \to 0} (fw)(x)$. When $\beta = 0$ ($\alpha = 0$), we denote by L^∞_w the space of all functions f continuous in $[-1, 1) \setminus \{0\}$ (respectively $(-1, 1] \setminus \{0\}$) and such that $\lim_{x \to -1} (fw)(x) = 0$ (respectively $\lim_{x \to 1} (fw)(x) = 0$).

The norm in L^∞_w is defined by

$$\|f\|_{L^\infty_w} := \|fw\|_\infty = \sup_{|x| \le 1} |(fw)(x)|.$$

For smoother functions we introduce the Sobolev type spaces W^p_r, $1 \le p \le +\infty$, $r \ge 1$, defined as follows

$$W^p_r(w) := W^p_r = \left\{ f \in L^p_w \mid f^{(r-1)} \in AC((-1,1) \setminus \{0\}), \|f^{(r)} \varphi^r w\|_p < +\infty \right\},$$

where $\varphi(x) = \sqrt{1-x^2}$ and $AC(A)$ is the set of all absolutely continuous functions in A. We equip W^p_r with the norm

$$\|f\|_{W^p_r} := \|fw\|_p + \|f^{(r)} \varphi^r w\|_p.$$

Notice that, in general, functions belonging to W^p_r are unbounded at 0 and/or at ± 1.

For a small t (say $t < t_0$), we define the following K-functional

$$K(f, t^r)_{w,p} = \inf \{ \|(f-g)w\|_p + t^r \|g^{(r)} \varphi^r w\|_p, \; g^{(r-1)} \in AC(-1,1) \}, \quad (2.5.46)$$

where $r \ge 1$ and $1 \le p \le +\infty$.

In order to define the modulus of smoothness we follow [311].

2.5 Weighted Polynomial Approximation

With $I_h = [-1 + 4r^2h^2, -4rh] \cup [4rh, 1 - 4r^2h^2]$, $h > 0$, $r \geq 1$ and $1 \leq p \leq +\infty$, for some "small" t (say $t < t_0$), we set

$$\Omega_\varphi^r(f,t)_{w,p}^* = \sup_{0 < h \leq t} \|(\Delta_{h\varphi}^r f)w\|_{L^p(I_h)}, \qquad (2.5.47)$$

where

$$\Delta_{h\varphi}^r f(x) = \sum_{k=0}^r (-1)^k \binom{r}{k} f\left(x + \frac{h\varphi(x)}{2}(r - 2k)\right).$$

Thus, we define the modulus of smoothness $\omega_\varphi^r(f,t)_{w,p}^*$ as

$$\omega_\varphi^r(f,t)_{w,p}^* = \Omega_\varphi^r(f,t)_{w,p}^* + \sum_{i=0}^{2} \inf_{q_i \in \mathcal{P}_{r-1}} \|(f - q_i)w\|_{L^p(I_i(t))}, \qquad (2.5.48)$$

where $I_0(t) = [-1, -1 + 4r^2t^2]$, $I_1(t) = [-4rt, 4rt]$, and $I_2(t) = [1 - 4r^2t^2, 1]$.

In some proofs, $4r^2t^2$ and $4rt$ are replaced by $4Ar^2t^2$ and $4Art$, respectively, where A is a positive constant. Since, after such a modification, the behavior of K, Ω_φ^r and ω_φ^r does not change for $t \to 0$, we will preserve the same notations.

For a discussion on the definition (2.5.48) the reader can consult [312]. Here we limit ourself to observe that the sum in (2.5.48) cannot be dropped. In fact, for $w(x) = |x|^\gamma$, $\gamma > 0$, and $f(x) = \mathrm{sgn}(x)$ we find, by a direct computation,

$$\Omega_\varphi^r(f,t)_{w,p}^* = 0 \quad \text{and} \quad \omega_\varphi^r(f,t)_{w,p}^* \sim t^{\gamma + 1/p}, \quad 1 \leq p \leq +\infty.$$

Now we can state the following result:

Theorem 2.5.3 *Let $w(x)$ be defined as in (2.5.45) and let $f \in L_w^p$, $1 \leq p \leq +\infty$. Let $r > 0$ be an integer and let t be a sufficiently small real number (say $t < t_0$). Then*

$$K(f,t^r)_{w,p} \leq C\omega_\varphi^r(f,t)_{w,p}^*. \qquad (2.5.49)$$

Moreover, if $0 < \gamma < 1 - 1/p$ and $1 < p < +\infty$, we have

$$\omega_\varphi^r(f,t)_{w,p}^* \leq CK(f,t^r)_{w,p}, \qquad (2.5.50)$$

while, if $\gamma + 1/p \neq j$, $j = 1, \ldots, r$, and $1 \leq p \leq +\infty$, we have

$$\omega_\varphi^r(f,t)_{w,p}^* \leq C\bigl[K(f,t^r)_{w,p} + t^r \|fw\|_p\bigr]. \qquad (2.5.51)$$

Here the constants C are independent of f and t.

Inequalities (2.5.49) and (2.5.50) seem to be more natural than (2.5.49) and (2.5.51), but the assumption on γ is very restrictive. On the other hand, the condition $\gamma < 1 - 1/p$ is sometimes the necessary condition for the boundedness of important projectors in some functional spaces.

Example 2.5.1 Let $u(x) := (1-x^2)^\alpha$ and let $\{p_n(u)\}_{n\in\mathbb{N}_0}$ be the associated system of orthonormal polynomials with positive leading coefficient. Denote by

$$S_n(u,f) = \sum_{k=0}^{n-1} c_k p_k(u)$$

the n-th Fourier sum of the function $f \in L_u^p$, $1 < p < +\infty$. Then it is well-known (see [24, 62]) that the bound $\|S_n(u,f)w\|_p \leq C\|fw\|_p$ is equivalent to the following conditions

$$\frac{w}{\sqrt{u\varphi}} \in L^p \quad \text{and} \quad \frac{u}{w}, \sqrt{\frac{u}{\varphi}\frac{1}{w}} \in L^q, \quad q^{-1} + p^{-1} = 1,$$

which imply $\gamma < 1 - 1/p$.

Example 2.5.2 We consider the Lagrange polynomial $L_n(u,f)$ interpolating a given function f on the zeros of orthonormal polynomials $p_n(u)$. It was proved in [306] that the inequality

$$\|L_n(u,f)w\|_p \leq C\left(\|fw\|_p + \frac{\|f'\varphi w\|_p}{n}\right), \quad 1 < p < +\infty,$$

holds if and only if

$$\frac{w}{\sqrt{u\varphi}} \in L^p \quad \text{and} \quad \frac{\sqrt{u\varphi}}{w} \in L^q, \quad q^{-1} + p^{-1} = 1.$$

These last conditions still imply $\gamma < 1 - 1/p$. We have considered this case also for this reason.

These two examples also show that if the parameter γ of the weight w does not satisfy $\gamma < 1 - 1/p$, then the weight u of the system $\{p_n(u)\}_{n\in\mathbb{N}_0}$ has to be chosen with inner zeros.

With $w \in GDT$ defined in (2.5.45) and $1 \leq p \leq +\infty$, we denote by $E_n(f)_{w,p}$ the error of the best approximation, i.e.,

$$E_n(f)_{w,p} = \inf_{P \in \mathcal{P}_n} \|(f-P)w\|_p.$$

Theorem 2.5.4 *Let $w(x)$ be defined as in (2.5.45) and let $1 \leq p \leq +\infty$. Then, for all $f \in W_r^p(w)$, we have*

$$E_n(f)_{w,p} \leq C\left(\frac{\|f^{(r)}\varphi^r w\|_p}{n^r} + \inf_{q \in \mathcal{P}_{r-1}} \|(f-q)w\|_{L^p([-4r/n, 4r/n])}\right). \quad (2.5.52)$$

2.5 Weighted Polynomial Approximation

Moreover, for all $f \in L_w^p$, we have

$$E_n(f)_{w,p} \leq C\omega_\varphi^r\left(f, \frac{1}{n}\right)^*_{w,p}$$

and

$$\omega_\varphi^r\left(f, \frac{1}{n}\right)^*_{w,p} \leq \frac{C}{n^r} \sum_{k=0}^n (1+k)^{r-1} E_k(f)_{w,p}. \tag{2.5.53}$$

The constant C is independent of n and f.

Notice that the quantity $\inf_{q \in \mathcal{P}_{r-1}} \|(f-q)w\|_{L^p([-4r/n, 4r/n])}$ can be estimated by using the following statement:

Proposition 2.5.1 *Let $0 < \gamma < 1 - 1/p$, $1 < p < +\infty$, $\sigma(x) = |x|^\gamma$, and let f be a function such that $f^{(r-1)}$ is absolutely continuous in $[t_0 - a, t_0 + a]$, $0 < a < 1$, and $\|f^{(r)}\sigma\|_{L^p(t_0-a, t_0+a)} < +\infty$. Then there exists a polynomial $p \in \mathcal{P}_{r-1}$, such that*

$$\|(f-p)\sigma\|_{L^p(t_0-t, t_0+t)} \leq Ct^r \|f^{(r)}\sigma\|_{L^p(t_0-t, t_0+t)}, \quad |t| < a, \tag{2.5.54}$$

where C is a positive constant independent of t and f.

Lemma 2.5.1 *Assume that f is a function such that $f^{(r-1)}$ is absolutely continuous in $[t_0 - a, t_0 + a] \setminus \{t_0\}$, $0 < a < 1$, and $\|f^{(r)}\sigma\|_{L^p(t_0-a, t_0+a)} < +\infty$, $1 \leq p \leq +\infty$. Let $\sigma(x) = |x|^\gamma$, $\gamma > 0$, and let $|t| < a$. If $\gamma + 1/p > r$, then we have*

$$\|f\sigma\|_{L^p(t_0-t, t_0+t)} \leq Ct^r \left[\|f^{(r)}\sigma\|_{L^p(t_0-a, t_0+a)} + \|f\sigma\|_{L^p(t_0-a, t_0+a)}\right], \tag{2.5.55}$$

while if

$$\gamma + \frac{1}{p} \leq r \quad \text{with} \quad \gamma + \frac{1}{p} \neq j, \quad j = 1, \ldots, r,$$

and, in addition, $f^{(r-\tau-1)}(t_0)$, where $\tau = [\gamma + 1/p]$ is the integer part of $\gamma + 1/p$, then there exist polynomials $p \in \mathcal{P}_{r-\tau-1}$ such that

$$\|(f-p)\sigma\|_{L^p(t_0-t, t_0+t)} \leq Ct^r \left[\|f^{(r)}\sigma\|_{L^p(t_0-a, t_0+a)} + \|f\sigma\|_{L^p(t_0-a, t_0+a)}\right]. \tag{2.5.56}$$

Finally, if $\gamma + 1/p = r$, then (2.5.55) holds with $t^r \log t^{-1}$ replaced by t^r and if $\gamma + 1/p = j$, $j = 1, \ldots, r-1$, then (2.5.56) holds with j replaced by τ and $t^r \log t^{-1}$ replaced by t^r.

Here C is a positive constant independent of f and t.

The above lemma has a "local" character and can be used in different contexts. Then we have the following statement:

Corollary 2.5.1 *Assume $f \in W_r^p(w)$, $1 \leq p \leq +\infty$. If either $\gamma + 1/p > r$ or $\gamma + 1/p \leq r$, $\gamma + 1/p \neq j$, $j = 1, \ldots, r$, and $f^{(r-[\gamma+1/p]-1)} \in AC([-1,1])$, we have*

$$E_n(f)_{w,p} \leq \frac{C}{n^r}\left(\|f^{(r)}\varphi^r w\|_p + \|fw\|_p\right), \qquad (2.5.57)$$

while if $\gamma < 1 - 1/p$ with $1 < p < +\infty$, then

$$E_n(f)_{w,p} \leq \frac{C}{n^r}\|f^{(r)}\varphi^r w\|_p. \qquad (2.5.58)$$

Here C is a positive constant independent of n and f.

Inequality (2.5.58) appeared for the first time in [62] and [310]. Now, we show that in general (2.5.57) is not true when $\gamma + 1/p$ is an integer. In [310], for $p = +\infty$ it has been proved that there exists a function $f \in W_r^\infty(w)$, $w(x) = |x|^\gamma$ (γ integer), such that

$$\limsup_{n \to +\infty} \frac{n^r}{\log n} E_n(f)_{w,\infty} > 0.$$

Now we consider the Sobolev space $W_2^p(w)$, where $p > 1$, $w(x) = |x|^\gamma$ and $\gamma + 1/p = 2$. The function $f(x) = (\log\log(e/|x|))\,\mathrm{sgn}(x)$ belongs to $W_2^p(w)$. In fact, we have for $x \neq 0$

$$|f''(x)| = \frac{1}{x^2}\left(\log^{-1}\frac{e}{|x|} + \log^{-2}\frac{e}{|x|}\right)$$

and then

$$|f''(x)||x|^\gamma = \frac{1}{|x|^{1/p}}\left(\log^{-1}\frac{e}{|x|} + \log^{-2}\frac{e}{|x|}\right).$$

However, an inequality of the type $E_n(f)_{w,p} \leq C/n^2$, with $C \neq C(f,n)$, is not possible. In fact, since f is odd together with its polynomial of the best approximation $P_n^*(x)$ $(= xR_{[\frac{n-1}{2}]}(x^2))$, we can write

$$E_n(f)_{w,p} = \left(\int_{-1}^{1} |f(x) - P_n^*(x)|^p |x|^{\gamma p} dx\right)^{1/p}$$

$$= 2\left(\int_0^1 |f(x) - xR_{[\frac{n-1}{2}]}(x^2)|^p |x|^{\gamma p} dx\right)^{1/p}$$

$$= \left(\int_0^1 |f(\sqrt{u}) - \sqrt{u}R_{[\frac{n-1}{2}]}(u)|^p u^{\gamma p/2} \frac{du}{\sqrt{u}}\right)^{1/p}$$

$$= \left(\int_0^1 \left|\frac{1}{\sqrt{u}}\log\log\frac{e}{\sqrt{u}} - R_{[\frac{n-1}{2}]}(u)\right|^p u^{(3/2-1/p)} du\right)^{1/p}$$

$$\geq E_n(g)_{L_\sigma^p([0,1])},$$

2.5 Weighted Polynomial Approximation

where $\sigma = u^{3/2 - 1/p}$ and $g(u) = (1/\sqrt{u}) \log \log (e/\sqrt{u})$. Notice that g is singular at the extremal point 0. Therefore, assuming $E_n(f)_{w,p} \leq C/n^2$, we have

$$E_n(g)_{L^p_\sigma([0,1])} \leq \frac{C}{n^2},$$

and, using the Stechkin inequality [98, p. 24], the inequality

$$\Omega^3_\varphi\left(g, \frac{1}{n}\right)_{L^p_\sigma([0,1])} \leq \frac{C}{n^2}, \qquad \varphi(x) = \sqrt{x(1-x)}, \tag{2.5.59}$$

holds true. Now we compute $\Omega^3_\varphi(g, 1/n)_{L^p_\sigma([0,1])}$. Neglecting the "smaller" terms in g''' (see also [98, p. 110]), we have

$$\Omega^3_\varphi\left(g, \frac{1}{n}\right)_{L^p_\sigma([0,1])} \geq \frac{C}{n^3} \left(\int_{C_1/n^2}^{C_2/n^2} \left(u^{-7/2} \left(\log \log \frac{e}{\sqrt{u}}\right) u^{3/2} u^{3/2 - 1/p} \right)^p du \right)^{1/p}$$

$$\sim \frac{1}{n^3} \left(\int_{C_1/n^2}^{C_2/n^2} \left(\frac{\log \log (e/\sqrt{u})}{u^{1/2 + 1/p}} \right)^p du \right)^{1/p} \sim \frac{\log \log n}{n^2}$$

for $n > e$, i.e., the estimate (2.5.59) is false.

A weaker version of the Jackson inequality is given in the following theorem.

Theorem 2.5.5 *For every function $f \in L^p_w$, $1 \leq p \leq +\infty$, we have*

$$E_m(f)_{w,p} \leq C \left(\int_0^{1/m} \frac{\Omega^r_\varphi(f,t)^*_{w,p}}{t} dt + \int_0^{2/m} \inf_{P \in \mathcal{P}_{r-1}} \|(f-P)w\|_{L^p(I(t))} \frac{dt}{t} \right),$$

where $r < m$, $I(t) = [-4rt, 4rt]$, and C is a positive constant independent of m and f.

As already observed $\Omega^r_\varphi(f,t)^*_{w,p}$ is not sufficient to characterize the behavior of the complete modulus $\omega^r_\varphi(f,t)^*_{w,p}$ and consequently of $E_m(f)_{w,p}$. However, letting

$$A(f,t) := \Omega^r_\varphi(f,t)^*_{w,p} + \inf_{P \in \mathcal{P}_{r-1}} \|(f-P)w\|_{L^p([-4rt, 4rt])}$$

and using (2.5.53) it is easy to verify that $A(f,t) \sim t^\lambda$, $0 < \lambda < r$, is equivalent to $\omega^r_\varphi(f,t)^*_{w,p} \sim t^\lambda$ and to $E_m(f)_{w,p} \sim m^{-\lambda}$. Notice that by the definition in (2.5.47) we have

$$\Omega^r_\varphi(f,t)^*_{w,p} \leq C \sup_{h \leq t} h^r \|f^{(r)} \varphi^r w\|_{L^p(I_h)},$$

with $I_h = [-1 + Ch^2, -4Ch] \cup [4Ch, 1 - Ch^2]$.

In the sequel, any polynomial $P_n \in \mathcal{P}_n$ such that $\|(f - P_n)w\|_p \leq C E_n(f)_{w,p}$, will be called the "near best approximant" polynomial of $f \in L^p_w$. The following theorem deals with the estimates of the derivatives of such "near best approximant" polynomials of $f \in L_w^p$.

Theorem 2.5.6 *Let $P_n \in \mathcal{P}_n$ be a "near best approximant" polynomial of $f \in L_w^p$ and let $1 \le p \le +\infty$. Then, if the parameter γ of the weight w satisfies $0 < \gamma < 1 - 1/p$, with $1 < p < +\infty$, we have*

$$\left\| P_n^{(r)} \left(\frac{\varphi}{n}\right)^r w \right\|_p \le C \omega_\varphi^r \left(f, \frac{1}{n}\right)_{w,p}^*, \qquad (2.5.60)$$

while if $\gamma + 1/p \ne j$, $j = 1, \ldots, r$, $1 \le p \le +\infty$, we get

$$\left\| P_n^{(r)} \left(\frac{\varphi}{n}\right)^r w \right\|_p \le C \left[\omega_\varphi^r \left(f, \frac{1}{n}\right)_{w,p}^* + \frac{\|fw\|_p}{n^r} \right]. \qquad (2.5.61)$$

In both cases the constant C is independent of n, P_n and f.

Using the K-functional defined before and Theorem 2.5.3, by (2.5.60) (with $\gamma < 1 - 1/p$) it follows that

$$\omega_\varphi^r \left(f, \frac{1}{n}\right)_{w,p}^* \sim E_n(f)_{w,p} + \frac{\|P_n^{(r)} \varphi^r w\|_p}{n^r}, \qquad 1 < p < +\infty,$$

while by (2.5.61) (with $\gamma + 1/p \ne j$, $j = 1, \ldots, r$) we deduce

$$\omega_\varphi^r \left(f, \frac{1}{n}\right)_{w,p}^* + \frac{\|fw\|_p}{n^r} \sim E_n(f)_{w,p} + \frac{\|P_n^{(r)} \varphi^r w\|_p}{n^r} + \frac{\|fw\|_p}{n^r}, \qquad 1 \le p \le +\infty.$$

We now consider the case of piecewise continuous functions in $(0, +\infty)$ and assume that $t_0 > 0$ is a singular point.

With $w_\alpha = x^\alpha e^{-x}$, $\alpha > -1$, $x > 0$, we define $w(x) = |x - t_0|^\gamma w_\alpha(x)$, $\gamma \ge 0$, $t_0 > 0$, as a generalized Laguerre weight ($w \in GL$). The properties and polynomial inequalities with such weights were given in [308]. Here we recall some functional spaces we use in the sequel. L_w^p is defined in a usual way if $1 \le p < +\infty$. When $p = +\infty$, then L_w^∞, $\alpha > 0$, is the collection of all functions $f \in C^0(0, +\infty)$ with the condition $(fw)(x) \to 0$ if $x \to 0$ or $+\infty$ and $x \to t_0$. For smoother functions we define the Sobolev type spaces

$$W_r^p(w_\alpha) = \left\{ f \in L_w^p \mid f^{(r-1)} \in AC(0, +\infty) \text{ and } \|f^{(r)} \varphi^r w\|_p < +\infty \right\}$$

and

$$W_r^p(w) = \left\{ f \in L_w^p \mid f^{(r-1)} \in AC[(0, +\infty) \setminus \{t_0\}] \text{ and } \|f^{(r)} \varphi^r w\|_p < +\infty \right\},$$

where $\varphi(x) = \sqrt{x}$, $r \ge 1$, $1 \le p \le +\infty$, $AC(I)$ is the set of all absolutely continuous functions on I and $\|g\|_{L^p(I)}^p = \int_I |g(t)|^p \, dt$.

2.5 Weighted Polynomial Approximation

The following modulus of smoothness was defined in [308]:

$$\omega_\varphi^r(f,t)^*_{w,p} = \Omega_\varphi^r(f,t)^*_{w,p} + \sum_{i=0}^{2} \inf_{q_i \in \mathcal{P}_{r-1}} \|(f-q_i)w\|_{L^p(A_i(t))},$$

where $r \geq 1$, $A_0(t) = (0, 4r^2t^2)$, $A_1(t) = [t_0 - 4rt, t_0 + 4rt]$, $A_2(t) = [t^{-2}, +\infty)$;

$$\Omega_\varphi^r(f,t)^*_{w,p} = \sup_{0 < h \leq t} \|w\Delta^r_{h\varphi} f\|_{I_h}, \quad I_h = [4r^2h^2, t_0 - 4rh] \cup [t_0 + 4rh, h^{-2}]$$

and

$$\Delta^r_{h\varphi} f(x) = \sum_{k=0}^{r} (-1)^k \binom{r}{k} f\left[x + \frac{h\sqrt{x}}{2}\left(\frac{r}{2} - k\right)\right].$$

Introducing the K-functional

$$K_r(f,t^r)_{w,p} = \inf_{g \in W^p_r(w_\alpha)} \left\{\|(f-g)w\|_p + t^r\|g^{(r)}\varphi^r w\|_p\right\},$$

we have the following result:

Lemma 2.5.2 *For all functions $f \in L^p_w$, $1 \leq p \leq +\infty$, and $w \in \mathrm{GL}$, we have*

$$K_r(f,t^r)_{w,p} \leq C\omega_\varphi^r(f,t)^*_{w,p}$$

and

$$\omega_\varphi^r(f,t)^*_{w,p} \leq C\left[K_r(f,t^r)_{w,p} + \inf_{q \in \mathcal{P}_{r-1}} \|(f-q)w\|_{L^p([t_0-4rt, t_0+4rt])}\right].$$

Denote by $E_n(f)_{w,p} = \inf_{P \in \mathcal{P}_n} \|(f-P)w\|_p$ the error of the best approximation of f in L^p_w. We can establish the following result:

Theorem 2.5.7 *For any $f \in L^p_w$, $1 \leq p \leq +\infty$, we have*

$$E_n(f)_{w,p} \leq C\omega_\varphi^r\left(f, \frac{1}{\sqrt{n}}\right)^*_{w,p}, \quad n > r > 1,$$

and

$$\omega_\varphi^r\left(f, \frac{1}{\sqrt{n}}\right)^*_{w,p} \leq \frac{C}{n^{r/2}} \sum_{k=0}^{n} (1+k)^{r/2-1} E_k(f)_{w,p}.$$

In order to establish a weak Jackson type inequality, we set

$$A_r(f,t) = \Omega_\varphi^r(f,t)^*_{w,p} + \inf_{q \in \mathcal{P}_{r-1}} \|(f-q)w\|_{L^p(t_0-4rt, t_0+4rt)},$$

where the second term on the right can be estimated by means of Lemma 2.5.1. We have the following statement:

Theorem 2.5.8 *For all functions $f \in L_w^p$,*

$$E_n(f)_{u,p} \leq C \int_0^{1/\sqrt{n}} \frac{A_r(f,t)}{t} dt.$$

Finally, for the near best approximating polynomials, we can establish the following result:

Theorem 2.5.9 *Let p_n be the near best approximating polynomial of $f \in L_w^p$. If $0 < \gamma \leq 1 - 1/p$, with $1 < p < +\infty$, then*

$$\left\| p_n^{(r)} \left(\frac{\varphi}{n}\right)^r w \right\|_p \leq C \omega_\varphi^r \left(f, \frac{1}{\sqrt{n}}\right)_{u,p}^*,$$

while if $\gamma + 1/p \neq j$, $j = 0, 1, \ldots, r$, and $1 \leq p \leq +\infty$, then

$$\left\| p_n^{(r)} \left(\frac{\varphi}{n}\right)^r w \right\|_p \leq C \left[\omega_\varphi^r \left(f, \frac{1}{\sqrt{n}}\right)_{w,p}^* + \frac{1}{n^{r/2}} \|fw\|_p \right].$$

Finally, we consider the case of piecewise continuous functions in $\mathbb{R}$ and assume that zero is a critical point.

To simplify notation, we assume $u(x) = |x|^\gamma e^{-|x|^\alpha} =: |x|^\gamma u_\alpha(x)$, where $\gamma \geq 0$, $\alpha > 1$, and consider the functions f that can have singularities at zero. The space L_u^p, $1 \leq p < +\infty$, is defined in the usual way, while, if $p = +\infty$, we assume $f \in C^0(\mathbb{R} \setminus \{0\})$ with the conditions $\lim(fu)(x) = 0$ if $|x| \to +\infty$ and $x \to 0$. In the sequel we also consider smoother functions and define the Sobolev spaces

$$W_r^p(u_\alpha) = \left\{ f \in L_u^p \mid f^{(r-1)} \in AC(\mathbb{R}) \text{ and } \|f^{(r)}u\|_p < +\infty \right\},$$

$$W_r^p(u) = \left\{ f \in L_u^p \mid f^{(r-1)} \in AC(\mathbb{R} \setminus \{0\}) \text{ and } \|f^{(r)}u\|_p < +\infty \right\},$$

with $1 \leq p \leq +\infty$ and $r \geq 1$.

To define a suitable modulus of smoothness, we follow [98] and [312] (see also [308]). With $t^* = c_\alpha/t^{1/(\alpha-1)}$, $c_\alpha = 1/\alpha^{1/(\alpha-1)}$ and t small (say $t < t_0$), we set

$$\Omega^r(f,t)_{u,p}^* = \sup_{h \leq t} \|u \Delta_h^r f\|_{L^p(I_h)}$$

where $I_h = [-h^*, -4rh] \cup [4rh, h^*]$ and $\Delta f(x) = f(x+h/2) - f(x-h/2)$, $\Delta^r = \Delta \Delta^{r-1}$. Then, we define the modulus of smoothness

$$\omega^r(f,t)_{u,p}^* = \Omega^r(f,t)_{u,p}^* + \sum_{i=0}^{2} \inf_{q_i \in \mathcal{P}_{r-1}} \|(f - q_i)u\|_{L^p(A_i(t))}, \qquad (2.5.62)$$

where $A_0(t) = (-\infty, -t^*)$, $A_1(t) = [-4rt, 4rt]$, $A_2(t) = [t^*, +\infty)$ (see [312] and [82] for a discussion about the motivation of definition (2.5.62)).

2.5 Weighted Polynomial Approximation

Now, we introduce the K-functional

$$K_r(f, t^r)_{u,p} = \inf\left\{\|(f-g)u\|_p + t^r\|g^{(r)}u\|,\ g \in W_r^p(u_\alpha)\right\}.$$

Observe that $K_r(f, t^r)_{u,p}$ is more closely connected to $W_r^p(u_\alpha)$ than to $W_r^p(u)$ and it is related to the previous modulus of smoothness by the following lemma:

Lemma 2.5.3 *For any function $f \in L_u^p$, with $1 \le p \le +\infty$, we have*

$$K_r(f, t^r)_{u,p} \le C\omega^r(f, t)_{u,p}^*$$

and

$$\omega^r(f, t)_{u,p}^* \le C\left[K_r(f, t^r)_{u,p} + \inf_{q \in \mathcal{P}_{r-1}} \|(f-q)u\|_{L^p(-4rt, 4rt)}\right].$$

Setting $E_n(f)_{u,p} = \inf_{P \in \mathcal{P}_n} \|(f-P)u\|_p$ and using the Mhaskar-Rakhmanov-Saff number $M_n = C(\alpha)n^{1/\alpha}$, where $C(\alpha) = \left(2^{\alpha-1}\Gamma(\alpha/2)^2/\Gamma(\alpha)\right)^{1/\alpha}$ (cf. (2.4.21)), we can establish the following result:

Theorem 2.5.10 *For all $f \in L_u^p$, $1 \le p \le +\infty$, we have*

$$E_n(f)_{u,p} \le C\omega^r\left(f, \frac{M_n}{n}\right)_{u,p}^*,\quad n > r \ge 1, \tag{2.5.63}$$

and

$$\omega^r(f, t)_{u,p}^* \le Ct^r \sum_{k=0}^{L}(k+1)^{r(1-1/\alpha)-1} E_k(f)_{u,p}, \tag{2.5.64}$$

where $L = [t^{\alpha/(1-\alpha)}]$. The constants C are independent of f and t.

Remark 2.5.1 In particular, if $f \in W_r^p(u)$, it follows by (2.5.63) that

$$E_n(f)_{u,p} \le C\left(\frac{M_n}{n}\right)^r \|f^{(r)}u\|_p + \inf_{q \in \mathcal{P}_{r-1}} \|(f-q)u\|_{L^p(-4rt, 4rt)}.$$

As in the previous cases, we can establish a weaker Jackson theorem, which is useful in several contexts. To this end, set

$$A_r(f, t) = \Omega^r(f, t)_{u,p}^* + \inf_{q \in \mathcal{P}_{r-1}} \|(f-q)u\|_{L^p(-4rt, 4rt)}.$$

Then, we get the following statement:

Theorem 2.5.11 *For all functions $f \in L_u^p$ we have*

$$E_n(f)_{u,p} \le C \int_0^{M_n/n} \frac{A_r(f, t)}{t}\,dt,\quad n > r > 1. \tag{2.5.65}$$

Obviously the use of (2.5.65) requires that this integral is finite. If, in this case, $A_r(f,t) \sim t^\lambda$, $0 < \lambda < r$, then $E_n(f)_{u,p} \sim (M_n/n)^\lambda$ and, using (2.5.64), also $\omega^r(f,t)_{u,p}^* \sim t^\lambda$. Therefore, for this class of functions, $\omega^r(f,t)_{u,p}^*$ is equivalent to $A_r(f,t)$. A natural question is: For which class of functions we have $\Omega^r(f,t)_{u,p}^* \sim \omega^r(f,t)_{u,p}$, i.e., when can we omit the term $\inf_{q \in \mathcal{P}_{r-1}} \|(f-q)u\|_p$?

We also observe that in order to estimate $\inf_{q \in \mathcal{P}_{r-1}} \|(f-q)u\|_p$ we can use Lemma 2.5.1 with $t_0 = 0$. The following theorem gives the estimates for derivatives of the "near best approximant" polynomials.

Theorem 2.5.12 *Let P_n be a near best polynomial approximant of degree n of $f \in L_u^p$. If $0 < \gamma < 1 - 1/p$, $1 < p < +\infty$, then*

$$\left(\frac{M_n}{n}\right)^r \|P_n^{(r)}u\|_p \le C \omega^r \left(f, \frac{M_n}{n}\right)_{u,p}^*,$$

while, if $\gamma + 1/p \ne j$, $j = 0, 1, \ldots, r$, and $1 \le p \le +\infty$, then

$$\left(\frac{M_n}{n}\right)^r \|P_n^{(r)}u\|_p \le C \left[\omega^r\left(f, \frac{M_n}{n}\right)_{u,p}^* + \left(\frac{M_n}{n}\right)^r \|fu\|_p\right].$$

The following statement is an immediate consequence of Theorem 2.5.12.

Corollary 2.5.2 *Let $f \in L_u^p$. If $0 < \gamma < 1 - 1/p$, $1 < p < +\infty$, then*

$$\omega^r\left(f, \frac{M_n}{n}\right)_{u,p}^* \sim \inf_{P \in \mathcal{P}_n} \left\{\|(f-P)u\|_p + \left(\frac{M_n}{n}\right)^r \|P^{(r)}u\|_p\right\},$$

while if $\gamma + 1/p \ne j$, $j = 0, 1, \ldots, r$, and $1 \le p \le +\infty$, then

$$\omega^r\left(f, \frac{M_n}{n}\right)_{u,p}^* + \|fu\|_p \left(\frac{M_n}{n}\right)^r$$
$$\sim \inf_{P \in \mathcal{P}_n} \left\{\|(f-P)u\|_p + \left(\frac{M_n}{n}\right)^r \left(\|P^{(r)}u\|_p + \|Pu\|_p\right)\right\},$$

where the constants in "$\sim$" are independent of f, P and n.

All results dealing with piecewise continuous functions in $\mathbb{R}^+$ and $\mathbb{R}$ have been proved with respect to more general weights in [308] and [309].

Chapter 3
Trigonometric Approximation

3.1 Approximating Properties of Operators

3.1.1 Approximation by Fourier Sums

We start this chapter with the Fourier operator and Fourier sums. First of all we note that the Fourier operator S_n, defined in (1.2.23), is a projector on $\mathcal{T}_n$, i.e., $S_n f \in \mathcal{T}_n$ for each $f \in L^1$ and $S_n f = f$ if $f \in \mathcal{T}_n$. Consequently, we can write for all $T \in \mathcal{T}_n$

$$\|f - S_n f\|_p \leq \|f - T\|_p + \|S_n f - T\|_p$$
$$= \|f - T\|_p + \|S_n(f - T)\|_p$$
$$\leq (1 + \|S_n\|_p)\|f - T\|_p, \tag{3.1.1}$$

where

$$\|S_n\|_p = \|S_n\|_{L^p \to L^p} := \sup_{\|f\|_p = 1} \|S_n f\|_p$$

denotes the usual norm of the operator S_n considered as a map from L^p in to L^p, $1 \leq p \leq +\infty$. This norm is known as the Lebesgue constant of the Fourier operator S_n in L^p and its behaviour is important in order to estimate the approximating properties of the Fourier partial sums. In fact, we get by (3.1.1), taking infimum with respect to $T \in \mathcal{T}_n$,

$$E_n^*(f)_p \leq \|f - S_n f\|_p \leq (1 + \|S_n\|_p) E_n^*(f)_p \tag{3.1.2}$$

and we have an approximation error comparable to the error of the best approximation, i.e., $\|f - S_n f\|_p \sim E_n^*(f)_p$ holds, if and only if the Lebesgue constants $\|S_n\|_p$ are uniformly bounded with respect to n, i.e., $\sup_n \|S_n\|_p < +\infty$. The behaviour of these Lebesgue constants is stated in the next theorem.

Theorem 3.1.1 *Let $n \in \mathbb{N}$, $1 \leq p \leq +\infty$, and $S_n : L^p \to L^p$. If $1 < p < +\infty$ then S_n is uniformly bounded in L^p, i.e.,*

$$\sup_n \|S_n f\|_p \leq c_p \|f\|_p \tag{3.1.3}$$

G. Mastroianni, G.V. Milovanović, *Interpolation Processes*,
© Springer 2008

holds, where the positive constant c_p is defined as

$$c_p := \begin{cases} 4\left(\dfrac{p}{p-1}\right)^{1/p} + 1 & \text{if } 1 < p < 2, \\ 1 & \text{if } p = 2, \\ 4p^{(p-1)/p} + 1 & \text{if } 2 < p < +\infty. \end{cases} \quad (3.1.4)$$

If $p = 1$ or $p = +\infty$, then we have

$$\|S_n f\|_p \leq C \log n \, \|f\|_p, \quad (3.1.5)$$

where C does not depend on n and f.

Proof The case $p = 2$ is obvious, since in the Hilbert space L^2 the Fourier projector $S_n : L^2 \to \mathcal{T}_n \subset L^2$ is an orthogonal projector.

In the more general case $1 < p < +\infty$ the proof is well known and was given by M. Riesz (cf. [508, Chap. VII], [402, Chap. II, p. 178]).

Finally, for $p = 1$ or $p = +\infty$, taking into account that

$$\|S_n\|_p = \frac{1}{\pi} \int_0^{2\pi} |D_n(\theta)| \, d\theta$$

holds, Theorem 1.2.1 immediately gives (3.1.5). $\square$

The previous theorem shows that S_n is a uniformly bounded operator in L^p, with $1 < p < +\infty$, while it is unbounded in L^1 as well as in L^∞. Consequently, we get by (3.1.2) in the general case $1 < p < +\infty$,

$$\|f - S_n f\|_p \sim E_n^*(f)_p, \quad 1 < p < +\infty, \quad (3.1.6)$$

where in the special case $p = 2$, the identity

$$\|f - S_n f\|_2 = E_n^*(f)_2$$

holds, since in the Hilbert space L^2 the Fourier sum $S_n f$ is the best approximation of f in $\mathcal{T}_n$.

The cases $p = 1$ and $p = +\infty$ are critical, since (3.1.2) and (3.1.5) give

$$\|f - S_n f\|_p \leq C \log n \, E_n^*(f)_p \quad (p = 1 \text{ and } p = +\infty) \quad (3.1.7)$$

and the convergence of the Fourier sums $S_n f$ to f is not always guaranteed. In particular, there is a counterexample, due to Kolmogorov [508], of an L^1 function for which the corresponding Fourier sum is everywhere divergent.

On the other hand, Lozinskiĭ (see e.g. [235]) proved that for each polynomial projector P_n (i.e., such that $\operatorname{Im} P_n \subset \mathcal{T}_n$ and $P_n f = f$, for each $f \in \mathcal{T}_n$) we have

$$\|S_n\|_\infty \leq \|P_n\|_\infty. \quad (3.1.8)$$

3.1 Approximating Properties of Operators

The same result can be extended in a natural way also to the L^1-norm (see e.g., [95, 475]), that means S_n is the projector of minimal norm in L^∞ as well as in L^1. In these spaces we cannot expect to obtain better approximating properties than those of the Fourier sum by using a polynomial projector.

Finally, we remark that from the behaviour of the Lebesgue constants we can deduce also results on the simultaneous approximation which generalize the estimates (3.1.6) and (3.1.7). In fact, using the property $S_n f' = (S_n f)'$, we obtain the following result from the previous theorem.

Corollary 3.1.1 *Let r, k be integers such that $0 \leq k \leq r$ and $f \in W_r^p$ with $1 \leq p \leq +\infty$. Then for all $n \in \mathbb{N}$ the estimates*

$$\|(f - S_n f)^{(k)}\|_p \leq c E_n^*(f^{(k)})_p, \quad 1 < p < +\infty, \tag{3.1.9}$$

$$\|(f - S_n f)^{(k)}\|_1 \leq c \log n \, E_n^*(f^{(k)})_1, \tag{3.1.10}$$

$$\|(f - S_n f)^{(k)}\|_\infty \leq c \log n \, E_n^*(f^{(k)})_\infty \tag{3.1.11}$$

hold, where in each case $c \neq c(n, f)$.

Proof We already remarked that S_n is a projector on $\mathcal{T}_n$ and that $(S_n f)' = S_n(f')$ holds. Thus using these facts, for $1 \leq p \leq +\infty$, $0 \leq k \leq n$, and for any $T \in \mathcal{T}_n$, we get

$$\|(f - S_n f)^{(k)}\|_p \leq \|f^{(k)} - S_n f^{(k)}\|_p$$
$$\leq \|f^{(k)} - T\|_p + \|S_n(f^{(k)} - T)\|_p$$
$$\leq (1 + \|S_n\|_p) \|f^{(k)} - T\|_p.$$

Hence the theorem follows if we take the infimum with respect to $T \in \mathcal{T}_n$ and recall the behaviour of $\|S_n\|_p$ (i.e., (3.1.3) and (3.1.5)). □

3.1.2 Approximation by Fejér and de la Vallée Poussin Means

In the previous section we saw that the Fourier sums $S_n f$ are not uniformly convergent to f for all functions $f \in C_{2\pi}$. Nevertheless, if we take particular means of the Fourier sums, we succeed in obtaining sequences of trigonometric polynomials that uniformly converge to an arbitrary function $f \in C_{2\pi}$. The first of these means is the Fejér mean $\sigma_n f$ defined in (1.2.25) by

$$\sigma_n f(x) := \frac{1}{n+1} \sum_{k=0}^{n} S_k f(x),$$

for which

$$\lim_{n \to \infty} \|\sigma_n f - f\|_\infty = 0,$$

when $f \in C_{2\pi}$ (cf. [125]). Concerning the rate of convergence of the Fejér means, if $f \in \text{Lip}\,\alpha$, $0 < \alpha \leq 1$, we have [32] (see also [8] and [507])

$$\|\sigma_n f - f\|_\infty \leq C \begin{cases} \frac{1}{n^\alpha} & \text{if } 0 < \alpha < 1, \\ \frac{\log n}{n} & \text{if } \alpha = 1, \end{cases}$$

where the constant $C \neq C(n, f)$. But, if f is more regular, also if f is an analytic function, the degree of approximation cannot be improved, since Hille [212] proved that $\|\sigma_n f - f\|_\infty = o(1/n)$ implies that f is a constant function. Alexits [8] proved the saturation:

$$\|\sigma_n f - f\|_\infty \leq \frac{C}{n} \iff \tilde{f} \in \text{Lip}\,1,$$

where $\tilde{f}$ is the corresponding conjugate function of f.

Thus, the Fejér means give a moderate degree of approximation; however their historical value is more important (for example, they can be used to prove the Weierstrass theorem mentioned in Chap. 1, Sect. 1.1.1).

A better approximation can be achieved if we take the means of the Fourier sums from n until $2n$, i.e., if we use the de la Vallée Poussin sums (see Sect. 1.2.2)

$$V_n f(x) := V_n^{2n} f(x) = \frac{1}{n+1} \sum_{k=n}^{2n} S_k f(x), \qquad n \geq 1. \tag{3.1.12}$$

Note that V_n is a *quasi-projector* in the sense that we have

$$V_n T = T \qquad (T \in \mathcal{T}_n), \tag{3.1.13}$$

but in general $V_n f \in \mathcal{T}_{2n}$ if $f \in L^1$.

The approximation error of this operator in the L^p spaces is estimated in the following way:

Theorem 3.1.2 *For all $1 \leq p \leq +\infty$ and $f \in L^p$, we have*

$$\|f - V_n f\|_p \leq C E_n^*(f)_p, \qquad C \neq C(n, f). \tag{3.1.14}$$

Proof First we observe that if $T^* \in \mathcal{T}_n$ is the polynomial of best approximation of f, i.e., $\|f - T^*\|_p = E_n^*(f)_p$, then we get by (3.1.13)

$$\|V_n f - f\|_p \leq \|V_n f - T^*\|_p + \|T^* - f\|_p = \|V_n(f - T^*)\|_p + E_n^*(f)_p$$
$$\leq (\|V_n\|_p + 1) E_n^*(f)_p.$$

Consequently, we estimate the norm of the operator $V_n : L^p \to L^p$ and prove that it is uniformly bounded with respect to n, i.e.,

$$\sup_n \|V_n\|_p < +\infty. \tag{3.1.15}$$

If $1 < p < +\infty$, (3.1.15) follows directly from the Riesz estimate (3.1.3). Finally, in the case $p = 1$ or $p = +\infty$, we have

$$\|V_n\|_p = \frac{1}{\pi} \int_0^{2\pi} |\mathcal{V}_n^{2n}(\theta)| \, d\theta,$$

where $\mathcal{V}_n^{2n}(\theta)$ is the de la Vallée Poussin kernel defined in (1.2.13). Then (3.1.15) follows directly from (1.2.15). □

As in the case of the Fourier sums, the previous theorem can be generalized and we can give estimates of the simultaneous approximation as specified in the following statement:

Theorem 3.1.3 *For all $n \in \mathbb{N}$, $1 \leq p \leq +\infty$, and $f \in W_k^p$, with $k \geq 0$, we have*

$$\|(V_n f - f)^{(k)}\|_p \leq C E_n^*(f^{(k)})_p, \qquad 1 \leq p \leq +\infty, \tag{3.1.16}$$

where $C \neq C(n, f)$.

For the sake of brevity we omit the proof of this theorem which is analogous to the proof of Corollary 3.1.1 and is based on the property $(V_n f)^{(k)} = V_n f^{(k)}$, as well as on (3.1.13) and (3.1.15).

In conclusion we point out that obviously the previous results are not much significant if $1 < p < +\infty$, since in this case we have the same results by using the simpler Fourier sums. But, in the cases $p = 1$ and $p = +\infty$, when the Fourier projector and more generally every polynomial projector is not uniformly bounded, we can use the quasi-projector V_n in order to obtain the uniform boundedness and the corresponding convergence with the best possible order.

3.2 Discrete Operators

3.2.1 A Quadrature Formula

In order to derive a quadrature rule for evaluating the integral $\int_0^{2\pi} f(t) \, dt$ of a periodic function f, we consider the trigonometric polynomial $L_n^* f$ interpolating f at the equispaced points

$$\tau_k := \frac{2k\pi}{2n+1}, \qquad k = 0, 1, \ldots, 2n. \tag{3.2.1}$$

We saw in Sect. 1.3.3 that

$$L_n^* f(t) = \frac{1}{2n+1} \sum_{k=0}^{2n} \frac{\sin(2n+1)(t - \tau_k)/2}{\sin(t - \tau_k)/2} f(\tau_k).$$

We have by (1.2.7)

$$L_n^* f(t) = \frac{2}{2n+1} \sum_{k=0}^{2n} D_n(t - \tau_k) f(\tau_k) \qquad (3.2.2)$$

and

$$L_n^* f(\tau_k) = f(\tau_k) \qquad k = 0, 1, \ldots, 2n. \qquad (3.2.3)$$

On the other hand, using (1.2.1) instead of (1.2.7), we get another representation of $L_n^* f$ in the form

$$L_n^* f(t) = \frac{A_0}{2} + \sum_{k=1}^{n} (A_k \cos kt + B_k \sin kt),$$

where

$$A_k = \frac{2}{2n+1} \sum_{\nu=0}^{2n} f(\tau_\nu) \cos k\tau_\nu, \qquad B_k = \frac{2}{2n+1} \sum_{\nu=0}^{2n} f(\tau_\nu) \sin k\tau_\nu.$$

Now, approximating f by $L_n^* f$, we get $\int_0^{2\pi} f(t)\,dt \approx \int_0^{2\pi} L_n f(t)\,dt = \pi A_0$, i.e.,

$$\int_0^{2\pi} f(t)\,dt \approx \frac{2\pi}{2n+1} \sum_{k=0}^{2n} f\left(\frac{2k\pi}{2n+1}\right).$$

More generally, we can consider the following quadrature sum

$$G_N(f) := \frac{2\pi}{N+1} \sum_{k=0}^{N} f\left(\frac{2k\pi}{N+1}\right). \qquad (3.2.4)$$

Such a quadrature sum is, in fact a periodic version of the well-known *trapezoidal rule*, since by the periodicity $f(2\pi) = f(0)$, we have

$$\frac{1}{2} f(0) + \frac{1}{2} f(2\pi) \equiv f(0).$$

For trigonometric polynomials of degree at most N, we can prove the following result:

Proposition 3.2.1 *For every $T \in \mathcal{T}_N$ the N-point quadrature formula G_N is exact, i.e.,*

$$\int_0^{2\pi} T(t)\,dt = \frac{2\pi}{N+1} \sum_{k=0}^{N} T\left(\frac{2k\pi}{N+1}\right). \qquad (3.2.5)$$

3.2 Discrete Operators

Proof Let $T \in \mathcal{T}_N$, i.e.,

$$T(t) = \frac{a_0}{2} + \sum_{v=1}^{N}(a_v \cos vt + b_v \sin vt).$$

Then

$$G_N(T) = \frac{2\pi}{N+1}\sum_{k=0}^{N} T\left(\frac{2k\pi}{N+1}\right)$$

$$= \pi a_0 + \frac{2\pi}{N+1}\left[\sum_{v=1}^{N} a_v\left(\sum_{k=0}^{N}\cos\frac{2kv\pi}{N+1}\right) + \sum_{v=1}^{N} b_v\left(\sum_{k=0}^{N}\sin\frac{2kv\pi}{N+1}\right)\right],$$

But using the identities (1.1.7) and (1.1.8), we get

$$\sum_{k=0}^{N}\cos\frac{2kv\pi}{N+1} = \sum_{k=0}^{N}\sin\frac{2kv\pi}{N+1} = 0.$$

Thus, for each $T \in \mathcal{T}_N$, we have $G_N(T) = \pi a_0$. On the other hand, it is easy to check that $\int_0^{2\pi} T(t)\,dt = a_0\pi$ holds too, which means that (3.2.5) holds. □

In the case $f \notin \mathcal{T}_n$, we have the quadrature error

$$e_N(f) := \int_0^{2\pi} f(t)dt - \frac{2\pi}{N+1}\sum_{k=0}^{N} f\left(\frac{2k\pi}{N+1}\right).$$

For functions f from $C^{2N+1}[0, 2\pi]$, this error can be expressed in the form (see [404] and [401])

$$e_N(f) = \frac{2\pi^2}{(N+1)(N!)^2}\Big[D(D^2+1)(D^2+2^2)\cdots(D^2+N^2)f(t)\Big]_{t=\xi},$$

where $D = d/dx$ and $\xi \in (0, 2\pi)$. Also, the following estimates of the error $e_N(f)$ hold:

Proposition 3.2.2 *For all $N \in \mathbb{N}$ and $f \in C_{2\pi}$, we have*

$$|e_N(f)| \leq 4\pi E_N^*(f)_\infty. \tag{3.2.6}$$

Moreover, if f is an absolutely continuous function, then

$$|e_N(f)| \leq C\frac{\|f'\|_1}{N}, \qquad C \neq C(N, f), \tag{3.2.7}$$

holds. In addition, if f satisfies

$$\int_0^1 \frac{\omega^k(f,t)_1}{t^2} dt < +\infty, \quad k \in \mathbb{N},$$

then

$$|e_N(f)| \le \frac{C}{N} \int_0^{1/N} \frac{\omega^k(f,t)_1}{t^2} dt, \quad C \ne C(N,f). \quad (3.2.8)$$

Proof First we prove (3.2.6). By Proposition 3.2.1, for all $T \in \mathcal{T}_N$, we have

$$|e_N(f)| = |e_N(f-T)| \le \int_0^{2\pi} |(f-T)(t)| dt + \frac{2\pi}{N+1} \sum_{k=0}^N \left|(f-T)\left(\frac{2k\pi}{N+1}\right)\right|$$

$$\le \|f-T\|_\infty (2\pi + 2\pi) = 4\pi \|f-T\|_\infty.$$

Taking infimum with respect to $T \in \mathcal{T}_N$ we get (3.2.6).

In order to prove (3.2.7), we note that integration by parts yields the following identity

$$(b-a)f(a) = \int_a^b f(t)dt + \int_a^b (t-a)f'(t)dt. \quad (3.2.9)$$

Using this equality for $k = 0, 1, \ldots, N$, we have

$$\frac{2\pi}{N+1}\left|f\left(\frac{2k\pi}{N+1}\right)\right| \le \int_{\frac{2k\pi}{N+1}}^{\frac{2(k+1)\pi}{N+1}} |f(t)| dt + \int_{\frac{2k\pi}{N+1}}^{\frac{2(k+1)\pi}{N+1}} \left(t - \frac{2k\pi}{N+1}\right) |f'(t)| dt$$

$$\le \int_{\frac{2k\pi}{N+1}}^{\frac{2(k+1)\pi}{N+1}} |f(t)| dt + \frac{2\pi}{N+1} \int_{\frac{2k\pi}{N+1}}^{\frac{2(k+1)\pi}{N+1}} |f'(t)| dt.$$

Hence we get

$$\frac{2\pi}{N+1} \sum_{k=0}^N \left|f\left(\frac{2k\pi}{N+1}\right)\right| \le C \int_0^{2\pi} |f(t)| dt + \frac{C}{n} \int_0^{2\pi} |f'(t)| dt \quad (3.2.10)$$

and then we have

$$|e_N(f)| \le \|f\|_1 + \frac{2\pi}{N+1} \sum_{k=0}^N \left|f\left(\frac{2k\pi}{N+1}\right)\right| \le C\left(\|f\|_1 + \frac{\|f'\|_1}{N}\right).$$

Consequently, we deduce from Proposition 3.2.1 that for all $T \in \mathcal{T}_N$

$$|e_N(f)| = |e_N(f-T)| \le C\left(\|f-T\|_1 + \frac{\|(f-T)'\|_1}{N}\right),$$

3.2 Discrete Operators

where, recalling Theorem 3.1.3 and using the Bernstein inequality (1.2.35) and the invariance $V_N T = T$, we can get the estimate

$$\|(f-T)'\|_1 \le \|(f-V_N f)'\|_1 + \|(V_N f - T)'\|_1$$
$$\le C E_N^*(f')_1 + Cn\|V_N(f-T)\|_1$$
$$\le C E_N^*(f')_1 + CN\|f-T\|_1, \qquad (3.2.11)$$

where in the last line we used the uniform boundedness of V_N in L^1 (see (3.1.15)).

So, for all $T \in \mathcal{T}_N$, we obtain

$$|e_N(f)| \le C\left[\|f-T\|_1 + \frac{\|(f-T)'\|_1}{N}\right] \le C\|f-T\|_1 + C E_N^*(f')_1.$$

Taking infimum with respect to $T \in \mathcal{T}_N$ and applying the Favard inequality (1.2.34), we obtain

$$|e_N(f)| \le C\left[E_N^*(f)_1 + \frac{E_N^*(f')_1}{N}\right] \le \frac{C}{N} E_N^*(f')_1 \le \frac{C}{N}\|f'\|_1.$$

This means that (3.2.7) holds.

Now let us prove (3.2.8). Denote by $T_N^* \in \mathcal{T}_N$ the polynomial of the best approximation of f, i.e., $\|f - T_N^*\|_1 = E_N^*(f)_1$. We get by (3.2.7)

$$|e_N(f)| = |e_N(f - T_N^*)| \le \frac{C}{n}\|(f - T_N^*)'\|_1. \qquad (3.2.12)$$

On the other hand, since $\lim_{N \to +\infty} E_N^*(f)_1 = 0$, we can write

$$f - T_N^* = \sum_{k=0}^{+\infty} T_{2^{k+1}N}^* - T_{2^k N}^* \qquad a.e. \qquad (3.2.13)$$

So let us examine the series $\sum_{k=0}^{+\infty} (T_{2^{k+1}N}^* - T_{2^k N}^*)'$. By the Bernstein inequality (1.2.35), we have

$$\sum_{k=0}^{+\infty} \|(T_{2^{k+1}N}^* - T_{2^k N}^*)'\|_1 \le \sum_{k=0}^{+\infty} 2^{k+1} N \|T_{2^{k+1}N}^* - T_{2^k N}^*\|_1$$
$$\le \sum_{k=0}^{+\infty} 2^{k+1} N \left(\|f - T_{2^{k+1}N}^*\|_1 + \|f - T_{2^k N}^*\|_1\right)$$
$$= \sum_{k=0}^{+\infty} 2^{k+1} N \left(E_{2^{k+1}N}^*(f)_1 + E_{2^k N}^*(f)_1\right)$$

$$\leq \sum_{k=0}^{+\infty} 2^{k+1} N E_{2^k N}^*(f)_1.$$

Hence, using the Jackson inequality (1.2.33) and recalling that $\omega^k(f,t)_1$ is a non-increasing function of t, we get

$$\sum_{k=0}^{+\infty} \|(T_{2^{k+1}N}^* - T_{2^k N}^*)'\|_1 \leq \sum_{k=0}^{+\infty} 2^{k+1} N E_{2^k N}^*(f)_1$$

$$\leq C \sum_{k=0}^{+\infty} 2^{k+1} N \omega^k(f, 2^{-k}/N)_1$$

$$= 2C \sum_{k=0}^{+\infty} \omega^k(f, 2^{-k}/N)_1 \int_{2^{-k-1}/N}^{2^{-k}/N} \frac{dt}{t^2}$$

$$\leq C \sum_{k=0}^{+\infty} \int_{2^{-k-1}/N}^{2^{-k}/N} \frac{\omega^k(f,t)_1}{t^2} dt$$

$$= C \int_0^{1/N} \frac{\omega^k(f,t)_1}{t^2} dt,$$

which guarantees the convergence of the series $\sum_{k=0}^{+\infty}(T_{2^{k+1}N}^* - T_{2^k N}^*)'$ in L^1. Consequently, we find by (3.2.13)

$$(f - T_N^*)' = \sum_{k=0}^{+\infty}(T_{2^{k+1}N}^* - T_{2^k N}^*)',$$

and then

$$\|(f - T_N^*)'\|_1 \leq \sum_{k=0}^{+\infty} \|(T_{2^{k+1}N}^* - T_{2^k N}^*)'\|_1 \leq C \int_0^{1/N} \frac{\omega^k(f,t)_1}{t^2} dt. \qquad (3.2.14)$$

In conclusion, using (3.2.14) the estimate (3.2.8) follows by (3.2.12). □

3.2.2 Discrete Versions of Fourier and de la Vallée Poussin Sums

When we use the Fourier sums or the de la Vallée Poussin means for the numerical approximation of a 2π-periodic function f, we encounter the problem of computing the Fourier coefficients (1.2.21) and (1.2.22), or equivalently integrals

$$S_n f(x) = \frac{1}{\pi} \int_0^{2\pi} D_n(x-t) f(t) dt,$$

3.2 Discrete Operators

$$V_n f(x) = \frac{1}{\pi} \int_0^{2\pi} \left[\frac{1}{n+1} \sum_{r=n}^{2n} D_r(x-t) \right] f(t)\,dt.$$

In the case when f is a continuous (or at most Riemman-integrable) 2π-periodic function, we can overcome these problems constructing discrete operators which are strictly connected with the Fourier and de la Vallée Poussin operators, but they use only a finite number of values of the function f.

In order to obtain such discrete Fourier and de la Vallée Poussin operators, we use the quadrature sum $G_N(f)$ defined by (3.2.4). Applying such a quadrature rule with $N = 2n$ nodes to the integral representation of $S_n f$,

$$S_n f(x) = \frac{1}{\pi} \int_0^{2\pi} D_n(x-t) f(t)\,dt, \tag{3.2.15}$$

we obtain the trigonometric polynomial

$$L_n^* f(x) = \frac{2}{2n+1} \sum_{k=0}^{2n} D_n(x - \tau_k) f(\tau_k) = \frac{1}{\pi} G_{2n}\{D_n(x - \cdot) f\}, \tag{3.2.16}$$

interpolating f at the equispaced points

$$\tau_k := \frac{2k\pi}{2n+1}, \quad k = 0, 1, \ldots, 2n. \tag{3.2.17}$$

Thus, we can consider the trigonometric polynomial $L_n^* f$ introduced in (1.3.16), as a discrete approximation of the Fourier sums. As we will see in the sequel, this different point of view permits us to study the approximation properties of the Lagrange operator by means of those already stated for the Fourier operator.

Another interpolation polynomial, different from the Lagrange one, can be obtained by applying a similar procedure for discretizing the de la Vallée Poussin means

$$V_n f(x) = \frac{1}{\pi} \int_0^{2\pi} \left[\frac{1}{n+1} \sum_{r=n}^{2n} D_r(x-t) \right] f(t)\,dt. \tag{3.2.18}$$

Note that for $f \in \mathcal{T}_n$, the integrand in (3.2.18) becomes a trigonometric polynomial of degree $3n$. If we want that the discrete approximation of $V_n f$, like its continuous version, preserves the polynomials of degree at most n, we have to take the quadrature rule with degree of exactness greater than or equal to $3n$. Therefore, we apply the quadrature rule (3.2.4) with $N = 3n$. In this way we get the following approximation of $V_n f$

$$\widetilde{V}_n f(x) := \frac{1}{\pi(n+1)} G_{3n}\left(\sum_{k=n}^{2n} D_k(x - \cdot) f \right),$$

i.e.,

$$\widetilde{V}_n f(x) := \frac{2}{(3n+1)(n+1)} \sum_{k=0}^{3n} \left[\sum_{r=n}^{2n} D_r(x - t_k) \right] f(t_k), \qquad (3.2.19)$$

with

$$t_k := \frac{2k\pi}{3n+1}, \qquad k = 0, 1, \ldots, 3n. \qquad (3.2.20)$$

By proper choice of the degree of exactness of the applied quadrature rule, we are sure that the discrete operator $\widetilde{V}_n$ satisfies the invariance property

$$(\forall T \in \mathfrak{T}_n) \qquad \widetilde{V}_n T = V_n T = T. \qquad (3.2.21)$$

Thus, $\widetilde{V}_n$ is a quasi-projector on $\mathfrak{T}_n$, like its continuous version. In addition, as in the approximation of the Fourier sums, the resulting discrete operator in this case satisfies an interpolation property on the same nodes of the applied quadrature rules. In fact, we have the following results:

Theorem 3.2.1 *For all integers $n > 1$ and for any 2π-periodic function f defined everywhere, the polynomial $\widetilde{V}_n f \in \mathfrak{T}_{2n}$ given by (3.2.19) interpolates f on the $3n + 1$ points t_k defined in (3.2.20), i.e.,*

$$\widetilde{V}_n f(t_k) = f(t_k), \qquad k = 0, 1, \ldots, 3n. \qquad (3.2.22)$$

Proof We deduce by (1.2.14)

$$\frac{1}{n+1} \sum_{r=n}^{2n} D_r(x) = \begin{cases} \dfrac{\sin \frac{(3n+1)x}{2} \sin \frac{(n+1)x}{2}}{2n \sin^2 \frac{x}{2}} & \text{if } x \neq 2\pi\nu, \quad \nu \in \mathbb{Z}, \\ \dfrac{3n+1}{2} & \text{if } x = 2\pi\nu, \quad \nu \in \mathbb{Z}, \end{cases}$$

which gives

$$\frac{2}{(3n+1)(n+1)} \sum_{r=n}^{2n} D_r(t_h - t_k) = \delta_{h,k}, \qquad h, k = 0, 1, \ldots, 3n,$$

and consequently

$$\widetilde{V}_n f(t_h) = \sum_{k=0}^{3n} \left[\frac{2}{(3n+1)(n+1)} \sum_{r=n}^{2n} D_r(t_h - t_k) \right] f(t_k) = \sum_{k=0}^{3n} \delta_{h,k} f(t_k) = f(t_h)$$

holds for all $h = 0, 1, \ldots, 3n$. □

3.2 Discrete Operators

We remark that the polynomial (3.2.19) was first considered by Szabados [460] (see also [231]). By the previous interpolation property, we refer to it as the *de la Vallée Poussin interpolating polynomial* in the sequel.

Moreover, we note that as in the case of the Lagrange operator, we can equivalently construct such a de la Vallée Poussin interpolating operator starting from the following expression of the continuous de la Vallée Poussin operator (see (1.2.26))

$$\widetilde{V}_n f(x) = \frac{a_0}{2} + \sum_{k=1}^{2n} \mu_k (a_k \cos kx + a_k \sin kx),$$

with

$$\mu_k := \begin{cases} 1 & \text{if } 1 \le k \le n, \\ \dfrac{2n - k + 1}{n + 1} & \text{if } n < k \le 2n, \end{cases}$$

and applying the quadrature rule G_{3n} to the Fourier coefficients a_k and b_k. In this way we obtain the following explicit form of the discrete de la Vallée Poussin operator

$$\widetilde{V}_n f(x) = \frac{A_0}{2} + \sum_{k=1}^{2n} \mu_k (A_k \cos kx + B_k \sin kx), \qquad (3.2.23)$$

where

$$A_k = \frac{2}{3n+1} \sum_{\nu=0}^{3n} f(t_\nu) \cos kt_\nu, \qquad B_k = \frac{2}{3n+1} \sum_{\nu=0}^{3n} f(t_\nu) \sin kt_\nu$$

and

$$t_\nu = \frac{2\nu\pi}{3n+1}, \qquad \nu = 0, 1, \ldots, 3n.$$

The approximation properties of the discrete Fourier and de la Vallée Poussin operators are strictly connected with their continuous versions. In order to recognize such connections we need some instruments that permit us to pass from continuous norms to discrete ones and vice versa. These tools are introduced in the next section.

3.2.3 Marcinkiewicz Inequalities

The so-called Marcinkiewicz inequalities constitute a basic tool for the study of the approximating properties of the Lagrange and de la Vallée Poussin operators. Generally speaking, such inequalities link the L^p-norm of a trigonometric polynomial to suitable quadrature sums of the same polynomial. Now we have by Proposition 3.2.1

$$\frac{2\pi}{N+1} \sum_{k=0}^{N} f(\theta_k) = \int_0^{2\pi} f(x)\,dx, \qquad \theta_k = \frac{2\pi k}{N+1},$$

for all $f = T \in \mathcal{T}_N$. Obviously, if we take $f = |T|^p$, with $T \in \mathcal{T}_N$ and $1 \le p < +\infty$ we cannot expect that the equality above still holds. The next theorem gives an inequality which holds in this case.

Theorem 3.2.2 *Let $N \in \mathbb{N}$ and $\theta_k = 2\pi k/(N+1)$, $k = 0, 1, \ldots, N$. Then, for each $T \in \mathcal{T}_{N\ell}$, with $\ell \in \mathbb{N}$ fixed, the inequality*

$$\left(\frac{2\pi}{N+1}\sum_{k=0}^{N}|T(\theta_k)|^p\right)^{1/p} \le C\left(\int_0^{2\pi}|T(x)|^p dx\right)^{1/p} \qquad (3.2.24)$$

holds, for all $1 \le p \le +\infty$ (for $p = +\infty$ (3.2.24) reduces to $\max_{0 \le k \le N}|T(\theta_k)| \le C\|T\|_\infty$). Here, C is an absolute positive constant ($C < 1 + 2\pi\ell$).

Proof First we observe that by the mean value theorem, we have

$$\|T\|_p = \left(\sum_{k=0}^{N}\int_{\theta_k}^{\theta_{k+1}}|T(t)|^p dt\right)^{1/p} = \left(\frac{2\pi}{N+1}\sum_{k=0}^{N}|T(\xi_k)|^p\right)^{1/p},$$

where $\theta_k \le \xi_k \le \theta_{k+1}$, $k = 0, 1, \ldots, N$. Then, using the Minkowski, Hölder and Bernstein inequalities, we get

$$\left(\frac{2\pi}{N+1}\sum_{k=0}^{N}|T(\theta_k)|^p\right)^{1/p} \le \left(\frac{2\pi}{N+1}\sum_{k=0}^{N}|T(\xi_k)|^p\right)^{1/p}$$

$$+ \left(\frac{2\pi}{N+1}\sum_{k=0}^{N}|T(\xi_k) - T(\theta_k)|^p\right)^{1/p}$$

$$= \|T\|_p + \left(\frac{2\pi}{N+1}\sum_{k=0}^{N}\left|\int_{\theta_k}^{\xi_k}T'(t)dt\right|^p\right)^{1/p}$$

$$\le \|T\|_p + \left(\frac{2\pi}{N+1}\sum_{k=0}^{N}\left(\int_{\theta_k}^{\theta_{k+1}}|T'(t)|dt\right)^p\right)^{1/p}$$

$$\le \|T\|_p + \left(\frac{2\pi}{N+1}\sum_{k=0}^{N}\left(\int_{\theta_k}^{\theta_{k+1}}dt\right)^{p-1}\right.$$

$$\left.\times \left(\int_{\theta_k}^{\theta_{k+1}}|T'(t)|^p dt\right)\right)^{1/p}$$

$$= \|T\|_p + \left(\frac{2\pi}{N+1}\left(\frac{2\pi}{N+1}\right)^{p-1}\int_0^{2\pi}|T'(t)|^p dt\right)^{1/p}$$

3.2 Discrete Operators

$$= \|T\|_p + \frac{2\pi}{N+1}\|T'\|_p \leq \|T\|_p + \frac{2\pi}{N+1} N\ell \|T\|_p$$

$$= \left(1 + \frac{2\pi N\ell}{N+1}\right)\|T\|_p,$$

i.e., (3.2.24) holds, with $C = 1 + 2\pi N\ell/(N+1) < 1 + 2\pi\ell$. Finally, in the case $p = +\infty$, (3.2.24) is trivial, since it reduces to

$$\max_{0 \leq k \leq n} |T(\theta_k)| \leq \max_{x \in [0, 2\pi]} |T(x)|. \qquad \square$$

Note that in (3.2.24) the inverse inequality does not hold in the general case. An "inverse" inequality, that gives the L^p-norm of a polynomial less than or equal to an absolute constant times a quadrature sum on equidistant points, could be achieved if we take more knots than the ones considered in (3.2.24). Precisely, we need at least as many nodes as the number of coefficients of the given polynomial. Equivalently, if the number of equidistant knots is fixed, in order to obtain the inverse inequality in (3.2.24), we have to decrease the degree of the trigonometric polynomial we take. Two inverse inequalities of this type are stated in the following result:

Theorem 3.2.3 *For every polynomial $T \in \mathcal{T}_n$ and $1 < p < +\infty$, we have*

$$\left(\int_0^{2\pi} |T(x)|^p dx\right)^{1/p} \leq C \left(\frac{2\pi}{2n+1} \sum_{k=0}^{2n} |T(\tau_k)|^p\right)^{1/p}, \qquad (3.2.25)$$

with $\tau_k = 2\pi k/(2n+1)$, and in the more general case $1 \leq p \leq +\infty$

$$\left(\int_0^{2\pi} |T(x)|^p dx\right)^{1/p} \leq C \left(\frac{2\pi}{3n+1} \sum_{k=0}^{3n} |T(t_k)|^p\right)^{1/p}, \qquad (3.2.26)$$

with $t_k = 2\pi k/(3n+1)$. For $p = +\infty$, (3.2.26) is given by

$$\|T\|_\infty \leq C \max_{0 \leq k \leq 3n} |T(t_k)|.$$

In both inequalities (3.2.25) and (3.2.26) C is a positive constant depending only on p (suitable values are $C = (1 + 2\pi) c_{p'}$ with $c_{p'}$ given by (3.1.4) and $p' = p/(p-1)$, in the case $1 < p < +\infty$, $C = (4/\pi^2) \log 2$ in the case $p = 1$ and $C = (4/\pi^2)(1 + 2\pi) \log 2$ in the case $p = +\infty$).

Proof First we observe that if $1 < p < +\infty$, then there exists a unique function $g \in L^{p'}$ with $p' = p/(p-1)$ and $\|g\|_{p'} = 1$, such that

$$\|T\|_p = \int_0^{2\pi} T(x) g(x) \, dx.$$

Then, if we take $T = L_n^* T$ and use the Hölder inequality, we get

$$\|T\|_p = \|L_n^* T\|_p = \int_0^{2\pi} L_n^* T(x) g(x) \, dx$$

$$= \frac{2}{2n+1} \sum_{k=0}^{2n} T(\tau_k) \int_0^{2\pi} D_n(x - \tau_k) g(x) \, dx$$

$$= \frac{2\pi}{2n+1} \sum_{k=0}^{2n} T(\tau_k) S_n g(\tau_k)$$

$$\leq \left(\frac{2\pi}{2n+1} \sum_{k=0}^{2n} |S_n g(\tau_k)|^{p'} \right)^{1/p'} \left(\frac{2\pi}{2n+1} \sum_{k=0}^{2n} |T(\tau_k)|^p \right)^{1/p}.$$

In order to obtain (3.2.25) we have to show that the first factor on the right hand side in the last inequality is bounded. Using (3.2.24) and (3.1.3), we obtain

$$\left(\frac{2\pi}{2n+1} \sum_{k=0}^{2n} |S_n g(\tau_k)|^{p'} \right)^{1/p'} \leq (1+2\pi) \|S_n g\|_{p'} \leq (1+2\pi) \|S_n\|_{p'} \leq (1+2\pi) c_{p'}.$$

Thus (3.2.25) holds.

Similarly if we consider $T = \widetilde{V}_n T$ and apply the Hölder inequality and (3.2.24), for $1 < p < +\infty$, we have

$$\|T\|_p = \|\widetilde{V}_n T\|_p = \int_0^{2\pi} \widetilde{V}_n T(x) g(x) \, dx = \frac{2\pi}{3n+1} \sum_{k=0}^{3n} T(t_k) V_n g(t_k)$$

$$\leq \left(\frac{2\pi}{3n+1} \sum_{k=0}^{3n} |V_n g(t_k)|^{p'} \right)^{1/p'} \left(\frac{2\pi}{3n+1} \sum_{k=0}^{3n} |T(t_k)|^p \right)^{1/p}$$

$$\leq (1+2\pi) \|V_n g\|_{p'} \left(\frac{2\pi}{3n+1} \sum_{k=0}^{3n} |T(t_k)|^p \right)^{1/p}$$

$$\leq (1+2\pi) \|V_n\|_{p'} \left(\frac{2\pi}{3n+1} \sum_{k=0}^{3n} |T(t_k)|^p \right)^{1/p}$$

$$\leq (1+2\pi) c_{p'} \left(\frac{2\pi}{3n+1} \sum_{k=0}^{3n} |T(t_k)|^p \right)^{1/p},$$

where in the last line we used

$$\|V_n\|_{p'} \leq \frac{1}{n+1} \sum_{k=n}^{2n} \|S_k\|_{p'} \leq c_{p'}.$$

3.2 Discrete Operators

Analogously if $p = 1$, setting $g = \text{sgn}(\widetilde{V}_n T)$,[1] we have

$$\|T\|_1 = \|\widetilde{V}_n T\|_1 = \int_0^{2\pi} \widetilde{V}_n T(x) g(x)\, dx = \frac{2\pi}{3n+1} \sum_{k=0}^{3n} T(t_k) V_n g(t_k)$$

$$\leq \max_{0 \leq k \leq 3n} |V_n g(t_k)| \left(\frac{2\pi}{3n+1} \sum_{k=0}^{3n} |T(t_k)| \right)$$

$$\leq \|V_n g\|_\infty \left(\frac{2\pi}{3n+1} \sum_{k=0}^{3n} |T(t_k)| \right)$$

$$\leq \|V_n\|_\infty \left(\frac{2\pi}{3n+1} \sum_{k=0}^{3n} |T(t_k)| \right) \leq \frac{4}{\pi^2} \log 2 \left(\frac{2\pi}{3n+1} \sum_{k=0}^{3n} |T(t_k)| \right)$$

since by (1.2.15),

$$\|V_n\|_\infty = \frac{1}{\pi} \int_0^{2\pi} |\mathcal{V}_n^{2n}(\theta)|\, d\theta \leq \frac{4}{\pi^2} \log 2.$$

Finally, in the case $p = +\infty$ a similar result can be achieved using (3.2.24) with $p = 1$ and noting that for all $x \in [0, 2\pi]$, we have

$$|T(x)| = |\widetilde{V}_n T(x)| \leq \frac{2}{(3n+1)(n+1)} \sum_{k=0}^{3n} \left[|T(t_k)| \left| \sum_{r=n}^{2n} D_r(x - t_k) \right| \right]$$

$$\leq \frac{\max_k |T(t_k)|}{n+1} \left[\frac{2}{3n+1} \sum_{k=0}^{3n} \left| \sum_{r=n}^{2n} D_r(x - t_k) \right| \right]$$

$$\leq (1 + 2\pi) \max_k |T(t_k)| \left[\frac{1}{\pi(n+1)} \int_0^{2\pi} \left| \sum_{r=n}^{2n} D_r(x - t) \right| dt \right]$$

$$\leq (1 + 2\pi) \max_k |T(t_k)| \|V_m\|_\infty$$

$$\leq (1 + 2\pi) \frac{4 \log 2}{\pi^2} \max_k |T(t_k)|.$$

If we take supremum with respect to x, the obtained inequality gives (3.2.26) for $p = +\infty$. □

---

[1] For all functions f, the symbol $\text{sgn}(f)$ denotes the sign function,

$$\text{sgn}(f)(x) := \begin{cases} 1 & \text{if } f(x) > 0, \\ 0 & \text{if } f(x) = 0, \\ -1 & \text{if } f(x) < 0. \end{cases}$$

Remark 3.2.1 Notice that in proving (3.2.26), we also stated that for all trigonometric polynomials $T \in \mathcal{T}_n$ and $1 \leq p \leq +\infty$, the inequality

$$\|\widetilde{V}_n T\|_p \leq C\left(\frac{2\pi}{3n+1}\sum_{k=0}^{2n}|T(t_k)|^p\right)^{1/p}, \qquad (3.2.27)$$

holds, where $C \neq C(n, T)$. The same inequality can be stated more generally. Namely, we have for all continuous and 2π-periodic functions f and $1 \leq p \leq +\infty$,

$$\|\widetilde{V}_n f\|_p \leq C\left(\frac{2\pi}{3n+1}\sum_{k=0}^{2n}|f(t_k)|^p\right)^{1/p}, \qquad (3.2.28)$$

where $C \neq C(n, f)$. This can be proved in a similar way as (3.2.27).

Furthermore, we remark that also in the cases $p = 1$ and $p = +\infty$, we can write

$$\|T\|_p \leq (1+2\pi)\|S_n\|_{p'}\left(\frac{2\pi}{2n+1}\sum_{k=0}^{2n}|T(\tau_k)|^p\right)^{1/p} \qquad (T \in \mathcal{T}_n), \qquad (3.2.29)$$

but obviously in these cases we cannot obtain a constant independent of n, since by (3.1.5), $\|S_n\|_{p'} \sim \log n$ holds when $p' = +\infty$ (case $p = 1$) or $p' = 1$ (case $p = +\infty$).

Thus, the inequalities (3.2.24) and (3.2.25), i.e., (3.2.26), state the equivalence between the L^p-norm of a trigonometric polynomial and a suitable quadrature sum of the same polynomial. The original proofs of such inequalities are due to Marcinkiewicz and they can be found in [508]. Here we proposed new and simpler proofs that can be also extended to the algebraic case.

3.2.4 Uniform Approximation

In this subsection we want to estimate the interpolation error of the discrete Fourier and de la Vallée Poussin operators constructed in Sect. 3.2.2, with respect to the uniform norm. With regard to this, we state the following result:

Theorem 3.2.4 *Let $f \in C_{2\pi}$ and $n \in \mathbb{N}$. Then*

$$\|L_n^* f - f\|_\infty \leq C \log n \, E_n^*(f)_\infty \qquad (3.2.30)$$

and

$$\|\widetilde{V}_n f - f\|_\infty \leq C E_n^*(f)_\infty \qquad (3.2.31)$$

hold, where C is an independent constant of f and n.

3.2 Discrete Operators

Proof At first we observe that by the reproducing property

$$(\forall T \in \mathcal{T}_n) \qquad L_n^* T = T = \widetilde{V}_n T,$$

we get

$$\|L_n^* f - f\|_\infty \leq \|L_n^*(f - T^*)\|_\infty + \|T^* - f\|_\infty \leq (\|L_n^*\|_\infty + 1) E_n^*(f)_\infty$$

and similarly

$$\|\widetilde{V}_n f - f\|_\infty \leq (\|\widetilde{V}_n\|_\infty + 1) E_n^*(f)_\infty.$$

Then, the estimate of the approximation errors is reduced to the study of the norms

$$\|L_n^*\|_\infty := \sup_{\|f\|_\infty = 1} \|L_n^* f\|_\infty \quad \text{and} \quad \|\widetilde{V}_n\|_\infty := \sup_{\|f\|_\infty = 1} \|\widetilde{V}_n f\|_\infty$$

of the discrete operators $L_n^* : C_{2\pi} \to C_{2\pi}$ and $\widetilde{V}_n : C_{2\pi} \to C_{2\pi}$, respectively. Note that for the Lagrange operator, we have

$$\|L_n^*\|_\infty \sim \|S_n\|_\infty \sim \log n. \tag{3.2.32}$$

In fact, the inequality $\|S_n\|_\infty \leq \|L_n^*\|_\infty$ follows from (3.1.8), while $\|L_n^* f\|_\infty \leq C \|S_n\|_\infty$ can be deduced by using (3.2.24) with the $2n+1$ equidistant points $\tau_k = 2\pi k/(2n+1)$, $p = 1$ and $T = \pi^{-1} D_n(x - \cdot) \in \mathcal{T}_n$ as follows

$$\|L_n^* f\|_\infty = \sup_x \frac{2}{2n+1} \sum_{k=0}^{2n} |D_n(x - \tau_k)| \leq \frac{C}{\pi} \sup_x \int_0^{2\pi} |D_n(x - t)| \, dt$$

$$= \frac{C}{\pi} \int_0^{2\pi} |D_n(t)| \, dt = C \|S_n\|_\infty.$$

Finally, the discrete and continuous de la Vallée Poussin operators are related by the following inequalities

$$\|\widetilde{V}_n\|_\infty \leq C \|V_n\|_\infty \leq C, \qquad C \neq C(n). \tag{3.2.33}$$

In fact, as in the case of the Lagrange operator, using (3.2.24) with $3n+1$ equidistant nodes $t_k = 2k\pi/(3n+1)$, we get

$$\|\widetilde{V}_n f\|_\infty = \sup_x \frac{2}{3n+1} \sum_{k=0}^{3n} \left| \sum_{r=n}^{2n} D_r(x - t_k) \right| \leq \frac{C}{\pi} \sup_x \int_0^{2\pi} \left| \sum_{r=n}^{2n} D_r(x - t) \right| dt$$

$$= \frac{C}{\pi} \int_0^{2\pi} \left| \sum_{r=n}^{2n} D_r(t) \right| dt = C \|V_n\|_\infty.$$

$\square$

Comparing (3.2.30) and (3.2.31), we can observe the advantage of the de la Vallée Poussin interpolation with respect to the trigonometric interpolation. Namely, while for all continuous and 2π-periodic functions f, the de la Vallée Poussin interpolating polynomials $\widetilde{V}_n f$ uniformly converge to f, as n tends to infinity, the same does not hold for the polynomials $L_n^* f$. Indeed, Grünwald (see [197, 475]) proved that for some continuous function f, $L_n^* f$ can be pointwise divergent almost everywhere. Similar results also hold more generally for the simultaneous approximation, as specified in the following corollary of Theorem 3.2.4.

Corollary 3.2.1 *Let $r \in \mathbb{N}$ and $f \in W_r^\infty$. Then for all positive integers $n, k \in \mathbb{N}$, with $k \leq r$, we have*

$$\|(f - L_n^* f)^{(k)}\|_\infty \leq C E_n^*(f^{(k)})_\infty \log n \qquad (3.2.34)$$

and

$$\|(f - \widetilde{V}_n f)^{(k)}\|_\infty \leq C E_n^*(f^{(k)})_\infty, \qquad (3.2.35)$$

where in each case C is a positive constant independent of n and f.

Proof The case $k = 0$ was stated in Theorem 3.2.4. For $k > 0$, since

$$\|(f - L_n^* f)^{(k)}\|_\infty \leq \|(f - S_n f)^{(k)}\|_\infty + \|(L_n^* f - S_n f)^{(k)}\|_\infty$$

holds, then (3.1.11) and the Bernstein inequality (1.2.35) give

$$\|(f - L_n^* f)^{(k)}\|_\infty \leq C E_n^*(f)_\infty \log n + n^k \|L_n^* f - f + f - S_n f\|_\infty$$
$$\leq C E_n^*(f)_\infty \log n + C E_n^*(f)_\infty n^k \log n.$$

Thus, (3.2.34) follows by iterating the Favard inequality (1.2.34). Finally, (3.2.35) can be proved, taking $V_n f$ instead of $S_n f$. □

In conclusion, we can observe that the uniform approximation properties of the interpolating Lagrange and de la Vallée Poussin operators are the same, which we have seen for the Fourier and de la Vallée Poussin continuous operators respectively (compare (3.1.11) with (3.2.34) and (3.1.16) with (3.2.35)). Thus, we can say that in the space $C_{2\pi}$ the discrete and continuous versions of the Fourier operator, as well as of the de la Vallée Poussin operator, are equivalent in the sense that they give the same order of approximation of a function $f \in C_{2\pi}$, but obviously the discrete operators, with respect to their continuous versions, are easier for computations and practical applications.

3.2.5 Lagrange Interpolation Error in L^p

This section is devoted to the study of the Lagrange interpolation error in the L^p-norm, with $1 < p < +\infty$. In dealing with $L_n^* f$, we need functions f everywhere

3.2 Discrete Operators

defined, so that even if we estimate the behaviour of the error $\|f - L_n^* f\|_p$ in the L^p-space, we assume f to be a continuous (or at least a Riemann-integrable) 2π-periodic function.

The starting point in our study is the Marcinkiewicz inequality

$$\|L_n^* f\|_p \leq C \left(\frac{2\pi}{2n+1} \sum_{k=0}^{2n} |f(\tau_k)|^p \right)^{1/p}, \quad 1 < p < +\infty, \qquad (3.2.36)$$

where $C \neq C(n, f)$, which follows directly from (3.2.25), taking into account that $L_n^* f$ belongs to $\mathcal{T}_n$ and satisfies the interpolation property

$$L_n^* f(\tau_k) = f(\tau_k), \quad \tau_k = \frac{2k\pi}{2n+1}, \quad k = 0, 1, \ldots, 2n.$$

Using (3.2.36), we can immediately deduce the following result:

Theorem 3.2.5 *For all $f \in C_{2\pi}$ and $n \in \mathbb{N}$, we have*

$$\|f - L_n^* f\|_p \leq C E_n^*(f)_\infty, \quad 1 < p < +\infty, \quad C \neq C(n, f). \qquad (3.2.37)$$

More generally, if $f \in W_r^\infty$, with $r \geq 0$, then for all positive integers $k \leq r$,

$$\|(f - L_n^* f)^{(k)}\|_p \leq C E_n^*(f^{(k)})_\infty, \quad 1 < p < +\infty, \qquad (3.2.38)$$

holds, where C is a positive constant independent of n and f.

Proof By (3.2.36) we obtain

$$\|L_n^* f\|_p \leq C \left(\frac{2\pi}{2n+1} \sum_{k=0}^{2n} |f(t_k)|^p \right)^{1/p} \leq C \|f\|_\infty, \quad C \neq C(n, f),$$

i.e., the Lagrange operator $L_n^* : C_{2\pi} \to L^p$ ($1 < p < +\infty$) is uniformly bounded with respect to n,

$$\sup_n \|L_n^*\|_{C_{2\pi} \to L^p} < +\infty, \quad 1 < p < +\infty. \qquad (3.2.39)$$

Consequently, if $T^* \in \mathcal{T}_n$ is the polynomial of the best approximation to f, i.e., $\|f - T^*\|_\infty = E_n^*(f)_\infty$, then we get

$$\|f - L_n^* f\|_p \leq \|f - T^*\|_p + \|L_n^*(f - T^*)\|_p$$
$$\leq \|f - T^*\|_\infty + \|L_n^*\|_{C_{2\pi} \to L^p} \|f - T^*\|_\infty$$
$$\leq \left(1 + \|L_n^*\|_{C_{2\pi} \to L^p}\right) E_n^*(f)_\infty \leq C E_n^*(f)_\infty$$

and (3.2.37) holds. In order to prove (3.2.38) with $k > 0$, we use (3.1.9), (3.2.37), and the Bernstein and Favard inequalities (1.2.35), (1.2.34), so that we have

$$\|(f - L_n^* f)^{(k)}\|_p \leq \|(f - S_n f)^{(k)}\|_p + \|(L_n^* f - S_n f)^{(k)}\|_p$$
$$\leq C E_n^*(f)_p + n^k \|L_n^* f - f + f - S_n f\|_p$$
$$\leq C E_n^*(f)_\infty + C n^k E_n^*(f)_\infty + C n^k E_n^*(f)_p$$
$$\leq C n^k E_n^*(f)_\infty \leq C E_n^*(f^{(k)})_\infty. \qquad \square$$

The estimates stated in the previous theorem are not homogeneous estimates because the presence of the uniform norm on the right-hand side. A sharp estimate of the L^p-error of the Lagrange interpolation is given in the following statement:

Theorem 3.2.6 Let $f \in C_{2\pi}$ such that

$$\int_0^1 \frac{\omega(f,t)_p}{t^{1+1/p}} dt < +\infty, \quad 1 < p < +\infty. \tag{3.2.40}$$

Then, for $k \geq 1$,

$$\|f - L_n^* f\|_p \leq \frac{c}{n^{1/p}} \int_0^{1/n} \frac{\omega^k(f,t)_p}{t^{1+1/p}} dt, \tag{3.2.41}$$

where the constant c is independent of n and f.

This theorem is a well-known result due to Hristov [215]. We will prove it in an alternative way, by using the following imbedding result (cf. [41]):

Lemma 3.2.1 Let the function g be defined a.e. on the interval $I = [a, a+\delta]$, $\delta > 0$, for which

$$\int_0^\delta \frac{\omega(g,t)_{L^p(I)}}{t^{1+1/p}} dt < +\infty.$$

Then g coincides a.e. with a continuous function G satisfying, for $1 < p < +\infty$,

$$\max_{x \in I} |G(x)| \leq C(p) \left[\delta^{-1/p} \|g\|_{L^p(I)} + \int_0^\delta \omega(g,t)_{L^p(I)} \frac{dt}{t^{1+1/p}} \right]. \tag{3.2.42}$$

The proof of Lemma 3.2.1 can be found in [219].

Proof of Theorem 3.2.6 We apply the Marcinkiewicz inequality

$$\|L_n^* f\|_p \leq c \left(\frac{2\pi}{2n+1} \sum_{k=0}^{2n} |f(\tau_k)|^p \right)^{1/p}, \quad \tau_k = \frac{2\pi k}{2n+1}, \tag{3.2.43}$$

3.2 Discrete Operators

as stated in (3.2.36). On the other hand, we set $I_j = (\tau_j, \tau_{j+1})$ and use (3.2.42), with $a = \tau_j$, $\delta = \tau_{j+1} - \tau_j$, so that

$$|f(\tau_j)| \leq c \left[\left(\frac{2n+1}{2\pi}\right)^{1/p} \|f\|_{L^p(I_j)} + \int_0^{2\pi/(2n+1)} \frac{\omega^k(f,t)_{L^p(I_j)}}{t^{1+1/p}} dt \right] \quad (3.2.44)$$

holds, since $f \in C_{2\pi}$. Consequently we get, by (3.2.43) and (3.2.44),

$$\|L_n^* f\|_p \leq c \left(\frac{2\pi}{2n+1} \sum_{j=0}^{2n} |f(\tau_j)|^p\right)^{1/p}$$

$$\leq c \left(\sum_{j=0}^{2n} \left[\|f\|_{L^p(I_j)}^p + \frac{2\pi}{2n+1} \left(\int_0^{2\pi/(2n+1)} \frac{\omega^k(f,t)_{L^p(I_j)}}{t^{1+1/p}} dt\right)^p\right]\right)^{1/p}$$

$$\leq c\|f\|_p + c \left[\frac{2\pi}{2n+1} \sum_{j=0}^{2n} \left(\int_0^{2\pi/(2n+1)} \frac{\omega^k(f,t)_{L^p(I_j)}}{t^{1+1/p}} dt\right)^p\right]^{1/p}$$

$$= c\|f\|_p + c \left[\frac{1}{n} \sum_{j=0}^{2n} \left(\int_0^{1/n} \frac{\omega^k\left(f, t\frac{2\pi n}{2n+1}\right)_{L^p(I_j)}}{t^{1+1/p}} dt\right)^p\right]^{1/p}$$

and using the well-known property $\omega^k(f, \lambda t)_p \leq ([\lambda]+1)^k \omega^k(f,t)_p$, for each positive λ, we obtain

$$\|L_n^* f\|_p \leq c\|f\|_p + \frac{c}{n^{1/p}} \left[\sum_{j=0}^{2n} \left(\int_0^{1/n} \frac{\omega^k(f,t)_{L^p(I_j)}}{t^{1+1/p}} dt\right)^p\right]^{1/p}. \quad (3.2.45)$$

An application of the generalized Minkowski inequality [206, Theorem 201] to the last sum gives

$$\left[\sum_{j=0}^{2n} \left(\int_0^{1/n} \frac{\omega^k(f,t)_{L^p(I_j)}}{t^{1+1/p}} dt\right)^p\right]^{1/p} < \int_0^{1/n} \left(\sum_{j=0}^{2n} \omega^k(f,t)_{L^p(I_j)}^p\right)^{1/p} \frac{dt}{t^{1+1/p}}. \quad (3.2.46)$$

Now, by Remark 1.2.2, we choose $g = g_t$ such that $\|f - g\|_p + t^k \|g^{(k)}\|_p \leq C\omega(f,t)_p$. For such a function g, using the properties 3° and 9° of the modulus of smoothness (Sect. 1.2.3), we find

$$\omega^k(f,t)_{L^p(I_j)} \leq \omega^k(f-g,t)_{L^p(I_j)} + \omega^k(g,t)_{L^p(I_j)}$$
$$\leq \|f-g\|_{L^p(I_j)} + \omega^k(g,t)_{L^p(I_j)}$$
$$\leq \|f-g\|_{L^p(I_j)} + t^k \|g^{(k)}\|_{L^p(I_j)}.$$

Thus, we get by (1.2.31)

$$\left(\sum_{j=0}^{2n}\omega^k(f,t)_{L^p(I_j)}^p\right)^{1/p} \le c \left(\sum_{j=0}^{2n}\left[\|f-g\|_{L^p(I_j)}^p + t^{kp}\|g^{(k)}\|_{L^p(I_j)}^p\right]\right)^{1/p}$$

$$\le c\left(\|f-g\|_p + t^k\|g^{(k)}\|_p\right) \le c\omega^k(f,t)_p,$$

i.e.,

$$\left(\sum_{j=0}^{2n}\omega^k(f,t)_{L^p(I_j)}^p\right)^{1/p} \le c\omega^k(f,t)_p. \tag{3.2.47}$$

Consequently, we have

$$\left[\sum_{j=0}^{2n}\left(\int_0^{1/n}\frac{\omega^k(f,t)_{L^p(I_j)}}{t^{1+1/p}}dt\right)^p\right]^{1/p} \le c \int_0^{1/n}\frac{\omega^k(f,t)}{t^{1+1/p}}dt$$

which, together with (3.2.45), gives

$$\|L_n^* f\|_p \le c\left[\|f\|_p + \frac{1}{n^{1/p}}\int_0^{1/n}\frac{\omega^k(f,t)_p}{t^{1+1/p}}dt\right]. \tag{3.2.48}$$

Now, recall (1.2.32) and let $T \in \mathcal{T}_n$ be such that

$$\|f-T\|_p + \frac{\|T^{(k)}\|_p}{n^k} \sim \omega^k\left(f,\frac{1}{n}\right)_p.$$

Applying (3.2.48) and the properties 3° and 9° of the modulus of smoothness (Sect. 1.2.3), we have

$$\|f - L_n^* f\|_p \le \|f-T\|_p + \|L_n^*(f-T)\|_p$$

$$\le c\left[\|f-T\|_p + \frac{1}{n^{1/p}}\int_0^{1/n}\frac{\omega^k(f-T,t)_p}{t^{1+1/p}}dt\right]$$

$$\le c\left[\|f-T\|_p + \frac{1}{n^{1/p}}\int_0^{1/n}\frac{\omega^k(T,t)_p}{t^{1+1/p}}dt\right.$$

$$\left. + \frac{1}{n^{1/p}}\int_0^{1/n}\frac{\omega^k(f,t)_p}{t^{1+1/p}}dt\right]$$

$$\le c\left[\|f-T\|_p + \frac{\|T^{(k)}\|_p}{n^k} + \frac{1}{n^{1/p}}\int_0^{1/n}\frac{\omega^k(f,t)_p}{t^{1+1/p}}dt\right]$$

3.2 Discrete Operators

$$\sim \omega^k(f, 1/n)_p + \frac{1}{n^{1/p}} \int_0^{1/n} \frac{\omega^k(f,t)_p}{t^{1+1/p}} \, dt.$$

Since $\omega^k(f,t)_p$ is an increasing function in t, we can write

$$\omega^k(f, 1/n)_p = n\omega^k(f, 1/n)_p \int_{1/n}^{2/n} dt \leq n \int_{1/n}^{2/n} \frac{\omega^k(f,t)_p}{t^{1+1/p}} t^{1+1/p} dt$$

$$\leq \frac{2^{1+1/p}}{n^{1/p}} \int_0^{2/n} \frac{\omega^k(f,t)_p}{t^{1+1/p}} dt \leq \frac{2^k}{n^{1/p}} \int_0^{1/n} \frac{\omega^k(f,t)_p}{t^{1+1/p}} dt,$$

so that (3.2.41) follows immediately. $\square$

Starting from the estimate (3.2.41) we can deduce the behaviour of the Lagrange operator in some subspaces of the L^p space. We remark that the Lagrange operator, because of its discrete nature, can only be defined on spaces of everywhere defined and bounded functions. Moreover, we point out that, even if it is defined, in general L_n^* is unbounded in the L^p spaces. This means that we have to consider L_n^* as a map in some suitable subspaces of L^p, containing continuous functions. The first result in this sense is given by the next theorem.

Theorem 3.2.7 *Let $1 < p < +\infty$ and $r \in \mathbb{N}$. Then for any $f \in W_r^p$ we have*

$$\sup_n \|L_n^* f\|_{W_r^p} \leq c \|f\|_{W_r^p} \tag{3.2.49}$$

where $c \neq c(f)$.

Proof First we write for any $f \in W_r^p$

$$\|L_n^* f\|_{W_r^p} \leq \|f\|_{W_r^p} + \|f - L_n^* f\|_{W_r^p}$$

$$= \|f\|_{W_r^p} + \|f - L_n^* f\|_p + \|(f - L_n^* f)^{(r)}\|_p. \tag{3.2.50}$$

We have by (1.2.30) and (1.2.29)

$$\omega^r(f,t)_p \leq cK^r(f,t)_p \leq ct^r \|f^{(r)}\|_p. \tag{3.2.51}$$

Thus, if we set $k = r \geq 1$ in (3.2.41) and use (3.2.51), we get

$$\|f - L_n^* f\|_p \leq \frac{c}{n^r} \|f^{(r)}\|_p, \quad c \neq c(n,f). \tag{3.2.52}$$

Moreover, recalling that $(S_n f)^{(r)} = S_n f^{(r)}$ and using (3.1.9) and the Bernstein inequality (1.2.35), we have

$$\|(f - L_n^* f)^{(r)}\|_p \leq \|f^{(r)} - S_n f^{(r)}\|_p + \|(S_n f - L_n^* f)^{(r)}\|_p$$

$$\leq c E_n^*(f^{(r)})_p + n^r \|f - S_n f\|_p + n^r \|f - L_n^* f\|_p$$

$$\leq c E_n^*(f^{(r)})_p + cn^r E_n^*(f)_p + n^r \|f - L_n^* f\|_p. \tag{3.2.53}$$

Consequently, by the Favard inequality (1.2.34), noting that $E_n^*(f^{(r)})_p \leq \|f^{(r)}\|_p$ and recalling (3.2.52), we obtain

$$\|(f - L_n^* f)^{(r)}\|_p \leq c E_n^*(f^{(r)})_p + n^r \|f - L_n^* f\|_p \leq c\|f^{(r)}\|_p, \qquad (3.2.54)$$

where c is independent of f and n.

Finally, (3.2.50), (3.2.52) and (3.2.54) give (3.2.49). □

Corollary 3.2.2 *Assume* $f \in W_s^p$, $1 < p < +\infty$ *and* $s \geq 1$. *Then for each integer* r *with* $0 \leq r \leq s$, *we have*

$$\|f - L_n^* f\|_{W_r^p} \leq \frac{c}{n^{s-r}} \|f\|_{W_s^p}, \qquad c \neq c(f, n). \qquad (3.2.55)$$

Proof Starting from (3.2.53) we find

$$\|(f - L_n^* f)^r\|_p \leq c \left[E_n^*(f^{(r)})_p + n^r E_n^*(f)_p + n^r \|f - L_n^* f\|_p \right].$$

Since $f \in W_s^p$ and $0 \leq r \leq s$, by iterating the Favard inequality (1.2.34), we deduce that

$$E_n^*(f^{(r)})_p \leq \frac{c}{n^{s-r}} \|f^{(s)}\|_p \quad \text{and} \quad n^r E_n^*(f)_p \leq \frac{c}{n^{s-r}} \|f^{(s)}\|_p.$$

Moreover, we get by (3.2.52)

$$n^r \|f - L_n^* f\|_p \leq \frac{c}{n^{s-r}} \|f^{(s)}\|_p.$$

Thus, we have

$$\|f - L_n^* f\|_{W_r^p} = \|f - L_n^* f\|_p + \|(f - L_n^* f)^{(r)}\|_p$$
$$\leq c \left[\frac{\|f^{(s)}\|_p}{n^s} + \frac{\|f^{(s)}\|_p}{n^{s-r}} \right] \leq \frac{c}{n^{s-r}} \|f\|_{W_s^p}$$

and the assertion follows. □

Theorem 3.2.7 and Corollary 3.2.2 give us the uniform boundedness of the Lagrange operator in the Sobolev spaces and the convergence estimate in the Sobolev norm, respectively. Similar results can also be achieved in the Besov spaces $B_{r,q}^p$. Indeed, we have:

Theorem 3.2.8 *Let* $1 < p < +\infty$, $1/p < r \in \mathbb{R}$ *and* $1 \leq q \leq +\infty$. *Then we have for any* $f \in B_{r,q}^p$

$$\sup_n \|L_n^* f\|_{B_{r,q}^p} \leq c \|f\|_{B_{r,q}^p}, \qquad (3.2.56)$$

where c *is independent of* f.

3.2 Discrete Operators

Proof Write
$$\|L_n^* f\|_{B_{r,q}^p} \leq \|f\|_{B_{r,q}^p} + \|f - L_n^* f\|_{B_{r,q}^p}$$
and consider the second term on the right-hand side.

Assuming $1 \leq q < +\infty$, we have

$$\|f - L_n^* f\|_{B_{r,q}^p} \sim \|f - L_n^* f\|_p + \left(\sum_{i \geq 1} (1+i)^{rq-1} E_i^* (f - L_n^* f)_p^q \right)^{1/q}$$

$$\leq cn^r \|f - L_n^* f\|_p + \left(\sum_{i \geq n} (1+i)^{rq-1} E_i^* (f)_p^q \right)^{1/q}, \quad (3.2.57)$$

since

$$E_i(f - L_n^* f)_p \begin{cases} = E_i(f)_p & \text{if } i \geq n, \\ \leq \|f - L_n^* f\|_p & \text{if } 1 \leq i < n. \end{cases} \quad (3.2.58)$$

On the other hand, by (3.2.41) and the Hölder inequality, we get

$$\|f - L_n^* f\|_p \leq \frac{c}{n^{1/p}} \int_0^{1/n} \frac{\omega^k(f,t)_p}{t^{r+1/q}} t^{r-1-1/p+1/q} \, dt$$

$$\leq \frac{c}{n^r} \left(\int_0^{1/n} \left[\frac{\omega^k(f,t)_p}{t^r}\right]^q \frac{dt}{t}\right)^{1/q}. \quad (3.2.59)$$

Hence we obtain by (3.2.57) and (3.2.59)

$$\|f - L_n^* f\|_{B_{r,q}^p} \leq c \left(\int_0^{1/n} \left[\frac{\omega^k(f,t)_p}{t^r}\right]^q \frac{dt}{t}\right)^{1/q} + c \left(\sum_{i \geq n} (1+i)^{rq-1} E_i^*(f)_p^q\right)^{1/q}$$

$$\leq c \left(\|f\|_{B_{r,q}^p} + \|f\|_{E_{r,q}^p}\right) \leq c \|f\|_{B_{r,q}^p}.$$

Thus, the theorem follows for $q < +\infty$.

The case $q = +\infty$ is similar. Namely, we have

$$\|f - L_n^* f\|_{B_{r,\infty}^p} \sim \|f - L_n^* f\|_p + \sup_{i \geq 1} i^r E_i^* (f - L_n^* f)_p$$

$$\leq cn^r \|f - L_n^* f\|_p + \sup_{i \geq n} i^r E_i^*(f)_p \quad (3.2.60)$$

having used (3.2.58). Since $f \in B_{r,\infty}^p$ then $\sup_{t>0} \dfrac{\omega^k(f,t)_p}{t^r} < +\infty$. Thus, we conclude from (3.2.41) that

$$\|f - L_n^* f\|_p \leq \frac{c}{n^{1/p}} \int_0^{1/n} \frac{\omega^k(f,t)_p}{t^{1+1/p}} \, dt$$

$$\leq \sup_{t>0} \frac{\omega^r(f,t)_p}{t^r} \left[\frac{c}{n^{1/p}} \int_0^{1/n} t^{r-1-1/p}\, dt \right] \leq \frac{c}{n^r} \|f\|_{B^p_{r,\infty}},$$

assuming $r > 1/p$. Consequently, the following inequalities

$$\|f - L_n^* f\|_{B^p_{r,q}} \leq c \left[\|f\|_{B^p_{r,q}} + \|f\|_{E^p_{r,q}} \right] \leq c \|f\|_{B^p_{r,q}}$$

also hold in the case $q = +\infty$ and the proof is complete. □

Corollary 3.2.3 *Let* $1 < p < +\infty$, $1/p < s \in \mathbb{R}$, $1 \leq q \leq +\infty$ *and* $f \in B^p_{s,q}$. *Then for any* $r \in \mathbb{R}$, *with* $0 \leq r \leq s$, *we have*

$$\|f - L_n^* f\|_{B^p_{r,q}} \leq \frac{c}{n^{s-r}} \|f\|_{B^p_{s,q}}, \qquad c \neq c(f,n). \tag{3.2.61}$$

Proof Assume $q < +\infty$ and start from (3.2.57). Since now $f \in B^p_{s,q}$, replacing r by s in (3.2.59), we get

$$\|f - L_n^* f\|_p \leq \frac{c}{n^s} \left(\int_0^{1/n} \left[\frac{\omega^k(f,t)_p}{t^s} \right]^q \frac{dt}{t} \right)^{1/q} \leq \frac{c}{n^s} \|f\|_{B^p_{s,q}}. \tag{3.2.62}$$

Moreover, we note that

$$\left(\sum_{i \geq n} (1+i)^{rq-1} E_i^*(f)_p^q \right)^{1/q} \leq \frac{1}{n^{s-r}} \left(\sum_{i \geq n} (1+i)^{sq-1} E_i^*(f)_p^q \right)^{1/q}$$

$$\leq c \frac{\|f\|_{E^p_{s,q}}}{n^{s-r}} \sim \frac{\|f\|_{B^p_{r,q}}}{n^{s-r}}. \tag{3.2.63}$$

Thus, the assertion follows from (3.2.57), (3.2.62), and (3.2.63), for $1 \leq q < +\infty$. In the case $q = +\infty$ we use a similar argument starting from (3.2.60). □

In conclusion, the Lagrange operator L_n^* is unbounded in the L^p-space, since it is a discrete operator, but it is bounded in some special subspaces of L^p, more precisely, in the Sobolev spaces by Theorem 3.2.7 and in the Besov space by Theorem 3.2.8.

Remark 3.2.2 All results, which we stated for the Lagrange operator, also hold for the de la Vallée Poussin interpolation. Their proofs are similar and based on (3.2.28). However, for $1 < p < +\infty$ we do not study the de la Vallée Poussin L^p interpolation error explicitly, since in this case (as for the continuous Fourier and de la Vallée Poussin operators) we have an optimal error estimate by using the simpler Lagrange interpolation and then the de la Vallée Poussin interpolation is not of much interest in the L^p-space for $1 < p < +\infty$.

3.2 Discrete Operators

3.2.6 Some Estimates of the Interpolation Errors in L^1-Sobolev Spaces

In the L^1-norm we have already stated (see Remark 3.2.1) that

$$\|\widetilde{V}_n f\|_1 \le \frac{4\log 2}{\pi^2} \left(\frac{2\pi}{3m+1} \sum_{k=0}^{3m} |f(t_k)| \right), \qquad t_k = \frac{2k\pi}{3m+1}, \qquad (3.2.64)$$

$$\|L_n^* f\|_1 \le \frac{4}{\pi^2} \log n \left(\frac{2\pi}{2m+1} \sum_{k=0}^{3m} |f(\tau_k)| \right), \qquad \tau_k = \frac{2k\pi}{2m+1}. \qquad (3.2.65)$$

Using such Marcinkiewicz-type inequalities, we can deduce the following L^1 error estimates for both the Lagrange and de la Vallée Poussin interpolation.

Theorem 3.2.9 *Let f be an absolutely continuous function. Then for all $n \in \mathbb{N}$, we have*

$$\|f - \widetilde{V}_n f\|_1 \le C \frac{E_n(f')_1}{n}, \qquad C \ne C(n, f), \qquad (3.2.66)$$

and

$$\|f - L_n^* f\|_1 \le C \log n \frac{\|f'\|_1}{n}, \qquad C \ne C(n, f). \qquad (3.2.67)$$

Proof Let us prove (3.2.66). We obtain by (3.2.64) and (3.2.10)

$$\|\widetilde{V}_n f\|_1 \le \frac{4\log 2}{\pi^2} \left(\frac{2\pi}{3n+1} \sum_{k=0}^{3n} |f(t_k)| \right) \le C \left(\|f\|_1 + \frac{\|f'\|_1}{n} \right).$$

Consequently, for all trigonometric polynomial $T \in \mathcal{T}_n$, using the invariance property $\widetilde{V}_n T = T$, we can write

$$\|f - \widetilde{V}_n f\|_1 \le \|f - T\|_1 + \|\widetilde{V}_n(f - T)\|_1 \le C\left[\|f - T\|_1 + \frac{\|f' - T'\|_1}{n} \right]$$

and as in (3.2.11), we have

$$\|f' - T'\|_1 \le C E_n^*(f')_1 + Cn\|f - T\|_1.$$

Thus, we get for all $T \in \mathcal{T}_n$

$$\|f - \widetilde{V}_n f\|_1 \le C \left[\|f - T\|_1 + \frac{\|f' - T'\|_1}{n} \right] \le C\|f - T\|_1 + C E_n^*(f')_1,$$

from which we deduce (3.2.66) taking the infimum with respect to $T \in \mathcal{T}_n$ and applying the Favard inequality (1.2.34), since

$$\|f - \widetilde{V}_n f\|_1 \leq C\left[E_n^*(f)_1 + \frac{E_n^*(f')_1}{n}\right] \leq \frac{C}{n} E_n^*(f')_1.$$

Finally, the proof of (3.2.67) is similar to that of (3.2.66) and based on (3.2.65) and (3.2.10). For the sake of brevity we omit this proof. $\square$

Starting from (3.2.66), we can deduce an error estimate similar to (3.2.41), but for $p = 1$. More precisely, we have the following statement:

Theorem 3.2.10 *If f is an absolutely continuous function such that*

$$\int_0^1 \frac{\omega^k(f,t)_1}{t^2} dt < +\infty, \qquad k > 1,$$

holds, then for all $n \in \mathbb{N}$ with $n \geq k$, we have

$$\|f - \widetilde{V}_n f\|_1 \leq \frac{C}{n} \int_0^{1/n} \frac{\omega^k(f,t)_1}{t^2} dt \tag{3.2.68}$$

and

$$\|f - L_n^* f\|_1 \leq \frac{C}{n} \log n \int_0^{1/n} \frac{\omega^k(f,t)_1}{t^2} dt \tag{3.2.69}$$

Proof Let $T_n^* \in \mathcal{T}_n$ be such that $\|f - T_n^*\|_1 = E_n^*(f)_1$. We get by (3.2.66) and (3.2.67)

$$\|f - \widetilde{V}_n f\|_1 = \|(f - T_n^*) - \widetilde{V}_n(f - T_n^*)\|_1 \leq \frac{C}{n}\|(f - T_n^*)'\|_1 \tag{3.2.70}$$

and

$$\|f - L_n^* f\|_1 = \|(f - T_n^*) - L_n^*(f - T_n^*)\|_1 \leq \frac{C}{n} \log n \|(f - T_n^*)'\|_1. \tag{3.2.71}$$

On the other hand, when proving Proposition 3.2.2 (see (3.2.14)), we stated

$$\|(f - T_n^*)'\|_1 \leq \sum_{k=0}^{+\infty} \|(T_{2^{k+1}n}^* - T_{2^k n}^*)'\|_1 \leq C \int_0^{1/n} \frac{\omega^k(f,t)_1}{t^2} dt \tag{3.2.72}$$

Thus, the estimates (3.2.68) and (3.2.69) follow by (3.2.70) and (3.2.71), respectively, using (3.2.72). $\square$

By Theorem 3.2.9 we can deduce the uniform boundedness of the discrete operator $\widetilde{V}_n$ in some subspaces of L^1 containing absolutely continuous functions. In fact, the following corollary holds:

3.2 Discrete Operators

Corollary 3.2.4 *Let $k \in \mathbb{N}$ and set $\widetilde{W}_k^1 := AC_{2\pi}^0 \cap W_k^1$. Then the discrete de la Vallée Poussin operator $\widetilde{V}_n$ is uniformly bounded in $\widetilde{W}_k^1$ with respect to n, i.e., we have*

$$\sup_n \|\widetilde{V}_n\|_{\widetilde{W}_k^1 \to \widetilde{W}_k^1} < +\infty. \tag{3.2.73}$$

Moreover, for all functions $f \in \widetilde{W}_k^1$ and for any $h \in \mathbb{N}$, such that $0 \le h \le k$,

$$\|f - \widetilde{V}_n f\|_{W_h^1} \le \frac{C}{n^{k-h}} \|f\|_{W_k^1}, \tag{3.2.74}$$

where $C \ne C(n, f)$.

Proof The uniform boundedness result (3.2.73) follows from (3.2.74) by taking $h = k$, since

$$\|\widetilde{V}_n f\|_{W_k^1} \le \|f - \widetilde{V}_n f\|_{W_k^1} + \|f\|_{W_k^1} \le (1+C)\|f\|_{W_k^1}.$$

Thus, we have to prove (3.2.74). First, we have

$$\|f - \widetilde{V}_n f\|_{W_h^1} = \|f - \widetilde{V}_n f\|_1 + \|(f - \widetilde{V}_n f)^{(h)}\|_1. \tag{3.2.75}$$

Using (3.2.66) and the well-known estimate

$$E_n^*(f)_1 \le \frac{C}{n^k} E_n^*(f^{(k)})_1, \tag{3.2.76}$$

we can estimate the first term in (3.2.75) as follows

$$\|f - \widetilde{V}_n f\|_1 \le \frac{C}{n} E_n^*(f')_1 \le \frac{C}{n^k} E_n^*(f^{(k)})_1 \le \frac{C}{n^k} \|f^{(k)}\|_1 \le \frac{C}{n^{k-h}} \|f\|_{W_k^1}.$$

For the second term in (3.2.75), we have

$$\|(f - \widetilde{V}_n f)^{(h)}\|_1 \le \|(f - V_n f)^{(h)}\|_1 + \|(V_n f - \widetilde{V}_n f)^{(h)}\|_1.$$

By (3.1.14), taking into account that $(V_n f)^h = V_n f^{(h)}$, we get

$$\|(f - V_n f)^{(h)}\|_1 = \|f^{(h)} - V_n f^{(h)}\|_1 \le C E_n^*(f^{(h)})_1. \tag{3.2.77}$$

On the other hand, iterating the Bernstein and Favard inequalities (1.2.35) and (1.2.34), and using (3.2.76), we obtain

$$\|(V_n f - \widetilde{V}_n f)^{(h)}\|_1 \le C n^h \|V_n f - \widetilde{V}_n f\|_1 \le C n^h \big(\|V_n f - f\|_1 + \|\widetilde{V}_n f - f\|_1\big)$$

$$\le C n^h \left[E_n^*(f)_1 + \frac{E_n^*(f')_1}{n} \right] \le C E_n^*(f^{(h)})_1.$$

Hence we have stated

$$\|(f - \widetilde{V}_n f)^{(h)}\|_1 \le \|(f - V_n f)^{(h)}\|_1 + \|(V_n f - \widetilde{V}_n f)^{(h)}\|_1 \le C E_n^*(f^{(h)})_1.$$

Finally, an iterative application of (1.2.34) gives

$$\|(f - \widetilde{V}_n f)^{(h)}\|_1 \le C E_n^*(f^{(h)})_1 \le \frac{C}{n^{k-h}} E_n^*(f^{(k)})_1 \le \frac{C}{n^{k-h}} \|f^{(k)}\|_1$$

$$\le \frac{C}{n^{k-h}} \|f\|_{W_k^1},$$

which concludes the proof. □

3.2.7 The Weighted Case

Here we give the weighted version of some of the results stated in the previous sections. The main arguments and theorems are extracted from [307].

We define "doubling" each integrable and 2π-periodic weight w (shortly $w \in D$) for which there exists a constant L such that

$$\int_{2I} w(x)dx \le L \int_I w(x)\,dx, \qquad (3.2.78)$$

for each interval I, being $2I$ the interval twice the length of I and with the same midpoint. The smallest constant L for which (3.2.78) holds is named "doubling constant".

Several properties of these weights can be found in [311, 456]. We recall here some inequalities we shall use in the sequel.

The first one is the Bernstein type inequality (see Theorem 3.1 in [311])

$$\int_0^{2\pi} |T'(x)|^p w(x)\,dx \le C n^p \int_0^{2\pi} |T(x)|^p w(x)\,dx \qquad (3.2.79)$$

that holds for every polynomial $T \in \mathcal{T}_n$, $1 \le p < +\infty$ and for each weight $w \in D$.

To state the second one we introduce the weight function w_n associated to w and defined as

$$w_n(x) = n \int_{x-1/n}^{x+1/n} w(t)\,dt, \qquad n \in \mathbb{N}. \qquad (3.2.80)$$

In the sequel, if $w = u^p$ we shall write $w_n = (u^p)_n$. w_n is a continuous function and moreover there exist two positive constants K and s, independent of n, such that for each x and y

$$w_n(x) \le K\big(1 + n|x - y|\big)^s w_n(y). \qquad (3.2.81)$$

This property is also equivalent [311] to definition (3.2.78).

Then we can state the following equivalence (see Theorem 2.1 in [311]):

For every $1 \le p < +\infty$ there is a constant C such that for every polynomial $T \in \mathcal{T}_n$ we have

3.2 Discrete Operators

$$\frac{1}{C}\int_0^{2\pi} |T(x)|^p w(x)\, dx \le \int_0^{2\pi} |T(x)|^p w_n(x)\, dx \le C\int_0^{2\pi} |T(x)|^p w(x)\, dx. \tag{3.2.82}$$

Moreover, we denote by $\lambda_n(w,t)$ the n-th Christoffel function related to the weight w and defined as

$$\lambda_n(w,t) = \inf_{T\in\mathcal{T}_n} \int_0^{2\pi} \left(\frac{T(x)}{T(t)}\right)^2 w(x)\, dx.$$

In [311, Theorem 3.3] the following estimate was proved

$$\frac{1}{Cn} w_n(t) \le \lambda_n(w,t) \le \frac{C}{n} w_n(t). \tag{3.2.83}$$

A special subclass of the doubling weights is the so called "A_p-class" of Muckenhoupt ([370]).

We say that a doubling weight w is an A_p weight ($w \in A_p$), $1 < p < +\infty$, if there exists a constant A such that for all intervals $I \subset [0, 2\pi)$

$$\left(\frac{1}{|I|}\int_I w(x)dx\right) \left(\frac{1}{|I|}\int_I w^{-p'/p}(x)dx\right)^{p/p'} \le A, \quad p' = \frac{p}{p-1}, \tag{3.2.84}$$

where $|I|$ is the measure of I. The smallest constant A for which (3.2.84) holds is named the "A_p constant" related to w.

The properties of this class of weights can be found for instance in [456, pp. 194–203].

We remark that in the general case a doubling weight is not also an A_p weight [370]. In fact, for example, a doubling weight can vanish on a set of positive measure (without being identically zero [456, Chap. 1, Sect. 8.8]). Moreover, a doubling weight can be unbounded on each subinterval of $[0, 2\pi)$.

The A_p weights give the class for which many singular and maximal operators are bounded in the related L^p weighted spaces. In particular, if

$$H(f,x) := \frac{1}{\pi} \text{P.V.} \int_{-\infty}^{+\infty} \frac{f(y)}{y-x}\, dy = \lim_{\varepsilon\to 0} \frac{1}{\pi}\int_{|y-x|\ge\varepsilon} \frac{f(y)}{y-x}\, dy$$

is the Hilbert transform of the function f, then

$$\int_0^{2\pi} |H(f,x)|^p w(x)dx \le C\int_0^{2\pi} |f(x)|^p w(x)\, dx \tag{3.2.85}$$

if and only if $w \in A_p$ (see [370]).

Moreover, (3.2.85) is equivalent to the boundedness of the Fourier operator (see [216, Theorem 8, p. 245]), i.e.,

$$\int_0^{2\pi} |S_n(f,x)|^p w(x)\, dx \le C\int_0^{2\pi} |f(x)|^p w(x)\, dx. \tag{3.2.86}$$

Sometimes in the sequel we assume $w = u^p$, $1 < p < +\infty$. Thus, the condition (3.2.84) becomes

$$\left(\int_I u^p\right)^{1/p} \left(\int_I u^{-p'}\right)^{1/p'} \leq C|I|, \quad p' = \frac{p}{p-1}, \tag{3.2.87}$$

for each interval $I \subset [0, 2\pi)$, where $|I|$ is the measure of I.

Moreover, we explicitly remark that if $u^p \in A_p$ then $u \in A_p$. Therefore, if $(u^p)_n$ satisfies (3.2.81) so does u_n.

First, we give the weighted version of (3.2.24).

Theorem 3.2.11 *Assume $u^p \in D$, $1 < p < +\infty$, and let $0 = \vartheta_0 < \vartheta_1 < \cdots < \vartheta_n < \vartheta_{n+1} = 2\pi$, with $\vartheta_{k+1} - \vartheta_k \sim n^{-1}$, $k = 0, 1, \ldots, n$. Then, there exists a constant C depending on the doubling constant such that, for each polynomial $T \in \mathcal{T}_{\ell n}$ (ℓ is a fixed integer),*

$$\sum_{k=0}^n \lambda_n(u^p, \vartheta_k)|T(\vartheta_k)|^p \leq C \int_0^{2\pi} |T(x)u(x)|^p\,dx. \tag{3.2.88}$$

Proof First we note that for $k = 0, 1, \ldots, n$, we have

$$|T(\vartheta_k)|^p (\vartheta_{k+1} - \vartheta_k) \leq 2^{p-1}\left[\int_{\vartheta_k}^{\vartheta_{k+1}} |T(x)|^p dx \right.$$
$$\left. + (\vartheta_{k+1} - \vartheta_k)^p \int_{\vartheta_k}^{\vartheta_{k+1}} |T'(x)|^p dx\right] \tag{3.2.89}$$

Multiply by $(u^p)_n(\vartheta_k)$ both sides of the previous inequality. Since $\vartheta_{k+1} - \vartheta_k \sim n^{-1}$, from (3.2.81) we have $(u^p)_n(\vartheta_k) \sim (u^p)_n(x)$, for each $x \in [\vartheta_k, \vartheta_{k+1}]$ and therefore,

$$|T(\vartheta_k)|^p \frac{(u^p)_n(\vartheta_k)}{n} \leq C\left[\int_{\vartheta_k}^{\vartheta_{k+1}} |T(x)|^p (u^p)_n(x)\,dx \right.$$
$$\left. + \frac{1}{n^p}\int_{\vartheta_k}^{\vartheta_{k+1}} |T'(x)|^p (u^p)_n(x)\,dx\right].$$

Then, by taking the sum with respect to k, we have

$$\sum_{k=0}^n |T(\vartheta_k)|^p \frac{(u^p)_n(\vartheta_k)}{n} \leq C\left[\int_0^{2\pi} |T(x)|^p (u^p)_n(x)dx \right.$$
$$\left. + \frac{1}{n^p}\int_0^{2\pi} |T'(x)|^p (u^p)_n(x)\,dx\right]$$

and (3.2.88) follows from (3.2.82), (3.2.79) and (3.2.83). □

3.2 Discrete Operators

Inequality (3.2.88) can be found in [311] with a different proof. Moreover, (3.2.88) with $u^p \in A_p$ and ϑ_k equispaced in $[0, 2\pi)$ can be found in [227, Theorem 1, p. 112]. However, the procedure followed in [227] is substantially based on (3.2.86) which in general does not hold for doubling weights. In the same paper [227], among other results, (3.2.88) is also generalized to the multivariate case.

Inequality (3.2.88) is not completely invertible under the same assumptions of the previous theorem, i.e., u^p is doubling, the degree of the polynomial is ℓn and there are exactly n knots ϑ_k that satisfy $\vartheta_{k+1} - \vartheta_k \sim n^{-1}$. If we do not make assumptions on the number of the knots we have the following result, proved in [311]:

Theorem 3.2.12 *Let u^p be a doubling weight and $1 \le p < +\infty$. Then there are two constants M and C such that for all m and $T \in \mathcal{T}_n$ we have*

$$\int_0^{2\pi} |T(x)u(x)|^p\, dx \le C \sum_{k=0}^{S} \lambda_n(u^p, \vartheta_k) |T(\vartheta_k)|^p$$

provided the points $\vartheta_0 < \vartheta_1 < \cdots < \vartheta_S$ satisfy $\vartheta_{k+1} - \vartheta_k \le \frac{1}{Mn}$ and $\vartheta_S \ge \vartheta_0 + 2\pi$.

As we can see, in the previous theorem the number of knots is not specified. This fact can produce a gap in the applications.

If we assume that $u^p \in A_p$, the points ϑ_k are equispaced and their number is exactly equal to the number of the coefficients of the trigonometric polynomial $T \in \mathcal{T}_n$, then (3.2.88) is invertible with $1 < p < +\infty$. In fact the following theorem holds.

Theorem 3.2.13 *Let $u^p \in A_p$, $1 < p < +\infty$ and $\tau_k = 2\pi k/(2n+1)$, $k = 0, 1, \ldots, 2n$. Then there exists a constant C, depending only on the weight u, such that for $T \in \mathcal{T}_n$ we have*

$$\int_0^{2\pi} |T(x)u(x)|^p\, dx \le C \sum_{k=0}^{2n} \lambda_n(u^p, \tau_k) |T(\tau_k)|^p. \tag{3.2.90}$$

Proof We can write

$$\int_0^{2\pi} |T(x)u(x)|^p\, dx = \sup_g \int_0^{2\pi} T(x)g(x)u(x)\, dx, \tag{3.2.91}$$

where g belongs to the set of all functions $\widetilde{g}$ such that

$$\|\widetilde{g}\|_{p'}^{p'} = \int_0^{2\pi} |\widetilde{g}(x)|^{p'}\, dx = 1, \qquad p' = \frac{p}{p-1}.$$

Now, we evaluate integrals at the right-hand side of (3.2.91). Since $T = L_n^* T$ and recalling that S_n denotes the Fourier operator, we get

$$\int_0^{2\pi} T(x)g(x)u(x)\,dx = \frac{1}{2n+1}\sum_{k=0}^{2n} T(\tau_k) \int_0^{2\pi} D_n(x - \tau_k)g(x)u(x)\,dx$$

$$= \frac{2\pi}{2n+1}\sum_{k=0}^{2n} T(\tau_k) S_n(gu, \tau_k)$$

$$= \frac{2\pi}{2n+1}\sum_{k=0}^{2n} T(\tau_k)(u^p)_n^{1/p}(\tau_k) \frac{S_n(gu, \tau_k)}{(u^p)_n^{1/p}(\tau_k)}$$

$$\leq \left(\frac{2\pi}{2n+1}\sum_{k=0}^{2n} |T(\tau_k)|^p (u^p)_n(\tau_k)\right)^{1/p} \times$$

$$\times \left(\frac{2\pi}{2n+1}\sum_{k=0}^{2n} ((u^p)_n(\tau_k))^{-p'/p} |S_n(gu, \tau_k)|^{p'}\right)^{1/p'},$$

where in the last step we applied the Hölder inequality with $p' = p/(p-1)$. Recalling that u^p is doubling, it follows from (3.2.83) that the first sum is equivalent to the right-hand side of (3.2.90). Therefore, we have to prove the uniform boundedness of the second sum. To achieve this, we set $I_k = [\tau_k - n^{-1}, \tau_k + n^{-1}]$. It follows from

$$1 \leq \left(\frac{1}{|I_k|}\int_{I_k} u^p(x)\,dx\right)^{1/p} \left(\frac{1}{|I_k|}\int_{I_k} u^{-p'}(x)\,dx\right)^{1/p'},$$

that

$$[(u^p)_n(\tau_k)]^{-p'/p} \leq \frac{C}{|I_k|}\int_{I_k} u^{-p'}(x)\,dx = C(u^{-p'})_n(\tau_k).$$

Moreover, since $u^p \in A_p$ then $u^{-p'} \in A_{p'}$. Thus, recalling (3.2.83), (3.2.88) and (3.2.86), we get

$$\left(\frac{2\pi}{2n+1}\sum_{k=0}^{2n} ((u^p)_n(\tau_k))^{-p'/p} |S_n(gu, \tau_k)|^{p'}\right)^{1/p'}$$

$$\leq C \left(\sum_{k=0}^{2n} \lambda_n(u^{-p'}, \tau_k)|S_n(gu, \tau_k)|^{p'}\right)^{1/p'}$$

$$\leq C \left(\int_0^{2\pi} |S_n(gu, x)u^{-1}(x)|^{p'}\,dx\right)^{1/p'} \leq C\|g\|_{p'} = C,$$

where C depends only on the doubling constant related to u. □

3.2 Discrete Operators

Theorem 3.2.13, which gives the weighted version of (3.2.25), appeared for the first time in [227] and was extended also to the multivariate case. However, the proof we presented drastically simplifies the one given in [227].

If we want to relax the assumptions on the weight u in Theorem 3.2.13, we can obtain a "weaker" version of (3.2.90), i.e., an inequality of the type (3.2.90) in which the Christoffel function, appearing on the right-hand side, is related to a weight different from u. In order to give a theorem in this direction we associate to the general weight U, its "singular part":

$$v(x) \equiv v(U, x) := \max\{U(x), 1\}. \tag{3.2.92}$$

With this notation we can state the following "weak" Marcinkiewicz inequality, whose proof can be found in [307].

Theorem 3.2.14 *Let u be an arbitrary weight in L^p, $1 < p < +\infty$. Assume $v \equiv v(u^p) = \max\{u^p, 1\} \in A_p$. Then there exists a constant C depending only on u such that for every polynomial $T \in \mathfrak{T}_n$*

$$\int_0^{2\pi} |T(x)u(x)|^p \, dx \leq C \sum_{k=0}^{2n} \lambda_n(v, \tau_k) |T(\tau_k)|^p \tag{3.2.93}$$

where $\tau_k = 2\pi k/(2n+1)$, $k = 0, 1, \ldots, 2n$.

Our aim now is to give some applications of the stated results to the weighted trigonometric interpolation.

We already saw that the Marcinkiewicz inequality (3.2.25) can be used to prove the boundedness of the interpolating operator L_n^* (defined in Sect. 3.2.5) as a map of C^0 into L^p.

An analogous result for the weighted case can be obtained using the "weak" Marcinkiewicz inequality (3.2.93). Let u be an arbitrary weight in L^p, $1 < p < +\infty$, $v = v(u^p) = \max\{u^p, 1\} \in A_p$ and f a continuous function. Then applying (3.2.93), we get

$$\|(L_n^* f)u\|_p \leq C \left(\sum_{k=0}^{2n} \lambda_n(v, \tau_k) |f(\tau_k)|^p \right)^{1/p} \leq C \|f\|_\infty \left(\int_0^{2\pi} v(t) dt \right)^{1/p},$$

i.e., L_n^* is a uniformly bounded operator as a map from C^0 into L_u^p.

Now, we want to investigate the behaviour of L_n^* in some Sobolev and Besov subspaces of the weighted L^p space.

Our first aim is to prove a weighted version of Theorems 3.2.7–3.2.8. To this end we need some additional definitions. With $r \in \mathbb{N}$ and $1 < p < +\infty$, we define the Sobolev space $W_r^p(u)$ as

$$W_r^p(u) = \left\{ f \in L_u^p \;\middle|\; \|f\|_{W_r^p(u)} := \|fu\|_p + \|f^{(r)}u\|_p < +\infty \right\},$$

where $u \in A_p$, $f \in L_u^p$ means that $fu \in L^p$ and $f^{(r)}$ denotes the r-th derivative of f in the sense of distributions.

Define the weighted K-functional of the function $f \in L_u^p$ as

$$K_k(f,t)_{u,p} = \inf_{g^{(k-1)} \in AC} \left\{ \|(f-g)u\|_p + t^k \|g^{(k)}u\|_p \right\},$$

where $k \in \mathbb{N}$ and AC denotes the space of the absolutely continuous functions.

Introduce now the seminorm

$$\|f\|_{p,q,r,u} := \begin{cases} \left(\int_0^1 \left[\dfrac{K_k(f,t)_{u,p}}{t^{r+1/q}} \right]^q dt \right)^{1/q}, & 1 \leq q < +\infty, \\ \sup_{t>0} \dfrac{K_k(f,t)_{u,p}}{t^r}, & q = +\infty, \end{cases}$$

associated to $K_k(f,t)_{u,p}$, $k > r$, where $0 \leq r \in \mathbb{R}$, $k \in \mathbb{N}$ and $1 \leq p \leq +\infty$. Define also the weighted Besov space with parameters p,q,r as

$$B_{r,q}^p(u) = \left\{ f \in L_u^p \mid \|f\|_{B_{r,q}^p(u)} := \|fu\|_p + \|f\|_{p,q,r,u} < +\infty \right\}.$$

We remark that in the non-weighted case, the K-functional $K_k(f,t)_p$ is equivalent to the modulus of smoothness $\omega_k(f,t)_p$ and the definition given here of the Besov spaces reduces to that of $B_{r,q}^p$.

In the weighted case we use the K-functional to define $B_{r,q}^p(u)$, since we do not have a satisfactory definition of a weighted modulus of smoothness for A_p weights. The theoretical problem is that, in the general case, the translation operator is unbounded in weighted spaces.

Now, let $E_n^*(f)_{u,p} = \inf_{T \in \mathcal{T}_n} \|(f-T)u\|_p$ be the error of the best weighted approximation in L^p by trigonometric polynomials. Then we can define the norm for $k > r$

$$\|f\|_{E_{r,q}^p(u)} := \begin{cases} \left(\displaystyle\sum_{i=0}^{+\infty} \left[(1+i)^{r-1/q} E_i^*(f)_{u,p}\right]^q \right)^{1/q}, & 1 \leq q < +\infty, \\ \sup_{i \geq 0}(1+i)^r E_i^*(f)_{u,p}, & q = +\infty, \end{cases}$$

where $1 \leq p \leq +\infty$, $0 \leq r \in \mathbb{R}$. Using the following two inequalities (see [246, Theorem 3])

$$E_n^*(f)_{u,p} \leq C K_k(f, n^{-1})_{u,p}, \qquad n \geq k, \tag{3.2.94}$$

and

$$K_k(f,t)_{u,p} \leq C t^k \sum_{i=0}^{[1/t]} (1+i)^{k-1} E_i^*(f)_{u,p}, \tag{3.2.95}$$

3.2 Discrete Operators

it is possible to prove that (see for instance [99])

$$\|f\|_{B^p_{r,q}(u)} \sim \|f\|_{E^p_{r,q}(u)}. \tag{3.2.96}$$

The following theorem gives two L^p_u estimates of the trigonometric interpolation error.

Theorem 3.2.15 *Assume that $u^p \in A_p$, $1 < p < +\infty$ and let f be a continuous function. Let $r \in \mathbb{N}$, $r \geq 1$. If $f \in W^p_r(u)$ we can write*

$$\|[f - L^*_n f]u\|_p \leq \frac{C}{n^r}\|f\|_{W^p_r(u)}, \tag{3.2.97}$$

where C is a positive constant independent of f and n.
Moreover, let $r \in \mathbb{R}$, $r > 1$, $1 \leq q \leq +\infty$. If $f \in B^p_{r,q}(u)$ then

$$\|[f - L^*_n f]u\|_p \leq \frac{C}{n^r}\|f\|_{B^p_{r,q}(u)}, \tag{3.2.98}$$

where C is a positive constant independent of f and n.

Proof We first prove (3.2.98). From (3.2.90), with $T = L^*_n f$ and (3.2.83), we have

$$\|L^*_n fu\|^p_p \leq C\sum_{i=0}^{2n}\lambda_n(u^p,\tau_i)|f(\tau_i)|^p \leq \frac{C}{n}\sum_{i=0}^{2n}(u^p)_n(\tau_i)|f(\tau_i)|^p. \tag{3.2.99}$$

Set $I_i = [\tau_i, \tau_{i+1}]$, $i = 0, 1, \ldots, 2n$, with $\tau_{2n+1} = 2\pi$. It is known that (cf. [41])

$$|f(\tau_i)| \leq \frac{1}{|I_i|}\int_{I_i}|f(t)|\,dt + \int_0^{1/n}\frac{\omega_k(f,t)_{L^1(I_i)}}{t^2}\,dt, \quad k > 1. \tag{3.2.100}$$

Now, let g be a function such that $g^{(k-1)} \in AC$. From the properties of the moduli of smoothness, we get

$$\omega_k(f,t)_{L^1(I_i)} \leq \omega_k(f-g,t)_{L^1(I_i)} + \omega_k(g,t)_{L^1(I_i)}$$
$$\leq 2^k \int_{I_i}|f(x) - g(x)|\,dx + t^k\int_{I_i}|g^{(k)}(x)|\,dx. \tag{3.2.101}$$

On the other hand, for every $f \in L^p_u$, we obviously have

$$\int_{I_i}|f(x)|\,dx \leq \left(\int_{I_i}|f(x)u(x)|^p\,dx\right)^{1/p}U_i, \tag{3.2.102}$$

where we put $U_i := \left(\int_{I_i}|u(x)|^{-p'}\,dx\right)^{1/p'}$ and $p' = p/(p-1)$. Then, using (3.2.102) in (3.2.101), after standard calculations, we get

$$|f(\tau_i)|^p (u^p)_n(\tau_i)|I_i| \leq C\left(\|fu\|^p_{L^p(I_i)} + |I_i|^p V^p_i\right), \tag{3.2.103}$$

where
$$V_i := \int_0^{1/n} \frac{\|(f-g)u\|_{L^p(I_i)} + t^k \|g^{(k)}u\|_{L^p(I_i)}}{t^2} dt.$$

According to (3.2.99), i.e.,
$$\|L_n^* f u\|_p \leq \left(\frac{C}{n} \sum_{i=0}^{2n} |f(\tau_i)|^p (u^p)_n(\tau_i)\right)^{1/p},$$

we find, after a summation in (3.2.103),
$$\|L_n^* f u\|_p \leq C \|f u\|_p + \frac{C}{n} \left(\sum_{i=0}^{2n} V_i^p\right)^{1/p}.$$

Applying now the Minkowski integral inequality [206, Theorem 201] to the second sum and taking the infimum over all $g^{(k-1)} \in AC$, we obtain
$$\|L_n^* f u\|_p \leq C \left(\|f u\|_p + \frac{1}{n} \int_0^{1/n} \frac{K_k(f,t)_{u,p}}{t^2} dt\right).$$

Then, for $1 \leq q < +\infty$, we get for $r \geq 2$
$$\|L_n^* f u\|_p \leq C \left[\|f u\|_p + \frac{1}{n^r} \left(\int_0^{1/n} \left[\frac{K_k(f,t)_{u,p}}{t^{r+1/q}}\right]^q dt\right)^{1/q}\right]. \qquad (3.2.104)$$

Moreover, for $q = +\infty$
$$\|L_n^* f u\|_p \leq C \left[\|f u\|_p + \frac{1}{n^r} \sup_{t>0} \frac{K_k(f,t)_{u,p}}{t^r}\right]. \qquad (3.2.105)$$

Now let $T \in \mathcal{T}_n$ and $1 \leq q < +\infty$. Then, from (3.2.104), we get
$$\|(f - L_n^* f)u\|_p \leq \|(f - T)u\|_p + \|L_n^*(f - T)u\|_p \qquad (3.2.106)$$
$$\leq C \|(f - T)u\|_p + \frac{C}{n^r} \|f - T\|_{B_{r,q}^p(u)}.$$

Taking the infimum over all $T \in \mathcal{T}_n$ we have
$$\|(f - L_n^* f)u\|_p \leq C E_n^*(f)_{u,p} + \frac{C}{n^r} \|f\|_{B_{r,q}^p(u)}. \qquad (3.2.107)$$

On the other hand, from (3.2.94) and (3.2.95), we deduce using the Hölder inequality
$$E_n^*(f)_{u,p} \leq C K_k(f, n^{-1})_{u,p} \leq \frac{C}{n^k} \sum_{i=0}^n (1+i)^{k-1} E_i^*(f)_{u,p} \qquad (3.2.108)$$

3.2 Discrete Operators

$$\leq \frac{C}{n^r}\left(\sum_{i=0}^{n}\left[(1+i)^{r-1/q}E_i^*(f)_{u,p}\right]^q\right)^{1/q} \leq \frac{C}{n^r}\|f\|_{E_{r,q}^p(u)}.$$

Thus, (3.2.98) follows from (3.2.107), (3.2.108) and (3.2.96). Analogously we can prove the case $q = +\infty$.

To prove (3.2.97) we can proceed similarly. We only have to replace (3.2.100) by the inequality (see [96] and [98])

$$|f(\tau_i)| \leq \frac{1}{|I_i|}\int_{I_i}|f(t)|\,dt + |I_i|^{r-1}\int_{I_i}|f^{(r)}(t)|\,dt, \quad r \geq 1. \qquad (3.2.109)$$

Using (3.2.102) we get

$$|f(\tau_i)|(u^p)_n^{1/p}(\tau_i) \leq C|I_i|^{1/p'}\left[\frac{1}{|I_i|}\|fu\|_{L^p(I_i)} + |I_i|^{r-1}\|f^{(r)}u\|_{L^p(I_i)}\right].$$

Consequently,

$$\|(f - L_n^*f)u\|_p \leq CE_n^*(f)_{u,p} + \frac{C}{n^r}\|f\|_{W_r^p(u)}$$

and (3.2.97) follows since $E_n^*(f)_{u,p} \leq C\|f^{(r)}u\|_p/n^r$. $\square$

A consequence of Theorem 3.2.15 is the uniform boundedness of the operator L_n^* in $B_{r,q}^p(u)$. We mention this result without proof.

Corollary 3.2.5 *Let* $u^p \in A_p$, $1 < p < +\infty$, $1 \leq q \leq +\infty$, $s \in \mathbb{R}$ *and* $s > 1$. *Set* $\widetilde{B}_{s,q}^p(u) = B_{s,q}^p(u) \cap C^0$. *Then we have*

$$\sup_n \|L_n^*\|_{\widetilde{B}_{s,q}^p(u) \to \widetilde{B}_{s,q}^p(u)} < +\infty. \qquad (3.2.110)$$

Moreover, for each function $f \in \widetilde{B}_{s,q}^p(u)$ *and for all* $r \in \mathbb{R}$ *such that* $0 \leq r \leq s$, *we get*

$$\|f - L_n^*f\|_{B_{r,q}^p(u)} \leq \frac{C}{n^{s-r}}\|f\|_{B_{s,q}^p(u)}, \qquad (3.2.111)$$

where C is a positive constant independent of f and n.

Chapter 4
Algebraic Interpolation in Uniform Norm

4.1 Introduction and Preliminaries

4.1.1 Interpolation at Zeros of Orthogonal Polynomials

The interpolation array and Lagrange operators in the general algebraic case were introduced in Sect. 1.4.2. As a sequence of polynomials $\{q_n\}_{n\in\mathbb{N}}$ ($q_n \in \mathcal{P}_n$), with zeros $x_{n,k}$ ($k = 1, \ldots, n$) as in (1.4.5) and the corresponding infinite triangular array $\mathcal{X}$ of these zeros (1.4.6), we consider here a sequence of orthonormal polynomials on $(-1, 1)$ with respect to the weight function w, i.e., $\{p_n(w)\}_{n\in\mathbb{N}}$, where $p_n(w; x) = \gamma_n x^n + \cdots$, with $\gamma_n > 0$. Then, instead of $L_n(\mathcal{X}, f)$ we will use the notation $L_n(w, f)$ for the corresponding Lagrange polynomials. In this particular case the Lagrange polynomials can also be expressed in the form

$$L_n(w, f)(x) = L_n(w, f; x) = \sum_{i=0}^{n-1} c_i p_i(w; x), \qquad (4.1.1)$$

where

$$c_i = \sum_{k=1}^{n} \lambda_{n,k}(w) p_i(w; x_{n,k}) f(x_{n,k}), \qquad (4.1.2)$$

$\lambda_{n,k}(w)$ are the Christoffel numbers and $x_{n,k}$ are the zeros of $p_n(w)$. The proof of (4.1.1) is easy. In fact, since $L_n(w, f) \in \mathcal{P}_{n-1}$ we can write

$$L_n(w, f; x) = \sum_{i=0}^{n-1} c_i p_i(w; x), \quad c_i = \int_{-1}^{1} L_n(w, f; x) p_i(w; x) w(x)\, dx.$$

Then, by the n-point Gaussian quadrature formula (2.2.43), we have exactly

$$c_i = \sum_{k=1}^{n} \lambda_{n,k}(w) L_n(w, f; x_{n,k}) p_i(w; x_{n,k}) = \sum_{k=1}^{n} \lambda_{n,k}(w) f(x_{n,k}) p_i(w; x_{n,k}),$$

because of $\deg(L_n(w, f) p_i(w)) \leq n - 1 + i \leq 2n - 2$. We deduce by (4.1.1) and (4.1.2)

$$L_n(w, f; x) = \sum_{k=1}^{n} \lambda_{n,k}(w) \left(\sum_{i=0}^{n-1} p_i(w; x_{n,k}) p_i(w; x) \right) f(x_{n,k}),$$

G. Mastroianni, G.V. Milovanović, *Interpolation Processes*,
© Springer 2008

i.e.,
$$L_n(w, f; x) = \sum_{k=1}^{n} \lambda_{n,k}(w) K_{n-1}(w; x, x_{n,k}) f(x_{n,k}), \quad (4.1.3)$$

where
$$K_{n-1}(w; x, y) := \sum_{i=0}^{n-1} p_i(w; x) p_i(w; y)$$

is the *Darboux kernel* introduced in Chap. 2.

By comparing this formula with the integral form of the Fourier partial sum

$$S_n(w, f; x) = \int_{-1}^{1} K_n(w; x, y) f(y) w(y) \, dy$$

it is easy to recognize that, as in the trigonometric case, we can derive the Lagrange polynomial $L_n(w, f)$ from the Fourier sum $S_n(w, f)$ applying a Gaussian quadrature rule. Moreover, comparing (4.1.3) with the standard Lagrange form

$$L_n(w, f; x) = \sum_{k=1}^{n} \ell_{n,k}(w; x) f(x_{n,k}),$$

we get the following form of the fundamental Lagrange polynomials $\ell_{n,k}(\mathcal{X}; x) = \ell_{n,k}(w; x)$ corresponding to the zeros of $p_n(w)$,

$$\ell_{n,k}(w; x) = \lambda_{n,k}(w) K_{n-1}(w; x, x_{n,k}),$$

i.e., recalling the Christoffel-Darboux formula (2.2.7), we obtain

$$\ell_{n,k}(w; x) = \frac{\gamma_{n-1}}{\gamma_n} \lambda_{n,k}(w) p_{n-1}(w; x_{n,k}) \frac{p_n(w; x)}{x - x_{n,k}}. \quad (4.1.4)$$

Finally, comparing (4.1.4) with the standard form (1.4.9), i.e.,

$$\ell_{n,k}(w; x) = \frac{p_n(w; x)}{p'_n(w; x_{n,k})(x - x_{n,k})}, \quad k = 1, \ldots, n,$$

we obtain the following useful formula

$$\frac{1}{p'_n(w; x_{n,k})} = \frac{\gamma_{n-1}}{\gamma_n} \lambda_{n,k}(w) p_{n-1}(w; x_{n,k}). \quad (4.1.5)$$

If f has a bounded n-th derivative, we can write according to (1.4.20)

$$|f(x) - L_n(w, f; x)| \leq \frac{|p_n(w; x)|}{\gamma_n n!} \max_{|x| \leq 1} |f^{(n)}(x)|. \quad (4.1.6)$$

4.1 Introduction and Preliminaries

Taking into account that we have $\gamma_n \sim 2^n$ for a large class of weights, the previous estimate turns out to be very useful in several cases when we consider sufficiently smooth functions. For example, $\gamma_n \sim 2^n$ holds for the Chebyshev weights and more generally for the Szegő class of weights w defined by (see Sect. 2.2.2)

$$\int_{-1}^{1} \left(\log w(x)/\sqrt{1-x^2}\right) dx > -\infty.$$

As we saw in Sect. 1.4.2, the behaviour of the Lebesgue constant

$$\Lambda_n(\mathcal{X}) = \|L_n(\mathcal{X})\|_\infty = \|L_n(w)\|_\infty$$

plays an important role in the study of the convergence of the Lagrange polynomials, because of (1.4.14), i.e.,

$$\|f - L_n(\mathcal{X}, f)\|_\infty \leq (1 + \|L_n(\mathcal{X})\|_\infty) E_{n-1}(f)_\infty, \tag{4.1.7}$$

where $E_{n-1}(f)_\infty$ is the error of the best uniform approximation, defined by $E_{n-1}(f)_\infty := \min_{P \in \mathcal{P}_{n-1}} \|f - P\|_\infty$.

According to (1.4.15) the Lebesgue constants are unbounded and for particular choices of the interpolation array $\mathcal{X}$, they can take very "large" values as the number n of the knots increases. For instance, this is the case if we use equidistant nodes in $[-1, 1]$ with the corresponding array $\mathcal{E}$ (see Sect. 1.4.6). Then we have the asymptotic expression (1.4.31) for the Lebesgue constants, i.e., $\|L_n(\mathcal{E})\|_\infty \sim 2^n/(en \log n)$, $n \to +\infty$.

This circumstance can strongly influence the numerical computation, where we deal with perturbed values of f and compute the polynomial $L_n(\mathcal{X}, f + \eta)$, where η is a perturbation of the function f appearing in the evaluation of $f(x_{n,k})$. Namely, instead of $f(x_{n,k})$ we always deal with $\tilde{f}_{n,k} = f(x_{n,k}) + \eta(x_{n,k})$, where $\eta(x_{n,k})$ is the corresponding error at the node $x_{n,k}$. In this case, we can estimate the actual error in the following form

$$\|f - L_n(\mathcal{X}, f + \eta)\|_\infty \leq \|f - L_n(\mathcal{X}, f)\|_\infty + \|L_n(\mathcal{X})\|_\infty \eta_n, \tag{4.1.8}$$

where $\eta_n := \max_{1 \leq k \leq n} |\eta(x_{n,k})|$. The first term on the right-hand side in (4.1.8) is the *theoretical error*. We have seen that it can be estimated by means of (4.1.7) or by using (4.1.6) if f is sufficiently regular. In the last case the theoretical error can be very small.

The second term in (4.1.8) represents the *numerical error* given by the amplification of the round-off error due to the numerical computation of f. Because of the magnitude of the Lebesgue constants, such a numerical error can be very large even if f is computed with machine precision. This fact is confirmed by the following example.

Table 4.1.1 Asymptotic values of the Lebesgue constant and actual errors in interpolation with equidistant nodes

n	$\|L_n(\mathcal{E})\|_\infty$	$\text{Err}_n(\mathcal{E}, \text{R})$	$\text{Err}_n(\mathcal{E}, \text{D})$	$\text{Err}_n(\mathcal{E}, \text{Q})$
5	1.46(0)	1.12(−3)	1.12(−3)	1.12(−3)
10	1.64(1)	2.38(−6)	3.85(−9)	3.85(−9)
15	2.97(2)	3.84(−5)	6.39(−14)	1.58(−15)
20	6.44(3)	7.32(−4)	1.99(−12)	1.43(−22)
30	3.87(6)	4.14(−1)	1.45(−9)	8.64(−28)
40	2.74(9)	4.44(2)	1.11(−6)	5.00(−25)
50	2.12(12)	2.75(5)	5.04(−4)	3.40(−22)
80	1.27(21)	2.50(14)	3.01(5)	3.53(−13)
100	1.01(27)		3.50(11)	2.22(−7)
120	8.51(32)		1.89(26)	1.13(−1)
150	6.99(41)			2.43(8)

Example 4.1.1 We take a function which is analytic in the hole complex plane, for example $f(x) = e^x$, and interpolate it at the equidistant nodes

$$x_{n,k} = -1 + 2\frac{k-1}{n-1}, \quad k = 1, \ldots, n \ (n \geq 2).$$

As we know, the interpolation process is uniformly convergent (see Sect. 1.4.4), i.e., the theoretical error tends to zero as $n \to +\infty$. The asymptotic values for $\|L_n(\mathcal{E})\|_\infty$ and the actual errors

$$\text{Err}_n(\mathcal{E}, \text{A}) := \|f - L_n(\mathcal{E}, f + \eta)\|_\infty$$

in the corresponding machine arithmetic A are given in Table 4.1.1 for some selected values of n. The all computation are performed on the WORKSTATION DIGITAL ULTIMATE ALPHA 533au2 in the R, D, and Q arithmetic, with the machine precisions 1.17(−7), 2.22(−16), and 1.93(−34), respectively. Numbers in parentheses indicate decimal exponents. The influence of the Lebesgue constant is evident.

On the other hand, taking the Chebyshev nodes $x_{n,k} = \cos(2k-1)\pi/(2n)$, $k = 1, \ldots, n$, we know that $\|L_n(\mathcal{T})\|_\infty \sim (2/\pi)\log n$, $n \to +\infty$ (see (1.4.32)). The actual errors $\text{Err}_n(\mathcal{T}, \text{R})$, $\text{Err}_n(\mathcal{T}, \text{D})$, and $\text{Err}_n(\mathcal{T}, \text{Q})$ are shown in Table 4.1.2. Now, the actual errors have the same order as the theoretical one, and they are limited by the machine precision.

According to the Faber inequality (1.4.15), the previous facts suggest the following definition of an optimal system of nodes:

Definition 4.1.1 We say that $\mathcal{X}$ is an *optimal system of nodes* if and only if there exists a constant $C \neq C(n)$ such that $\|L_n(\mathcal{X})\|_\infty \leq C \log n \ (n > 1)$.

4.1 Introduction and Preliminaries

Table 4.1.2 Asymptotic values of the Lebesgue constant and actual errors in interpolation with Chebyshev nodes

n	$\|L_n(T)\|_\infty$	$\mathrm{Err}_n(T,\mathrm{R})$	$\mathrm{Err}_n(T,\mathrm{D})$	$\mathrm{Err}_n(T,\mathrm{Q})$
5	1.02	6.40(−4)	6.40(−4)	6.40(−4)
10	1.47	1.43(−6)	6.03(−10)	6.03(−10)
15	1.72	1.19(−6)	2.66(−15)	5.05(−17)
20	1.91	1.67(−6)	3.11(−15)	8.32(−25)
30	2.17	1.91(−6)	3.55(−15)	2.70(−33)
40	2.35	2.62(−6)	4.44(−15)	4.24(−33)
50	2.49	2.86(−6)	4.00(−15)	5.39(−33)
80	2.79	9.88(−4)	8.88(−15)	5.39(−33)
100	2.93		7.55(−15)	5.39(−33)
120	3.05		7.55(−15)	6.55(−33)
150	3.19		7.11(−15)	1.04(−32)
200	3.37		8.88(−15)	6.93(−33)

In Sect. 4.2 we consider some optimal systems of nodes. Before, we need some auxiliary results which are important in the further investigation of the algebraic interpolation.

4.1.2 Some Auxiliary Results

In this section we give certain estimates which are connected with the zeros of orthogonal polynomials. Separately, we consider the cases of the intervals $(-1, 1)$, $(0, +\infty)$, and $(-\infty, +\infty)$.

Case $(-1, 1)$ Let $v^{\gamma,\delta}(x) = (1-x)^\gamma (1+x)^\delta$ be the Jacobi weight with parameters $\gamma, \delta > -1$.

Lemma 4.1.1 *Let $x \in [-1, 1]$ and $-1 = x_0 < x_1 < x_2 < \cdots < x_n < x_{n+1} = 1$ be a set of points with an arc sine distribution (i.e., $x_{k+1} - x_k \sim \sqrt{1 - x_k^2}/n$). If d is the index of a point x_d closest to x, i.e., $|x - x_d| = \min_{1 \le k \le n} |x - x_k|$, and $\Delta x_k := x_{k+1} - x_k$ for $k = 1, \ldots, n$, then*

$$\sum_{\substack{k=1 \\ k \ne d}}^n \frac{\Delta x_k}{|x - x_k|} \sim \log n \qquad \text{for } |x| \le 1; \tag{4.1.9}$$

$$\sum_{\substack{k=1 \\ k \ne d}}^n \frac{v^{\gamma,\delta}(x_k)\Delta x_k}{|x - x_k|} \sim \log n \qquad \text{for } |x| \le a \ (\gamma, \delta \ge -1); \tag{4.1.10}$$

$$\sum_{\substack{k=1\\k\neq d}}^{n}\frac{(1+x_k)^\delta \Delta x_k}{|x-x_k|} \sim \left(\sqrt{1+x}+\frac{1}{n}\right)^{2\delta} \quad \text{for } -1 \le x \le -a \ (-1 \le \delta < 0);$$

(4.1.11)

$$\sum_{\substack{k=1\\k\neq d}}^{n}\frac{(1-x_k)^\gamma \Delta x_k}{|x-x_k|} \sim \left(\sqrt{1-x}+\frac{1}{n}\right)^{2\gamma} \quad \text{for } a \le x \le 1 \ (-1 \le \gamma < 0); \quad (4.1.12)$$

where a is an arbitrary fixed point in $(1/2, 1)$.

Proof First we observe that the exclusion of the point x_d from the previous sums assures that the distance of x from the other points is greater than or equal to Δx_d.

1° We prove (4.1.9) for $d \neq 1, n$, i.e., when $-1 < x_{d-1} < x_d \le x < x_{d+1} < 1$, for instance. The case $d \in \{1, n\}$ is similar. We write

$$\sum_{\substack{k=1\\k\neq d}}^{n}\frac{\Delta x_k}{|x-x_k|} = \sum_{k=1}^{d-1}\frac{\Delta x_k}{x-x_k} + \sum_{k=d+1}^{n}\frac{\Delta x_k}{x_k-x} =: I_1 + I_2.$$

For I_1 we have

$$I_1 = \sum_{k=1}^{d-1}\frac{\Delta x_k}{x-x_k} = \sum_{k=2}^{d-2}\frac{\Delta x_k}{x-x_k} + \frac{\Delta x_1}{x-x_1} + \frac{\Delta x_{d-1}}{x-x_{d-1}},$$

where the last two terms are bounded since $\Delta x_{d-1} \sim x - x_{d-1}$, because x_d is the point closest to x, and

$$\frac{\Delta x_1}{x-x_1} \sim \frac{\Delta x_1}{\sum_{i=1}^{d-1}\Delta x_i} \le 1.$$

On the other hand, $\Delta x_k = \int_{x_k}^{x_{k+1}} dt \sim \Delta x_{k\pm 1}$ gives

$$C\int_{x_{k-1}}^{x_k}\frac{dt}{x-t} < \frac{\Delta x_k}{x-x_k} < \int_{x_k}^{x_{k+1}}\frac{dt}{x-t}, \quad k = 2, \ldots, d-2.$$

Hence, taking the sum with respect to k, we get

$$C\log n \le C\int_{x_1}^{x_{d-2}}\frac{dt}{x-t} < \sum_{k=2}^{d-2}\frac{\Delta x_k}{x-x_k} < C\int_{x_2}^{x_{d-1}}\frac{dt}{x-t} \le C\log n,$$

4.1 Introduction and Preliminaries

i.e.,

$$\log n \sim \sum_{k=2}^{d-2} \frac{\Delta x_k}{x - x_k} \leq I_1 \leq C \log n.$$

Analogously we prove $I_2 \sim \log n$ and, therefore, (4.1.9) follows.

2° In order to prove (4.1.10) we set

$$\sum_{\substack{k=1 \\ k \neq d}}^{n} \frac{v^{\gamma,\delta}(x_k) \Delta x_k}{|x - x_k|} = \left(\sum_{|x - x_k| > (1-a)/4} + \sum_{|x - x_k| \leq (1-a)/4} \right) =: A_1 + A_2.$$

For A_1 we have

$$A_1 \leq \frac{4}{1-a} \sum_{k=1}^{n} v^{\gamma,\delta}(x_k) \Delta x_k \sim \int_{x_1}^{x_n} (1-t)^{\gamma}(1+t)^{\delta} \, dt$$

$$\sim \begin{cases} 1, & \text{if } \gamma, \delta > -1, \\ \log n, & \text{if } \min\{\gamma, \delta\} = -1. \end{cases}$$

For A_2 we write

$$A_2 \leq v^{\gamma,\delta}(x) \sum_{|x - x_k| \leq (1-a)/4} \frac{\Delta x_k}{|x - x_k|} + \sum_{|x - x_k| \leq (1-a)/4} \left| \frac{v^{\gamma,\delta}(x_k) - v^{\gamma,\delta}(x)}{x_k - x} \right| \Delta x_k$$

$$\leq C \sum_{|x - x_k| \leq (1-a)/4} \frac{\Delta x_k}{|x - x_k|} + C \sum_{|x - x_k| \leq (1-a)/4} \Delta x_k,$$

since $v^{\gamma,\delta}(x)$ is bounded for $|x| \leq a$, $x_k \in I := \left[-\frac{3a+1}{4}, \frac{3a+1}{4} \right] \supset [-a, a]$ and $v^{\gamma,\delta}(x)$ has a bounded derivative in I.

Hence we get by (4.1.9)

$$A_2 \leq C \sum_{|x - x_k| \leq (1-a)/4} \left(\frac{\Delta x_k}{|x - x_k|} + \Delta x_k \right) \leq C \sum_{\substack{k=1 \\ k \neq d}} \frac{\Delta x_k}{|x - x_k|} + C \sum_{k=1} \Delta x_k \leq C \log n.$$

Moreover

$$A_2 \geq \min_{x \in I} v^{\gamma,\delta}(x) \sum_{|x - x_k| \leq (1-a)/4} \frac{\Delta x_k}{|x - x_k|} \sim \log n,$$

since (4.1.9) also holds if the sum is extended to the indices $k \neq d$ such that $|x - x_k|$ is less or equal to a suitable constant.

Thus, we have

$$C\log n \leq A_2 \leq \sum_{\substack{k=1\\k\neq d}}^{n} \frac{v^{\gamma,\delta}(x_k)\Delta x_k}{|x-x_k|} = A_1 + A_2 \leq C\log n,$$

which gives (4.1.10).

3° In order to prove (4.1.11), we assume $x > -1$ and write

$$\sum_{\substack{k=1\\k\neq d}}^{n} \frac{(1+x_k)^\delta \Delta x_k}{|x-x_k|} = \left(\sum_{x_1 \leq x_k \leq 2x+1} + \sum_{x_k > 2x+1}\right) =: J_1 + J_2.$$

Since $x_k > 2x + 1$ implies $|x - x_k| > 2(1 + x_k)$, for J_2 we get

$$J_2 \leq 2 \sum_{x_k > 2x+1} (1+x_k)^{\delta-1} \Delta x_k \leq \int_{2x}^{1} (1+t)^{\delta-1} dt$$

$$\leq C \int_{x}^{+\infty} (1+t)^{\delta-1} dt = C(1+x)^\delta,$$

where we used $\delta < 0$.

Assuming $d \neq 1, n$, for J_1 we can write

$$J_1 = \left(\sum_{x_1 \leq x_k \leq (x-1)/2} + \sum_{(x-1)/2 < x_k \leq x_{d-1}} + \sum_{x_{d+1} \leq x_k \leq 2x+1}\right) \frac{(1+x_k)^\delta \Delta x_k}{|x-x_k|}$$

$$\leq \sum_{x_1 \leq x_k \leq (x-1)/2} (1+x_k)^{\delta-1} \Delta x_k$$

$$+ \left(\sum_{(x-1)/2 < x_k \leq x_{d-1}} + \sum_{x_{d+1} \leq x_k \leq 2x+1}\right) (x-x_k)^{\delta-1} \Delta x_k$$

$$=: J' + J'' + J''',$$

since $x - x_k \geq 1 + x_k$ if and only if $x_k \leq (x-1)/2$, while in the other cases $|x - x_k| < 1 + x_k$.

Now we have

$$J' = \sum_{x_1 \leq x_k \leq (x-1)/2} (1+x_k)^{\delta-1} \Delta x_k \leq \int_{x_1}^{+\infty} (1+t)^{\delta-1} dt \leq \frac{C}{n^{2\delta}}.$$

Moreover, using $x - x_{d-1} \sim \Delta x_{d-1}$, we get

$$J'' = \sum_{(x-1)/2 < x_k \leq x_{d-2}} (x-x_k)^{\delta-1} \Delta x_k + (x-x_{d-1})^{\delta-1} \Delta x_{d-1}$$

$$\leq \sum_{(x-1)/2 < x_k \leq x_{d-2}} \int_{x_k}^{x_{k+1}} (x-t)^{\delta-1} dt + (\Delta x_{d-1})^\delta \leq \frac{C}{n^{2\delta}}.$$

4.1 Introduction and Preliminaries

Finally, J''' can be estimated as J'' by considering separately the term corresponding to $k - d + 1$ and estimating the remainder sum by the integral. In this way we obtain that $J''' \leq Cn^{-2\delta}$ also holds and then we have $J_2 = J' + J'' + J''' \leq Cn^{-2\delta}$. Thus, we obtain

$$\sum_{\substack{k=1 \\ k \neq d}}^{n} \frac{(1+x_k)^\delta \Delta x_k}{|x - x_k|} = J_1 + J_2 \leq C(1+x)^\delta + \frac{C}{n^{2\delta}} \leq C\left(\sqrt{1-x} + \frac{1}{n}\right)^{2\delta}$$

for all $-1 < x < -a$. Finally, in the case $x = -1$, it is easy to prove that

$$\sum_{\substack{k=1 \\ k \neq d}}^{n} \frac{(1+x_k)^\delta \Delta x_k}{1 + x_k} \sim n^{-2\delta}.$$

Hence, we conclude that for all $x \in [-1, -a]$, we have

$$\sum_{\substack{k=1 \\ k \neq d}}^{n} \frac{(1+x_k)^\delta \Delta x_k}{|x - x_k|} \sim \max\left\{\left(\sqrt{1+x}\right)^{2\delta}, \frac{1}{n^{2\delta}}\right\} \sim \left(\sqrt{1-x} + \frac{1}{n}\right)^{2\delta},$$

i.e., (4.1.11) holds.

4° The estimate (4.1.12) can be proved in a similar way as (4.1.11). □

Remark 4.1.1 By (4.1.9)–(4.1.12), we can easily deduce the following useful estimate

$$\sum_{\substack{k=1 \\ k \neq d}}^{n} \frac{v^{\gamma, \delta}(x_k) \Delta x_k}{|x - x_k|} \leq C \left(\sqrt{1-x} + \frac{1}{n}\right)^{2\gamma} \left(\sqrt{1+x} + \frac{1}{n}\right)^{2\delta} \log n, \qquad (4.1.13)$$

which holds for all $|x| \leq 1$ and for $-1 \leq \gamma, \delta \leq 0$. Obviously the constant C in (4.1.13) is independent of n and x.

Lemma 4.1.2 *Let $t_0 \in (-1, 1)$ be fixed and let n points $x_1, \ldots, x_n$ be given with an arc sine distribution on $[-1, 1]$ and such that*

$$-1 < x_1 < \cdots < x_s < t_0 < x_{s+1} < \cdots < x_n < 1$$

holds, with

$$|x_s - t_0| \sim \frac{1}{n} \sim |x_{s+1} - t_0|.$$

Then, for all $x \in [-1, 1] \setminus (t_0 - c/n, t_0 + c/n)$ and any $0 \leq \gamma \leq 1$, we have

$$\sum_{\substack{k=1 \\ k \neq d}}^{n} \frac{\Delta x_k}{|x - x_k||t_0 - x_k|^\gamma} \leq C \frac{\log n}{|t_0 - x|^\gamma}, \qquad C \neq C(n, x), \qquad (4.1.14)$$

where d is the index of a point x_d closest to x.

Proof For the sake of brevity we consider only the case $x > t_0$. The other case being similar. We write

$$\sum_{\substack{k=1 \\ k \neq d}}^{n} \frac{\Delta x_k}{|x - x_k||t_0 - x_k|^\gamma}$$

$$= I_1 + I_2 + I_3 + I_4$$

$$= \left\{ \sum_{x_k < 2t_0 - x} + \sum_{2t_0 - x \leq x_k \leq x_s} + \sum_{x_{s+1} \leq x_k < (t_0 + x)/2} + \sum_{x_k \geq (t_0 + x)/2} \right\} \frac{\Delta x_k}{|x - x_k||t_0 - x_k|^\gamma}.$$

We observe that by $x_k < 2t_0 - x$ we have $t_0 - x_k > x - t_0$ and then by (4.1.9) for I_1 we get

$$I_1 = \sum_{x_k < 2t_0 - x} \frac{\Delta x_k}{(x - x_k)(t_0 - x_k)^\gamma} \leq \frac{C}{(x - t_0)^\gamma} \sum_{\substack{k=1 \\ k \neq d}}^{n} \frac{\Delta x_k}{|x - x_k|} \leq \frac{C}{(x - t_0)^\gamma} \log n.$$

In estimating I_2 we take into account that $x - x_k > x - t_0$, so that we have

$$I_2 = \sum_{2t_0 - x \leq x_k \leq x_s} \frac{\Delta x_k}{(x - x_k)(t_0 - x_k)^\gamma} \leq \frac{C}{(x - t_0)^\gamma} \int_{2t_0 - x}^{x_s} \frac{dt}{(t_0 - t)^\gamma}$$

$$\leq \frac{C}{(x - t_0)^\gamma} \log n.$$

For I_3 we note that from $x_k < (t_0 + x)/2$ we get $x - x_k > (x - t_0)/2$. Therefore,

$$I_3 = \sum_{x_{s+1} \leq x_k \leq (t_0 + x)/2} \frac{\Delta x_k}{(x - x_k)(x_k - t_0)^\gamma} \leq \frac{2}{x - t_0} \sum_{x_{s+1} \leq x_k \leq (t_0 + x)/2} \frac{\Delta x_k}{(x_k - t_0)^\gamma}$$

$$\leq \frac{C}{x - t_0} \int_{x_{s+1}}^{(t_0 + x)/2} \frac{dt}{(t - t_0)^\gamma} \leq \frac{C}{(x - t_0)^\gamma} \log n.$$

Finally, for I_4, since $x_k > (t_0 + x)/2$ implies $x_k - t_0 > (x - t_0)/2$, using (4.1.9) we have

$$I_4 = \sum_{x_k > (t_0 + x)/2} \frac{\Delta x_k}{(x - x_k)(x_k - t_0)^\gamma} \leq \frac{2^\gamma}{(x - t_0)^\gamma} \sum_{\substack{k=1 \\ k \neq d}}^{n} \frac{\Delta x_k}{|x - x_k|} \leq \frac{C}{(x - t_0)^\gamma} \log n. \quad \square$$

4.1 Introduction and Preliminaries

Case $(0, +\infty)$ Let $w_\alpha(x) := x^\alpha e^{-x}$ be a Laguerre weight on $(0, +\infty)$, with the parameter $\alpha > -1$.

Lemma 4.1.3 *Let $x_1, x_2, \ldots, x_n$ be the zeros of the Laguerre polynomial $p_n(w_\alpha)$ such that $0 < x_1 < x_2 < \cdots < x_n < 4n$. If $0 \leq \theta, \tau \leq 1$, then for all $x \in [a/n, 4n + b]$, with fixed constants $a, b > 0$, and for n sufficiently large (say $n > n_0$) we have*

$$\sum_{\substack{k=1 \\ k \neq d}}^{n} \left(\frac{x}{x_k}\right)^\tau \left(\frac{4n - x}{4n - x_k}\right)^\theta \frac{\Delta x_k}{|x - x_k|} \leq C \log n, \quad C \neq C(n, x), \quad (4.1.15)$$

where x_d is a zero closest to x and $\Delta x_k = x_{k+1} - x_k$.

Proof First we assume $x_{d-1} < x \leq x_d < x_{d+1}$ and $2 \leq d \leq n - 1$ and decompose the sum on the right-hand side in (4.1.15) as

$$\left\{ \sum_{x_k \in [x_1, x/2]} + \sum_{x_k \in [x/2, x_{d-1}]} + \sum_{x_k \in [x_{d+1}, (x+4n)/2]} + \sum_{x_k \in [(x+4n)/2, x_n]} \right\}$$

$$\times \left(\frac{x}{x_k}\right)^\tau \left(\frac{4n - x}{4n - x_k}\right)^\theta \frac{\Delta x_k}{|x - x_k|}$$

$$=: J_1(x) + J_2(x) + J_3(x) + J_4(x).$$

We only estimate $J_1(x)$ and $J_2(x)$, since the estimations of $J_3(x)$ and $J_4(x)$ are similar.

Since $(4n - x_k)^\theta > (4n - x)^\theta$ and $x - x_k > x/2$, we have for J_1

$$J_1(x) = \sum_{x_1 \leq x_k \leq x/2} \left(\frac{x}{x_k}\right)^\tau \left(\frac{4n - x}{4n - x_k}\right)^\theta \frac{\Delta x_k}{x - x_k}$$

$$\leq Cx^{\tau-1} \sum_{x_1 \leq x_k \leq x/2} \frac{\Delta x_k}{x_k^\tau} = Cx^{\tau-1} \left\{ \frac{x_2 - x_1}{x_1^\tau} + \sum_{x_2 \leq x_k \leq x/2} \frac{\Delta x_{k-1}}{x_k^\tau} \right\}.$$

Then, using the following estimate of the Laguerre zeros (see (2.3.50))

$$x_k \sim \frac{k^2}{n}, \quad k = 1, \ldots, n, \quad (4.1.16)$$

and taking into account that $n^{-1} \leq Cx$, we get

$$J_1(x) \leq Cx^{\tau-1}\left\{\frac{x_2-x_1}{x_1^\tau} + \sum_{x_2 \leq x_k \leq x/2}\frac{\Delta x_{k-1}}{x_k^\tau}\right\}$$

$$\leq Cx^{\tau-1}\left\{\left(\frac{C}{n}\right)^{1-\tau} + \sum_{x_2 \leq x_k \leq x/2}\frac{\Delta x_{k-1}}{x_k^\tau}\right\}$$

$$\leq Cx^{\tau-1}\left\{x^{1-\tau} + \int_{x_2}^{x/2}\frac{dt}{t^\tau}\right\} \leq C\begin{cases}1, & \text{if } 0 \leq \tau < 1,\\ \log n, & \text{if } \tau = 1.\end{cases}$$

In order to estimate $J_2(x)$, we use $x_k > x/2$ and that by (4.1.16) we have $x \sim x_d$. Then

$$J_2(x) := \sum_{x/2 \leq x_k \leq x_{d-1}}\left(\frac{x}{x_k}\right)^\tau\left(\frac{4n-x}{4n-x_k}\right)^\theta\frac{\Delta x_k}{x-x_k}$$

$$\leq C\sum_{x/2 \leq x_k \leq x_{d-1}}\frac{\Delta x_k}{x-x_k} = C\sum_{x/2 \leq x_k \leq x_{d-2}}\frac{\Delta x_k}{x-x_k} + C\frac{\Delta x_{d-1}}{x-x_{d-1}}$$

$$\leq C\int_{x/2}^{x_{d-1}}\frac{dt}{x-t} + C \leq C\log n.$$

Now, we assume $0 < x < x_1$. In this case we set

$$\left\{\sum_{x_2 \leq x_k \leq (x+4n)/2} + \sum_{(x+4n)/2 \leq x_k \leq x_n}\right\}\left(\frac{x}{x_k}\right)^\tau\left(\frac{4n-x}{4n-x_k}\right)^\theta\frac{\Delta x_k}{x-x_k} =: I_1(x) + I_2(x).$$

Since $x < x_k$, $\tau \geq 0$ and $(4n-x_k)^\theta > ((4n-x)/2)^\theta$, we get

$$I_1(x) := \left(\frac{x}{x_k}\right)^\tau\left(\frac{4n-x}{4n-x_k}\right)^\theta\frac{\Delta x_k}{x-x_k}$$

$$\leq C\sum_{x_2 \leq x_k \leq (x+4n)/2}\frac{\Delta x_k}{(x_k-x)} \leq C\int_{x_2}^{(x+4n)/2}\frac{dt}{t-x} \leq C\log n.$$

Also, because of $x < x_k$, $\tau \geq 0$ and $x_k - x > ((4n-x)/2)$, we obtain for I_2

$$I_2(x) := \sum_{(x+4n)/2 \leq x_k \leq x_n}\left(\frac{x}{x_k}\right)^\tau\left(\frac{4n-x}{4n-x_k}\right)^\theta\frac{\Delta x_k}{x-x_k}$$

$$\leq C(4n-x)^{\theta-1}\sum_{(x+4n)/2 \leq x_k \leq x_n}\frac{\Delta x_k}{(4n-x_k)^\theta}$$

$$\leq C(4n-x)^{\theta-1}\int_{(x+4n)/2}^{x_n}\frac{dt}{(4n-t)^\theta} \leq C\begin{cases}1, & \text{if } 0 \leq \theta < 1,\\ \log n, & \text{if } \theta = 1.\end{cases}$$

Finally, the case $x_n < x < 4n$ is similar to the previous one. □

4.1 Introduction and Preliminaries

Case $(-\infty, +\infty)$ Let $w(x) := e^{-x^2}$ be the Hermite weight function on $(-\infty, |\infty)$.

Lemma 4.1.4 *Let $0 < \theta < 1$ and $x_1, x_2, \ldots, x_n$ be the zeros of the Hermite polynomial $p_n(w)$ in an increasing order $x_1 < x_2 < \cdots < x_n$. If $x \in [-\sqrt{2n}, \sqrt{2n}]$ and if x_d is a zero closest to x, then*

$$\sum_{\substack{k=1 \\ k \neq d}}^{n} \left(\frac{2n - x^2}{2n - x_k^2}\right)^\theta \frac{\Delta x_k}{|x - x_k|} \leq C \log n, \qquad C \neq C(n, x) \qquad (4.1.17)$$

holds, where we set $\Delta x_k := x_{k+1} - x_k$, $k = 1, \ldots, n$ $(x_{n+1} := \sqrt{2n})$.

Proof For the sake of simplicity, we assume that $x_1 \leq x_{d-1} < x < x_{d+1} \leq x_n$; the cases $d = 1, n$ are similar.

Now, we decompose the sum on the right-hand side in (4.1.17) as follows

$$\left\{ \sum_{x_k \in \left[x_1, \frac{x-\sqrt{2n}}{2}\right]} + \sum_{x_k \in \left[\frac{x-\sqrt{2n}}{2}, x_{d-1}\right]} + \sum_{x_k \in \left[x_{d+1}, \frac{x+\sqrt{2n}}{2}\right]} + \sum_{x_k \in \left[\frac{x+\sqrt{2n}}{2}, x_n\right]} \right\}$$

$$\times \left(\frac{2n - x^2}{2n - x_k^2}\right)^\theta \frac{\Delta x_k}{|x - x_k|}$$

$$=: I_1(x) + I_2(x) + I_3(x) + I_4(x).$$

The estimates of $I_3(x)$ and $I_4(x)$ are similar to the ones of $I_2(x)$ and $I_1(x)$, respectively. Therefore, we examine only these last two terms for brevity.

In the case $x_1 \leq x_k \leq (x - \sqrt{2n})/2$, we have $\sqrt{2n} - x < \sqrt{2n} - x_k$ and $x - x_k \geq (x + \sqrt{2n})/2$. Hence the sum $I_1(x)$ reduces to

$$I_1(x) := \sum_{x_k \in \left[x_1, (x-\sqrt{2n})/2\right]} \left(\frac{2n - x^2}{2n - x_k^2}\right)^\theta \frac{\Delta x_k}{x - x_k}$$

$$\leq C \left(\sqrt{2n} + x\right)^{\theta - 1} \sum_{x_k \in \left[x_1, (x-\sqrt{2n})/2\right]} \frac{\Delta x_k}{\left(\sqrt{2n} + x\right)^\theta}$$

$$\leq C \left(\sqrt{2n} + x\right)^{\theta - 1} \int_{x_1}^{(x-\sqrt{2n})/2} \frac{dt}{\left(\sqrt{2n} + t\right)^\theta} \leq C.$$

Finally, if $(x - \sqrt{2n})/2 \leq x_k \leq x_{d-1}$, then we have $\sqrt{2n} - x_k \geq \sqrt{2n} - x$ and $\sqrt{2n} + x_k \geq (\sqrt{2n} + x)/2$. Consequently, we get

$$I_2(x) := \sum_{x_k \in \left[(x-\sqrt{2n})/2, x_{d-1}\right]} \left(\frac{2n - x^2}{2n - x_k^2}\right)^\theta \frac{\Delta x_k}{x - x_k}$$

$$\leq C \sum_{x_k \in \left[(x-\sqrt{2n})/2, x_{d-1}\right]} \frac{\Delta x_k}{x - x_k}$$

$$\leq C \left(\sum_{x_k \in \left[(x-\sqrt{2n})/2, x_{d-2}\right]} \frac{\Delta x_k}{x - x_k} + \frac{\Delta x_{d-1}}{x - x_{d-1}} \right)$$

$$\leq C \int_{(x-\sqrt{2n})/2}^{x_{d-1}} \frac{dt}{x - t} + C \leq C \log n,$$

where we used $\Delta x_{d-1} \sim x - x_{d-1}$. $\square$

4.2 Optimal Systems of Nodes

4.2.1 Optimal Systems of Knots on $[-1, 1]$

We start this section with two classical examples of well known optimal systems of nodes. We briefly cite these results without their proofs, but in the sequel we will see that they are particular cases of more general results, which will be proved.

4.2.1.1 Interpolation at Jacobi Abscissas

For $\mathcal{X} = \{p_n(v^{\alpha,\beta})\}_{n=1,2,...}$, the behaviour of the Lebesgue constants $\|L_n(\mathcal{X})\|_\infty = \|L_n(v^{\alpha,\beta})\|_\infty$ is described by the following classical result due to Szegő, whose proof can be found in [470, Theorem 14.4, p. 335].

Theorem 4.2.1 (Szegő) *Let $\gamma = \max(\alpha, \beta)$. For all $n \in \mathbb{N}$*

$$\|L_n(v^{\alpha,\beta})\|_\infty \sim \begin{cases} \log n, & \text{if } -1 < \alpha, \beta \leq -1/2, \\ n^{\gamma+1/2}, & \text{otherwise}, \end{cases} \quad (4.2.1)$$

holds, where the constants in "$\sim$" are independent of n.

Hence the optimal Lebesgue constants $\|L_n(v^{\alpha,\beta})\|_\infty \sim \log n$ can be obtained if and only if $\alpha \leq -1/2$ and also $\beta \leq -1/2$ holds. In the other cases the Lebesgue constants corresponding to the Jacobi abscissas grow algebraically as $n \to +\infty$.

4.2.1.2 Interpolation at the "Practical Abscissas"

The "practical abscissas", also known as Clenshaw's abscissas, are the following system of knots

$$x_{n,k} = \cos \frac{(n-k)\pi}{n-1}, \quad k = 1, \ldots, n.$$

Such points are the zeros of the polynomials $q_n(x) = p_{n-2}(v^{1/2,1/2}; x)(1-x^2)$, i.e. they are zeros the Chebyshev polynomial of the second kind to which we add the endpoints ± 1.

The following statement is easy to prove.

Theorem 4.2.2 *If $\mathcal{X} = \{U_{n-2}(x)(1-x^2)\}_n$ is the system of the practical abscissas, then the corresponding Lebesgue constants $\|L_n(\mathcal{X})\|_\infty$ are optimal, i.e.,*

$$\|L_n(\mathcal{X})\|_\infty \sim \log n \tag{4.2.2}$$

holds.

For a long time these two examples of optimal systems of nodes were the only ones known. On the other hand the Lagrange interpolation polynomials are easily computable and, also for this reason, they are useful in the approximation of functions and their derivatives, as well as in the numerical integration and in the projection method for the numerical treatment of functional equations. Consequently this leads to the necessity to investigate the existence of more optimal matrices of knots.

With regard to this we want to give some preliminary and simple observations on two necessary "ingredients" for the construction of good systems of nodes.

The first one derives from the trigonometric interpolation and concerns the distribution of the interpolation knots. We recall that the set $\{x_{n,k} = \cos\theta_{n,k}, \ k = 1, \ldots, n\}$ of nodes in $[-1, 1]$ has an *arc sine distribution* if and only if

$$|\theta_{n,k} - \theta_{n,k+1}| \sim \frac{1}{n}, \quad k = 0, 1, \ldots, n, \tag{4.2.3}$$

holds, where we set $\theta_{n,0} = \pi$ and $\theta_{n,n+1} = 0$. Obviously the condition (4.2.3) is equivalent to the following condition on the points $x_{n,k} = \cos\theta_{n,k}$,

$$|x_{n,k+1} - x_{n,k}| \sim \frac{\sqrt{1-x^2}}{n}, \quad x_{n,k} \leq x \leq x_{n,k+1}, \tag{4.2.4}$$

i.e., the distance of the points $x_{n,k}$ closest to the extremes ± 1 is $O(n^{-2})$, while the distance between two "internal" points $x_{n,k}$ is $O(n^{-1})$.

In [465] Szabados and Vértesi stated the next proposition. It shows a case in which the nodes are not *arc sine distributed* and this implies the Lebesgue constants grow in an algebraic way.

Proposition 4.2.1 Let be $n \in \mathbb{N}$ and $-1 = x_{n,0} < x_{n,1} < \cdots < x_{n,n} < x_{n,n+1} = 1$ such that $x_{n,k} = \cos\theta_{n,k}$ for $k = 0, 1, \ldots, n+1$.

If for some k and $n > n_0$ we have

$$\theta_{n,k} - \theta_{n,k+1} \sim \frac{1}{n^{1+\alpha}}, \quad \alpha > 0,$$

then $\|L_n(\mathcal{X})\| \geq Cn^\alpha$ holds.

Proof Let f be the following function

$$f(x) := \begin{cases} \dfrac{x - x_{n,k-1}}{x_{n,k} - x_{n,k-1}} & \text{if } x_{n,k-1} \leq x \leq x_{n,k}, \\ \dfrac{x - x_{n,k+1}}{x_{n,k} - x_{n,k+1}} & \text{if } x_{n,k} \leq x \leq x_{n,k+1}, \\ 0 & \text{otherwise}, \end{cases}$$

and note that $0 \leq f(x) \leq 1$, for all $|x| \leq 1$. Moreover, denote by $P = L_n(\mathcal{X}, f)$ the Lagrange polynomial interpolating f at the points $x_{n,k}$, $k = 1, \ldots, n$.

Observing that $P(\cos t)$ is a trigonometric polynomial of degree at most $n - 1$ and then using the Bernstein inequality (1.2.35), we get

$$n^{1+\alpha} \sim \frac{1}{\theta_{n,k} - \theta_{n,k+1}} = \frac{P(\cos\theta_{n,k}) - P(\cos\theta_{n,k+1})}{\theta_{n,k} - \theta_{n,k+1}}$$

$$= \left[\frac{d}{dt}P(\cos t)\right]_{t=\vartheta} \leq n\|P\|_\infty = n\|L_n(\mathcal{X}, f)\|_\infty$$

$$\leq n\|L_n(\mathcal{X})\|_\infty$$

where $\theta_{n,k+1} \leq \vartheta \leq \theta_{n,k+1}$. Thus, we have

$$n^{1+\alpha} \leq Cn\|L_n(\mathcal{X})\|_\infty$$

i.e., $\|L_n(\mathcal{X})\|_\infty \geq Cn^\alpha$. □

In view of the previous proposition, the *arc sine distribution* of the interpolation knots is recommended in order to get optimal Lebesgue constants. According to this fact, we note that in the examples we have shown at the beginning, the zeros of the polynomials $q_n(x) = p_n(v^{\alpha,\beta}; x)$, with parameters $-1 < \alpha, \beta \leq -1/2$, and $q_n(x) = (1 - x^2)p_n(v^{1/2,1/2}; x)$ have the *arc sine distribution*.

The second "ingredient" concerns the magnitude of the sequence $\{\|q_n\|_\infty\}_n$, where $\mathcal{X} = \{q_n\}_n$. We have the following result from a theorem stated in [313]:

Theorem 4.2.3 Let u and w be two weight functions on $[-1, 1]$ and let $\mathcal{X} = \{q_n\}$ be such that every interval $I \subset [-1, 1]$, with $\int_I u(x)dx > 0$, contains at least one

4.2 Optimal Systems of Nodes

zero of q_n, whenever n is sufficiently large ($\mathcal{X}$ is u-regular). Then

$$\|q_n\|_\infty \leq C \left(\int_{-1}^{1} |q_n(x)| u(x) dx \right) \|L_n(\mathcal{X})\|_\infty \tag{4.2.5}$$

holds, where C depends only on u and $\mathcal{X}$.

Taking into account that if the zeros of q_n are *arc sine distributed*, then $\mathcal{X} = \{q_n\}$ is u-regular for every weight function u on $[-1, 1]$, we can deduce the following immediate consequence of the previous result.

Proposition 4.2.2 *If the zeros of $\mathcal{X} = \{q_n\}_n$ are arc sine distributed on $[-1, 1]$ and if the condition*

$$\sup_n \int_{-1}^{1} |q_n(x)| u(x) dx < +\infty \tag{4.2.6}$$

is satisfied for some weight function u, then we have

$$\|L_n(\mathcal{X})\|_\infty \geq C \|q_n\|_\infty, \quad C \neq C(n). \tag{4.2.7}$$

In order to get an optimal sequence of nodes, (4.2.7) suggests us to consider, as in the trigonometric case, sequences of polynomials that are uniformly bounded with respect to n, i.e., such that

$$\sup_n \|q_n\|_\infty < +\infty \tag{4.2.8}$$

holds, which implies (4.2.6) holds with $u = 1$.

The Szegő theorem and also the interpolation at the practical abscissas, confirm this fact. Namely, we recall the following bound (see (2.3.41))

$$|p_n(v^{\alpha,\beta}; x)| \leq C \left(\sqrt{1-x} + \frac{1}{n} \right)^{-\alpha-1/2} \left(\sqrt{1+x} + \frac{1}{n} \right)^{-\beta-1/2}, \quad |x| \leq 1, \tag{4.2.9}$$

where $C \neq C(n, x)$, which constitutes a precise estimate, since

$$\|p_n(v^{\alpha,\beta})\|_\infty \sim n^{\max\{\alpha,\beta\}+1/2} \tag{4.2.10}$$

holds, where the maximum of $|p_n(v^{\alpha,\beta}; x)|$ is assumed near the end points ± 1. Then, using the previous bound, we note that the sequence $q_n(x) = p_n(v^{\alpha,\beta}; x)$ with $\alpha, \beta > -1/2$ satisfies (4.2.6), but not (4.2.8), while the two sequences $q_n(x) = p_n(v^{\alpha,\beta}; x)$, with $-1 < \alpha, \beta \leq -1/2$, and $q_n(x) = (1-x^2) p_n(v^{1/2,1/2}; x)$ are uniformly bounded with respect to n. Also, their zeros are *arc sine distributed*.

We point out that in general the uniform boundedness of the polynomials q_n and the *arc sine distribution* of their zeros are two independent conditions and both are necessary for having optimal $\mathcal{X} = \{q_n\}_n$. The next two examples show two different cases in which only one of the previous conditions is satisfied. In both cases the Lebesgue constants grow in an algebraic way.

Example 4.2.1 Consider the sequence $q_{2n+1} = T_n T_{n+1}$, with $T_n(x) = \cos(n \arccos x)$ for all $|x| \leq 1$. It is easy to check that $\|q_{2n+1}\|_\infty = 1$, but the zeros of q_{2n+1} have not an *arc sine distribution*. In fact, for instance, if we take the first zero $x_{n,1} = \cos(\pi/(2n))$ of T_n and the first zero $x_{n+1,1} = \cos(\pi/(2n+2))$ of T_{n+1}, we have

$$|\theta_{n,1} - \theta_{n+1,1}| = \frac{\pi}{2n} - \frac{\pi}{2n+2} = \frac{\pi}{2n(n+1)}.$$

Consequently, Proposition 4.2.1 (with $\alpha = 1$) gives $\|L_n(\mathcal{X})\|_\infty \geq Cn$, with $C \neq C(n)$.

Example 4.2.2 The zeros $x_{n,k} = \cos(k\pi/(n+1))$, $k = 1, \ldots, n$, of the orthonormal Chebyshev polynomials of the second kind $\{p_n(v^{1/2,1/2}; x)\}$ satisfy (4.2.3), but since $p_n(v^{1/2,1/2}; x) = U_n(x)/\sqrt{\pi}$, with $U_n(\cos\theta) = \sin(n+1)\theta/\sin\theta$, then it is easy to see that

$$\|p_n(v^{1/2,1/2})\|_\infty = \frac{\|U_n\|_\infty}{\sqrt{\pi}} = \frac{n+1}{\sqrt{\pi}}.$$

Thus, (4.2.8) does not hold, but (4.2.6) is satisfied since, for instance, we have

$$\sup_n \int_{-1}^1 |p_n(v^{1/2,1/2}; x)| \sqrt{1-x^2}\, dx < +\infty.$$

Thus, by Proposition 4.2.2 we can apply (4.2.7) which gives $\|L_n(\mathcal{X})\|_\infty \geq Cn$, with $C \neq C(n)$.

From the reasoning above we can conclude that uniformly bounded sequences $\mathcal{X}$ with an *arc sine distribution* of their zeros are the favorite candidates in order to construct interpolation processes $\{L_n(\mathcal{X})\}$ that have Lebesgue constants of order $\log n$.

While many sequences of orthonormal polynomials in $[-1, 1]$ have *arc sine distributed* zeros, only a "few" of them are uniformly bounded.

Using Propositions 4.2.2 and 4.2.1, sometimes it is possible to slightly modify the sequence $\mathcal{X}$ to obtain $\widetilde{\mathcal{X}}$ such that $\|L_n(\widetilde{\mathcal{X}})\|_\infty \sim \log n$, while $\|L_n(\mathcal{X})\|_\infty \not\sim \log n$. In the next section we will give an important example.

4.2.2 Additional Nodes Method with Jacobi Zeros

By the Szegő theorem we only have a restricted class of the Jacobi polynomials $p_n(v^{\alpha,\beta})$ with $\alpha, \beta \leq -1/2$ which gives an optimal interpolation process. If we want to use the zeros of $p_n(v^{\alpha,\beta})$ with $\alpha, \beta > -1/2$, we may modify the system of knots $\mathcal{X} = \{p_n(v^{\alpha,\beta})\}$ as follows.

Since in the case $\max\{\alpha, \beta\} > -1/2$, the polynomials $p_n(v^{\alpha,\beta}; x)$ are not bounded with respect to n eventually at the ends of the interval $[-1, 1]$ (see

4.2 Optimal Systems of Nodes

(4.2.10)), we construct two sequences of polynomials $\{Z_r = Z_{n,r}\}$ and $\{Y_s = Y_{n,s}\}$ of fixed degree r and s respectively, with their zeros "close" to the endpoints ± 1 and such that the new system of polynomials $\{Q_{n+r+s} := Y_{n,s} Z_{n,r} p_n(v^{\alpha,\beta})\}_n$ is uniformly bounded with respect to n and the zeros of Q_{n+r+s} have an *arc sine distribution*.

In order to construct the polynomials $Z_{n,r}$ and $Y_{n,s}$, denote by $x_1 < x_2 < \cdots < x_n$ the zeros of $p_n(v^{\alpha,\beta})$ and define

$$y_j = -1 + j\frac{1+x_1}{1+s}, \qquad j = 1, 2, \ldots, s, \tag{4.2.11}$$

$$z_i = x_n + i\frac{1-x_n}{1+r}, \qquad i = 1, 2, \ldots, r, \tag{4.2.12}$$

where r, s are fixed positive integers (i.e., the points y_j and z_i are dependent on n, but the numbers r and s of such points are independent of n).

Using the points $\{y_j\}_{j=1}^s$ and $\{z_i\}_{i=1}^r$ we set

$$Y_s(x) = Y_{n,s}(x) = \prod_{j=1}^s (x - y_j), \qquad Z_r(x) = Z_{n,r}(x) = \prod_{i=1}^r (x - z_i). \tag{4.2.13}$$

In this way the set $\{y_j\}_{j=1}^s \cup \{x_j\}_{j=1}^n \cup \{z_j\}_{j=1}^r$ of the zeros of the polynomial

$$Q_{n+r+s}(x) = Y_{n,s}(x) Z_{n,r}(x) p_n(v^{\alpha,\beta}; x)$$

has an *arc sine distribution*, since

$$x_1 - y_s \sim \frac{1}{n^2} \sim z_1 - x_n, \tag{4.2.14}$$

$$y_{j+1} - y_j \sim \frac{1}{n^2}, \quad j = 1, \ldots, s-1, \tag{4.2.15}$$

$$z_{j+1} - z_j \sim \frac{1}{n^2}, \quad j = 1, \ldots, r-1, \tag{4.2.16}$$

hold uniformly with respect to n.

On the other hand, it is easy to verify that Q_{n+r+s} satisfies (4.2.6) for some weight function u (for instance we can take $u = v^{\alpha,\beta}$). Moreover, since

$$|Y_s(x)| \le C\left(\sqrt{1+x} + \frac{1}{n}\right)^{2s}, \qquad |Z_r(x)| \le C\left(\sqrt{1-x} + \frac{1}{n}\right)^{2r} \tag{4.2.17}$$

hold for all $|x| \le 1$, with $C \ne C(n, x)$, we get using (4.2.9)

$$|Y_s(x) Z_r(x) p_n(v^{\alpha,\beta}; x)| \le C\left(\sqrt{1-x} + \frac{1}{n}\right)^{-\alpha - \frac{1}{2} + 2r} \left(\sqrt{1+x} + \frac{1}{n}\right)^{\beta - \frac{1}{2} + 2s},$$

i.e.,
$$|Y_s(x)Z_r(x)p_n(v^{\alpha,\beta};x)| \leq C \neq C(n,x), \qquad (4.2.18)$$
for each $x \in [-1,1]$, whenever
$$r \geq \frac{\alpha}{2} + \frac{1}{4} \quad \text{and} \quad s \geq \frac{\beta}{2} + \frac{1}{4}.$$

Then, requiring at most some additional conditions on r and s, we can expect that the Lebesgue constants corresponding to $\mathcal{X} = \{Q_{n+r+s}\}_n$ are of order $\log n$.

In fact, let us denote by $L_{n,r,s}(v^{\alpha,\beta}, f) \in \mathcal{P}_{n+r+s-1}$ the Lagrange polynomial interpolating f at the points
$$-1 < y_1 < \cdots < y_s < x_1 < \cdots < x_n < z_1 < \cdots < z_r < 1.$$

Such a polynomial can be written as

$$L_{n,r,s}(v^{\alpha,\beta}, f; x)$$
$$= Y_s(x)Z_r(x) \sum_{k=1}^{n} \frac{\ell_{n,k}(v^{\alpha,\beta};x)}{Y_s(x_k)Z_r(x_k)} f(x_k)$$
$$+ Y_s(x)p_n(v^{\alpha,\beta};x) \sum_{k=1}^{r} \frac{1}{Y_s(z_k)p_n(v^{\alpha,\beta};z_k)} \prod_{\substack{i=1 \\ i \neq k}}^{r} \frac{x-z_i}{z_k - z_i} f(z_k)$$
$$+ Z_r(x)p_n(v^{\alpha,\beta};x) \sum_{k=1}^{s} \frac{1}{Z_r(y_k)p_n(v^{\alpha,\beta};y_k)} \prod_{\substack{i=1 \\ i \neq k}}^{s} \frac{x-y_i}{y_k - y_i} f(y_k). \qquad (4.2.19)$$

The behaviour of the Lebesgue constants $\|L_{n,r,s}(v^{\alpha,\beta})\|_\infty$ corresponding to the interpolation process $L_{n,r,s}(v^{\alpha,\beta}, f)$, is described by the following theorem.

Theorem 4.2.4 *Let $\alpha, \beta > -1$ and r, s be non negative integers. We have optimal Lebesgue constants*
$$\|L_{n,r,s}(v^{\alpha,\beta})\|_\infty \sim \log n \qquad (4.2.20)$$
if and only if the parameters α, β, r, s satisfy the relations
$$\frac{\alpha}{2} + \frac{1}{4} \leq r \leq \frac{\alpha}{2} + \frac{5}{4}, \qquad (4.2.21)$$
$$\frac{\beta}{2} + \frac{1}{4} \leq s \leq \frac{\beta}{2} + \frac{5}{4}. \qquad (4.2.22)$$

Proof Sufficiency of the conditions. We assume that (4.2.21) and (4.2.22) hold and prove
$$\|L_{n,r,s}(v^{\alpha,\beta}, f)\|_\infty \leq C \log n \|f\|_\infty, \qquad C \neq C(n, f).$$

4.2 Optimal Systems of Nodes

To this end it is sufficient by (4.2.19) to state that

$$\mathcal{L}_1(x) := |Y_s(x) Z_r(x)| \sum_{k=1}^{n} \frac{|\ell_{n,k}(v^{\alpha,\beta}; x)|}{|Y_s(x_k) Z_r(x_k)|} \leq C \log n, \tag{4.2.23}$$

$$\mathcal{L}_2(x) := |Y_s(x) p_n(v^{\alpha,\beta}; x)| \sum_{k=1}^{r} \frac{1}{|Y_s(z_k) p_n(v^{\alpha,\beta}; z_k)|} \prod_{\substack{i=1 \\ i \neq k}}^{r} \left| \frac{x - z_i}{z_k - z_i} \right| \leq C, \tag{4.2.24}$$

$$\mathcal{L}_3(x) := |Z_r(x) p_n(v^{\alpha,\beta}; x)| \sum_{k=1}^{s} \frac{1}{|Z_r(y_k) p_n(v^{\alpha,\beta}; y_k)|} \prod_{\substack{i=1 \\ i \neq k}}^{s} \left| \frac{x - y_i}{y_k - y_i} \right| \leq C, \tag{4.2.25}$$

where in all cases $C \neq C(n, x)$.

Now we use the following estimates for $\mathcal{L}_1$

$$|\ell_{n,d}(v^{\alpha,\beta}; x)| \sim 1, \tag{4.2.26}$$

$$|\ell_{n,k}(v^{\alpha,\beta}; x)| \sim |p_n(v^{\alpha,\beta}; x)| \frac{v^{\frac{\alpha}{2}+\frac{1}{4}, \frac{\beta}{2}+\frac{1}{4}}(x_k)}{|x_k - x|} \Delta x_k, \quad x \neq x_k, \tag{4.2.27}$$

$$|Y_s(x)| \sim (1+x)^s, \quad |Z_r(x)| \sim (1-x)^r, \quad -1 + \frac{C}{n^2} \leq x \leq 1 - \frac{C}{n^2}, \tag{4.2.28}$$

where d in (4.2.26) denotes the index of a Jacobi zero x_d closest to x (i.e., $|x - x_d| = \min_{1 \leq k \leq n} |x - x_k|$). Consequently, using also (4.2.18), we get

$$\mathcal{L}_1(x) \sim 1 + \left| Y_s(x) Z_r(x) p_n(v^{\alpha,\beta}; x) \right| \sum_{\substack{k=1 \\ k \neq d}}^{n} \frac{v^{\frac{\alpha}{2}+\frac{1}{4}-r, \frac{\beta}{2}+\frac{1}{4}-s}(x_k)}{|x_k - x|} \Delta x_k$$

$$\leq C \left(1 + \sum_{\substack{k=1 \\ k \neq d}}^{n} \frac{v^{\frac{\alpha}{2}+\frac{1}{4}-r, \frac{\beta}{2}+\frac{1}{4}-s}(x_k)}{|x_k - x|} \Delta x_k \right) \leq C \log n,$$

where the last sum is estimated by means of (4.1.13) which we can apply since (4.2.21) and (4.2.22) hold.

The proofs of (4.2.24) and (4.2.25) are similar and based on the estimates (4.2.17), (4.2.9) and

$$|p_n(v^{\alpha,\beta}; x)| \sim n^{\alpha+1/2}, \quad 1 - \frac{C}{n^2} \leq x \leq 1, \tag{4.2.29}$$

$$|p_n(v^{\alpha,\beta}; x)| \sim n^{\beta+1/2}, \quad -1 \leq x \leq -1 + \frac{C}{n^2}. \tag{4.2.30}$$

Let us estimate for instance only the sum $\mathcal{L}_2(x)$. Taking into account that

$$|Y_s(x)p_n(v^{\alpha,\beta};x)| \leq C\left(\sqrt{1-x}+\frac{1}{n}\right)^{-\alpha-\frac{1}{2}}\left(\sqrt{1+x}+\frac{1}{n}\right)^{-\beta-\frac{1}{2}+2s}, \quad (4.2.31)$$

$$|Y_s(z_k)| \sim 1, \qquad |p_n(v^{\alpha,\beta};z_k)| \sim n^{\alpha+\frac{1}{2}}, \quad (4.2.32)$$

$$\prod_{\substack{i=1\\i\neq k}}^{r}|x-z_i| \leq C\left(\sqrt{1-x}+\frac{1}{n}\right)^{2r-2}, \qquad \prod_{\substack{i=1\\i\neq k}}^{r}|z_k-z_i| \sim n^{2-2r}, \quad (4.2.33)$$

then we have

$$\mathcal{L}_2(x) := |Y_s(x)p_n(v^{\alpha,\beta};x)|\sum_{k=1}^{s}\frac{1}{|Y_s(z_k)p_n(v^{\alpha,\beta},z_k)|}\prod_{\substack{i=1\\i\neq k}}^{r}\left|\frac{x-z_i}{z_k-z_i}\right|$$

$$\leq Cm^{-\alpha-\frac{5}{2}+2r}\left(\sqrt{1-x}+\frac{1}{n}\right)^{-\alpha-\frac{5}{2}+2r}\left(\sqrt{1+x}+\frac{1}{n}\right)^{-\beta-\frac{1}{2}+2s}$$

$$\leq C\left(\sqrt{1-x}+\frac{1}{n}\right)^{\alpha+\frac{5}{2}-2r}\left(\sqrt{1-x}+\frac{1}{n}\right)^{-\alpha-\frac{5}{2}+2r}\left(\sqrt{1+x}+\frac{1}{n}\right)^{-\beta-\frac{1}{2}+2s}$$

$$\leq C \neq C(n,x),$$

since $\alpha + 5/2 - 2r \geq 0$ and also $-\beta - 1/2 + 2s \geq 0$ by (4.2.21) and (4.2.22).

Necessity of the conditions. Now let us prove that (4.2.21) and (4.2.22) are also necessary conditions for the Lebesgue constants to be optimal. First, we observe that if

$$0 \leq r < \frac{\alpha}{2} + \frac{1}{4} \quad \text{or} \quad 0 \leq s < \frac{\beta}{2} + \frac{1}{4},$$

then the system $\{q_n := Y_s Z_r p_n(v^{\alpha,\beta})\}$ is unbounded with respect to n. In fact, if we take $x = (y_1 - 1)/2$ or $x = (1+z_r)/2$, then we have $|q_n(x)| \sim n^{\beta+1/2-2s}$ or $|q_n(x)| \sim n^{\alpha+1/2-2r}$, respectively, and consequently

$$\|q_n\|_\infty \geq Cn^\mu, \qquad \mu = \max\{\alpha + 1/2 - 2r, \beta + 1/2 - 2s\}.$$

On the other hand, for some weight u, the L^1-condition (4.2.6) is satisfied by $q_n := Y_s Z_r p_n(v^{\alpha,\beta})$. Hence, we can apply (4.2.7) which gives

$$\|L_{n,r,s}\|_\infty \geq C\|q_n\|_\infty \geq Cn^{\max\{\alpha+1/2-2r,\beta+1/2-2s\}},$$

i.e., the Lebesgue constants grow algebraically.

Now we assume that the inequalities on the right-hand side of (4.2.21) and (4.2.22) do not hold and prove that also in this case the Lebesgue constants are not optimal.

4.2 Optimal Systems of Nodes

Suppose $r > \alpha/2 + 5/4$; the case $s > \beta/2 + 5/4$ is similar.

Taking two fixed Jacobi zeros x_d and x_{d+1} "close" to the origin (e.g., $x_d, x_{d+1} \in [-a, a]$ with $0 < a < 1/2$), we consider the midpoint $\bar{x} = (x_d + x_{d+1})/2$. We get by (4.2.19)

$$\|L_{n,r,s}(v^{\alpha,\beta})\|_\infty \geq \mathcal{L}_2(\bar{x})$$

$$= |Y_s(\bar{x})p_n(v^{\alpha,\beta}; \bar{x})| \sum_{k=1}^{s} \frac{1}{|Y_s(z_k)p_n(v^{\alpha,\beta}; z_k)|} \prod_{\substack{i=1 \\ i \neq k}}^{r} \left|\frac{\bar{x} - z_i}{z_k - z_i}\right|.$$

On the other hand we have by (4.2.28)

$$|Y_s(\bar{x})| \sim (1+\bar{x})^s \geq C, \qquad \prod_{\substack{i=1 \\ i \neq k}}^{r} |\bar{x} - z_i| \sim (1-\bar{x})^{r-1} \geq C, \quad 0 < C \neq C(n),$$

and from (4.2.26) and (4.2.27) we deduce

$$C \leq \ell_{n,d}(v^{\alpha,\beta}; \bar{x}) \sim |p_n(v^{\alpha,\beta}; \bar{x})| \frac{v^{\frac{\alpha}{2}+\frac{1}{4}, \frac{\beta}{2}+\frac{1}{4}}(x_d)}{\bar{x} - x_d} \Delta x_d \leq C|p_n(v^{\alpha,\beta}; \bar{x})|$$

since $\Delta x_d = 2(\bar{x} - x_d)$.

So, using these estimates, (4.2.32) and (4.2.33), we obtain

$$\|L_{n,r,s}(v^{\alpha,\beta})\|_\infty \geq |Y_s(\bar{x})p_n(v^{\alpha,\beta}; \bar{x})| \sum_{k=1}^{s} \frac{1}{|Y_s(z_k)p_n(v^{\alpha,\beta}; z_k)|} \prod_{\substack{i=1 \\ i \neq k}}^{r} \left|\frac{\bar{x} - z_i}{z_k - z_i}\right|$$

$$\geq C \sum_{k=1}^{s} \frac{1}{|Y_s(z_k)p_n(v^{\alpha,\beta}; z_k)|} \frac{1}{\prod_{\substack{i=1 \\ i \neq k}}^{r} |z_k - z_i|}$$

$$\geq Cn^{-\alpha - 5/2 - 2r},$$

i.e., the Lebesgue constants grow algebraically. □

Theoretically, Theorem 4.2.4 assures that it is always possible to get an optimal interpolation process also using the zeros of the Jacobi polynomials $\{p_n(v^{\alpha,\beta})\}$ with $\alpha, \beta > -1/2$, since for any $\alpha, \beta > -1/2$, there exists at least one integer r that satisfies (4.2.21) and one integer s that satisfies (4.2.22).

We remark that the previous theorem still holds with other definitions of the additional points y_j and z_j, under the condition that (4.2.14)–(4.2.16) are verified (in case y_1 and z_r can be replaced by -1 and 1, respectively).

Special cases of Theorem 4.2.4 are the Szegő theorem which corresponds to the choice $r = s = 0$ and the interpolation at the practical abscissas

$$\left\{\cos\frac{(n-k)\pi}{n-1}, \quad k = 1, \ldots, n\right\},$$

which corresponds to the case $r = s = 1$ and $\alpha = \beta = 1/2$. But we can construct many other examples, since because of Theorem 4.2.4 we can easily change any "bad" array of nodes into an "optimal" one.

Example 4.2.3 Let us consider the Chebyshev weights of the third and fourth kind

$$v^{1/2,-1/2}(x) := \sqrt{\frac{1-x}{1+x}}, \qquad v^{-1/2,1/2}(x) := \sqrt{\frac{1+x}{1-x}}. \qquad (4.2.34)$$

The zeros of the corresponding systems of polynomials are explicitly known

$$\mathcal{X}_1 := \{p_n(v^{1/2,-1/2})\} = \left\{-\cos\frac{2k\pi}{2n+1}, \quad k=1,\ldots,n\right\}_n,$$

$$\mathcal{X}_2 := \{p_n(v^{-1/2,1/2})\} = \left\{-\cos\frac{k\pi}{2n+1}, \quad k=1,\ldots,n\right\}_n.$$

The weights in (4.2.34) do not satisfy the Szegő theorem. Hence we have "bad" Lebesgue constants corresponding to the matrices of nodes $\mathcal{X}_1$ and $\mathcal{X}_2$. But, if we add one of the endpoints ± 1 to the previous matrices, i.e., if we consider

$$\mathcal{X}_1^* = \{(1-x)p_n(v^{1/2,-1/2}; x)\}_n, \qquad \mathcal{X}_2^* = \{(1+x)p_n(v^{-1/2,1/2}; x)\}_n,$$

we get "optimal" matrices of nodes by Theorem 4.2.4 (taking $r = 1$, $s = 0$ for $v^{1/2,-1/2}$ and $r = 0$, $s = 1$ for $v^{-1/2,1/2}$, the conditions (4.2.21) and (4.2.22) are satisfied).

Example 4.2.4 Another simple application of Theorem 4.2.4 can be achieved with the Legendre polynomials $\{p_n(x)\}$, which are special Jacobi polynomials orthonormal with respect to the weight $v(x) = 1$.

In this case $\alpha = \beta = 0$ and the Szegő theorem assures the Lagrange interpolation at the Legendre zeros has Lebesgue constants of order $\sqrt{n}$. On the other hand, the conditions in (4.2.21) and (4.2.22) are satisfied with $r = s = 1$. Hence by Theorem 4.2.4, the Lebesgue constants become of order $\log n$ if, for instance, we add the endpoints ± 1 to the Legendre zeros.

Now, we consider a Lagrange-Hermite interpolation process that uses the values of the function f in $[-1, 1]$ and its derivatives at the end-points ± 1. The problem is the following: assume that the values $f^{(i)}(-1)$, $i = 0, 1, \ldots, s-1$, and $f^{(j)}(1)$, $j = 0, 1, \ldots, r-1$, are known (we set $f^{(0)} = f$); then find n nodes $x_1 < x_2 < \cdots < x_n$ in $(-1, 1)$ such that the Hermite interpolating polynomial P defined by

$$P(x_k) = f(x_k), \qquad k = 1, \ldots, n, \qquad (4.2.35)$$

$$P^{(i)}(-1) = f^{(i)}(-1), \qquad i = 0, 1, \ldots, s-1, \qquad (4.2.36)$$

$$P^{(j)}(1) = f^{(j)}(1), \qquad j = 0, 1, \ldots, r-1, \qquad (4.2.37)$$

4.2 Optimal Systems of Nodes

satisfies the error estimate

$$\|f - P\|_\infty \leq C E_{n-1}(f) \log n, \qquad C \neq C(n, f).$$

Such an interpolation process turns out to be very useful in the numerical solution of boundary value problems.

For the sake of simplicity we assume $f \in C^N[-1, 1]$, with $N = \max\{r - 1, s - 1\}$, but we could also suppose that f is simply continuous in $[-1, 1]$ and has continuous derivatives in a neighborhood of ± 1.

Moreover, we assume that the knots x_k are the zeros of the Jacobi polynomial $p_n(v^{\alpha,\beta})$ and then the question is how to choose the parameters $\alpha, \beta > -1$, once the positive integers r and s have been given.

The case $r = s = 1$ is an accordance with Theorem 4.2.4. The case $r, s > 1$ can be interpreted as a limiting case of the additional knots method, setting

$$y_1 = y_2 = \cdots = y_s = -1, \qquad z_1 = z_2 = \cdots = z_r = 1, \tag{4.2.38}$$

i.e., considering only two additional points $1, -1$ of multiplicity r and s, respectively. Therefore, we have

$$Y_s(x) := (1 + x)^s, \qquad Z_r(x) := (1 - x)^r. \tag{4.2.39}$$

If we write the last two sums in (4.2.19), i.e., the Lagrange polynomials interpolating $f/(Y_s p_n(v^{\alpha,\beta}))$ at the points $\{z_j\}$ and $f/(Z_r p_n(v^{\alpha,\beta}))$ at the points $\{y_i\}$, using the Newton formula, we have

$$L_{n,r,s}(v^{\alpha,\beta}, f; x) = Y_s(x) Z_r(x) \sum_{k=1}^{n} \ell_{n,k}(v^{\alpha,\beta}; x) \frac{f(x_k)}{Y_s(x_k) Z_r(x_k)}$$

$$+ Z_r(x) p_n(v^{\alpha,\beta}; x) \left(\frac{f(-1)}{Z_r(-1) p_n(v^{\alpha,\beta}; -1)} \right.$$

$$+ \sum_{i=1}^{s-1} (x - y_1) \cdots (x - y_i) \left[y_1, \ldots, y_{i+1}; \frac{f}{Z_r p_n(v^{\alpha,\beta})} \right] \right)$$

$$+ Y_s(x) p_n(v^{\alpha,\beta}; x) \left(\frac{f(1)}{Y_s(1) p_n(v^{\alpha,\beta}; 1)} \right.$$

$$+ \sum_{j=1}^{r-1} (x - z_1) \cdots (x - z_j) \left[z_1, \ldots, z_{j+1}; \frac{f}{Y_s p_n(v^{\alpha,\beta})} \right] \right).$$

Then, taking the limits for $y_i \to -1$, $i = 1, \ldots, s$, and $z_j \to 1$, $j = 1, \ldots, r$, and recalling the property

$$\xi_1 = \xi_2 = \cdots = \xi_{i+1} = \xi \quad \Rightarrow \quad [\xi_1, \xi_2, \ldots, \xi_{i+1}; f] = \frac{f^{(i)}(\xi)}{i!},$$

we get

$$\mathcal{L}_{n,r,s}(v^{\alpha,\beta}, f; x)$$
$$:= Y_s(x) Z_r(x) \sum_{k=1}^{n} \ell_{n,k}(v^{\alpha,\beta}; x) \frac{f(x_k)}{Y_s(x_k) Z_r(x_k)}$$
$$+ Z_r(x) p_n(v^{\alpha,\beta}; x) \sum_{i=0}^{s-1} \frac{(1+x)^i}{i!} \left(\frac{f}{Z_r p_n(v^{\alpha,\beta})} \right)^{(i)} (-1)$$
$$+ Y_s(x) p_n(v^{\alpha,\beta}; x) \sum_{j=0}^{r-1} (-1)^j \frac{(1-x)^j}{j!} \left(\frac{f}{Y_s p_n(v^{\alpha,\beta})} \right)^{(j)} (1), \quad (4.2.40)$$

with the polynomials Y_s and Z_r defined by (4.2.39).

It is easy to check that the polynomial (4.2.40) satisfies the conditions (4.2.35), (4.2.36) and (4.2.37).

According to the error estimate for the approximation $f \approx \mathcal{L}_{n,r,s}(v^{\alpha,\beta}, f)$, we recall that, by Theorem 4.2.4, the conditions

$$\frac{\alpha}{2} + \frac{1}{4} \leq r \leq \frac{\alpha}{2} + \frac{5}{4}, \quad \frac{\beta}{2} + \frac{1}{4} \leq s \leq \frac{\beta}{2} + \frac{5}{4},$$

are necessary and sufficient for

$$\|f - \mathcal{L}_{n,r,s}(v^{\alpha,\beta}, f)\|_\infty \leq C E_{n-1}(f)_\infty \log n, \qquad C \neq C(n, f). \quad (4.2.41)$$

Then, since the polynomial $\mathcal{L}_{n,r,s}(v^{\alpha,\beta}, f)$ can be deduced from $L_{n,r,s}(v^{\alpha,\beta}, f)$ when the additional knots tend to the endpoints ± 1, we expect the same result to hold for $\mathcal{L}_{n,r,s}(v^{\alpha,\beta}, f)$, too. In fact, the following statement holds.

Theorem 4.2.5 *Let $\alpha, \beta > -1$ and r, s be positive integers. For all $n \in \mathbb{N}$ and $f \in C^N[-1, 1]$, with $N = \max\{r-1, s-1\}$, we have*

$$\|f - \mathcal{L}_{n,r,s}(v^{\alpha,\beta}, f)\|_\infty \leq C E_{n-1}(f)_\infty \log n, \qquad C \neq C(n, f), \quad (4.2.42)$$

if and only the relations

$$2r - \frac{5}{2} \leq \alpha \leq 2r - \frac{1}{2}, \quad (4.2.43)$$

$$2s - \frac{5}{2} \leq \beta \leq 2s - \frac{1}{2}, \quad (4.2.44)$$

are satisfied.

Proof Sufficiency of the conditions. Assume (4.2.43) and (4.2.44) hold and prove (4.2.42).

4.2 Optimal Systems of Nodes

Taking into account that for all $P \in \mathcal{P}_{n+r+s-1}$ we have $\mathcal{L}_{n,r,s}(v^{\alpha,\beta}, P) = P$, we can write

$$f - \mathcal{L}_{n,r,s}(v^{\alpha,\beta}, f) = (f - P) - \mathcal{L}_{n,r,s}(v^{\alpha,\beta}, f - P), \qquad P \in \mathcal{P}_{n+r+s-1}. \tag{4.2.45}$$

On the other hand, by a result of Gopengauz [187], there exists a polynomial $Q \in \mathcal{P}_{n+r+s-1}$ such that

$$|f^{(i)}(x) - Q^{(i)}(x)| \leq C \left(\frac{\sqrt{1-x^2}}{n} \right)^{N-i} \omega\left(f^{(i)}, \frac{\sqrt{1-x^2}}{n} \right) \tag{4.2.46}$$

holds, for all $x \in [-1, 1]$, and $i = 0, 1, \ldots, N$, with $C \neq C(n, f, x)$.

In particular, we have by (4.2.46)

$$f^{(i)}(-1) - Q^{(i)}(-1) = 0, \qquad i = 0, 1, \ldots, s-1,$$
$$f^{(i)}(1) - Q^{(i)}(1) = 0, \qquad i = 0, 1, \ldots, r-1,$$

and then, by (4.2.45) and (4.2.40), for all $x \in [-1, 1]$, we get

$$|f(x) - \mathcal{L}_{n,r,s}(v^{\alpha,\beta}, f; x)|$$
$$\leq |f(x) - Q(x)| + |\mathcal{L}_{n,r,s}(v^{\alpha,\beta}, f - Q; x)|$$
$$\leq |f(x) - Q(x)| + |Y_s(x) Z_r(x)| \sum_{k=1}^{n} |f(x_k) - Q(x_k)| \left| \frac{\ell_{n,k}(v^{\alpha,\beta}; x)}{Y_s(x_k) Z_r(x_k)} \right|$$
$$\leq \|f - Q\|_{\infty} \left(1 + |Y_s(x) Z_r(x)| \sum_{k=1}^{n} \left| \frac{\ell_{n,k}(v^{\alpha,\beta}; x)}{Y_s(x_k) Z_r(x_k)} \right| \right).$$

But, following the proof of (4.2.23), we have

$$|Y_s(x) Z_r(x)| \sum_{k=1}^{n} \left| \frac{\ell_{n,k}(v^{\alpha,\beta}; x)}{Y_s(x_k) Z_r(x_k)} \right| \leq C \log n, \qquad C \neq C(n, x).$$

Thus, we get

$$|f(x) - \mathcal{L}_{n,r,s}(v^{\alpha,\beta}, f; x)| \leq C \|f - Q\|_{\infty} \log n.$$

On the other hand, we deduce by (4.2.46)

$$\|f - Q\|_{\infty} \leq C \omega\left(f, \frac{1}{n} \right), \qquad C \neq C(n, f),$$

and then

$$\|f - \mathcal{L}_{n,r,s}(v^{\alpha,\beta}, f)\|_{\infty} \leq C \omega\left(f, \frac{1}{n} \right) \log n, \qquad C \neq C(n, f),$$

holds, for all $f \in C^N[-1, 1]$. Consequently, for any $P \in \mathcal{P}_{n-1}$, we have

$$\|f - \mathcal{L}_{n,r,s}(v^{\alpha,\beta}, f)\|_\infty = \|(f - P) - \mathcal{L}_{n,r,s}(v^{\alpha,\beta}, f - P)\|_\infty$$

$$\leq C\omega\left(f - P, \frac{1}{n}\right)\log n$$

$$\leq C\|f - P\|_\infty \log n,$$

which, by taking the infimum with respect to $P \in \mathcal{P}_{n-1}$, gives (4.2.42).

Necessity of the conditions. We firstly point out that (4.2.42) is equivalent to the following bound of the Lebesgue constants

$$\|\mathcal{L}_{n,r,s}(v^{\alpha,\beta})\|_\infty := \sup_{\|f\|_\infty \leq 1} \|\mathcal{L}_{n,r,s}(v^{\alpha,\beta}, f)\|_\infty \leq C\log n, \qquad (4.2.47)$$

where $C \neq C(n)$. In fact, from (4.2.42) we deduce for all continuous functions

$$\|\mathcal{L}_{n,r,s}(v^{\alpha,\beta}, f)\|_\infty \leq \|f\|_\infty + \|f - \mathcal{L}_{n,r,s}(v^{\alpha,\beta}, f)\|_\infty \leq C\|f\|_\infty \log n,$$

which gives (4.2.47). Conversely, (4.2.47) implies (4.2.42), since for all $P \in \mathcal{P}_{n-1}$

$$\|f - \mathcal{L}_{n,r,s}(v^{\alpha,\beta}, f)\|_\infty \leq \|f - P\|_\infty + \|\mathcal{L}_{n,r,s}(v^{\alpha,\beta}, f - P)\|_\infty$$

$$\leq \|f - P\|_\infty \left(1 + \|\mathcal{L}_{n,r,s}(v^{\alpha,\beta})\|_\infty\right)$$

$$\leq C\|f - P\|_\infty \log n,$$

which gives (4.2.42) by taking the infimum with respect to $P \in \mathcal{P}_{n-1}$.

This means, we assume that (4.2.43) or (4.2.44) does not hold and then prove that in this case

$$\|\mathcal{L}_{n,r,s}(v^{\alpha,\beta})\|_\infty \geq Cn^\mu$$

holds for some $\mu > 0$.

For example, assume $\beta > 2s - 1/2$. In this case we fix $\bar{x} := (x_1 - 1)/2$, i.e., we set $\bar{x}$ to be the midpoint between -1 and the first zero x_1. Then we get by (4.2.40)

$$\|\mathcal{L}_{n,r,s}(v^{\alpha,\beta})\|_\infty \geq Y_s(\bar{x})Z_r(\bar{x}) \sum_{k=1}^n \frac{|\ell_{n,k}(v^{\alpha,\beta}; \bar{x})|}{Y_s(x_k)Z_r(x_k)} = v^{r,s}(\bar{x}) \sum_{k=1}^n \frac{|\ell_{n,k}(v^{\alpha,\beta}; \bar{x})|}{v^{r,s}(x_k)}$$

$$\geq v^{r,s}(\bar{x}) \sum_{|x_k| \leq 1/2} \frac{|\ell_{n,k}(v^{\alpha,\beta}; \bar{x})|}{v^{r,s}(x_k)}$$

$$\sim v^{r,s}(\bar{x})|p_n(v^{\alpha,\beta}; \bar{x})| \sum_{|x_k| \leq 1/2} \frac{v^{\frac{\alpha}{2}+\frac{1}{4}-r, \frac{\beta}{2}+\frac{1}{4}-s}(x_k)}{|x_k - \bar{x}|} \Delta x_k,$$

where we used (4.2.27).

4.2 Optimal Systems of Nodes

But, recalling that $\bar{x} = (x_1 - 1)/2$, we have

$$v^{r,s}(\bar{x}) \sim (1+\bar{x})^s = \left(\frac{1+x_1}{2}\right)^s \sim \frac{1}{n^{2s}}$$

and consequently we get by (4.2.30)

$$v^{r,s}(\bar{x})|p_n(v^{\alpha,\beta};\bar{x})| \sim n^{\beta+1/2-2s}.$$

Using this fact and taking into account that $|x_k - \bar{x}| \leq 2$ and $v^{\frac{\alpha}{2}+\frac{1}{4}-r,\frac{\beta}{2}+\frac{1}{4}-s}(x_k) \sim 1$ for $|x_k| \leq 1/2$, we obtain

$$\|\mathcal{L}_{n,r,s}(v^{\alpha,\beta})\|_\infty \geq Cv^{r,s}(\bar{x})|p_n(v^{\alpha,\beta};\bar{x})| \sum_{|x_k|\leq 1/2} \frac{v^{\frac{\alpha}{2}+\frac{1}{4}-r,\frac{\beta}{2}+\frac{1}{4}-s}(x_k)}{|x_k - \bar{x}|} \Delta x_k$$

$$\geq Cn^{\beta+1/2-2s} \sum_{|x_k|\leq 1/2} \Delta x_k \geq Cn^{\beta+1/2-2s}.$$

In the case when (4.2.43) is not satisfied because $\alpha > 2r - 1/2$, we can proceed analogously, taking $\bar{x} = (x_n + 1)/2$.

Now we study the case $\beta < 2s - 5/2$. Similarly as in the proof of Theorem 4.2.4, in this case we consider the midpoint of two zeros of $p_n(v^{\alpha,\beta})$ which are "near" to the origin, i.e., we take $\bar{x} = (x_d + x_{d+1})/2$ with $x_d, x_{d+1} \in [-1/2, 1/2]$ for example. Then we get by (4.2.40) and (4.2.27)

$$\|\mathcal{L}_{n,r,s}(v^{\alpha,\beta})\|_\infty \geq v^{r,s}(\bar{x}) \sum_{k=1}^n \frac{|\ell_{n,k}(v^{\alpha,\beta};\bar{x})|}{v^{r,s}(x_k)} \geq v^{r,s}(\bar{x}) \frac{|\ell_{n,1}(v^{\alpha,\beta};\bar{x})|}{v^{r,s}(x_1)}$$

$$\sim v^{r,s}(\bar{x})|p_n(v^{\alpha,\beta};\bar{x})| \frac{v^{\frac{\alpha}{2}+\frac{1}{4}-r,\frac{\beta}{2}+\frac{1}{4}-s}(x_1)\Delta x_1}{|\bar{x} - x_1|}.$$

Now, for the opposite choice of the point $\bar{x}$, we have $|p_n(v^{\alpha,\beta};\bar{x})| \sim 1$, $v^{r,s}(\bar{x}) \sim 1$ and also $|\bar{x} - x_1| \leq 2$. Moreover, by the *arc sine distribution* of the knots x_k, we observe that

$$v^{\frac{\alpha}{2}+\frac{1}{4}-r,\frac{\beta}{2}+\frac{1}{4}-s}(x_1) \sim (1+x_1)^{\frac{\beta}{2}+\frac{1}{4}-s} \sim n^{2s-\beta-1/2}$$

and $\Delta x_1 \sim n^{-2}$ hold. Hence, collecting these facts, we get

$$\|\mathcal{L}_{n,r,s}(v^{\alpha,\beta})\|_\infty \geq Cv^{r,s}(\bar{x})|p_n(v^{\alpha,\beta};\bar{x})| \frac{v^{\frac{\alpha}{2}+\frac{1}{4}-r,\frac{\beta}{2}+\frac{1}{4}-s}(x_1)\Delta x_1}{|\bar{x} - x_1|} \geq Cn^{2s-\beta-5/2}.$$

Finally in the case $\alpha < 2r - 5/2$, we can proceed analogously as in the previous case, taking the same point $\bar{x}$ and noting that

$$\|\mathcal{L}_{n,r,s}(v^{\alpha,\beta})\|_\infty \geq v^{r,s}(\bar{x}) \frac{|\ell_{n,n}(v^{\alpha,\beta};\bar{x})|}{v^{r,s}(x_n)}.$$

□

Example 4.2.5 Let f be a given function belonging to $C^2[-1, 1]$. We consider the following interpolation problem: find a good polynomial P such that

$$P(\pm 1) = f(\pm 1), \quad P(x_k) = f(x_k), \quad k = 1, \ldots, n, \quad P'(1) = f'(1),$$

where $\{x_k\}$ is the set of zeros of the Jacobi polynomial $p_n(v^{\alpha,\beta})$.

If the points x_k are the Chebyshev zeros of the first kind, then we have by Theorem 4.2.5

$$\|f - P\|_\infty \leq C n^2 E_{n-1}(f)_\infty, \quad C \neq C(n, f),$$

and only if we choose the zeros of $p_n(v^{\alpha,\beta})$ with

$$-\frac{1}{2} \leq \alpha \leq \frac{3}{2}, \quad \frac{3}{2} \leq \beta \leq \frac{7}{2},$$

we obtain the optimal estimate

$$\|f - P\|_\infty \leq C E_{n-1}(f)_\infty \log n, \quad C \neq C(n, f).$$

A brief history of the *additional nodes method* may be useful. According to our knowledge, in 1958 Egerváry and Turán [106] were the first to use the points ± 1. They proved that the sequence of the Hermite-Fejér polynomials based on the Legendre zeros together with ± 1 is uniformly convergent (this result is false if we drop the points ± 1). The first use of the additional points in the Lagrange-Hermite interpolation is due to Szász [467] in 1959, while the use of the points ± 1 appeared in some papers by Freud [132, 133] and Vértesi [494, 495]. In 1987, Szabados [461, 462] was the first who successfully used not only ± 1, but other additional points to minimize the norm of the derivatives of the Lagrange polynomials based on the Chebyshev zeros of the first kind. This problem was thoroughly investigated in some papers by Szabados and Vértesi [464], and by Halász in [205]. In [421], and subsequently in [289], simultaneous interpolation processes based on the zeros of Jacobi polynomials were constructed. This procedure was then extensively used by several authors and in different contexts, and nowadays is referred to as the "*additional nodes method*". For an exhaustive bibliography, the interested reader can consult [465, p. 279] and [65] and the references therein.

4.2.3 Other "Optimal" Interpolation Processes

4.2.3.1 Interpolation with Associated Polynomials

Consider the Jacobi weight $v^{\alpha,\beta}(x) = (1-x)^\alpha (1+x)^\beta$ and associate with it a new weight function (cf. (2.2.32))

$$w^{\alpha,\beta}(x) = \frac{v^{\alpha,\beta}(t)}{\pi^2 v^{2\alpha,2\beta}(t) + H^2(v^{\alpha,\beta}; t)},$$

4.2 Optimal Systems of Nodes

where $H(g)$ is the finite Hilbert transform defined by

$$H(g;t) = \text{P.V.} \int_{-1}^{1} \frac{g(x)}{x-t} dx = \lim_{\varepsilon \to 0} \int_{|x-t| \geq \varepsilon} \frac{g(x)}{x-t} dx.$$

The orthonormal polynomials $\{p_n(w^{\alpha,\beta})\}_n$ corresponding to the weight $w^{\alpha,\beta}$ are called the associated polynomials and they are strictly connected with the Jacobi polynomials $\{p_n(v^{\alpha,\beta})\}_n$. In fact, if

$$\begin{cases} xp_n(v^{\alpha,\beta};x) = a_{n+1}p_{n+1}(v^{\alpha,\beta};x) + b_n p_n(v^{\alpha,\beta};x) + a_n p_{n-1}(v^{\alpha,\beta};x), \\ p_0(v^{\alpha,\beta};x) = \left(\int_{-1}^{1} v^{\alpha,\beta}(x)\,dx \right)^{-1/2}, \\ p_{-1}(v^{\alpha,\beta};x) = 0 \end{cases}$$

is the three-term recurrence relation for the Jacobi polynomials, then the sequence $\{p_n(w^{\alpha,\beta})\}$ satisfies the following recurrence relation

$$\begin{cases} xp_{n-1}(w^{\alpha,\beta};x) = a_{n+1}p_n(w^{\alpha,\beta};x) + b_n p_{n-1}(w^{\alpha,\beta};x) + a_n p_{n-2}(w^{\alpha,\beta};x), \\ p_0(w^{\alpha,\beta};x) = \left(p_0^2(v^{\alpha,\beta})a_1 \right)^{-1}, \\ p_{-1}(w^{\alpha,\beta};x) = 0, \end{cases}$$

with the same coefficients a_n and b_n. From this link, the zeros of $p_n(w^{\alpha,\beta})$ and the Christoffel numbers related to $w^{\alpha,\beta}$ can be computed by solving the eigenvalue problem of the corresponding Jacobi matrix.

In particular, it is possible to prove that the zeros of $p_n(w^{\alpha,\beta})$ interlace with the zeros of $p_{n+1}(v^{\alpha,\beta})$ and they have an *arc sine distribution*. For more details on the associated polynomials see Sect. 2.2.3.

Now, we consider the sequence

$$\mathcal{X} = \left\{ Y_s(x) Z_r(x) p_n(w^{\alpha,\beta};x) \right\}_{n=1,2,\ldots},$$

where Y_s and Z_r are defined in (4.2.13), and denote by $L_{n,r,s}(w^{\alpha,\beta},f)$ the Lagrange polynomial that interpolates a given function f at the zeros of $Y_s Z_r p_n(w^{\alpha,\beta})$. The following theorem holds.

Theorem 4.2.6 *We have*

$$\|L_{n,r,s}(w^{\alpha,\beta})\|_\infty \sim \log n, \qquad (4.2.48)$$

whenever the parameters α, β, r, s *satisfy*

$$\frac{|\alpha|}{2} + \frac{1}{4} \leq r < \frac{|\alpha|}{2} + \frac{5}{4},$$

$$\frac{|\beta|}{2} + \frac{1}{4} \leq s < \frac{|\beta|}{2} + \frac{5}{4}.$$

The proof of this theorem can be found in [300]. The special case $\alpha = \beta = 0$ was separately considered in [299].

We remark that, since $|\alpha|$ and $|\beta|$ assume non-negative values, the number of additional points r and s is greater than or equal to 1.

4.2.3.2 Interpolation at Stieltjes Zeros

Now, we consider the Stieltjes polynomials $E_{n+1}(x)$ defined (up to a multiplicative constant) by (see Sect. 2.2.4; in particular (2.2.56), for the Legendre measure)

$$\int_{-1}^{1} E_{n+1}(x) p_n(x) x^k dx = 0, \quad k = 0, 1, \ldots, n, \quad n \geq 1. \tag{4.2.49}$$

The zeros of E_{n+1} were used by Kronrod to construct the well-known extended quadrature formula, which has later been extensively studied by several authors [107, 108, 152, 173, 366–369, 389].

Such zeros have an *arc sine distribution* and they generate an optimal interpolatory process. In fact, if $L_{n+1} f$ denotes the Lagrange polynomial that interpolates a given function f at the zeros of $E_{n+1}(x)$, then the following theorem holds for the corresponding Lebesgue constant $\|L_{n+1}\|_\infty$.

Theorem 4.2.7 *For all $n \in \mathbb{N}$, we have*

$$\|L_{n+1}\|_\infty \sim \log n.$$

4.2.3.3 Extended Interpolation

The underlying idea of the "extended interpolation" is to interpolate a function at the zeros of the sequence $\{q_N\} = \{p_n(v^{\alpha,\beta}) p_m(v^{\gamma,\delta}) Y_s Z_r\}$, where Y_s and Z_r are defined as in (4.2.13) and the parameters $\alpha, \beta, \gamma, \delta, r, s$ (and analogously n and m) are suitably related.

An extended interpolation turns out to be useful for the numerical evaluation of the interpolation error based on the zeros of orthogonal polynomials. More precisely, if $L_n(w, f)$ is the Lagrange polynomial interpolating f at the zeros of $p_n(w)$, the difference $|L_m(w, f) - L_n(w, f)|$, $m > n$, is assumed to be a numerical evaluation of the error of $L_n(w, f)$. Thus, if $m = n + 1$, following this procedure, we need $2n + 1$ evaluations of the function in order to compare $L_n(w, f)$ with $L_{n+1}(w, f)$. But if we consider the extended interpolation polynomial $L_{2n}(w, u; f)$ based on the zeros of $p_n(w) p_n(u)$, by using only $2n$ evaluations of the function f, we can compare $L_n(w, f)$ with $L_{2n}(w, u, f)$. Then, we have the difference $|L_{2n}(w, u, f) - L_n(w, f)|$ which is more precise than $|L_{n+1}(w, f) - L_n(w, f)|$, when both polynomials $L_n(w, f)$ and $L_{2n}(w, u, f)$ have the same order of convergence to f. According to the quality of the Lebesgue constants of the extended interpolation process, following some indications given in Sect. 4.2.2, it is necessary that

4.2 Optimal Systems of Nodes

the interpolation points have an *arc sine distribution* and that $\sup_N \|q_N\|_\infty < +\infty$. This last condition is not difficult to satisfy. In fact, from the pointwise estimate for the Jacobi polynomials, it is possible to determine r and s (the number of additional nodes) in such a way that the sequence $\{q_N\}$ becomes uniformly bounded with respect to N.

On the other hand, the choice of the parameters $\alpha, \beta, \gamma, \delta, r, s, n$ and m, such that the zeros of q_N have an *arc sine distribution*, is still an open problem. However, some important examples are known. For instance, we consider

$$\mathcal{X} = \{q_N\} = \left\{ p_{n+1}(v^{\alpha,\beta}) p_n(v^{\alpha+1,\beta+1}) Y_s Z_r \right\}_n, \qquad (4.2.50)$$

where Y_s and Z_r are defined as in (4.2.13), but replacing in those definitions x_1 and x_n by the first and the last zero of the polynomial

$$Q_{2n+1} := p_{n+1}(v^{\alpha,\beta}) p_n(v^{\alpha+1,\beta+1}),$$

respectively. It is possible to prove (see [63, 65]) that the zeros $x_{n,k}(v^{\alpha+1,\beta+1})$ of $p_n(v^{\alpha+1,\beta+1})$ interlace with the zeros $x_{n+1,k}(v^{\alpha,\beta})$ of $p_{n+1}(v^{\alpha,\beta})$, i.e.

$$x_{n+1,k}(v^{\alpha,\beta}) < x_{n,k}(v^{\alpha+1,\beta+1}) < x_{n+1,k+1}(v^{\alpha,\beta}), \qquad k = 1, \ldots, n$$

holds. Moreover, the set $\{x_{n+1,k}(v^{\alpha,\beta})\} \cup \{x_{n,k}(v^{\alpha+1,\beta+1})\}$ of the zeros of Q_{2n+1} has an *arc sine distribution*.

On the other hand, by (4.2.9) we have

$$|Q_{2n+1}(x)| \leq C \left(\sqrt{1-x} + \frac{1}{n} \right)^{-2\alpha-2} \left(\sqrt{1+x} + \frac{1}{n} \right)^{-2\beta-2}.$$

Hence, if we want sequences of polynomials that are uniformly bounded with respect to n, as suggested in Sect. 4.2.2, we are forced to consider the product

$$q_N(x) := Q_{2n+1}(x) Y_s(x) Z_r(x),$$

adding s equispaced knots between -1 and the first zero of Q_{2n+1} and r equispaced nodes between the last zero of Q_{2n+1} and $+1$.

In this way the *arc sine distribution* is preserved and we have

$$|q_N(x)| \leq C \left(\sqrt{1-x} + \frac{1}{n} \right)^{-2\alpha-2+2r} \left(\sqrt{1+x} + \frac{1}{n} \right)^{-2\beta-2+2s}, \qquad (4.2.51)$$

so that (4.2.8) is true, if

$$\alpha + 1 \leq r \leq \alpha + 2 \quad \text{and} \quad \beta + 1 \leq s \leq \beta + 2 \qquad (4.2.52)$$

hold. The next theorem proves that these are the sufficient conditions for the parameters α, β, r and s to have optimal Lebesgue constants with $\mathcal{X}$ given by (4.2.50) (see [66]).

Theorem 4.2.8 Let $\mathcal{X}$ be defined by (4.2.50) and let $L_{2n+1,r,s} f$ be the corresponding Lagrange polynomials. If the parameters α, β, r and s satisfy (4.2.52), then we have

$$\|L_{2n+1,r,s}\|_\infty \sim \log n.$$

Another optimal sequence of polynomials is given by

$$\{\widetilde{Q}_{2n+r+s}\}_n = \{p_n(v^{\alpha+1,\beta})p_n(v^{\alpha,\beta+1})Y_s Z_r\}_n,$$

under the same conditions (4.2.52) for the parameters α, β, r and s.

Finally, the following two sequences of polynomials constitute additional significant examples:

$$\{q_{2n+2}(x)\}_n = \{(1-x^2)p_n(v^{\alpha,-\alpha};x)p_n(v^{-\alpha,\alpha};x)\}_n$$

and

$$\{\tilde{q}_{2n+3}(x)\}_n = \{(1-x^2)p_n(v^{\alpha,\beta};x)p_{n+1}(v^{-\alpha,-\beta};x)\}_n,$$
$$0 < \alpha, \beta < 1, \ \alpha + \beta = 1.$$

The zeros of q_{2n+2} and/or $\tilde{q}_{2n+3}$ are used in some quadrature methods and in the numerical treatment of singular integral equations (cf. [304]). If we denote by $L_{2n+2} f$ and $L_{2n+3} f$ the Lagrange polynomials based on the zeros of q_{2n+2} and $\tilde{q}_{2n+3}$, respectively, the following theorem holds (see [304]):

Theorem 4.2.9 The zeros of q_{2n+2} and $\tilde{q}_{2n+3}$ have an arc sine distribution and, moreover,

$$\|L_{2n+2}\|_\infty \sim \log n \sim \|L_{2n+3}\|_\infty$$

holds.

In conclusion, we want to mention a result that recently appeared in [108]. Namely, the authors considered the sequence $\{R_{2n+1}\} = \{p_n E_{n+1}\}_{n \geq 1}$, where p_n is the n-th Legendre polynomial and E_{n+1} is the $(n+1)$-th Stieltjes polynomial (see (4.2.49)). They studied the distribution of the zeros of R_{2n+1} and the sequence $\{\mathcal{L}_{2n+1} f\}$ of the Lagrange polynomials based on these zeros. The result is as follows.

Theorem 4.2.10 The zeros of R_{2n+1} have an arc sine distribution and, moreover,

$$\|\mathcal{L}_{2n+1}\|_\infty \sim \log n.$$

4.2.4 Some Simultaneous Interpolation Processes

It is very useful to approximate the derivatives of a smooth function f by using only the values of f at some suitable points. In the particular case $\mathcal{X} = \{p_n(v^{\alpha,\beta})\}_n$, using the additional nodes, we state the following result (see [289, 421, 461, 462]):

4.2 Optimal Systems of Nodes

Theorem 4.2.11 Let $f \in C^q[-1, 1]$ and let $L_{n,r,s}(v^{\alpha,\beta}, f)$ be the Lagrange polynomial defined in (4.2.19). Then, for $i = 0, 1, \ldots, q$, we have

$$\|f^{(i)} - L_{n,r,s}^{(i)}(v^{\alpha,\beta}, f)\|_\infty \leq C \frac{E_{n-1-q}(f^{(q)})_\infty}{n^{q-i}} \log n, \quad (4.2.53)$$

with $C \neq C(n, f)$, if the parameters $\alpha, \beta > -1$ and $r, s \in \mathbb{N}$ satisfy the following conditions

$$\frac{\alpha + i}{2} + \frac{1}{4} \leq r \leq \frac{\alpha + i}{2} + \frac{5}{4},$$

$$\frac{\beta + i}{2} + \frac{1}{4} \leq s \leq \frac{\beta + i}{2} + \frac{5}{4}.$$

Proof By the Gopengauz theorem, there exists a polynomial $Q \in \mathcal{P}_{n-1}$ such that

$$|f^{(i)}(x) - Q^{(i)}(x)| \leq C \left(\frac{\sqrt{1-x^2}}{n}\right)^{q-i} E_{n-1-q}(f^{(q)})_\infty \quad (4.2.54)$$

holds, for all $x \in [-1, 1]$ and for $i = 0, 1, \ldots, q$.

Taking into account that $L_{n,r,s}(v^{\alpha,\beta}, Q) = Q$, we have

$$\|f^{(i)} - L_{n,r,s}^{(i)}(v^{\alpha,\beta}, f)\|_\infty \leq \|f^{(i)} - Q^{(i)}\|_\infty + \|L_{n,r,s}^{(i)}(v^{\alpha,\beta}, f - Q)\|_\infty. \quad (4.2.55)$$

For the first term, by (4.2.54) we conclude that

$$\|f^{(i)} - Q^{(i)}\|_\infty \leq C \frac{E_{n-1-q}(f^{(q)})_\infty}{n^{q-i}}, \quad C \neq C(n, f). \quad (4.2.56)$$

For the second term in (4.2.55), set $\varphi_n(x) := \sqrt{1-x^2} + 1/n$. Then, the Bernstein inequality in the form

$$\|P^{(i)} u \varphi_N^i\|_\infty \leq C N^i \|Pu\|_\infty, \quad P \in \mathcal{P}_N, \quad (4.2.57)$$

yields

$$\|L_{n,r,s}^{(i)}(v^{\alpha,\beta}, f - Q)\|_\infty = \|L_{n,r,s}^{(i)}(v^{\alpha,\beta}, f - Q)\varphi_n^{-i} \varphi_n^i\|_\infty$$

$$\leq C n^i \|L_{n,r,s}(v^{\alpha,\beta}, f - Q)\varphi_n^{-i}\|_\infty.$$

On the other hand, by (4.2.19) and (4.2.54), we have for all $x \in [-1, 1]$

$$|L_{n,r,s}(v^{\alpha,\beta}, f - Q; x)\varphi_n^{-i}(x)|$$

$$\leq |Y_s(x) Z_r(x) \varphi_n^{-i}(x)| \sum_{k=1}^n \frac{|\ell_{n,k}(v^{\alpha,\beta}; x)|}{|Y_s(x_k) Z_r(x_k) \varphi_n^{-i}(x_k)|} \frac{|f(x_k) - Q(x_k)|}{\left[\sqrt{1-x_k^2} + \frac{1}{n}\right]^i}$$

$$+ \sum_{k=1}^{r} \frac{|Y_s(x) p_n(v^{\alpha,\beta}; x) \varphi_n^{-i}(x)|}{|Y_s(z_k) p_n(v^{\alpha,\beta}; z_k) \varphi_n^{-i}(z_k)|} \prod_{\substack{i=1 \\ i \neq k}}^{r} \left|\frac{x - z_i}{z_k - z_i}\right| \frac{|f(z_k) - Q(z_k)|}{\left[\sqrt{1 - z_k^2} + \frac{1}{n}\right]^i}$$

$$+ \sum_{k=1}^{s} \frac{|Z_r(x) p_n(v^{\alpha,\beta}; x) \varphi_n^{-i}(x)|}{|Z_r(y_k) p_n(v^{\alpha,\beta}; y_k) \varphi_n^{-i}(y_k)|} \prod_{\substack{i=1 \\ i \neq k}}^{s} \left|\frac{x - y_i}{y_k - y_i}\right| \frac{|f(y_k) - Q(y_k)|}{\left[\sqrt{1 - y_k^2} + \frac{1}{n}\right]^i}$$

$$\leq \frac{C}{n^q} E_{n-1-q}(f^{(q)})_\infty \left(|Y_s(x) Z_r(x) \varphi_n^{-i}(x)| \sum_{k=1}^{n} \frac{|\ell_{n,k}(v^{\alpha,\beta}; x)|}{|Y_s(x_k) Z_r(x_k) \varphi_n^{-i}(x_k)|} \right.$$

$$+ |Y_s(x) p_n(v^{\alpha,\beta}; x) \varphi_n^{-i}(x)| \sum_{k=1}^{r} \frac{1}{|Y_s(z_k) p_n(v^{\alpha,\beta}; z_k) \varphi_n^{-i}(z_k)|} \prod_{\substack{i=1 \\ i \neq k}}^{r} \left|\frac{x - z_i}{z_k - z_i}\right|$$

$$+ |Z_r(x) p_n(v^{\alpha,\beta}; x) \varphi_n^{-i}(x)|$$

$$\left. \times \sum_{k=1}^{s} \frac{1}{|Z_r(y_k) p_n(v^{\alpha,\beta}; y_k) \varphi_n^{-i}(y_k)|} \prod_{\substack{i=1 \\ i \neq k}}^{s} \left|\frac{x - y_i}{y_k - y_i}\right| \right)$$

and by means of the same arguments used to state (4.2.23), (4.2.24) and (4.2.25), it is easy to prove that the inequalities

$$|Y_s(x) Z_r(x) \varphi_n^{-i}(x)| \sum_{k=1}^{n} \frac{|\ell_{n,k}(v^{\alpha,\beta}; x)|}{|Y_s(x_k) Z_r(x_k) \varphi_n^{-i}(x_k)|} \leq C \log n,$$

$$|Y_s(x) p_n(v^{\alpha,\beta}; x) \varphi_n^{-i}(x)| \sum_{k=1}^{r} \frac{1}{|Y_s(z_k) p_n(v^{\alpha,\beta}, z_k) \varphi_n^{-i}(z_k)|} \prod_{\substack{i=1 \\ i \neq k}}^{r} \left|\frac{x - z_i}{z_k - z_i}\right| \leq C,$$

$$|Z_r(x) p_n(v^{\alpha,\beta}; x) \varphi_n^{-i}(x)| \sum_{k=1}^{s} \frac{1}{|Z_r(y_k) p_n(v^{\alpha,\beta}, y_k) \varphi_n^{-i}(y_k)|} \prod_{\substack{i=1 \\ i \neq k}}^{s} \left|\frac{x - y_i}{y_k - y_i}\right| \leq C$$

hold, for all $x \in [-1, 1]$, with $C \neq C(n, x)$.

Thus, we conclude

$$\|L_{n,r,s}^{(i)}(v^{\alpha,\beta}, f - Q)\|_\infty \leq Cn^i \|L_{n,r,s}(v^{\alpha,\beta}, f - Q)\varphi_n^{-i}\|_\infty$$

$$\leq C \frac{E_{n-1-q}(f^{(q)})_\infty}{n^{q-i}} \log n$$

and the theorem follows by (4.2.55) and (4.2.56). □

4.3 Weighted Interpolation

4.3.1 Weighted Interpolation at Jacobi Zeros

Until now we only considered the Lagrange interpolation for continuous functions in $[-1, 1]$, but in some applications, as well as from a theoretical point of view, it is interesting to consider interpolatory processes also for locally continuous functions, i.e., functions that are continuous on each compact subinterval $[a, b] \subset (-1, 1)$, and that may tend to infinity with a known behaviour at the endpoints ± 1. This section is devoted to the study of the Lagrange interpolation with the Jacobi abscissas for this kind of functions. More precisely, if w is a Jacobi weight having positive exponents, we consider functions belonging to the weighted uniform space

$$C_w := \left\{ f \in C_{\text{loc}} \;\middle|\; \lim_{|x| \to 1} f(x)w(x) = 0 \right\}, \qquad (4.3.1)$$

equipped with the norm

$$\|f\|_{C_w} := \|fw\|_\infty = \max_{|x| \leq 1} |f(x)w(x)|.$$

Of course, in the case $w(x) := 1$ we set $C_w := C^0 = C[-1, 1]$, while in the case $w(x) := (1-x)^\rho$ (or $w(x) := (1+x)^\sigma$) C_w consists of the set of all functions that are continuous on each interval $[a, b] \subset [-1, 1)$ (respectively, $[a, b] \subset (-1, 1]$) and such that $\lim_{x \to 1} f(x)w(x) = 0$ (respectively, $\lim_{x \to -1} f(x)w(x) = 0$).

The norm of the Lagrange operator $L_n(v^{\alpha,\beta}) : C_{v^{\rho,\sigma}} \to C_{v^{\rho,\sigma}}$, considered as a map from $C_{v^{\rho,\sigma}}$ into itself, is given by

$$\|L_n(v^{\alpha,\beta})\|_{v^{\rho,\sigma},\infty} := \sup_{\|fv^{\rho,\sigma}\|_\infty \leq 1} \|L_n(v^{\alpha,\beta}, f)v^{\rho,\sigma}\|_\infty$$

$$= \max_{|x| \leq 1} \left\{ v^{\rho,\sigma}(x) \sum_{k=1}^{n} \frac{|\ell_{n,k}(v^{\alpha,\beta}; x)|}{v^{\rho,\sigma}(x_k)} \right\}, \qquad (4.3.2)$$

where x_k are the zeros of $p_n(v^{\alpha,\beta})$, and $\ell_{n,k}(v^{\alpha,\beta}; x)$ are the fundamental Lagrange polynomials corresponding to the Jacobi abscissas (see e.g. (4.1.4)).

The numbers $\|L_n(v^{\alpha,\beta})\|_{v^{\rho,\sigma},\infty}$ are known as the *weighted Lebesgue constants* and, as in the non-weighted case, they appear in the estimate of the Lagrange interpolation error:

$$\|(f - L_n(v^{\alpha,\beta}, f))v^{\rho,\sigma}\|_\infty \leq \left(1 + \|L_n(v^{\alpha,\beta})\|_{v^{\rho,\sigma},\infty}\right) E_{n-1}(f)_{v^{\rho,\sigma},\infty}, \qquad (4.3.3)$$

where

$$E_n(f)_{v^{\rho,\sigma},\infty} := \min_{P \in \mathcal{P}_n} \|(f - P)v^{\rho,\sigma}\|_\infty$$

denotes the error of the best weighted uniform approximation.

For the behaviour of this weighted Lebesgue constant it is not difficult to prove the lower bound

$$\|L_n(v^{\alpha,\beta})\|_{v^{\rho,\sigma},\infty} \geq C\log n, \qquad C \neq C(n). \tag{4.3.4}$$

Hence, as in the non-weighted case, the Lebesgue constants grow at least like $\log n$, as $n \to +\infty$. The next theorem gives the necessary and sufficient conditions in order to obtain the "optimal" Lebesgue constants.

Theorem 4.3.1 *Given two Jacobi weights $v^{\alpha,\beta}$ with $\alpha, \beta > -1$, and $v^{\rho,\sigma}$ with $\rho, \sigma \geq 0$, we have*

$$\|L_n(v^{\alpha,\beta})\|_{v^{\rho,\sigma},\infty} \sim \log n, \tag{4.3.5}$$

if and only if the following conditions

$$\frac{\alpha}{2} + \frac{1}{4} \leq \rho \leq \frac{\alpha}{2} + \frac{5}{4}, \quad \frac{\beta}{2} + \frac{1}{4} \leq \sigma \leq \frac{\beta}{2} + \frac{5}{4} \tag{4.3.6}$$

are satisfied.

Proof Let x be an arbitrary fixed number in $[-1, 1]$ ($x \neq x_k$, $k = 1, \ldots, n$) and let x_d be a Jacobi zero closest to x. Taking into account that $(1 \pm x) \sim (1 \pm x_d)$ and using the estimates (4.2.26) and (4.2.27), we get

$$v^{\rho,\sigma}(x)\frac{|\ell_{n,d}(v^{\alpha,\beta};x)|}{v^{\rho,\sigma}(x_d)} \sim 1, \tag{4.3.7}$$

$$v^{\rho,\sigma}(x)\frac{|\ell_{n,k}(v^{\alpha,\beta};x)|}{v^{\rho,\sigma}(x_k)} \sim v^{\rho,\sigma}(x)|p_n(v^{\alpha,\beta};x)|\frac{v^{\frac{\alpha}{2}+\frac{1}{4}-\rho,\frac{\beta}{2}+\frac{1}{4}-\sigma}(x_k)}{|x_k-x|}\Delta x_k. \tag{4.3.8}$$

Thus, if (4.3.6) holds, applying (4.1.13) and (4.2.9), for each $x \in [-1, 1]$, we get

$$v^{\rho,\sigma}(x)|p_n(v^{\alpha,\beta};x)|\sum_{\substack{k=1\\k\neq d}}^{n}\frac{v^{\frac{\alpha}{2}+\frac{1}{4}-\rho,\frac{\beta}{2}+\frac{1}{4}-\sigma}(x_k)}{|x_k-x|}\Delta x_k \leq C\log n.$$

Consequently, we have by (4.3.2)

$$\|L_n(v^{\alpha,\beta})\|_{v^{\rho,\sigma},\infty} = \max_{|x|\leq 1}\left(v^{\rho,\sigma}(x)\sum_{k=1}^{n}\frac{|\ell_{n,k}(v^{\alpha,\beta};x)|}{v^{\rho,\sigma}(x_k)}\right) \leq C\log n$$

and recalling (4.3.4), we get (4.3.5).

To prove that the conditions (4.3.6) are also necessary for (4.3.5) to hold, we can proceed as in the proof of Theorem 4.2.5 and state that

$$\|L_n(v^{\alpha,\beta})\|_{v^{\rho,\sigma},\infty} \geq Cn^{\mu}, \qquad \mu > 0,$$

holds, when the conditions in (4.3.6) are not satisfied.

4.3 Weighted Interpolation

In fact, in the case $\sigma < \beta/2 + 1/4$, if we fix $\bar{x} := (x_1 - 1)/2$, then we get by (4.3.2)

$$\|L_n(v^{\alpha,\beta})\|_{v^{\rho,\sigma},\infty} = \max_{|x|\leq 1} v^{\rho,\sigma}(x) \sum_{k=1}^{n} \frac{|\ell_{n,k}(v^{\alpha,\beta};x)|}{v^{\rho,\sigma}(x_k)}$$

$$\geq v^{\rho,\sigma}(\bar{x}) \sum_{k=1}^{n} \frac{|\ell_{n,k}(v^{\alpha,\beta};\bar{x})|}{v^{\rho,\sigma}(x_k)} \geq v^{\rho,\sigma}(\bar{x}) \sum_{|x_k|\leq 1/2} \frac{|\ell_{n,k}(v^{\alpha,\beta};\bar{x})|}{v^{\rho,\sigma}(x_k)}$$

$$\geq Cn^{\beta+1/2-2\sigma},$$

as in the proof of Theorem 4.2.5.

In the case $\sigma > \beta/2 + 5/4$, we consider for instance the point $\bar{x} = (x_d + x_{d+1})/2$ with $x_d, x_{d+1} \in [-1/2, 1/2]$. According to (4.3.8), we get

$$\|L_n(v^{\alpha,\beta})\|_{v^{\rho,\sigma},\infty} = \max_{|x|\leq 1} \left(v^{\rho,\sigma}(x) \sum_{k=1}^{n} \frac{|\ell_{n,k}(v^{\alpha,\beta};x)|}{v^{\rho,\sigma}(x_k)} \right)$$

$$\geq v^{\rho,\sigma}(\bar{x}) \frac{|\ell_{n,1}(v^{\alpha,\beta};\bar{x})|}{v^{\rho,\sigma}(x_1)} \geq Cn^{2\sigma-\beta-5/2},$$

as we already stated in proving Theorem 4.2.5.

Finally, in the cases when $\rho < \alpha/2 + 1/4$ or $\rho > \alpha/2 + 5/4$, we can proceed analogously as in the previous cases. □

Comparing (4.3.6) with (4.2.21) and (4.2.22), we can observe that the required conditions for having optimal Lebesgue constants are the same as in the case of interpolation with additional knots as well as in the case of weighted interpolation. In some sense these two interpolation processes are equivalent. The only difference is that while the numbers r and s of the additional knots have to be integers, the $\rho, \sigma \geq 0$ may be not integers, since they are the exponents of a bounded Jacobi weight.

If the conditions in (4.3.6) are satisfied, then by (4.3.3), we get

$$\|[f - L_n(v^{\alpha,\beta}, f)]v^{\rho,\sigma}\|_\infty \leq CE_{n-1}(f)_{v^{\rho,\sigma},\infty} \log n, \qquad f \in C_{v^{\rho,\sigma}}.$$

Example 4.3.1 For $f(x) = \log(1+x)$ and $v^{\rho,\sigma}(x) = \sqrt{1+x}$ ($\rho = 0$ and $\sigma = 1/2$), we have $f \in C_{v^{\rho,\sigma}}$ and $E_n(f)_{v^{\rho,\sigma},\infty} \leq C/n$. Then, using (4.3.6), we can choose $\alpha = -1/2, \beta = 1/2$ and obtain

$$\|[f - L_n(v^{\alpha,\beta}, f)]v^{\rho,\sigma}\|_\infty \leq C \frac{\log n}{n}, \qquad C \neq C(n).$$

Example 4.3.2 Let $L_n(v^{\alpha,\beta}, f)$ be the Lagrange interpolation polynomial for the function $f(x) = \sqrt[4]{1-x^2}$ at the Chebyshev zeros of the first kind, i.e., when

$v^{\alpha,\beta}(x) = (1-x^2)^{-1/2}$. Then, we have

$$\|f - L_n(v^{\alpha,\beta}, f)\|_\infty \leq C \frac{\log n}{\sqrt{n}}, \quad C \neq C(n),$$

and $E_n(f)_\infty \leq C/\sqrt{n}$.

On the other hand, if we take ρ, σ such that (4.3.6) and $\rho, \sigma > k/2 - 1/4$, $k \in \mathbb{N}$, hold, then we obtain

$$\|[f - L_n(v^{\alpha,\beta}, f)]v^{\rho,\sigma}\|_\infty \leq C \frac{\log n}{n^k}, \quad C \neq C(n).$$

As the previous examples show, once the weight $v^{\rho,\sigma}$ of the norm is fixed, the conditions in (4.3.6) give us a rule for choosing the interpolation knots (we can find infinitely many values for α and β such that (4.3.6) holds). Conversely, if we fix the interpolation nodes, i.e., if we have the Lagrange polynomial $L_n(v^{\alpha,\beta}, f)$, then (4.3.6) tells us how to choose the weight $v^{\rho,\sigma}$ of the norm.

Nevertheless in many applications, we have fixed both the weighted norm and the interpolation knots. In these cases, (4.3.6) could be not satisfied, but we can still obtain an optimal interpolation process by means of the additional knots method. In fact, if $L_{n,r,s}(v^{\alpha,\beta}, f)$ is the Lagrange polynomial interpolating f at the zeros of $p_n(v^{\alpha,\beta}) Y_s Z_r$, with Y_s and Z_r defined by (4.2.13), then the following theorem holds.

Theorem 4.3.2 *Let $r, s \in \mathbb{N}$, $\alpha, \beta > -1$ and $\rho, \sigma \geq 0$. Then the necessary and sufficient conditions for*

$$\|L_{n,r,s}(v^{\alpha,\beta})\|_{v^{\rho,\sigma},\infty} \leq C \log n, \quad C \neq C(n),$$

are the following

$$\frac{\alpha}{2} + \frac{1}{4} \leq \rho + r \leq \frac{\alpha}{2} + \frac{5}{4}, \quad \frac{\beta}{2} + \frac{1}{4} \leq \sigma + s \leq \frac{\beta}{2} + \frac{5}{4}. \quad (4.3.9)$$

For the sake of brevity we omit the proof of this theorem, since it can easily be deduced from the proofs of Theorems 4.2.4 and 4.3.1. In conclusion, we give a generalization of Theorem 4.3.1 which treats the Lagrange operator as a map between two different uniform weighted spaces, i.e., $L_n(v^{\alpha,\beta}) : C_{v^{\gamma,\delta}} \to C_{v^{\rho,\sigma}}$.

Theorem 4.3.3 *Let $\rho \geq \gamma \geq 0$ and $\sigma \geq \delta \geq 0$. For all $n \in \mathbb{N}$ and $f \in C_{v^{\gamma,\delta}}$ we have*

$$\|L_n(v^{\alpha,\beta}, f)v^{\rho,\sigma}\|_\infty \leq C \log n \|fv^{\gamma,\delta}\|_\infty, \quad C \neq C(n, f), \quad (4.3.10)$$

if and only if

$$\rho \geq \frac{\alpha}{2} + \frac{1}{4} \quad \text{and} \quad \gamma \leq \frac{\alpha}{2} + \frac{5}{4}, \quad (4.3.11)$$

$$\sigma \geq \frac{\beta}{2} + \frac{1}{4} \quad \text{and} \quad \delta \leq \frac{\beta}{2} + \frac{5}{4}. \quad (4.3.12)$$

4.3 Weighted Interpolation

Proof For all $f \in C_{v^{\gamma,\delta}}$ and each $x \in [-1, 1]$, we can write

$$|L_n(v^{\alpha,\beta}, f; x)v^{\rho,\sigma}(x)| = \left| v^{\rho,\sigma}(x) \sum_{k=1}^{n} \ell_{n,k}(v^{\alpha,\beta}; x) f(x_k) \right|$$

$$\leq \|fv^{\gamma,\delta}\|_\infty \left(v^{\rho,\sigma}(x) \sum_{k=1}^{n} \frac{|\ell_{n,k}(v^{\alpha,\beta}; x)|}{v^{\gamma,\delta}(x_k)} \right).$$

On the other hand, if d denotes the index of a knot x_d closest to x, then by (4.2.26) and (4.2.27), we get

$$v^{\rho,\sigma}(x) \frac{|\ell_{n,d}(v^{\alpha,\beta}; x)|}{v^{\gamma,\delta}(x_d)} \sim 1,$$

$$v^{\rho,\sigma}(x) \frac{|\ell_{n,k}(v^{\alpha,\beta}; x)|}{v^{\gamma,\delta}(x_k)} \sim v^{\rho,\sigma}(x)|p_n(v^{\alpha,\beta}; x)| \frac{v^{\frac{\alpha}{2}+\frac{1}{4}-\gamma, \frac{\beta}{2}+\frac{1}{4}-\delta}(x_k)}{|x_k - x|} \Delta x_k,$$

where in the first '$\sim$' we used $(1 \pm x_d) \sim (1 \pm x)$ and $\sigma - \gamma \geq 0$, $\rho - \delta \geq 0$.

Moreover, assuming that (4.3.11) and (4.3.12) hold, we obtain by (4.1.13)

$$\sum_{\substack{k=1 \\ k \neq d}}^{n} \frac{v^{\frac{\alpha}{2}+\frac{1}{4}-\gamma, \frac{\beta}{2}+\frac{1}{4}-\delta}(x_k)}{|x_k - x|} \Delta x_k$$

$$\leq C \left(\sqrt{1-x} + \frac{1}{n} \right)^{\alpha+1/2-2\gamma} \left(\sqrt{1+x} + \frac{1}{n} \right)^{\beta+1/2-2\delta} \log n,$$

and by (4.2.9), we have

$$v^{\rho,\sigma}(x)|p_n(v^{\alpha,\beta}; x)| \leq C \left(\sqrt{1-x} + \frac{1}{n} \right)^{-\alpha-1/2+2\rho} \left(\sqrt{1+x} + \frac{1}{n} \right)^{-\beta-1/2+2\sigma}.$$

Thus, using the previous estimates, we obtain

$$|L_n(v^{\alpha,\beta}, f; x)v^{\rho,\sigma}(x)|$$

$$\leq \|fv^{\gamma,\delta}\|_\infty v^{\rho,\sigma}(x) \sum_{k=1}^{n} \frac{|\ell_{n,k}(v^{\alpha,\beta}; x)|}{v^{\gamma,\delta}(x_k)}$$

$$\leq C\|fv^{\gamma,\delta}\|_\infty \left(\sqrt{1-x} + \frac{1}{n} \right)^{2\rho-2\gamma} \left(\sqrt{1+x} + \frac{1}{n} \right)^{2\sigma-2\delta} \log n$$

$$\leq C\|fv^{\gamma,\delta}\|_\infty \log n.$$

Taking supremum with respect to $x \in [-1, 1]$, for each $f \in C_{v^{\gamma,\delta}}$, this gives

$$\|L_{n,r,s}(v^{\alpha,\beta}, f)v^{\rho,\sigma}\|_\infty \leq C\|fv^{\gamma,\delta}\|_\infty \log n,$$

i.e., (4.3.10) holds.

Finally, the proof that (4.3.11) and (4.3.12) are also necessary conditions for (4.3.10) is similar to the one in Theorem 4.3.1. □

At the end of this subsection, we state a result on the behaviour of the Fourier sums in the Jacobi polynomials in the space $C_{v^{\rho,\sigma}}$. Namely, with $f \in C_{v^{\rho,\sigma}}$ and

$$S_n(v^{\alpha,\beta}, f; x) = \sum_{k=0}^{n-1} c_k p_k(v^{\alpha,\beta}; x), \quad c_k = \int_0^{+\infty} f p_k(v^{\alpha,\beta}) v^{\alpha,\beta},$$

we have:

Theorem 4.3.4 *For every* $f \in C_{v^{\rho,\sigma}}$ *the estimate*

$$\|S_m(v^{\alpha,\beta}, f) v^{\rho,\sigma}\|_\infty \le C \|f v^{\rho,\sigma}\|_\infty \log n$$

holds, with $C \ne C(n, f)$, *if and only if*

$$\max\left(0, \frac{\alpha}{2} + \frac{1}{4}\right) \le \rho \le \min\left(\frac{\alpha}{2} + \frac{3}{4}, \alpha + 1\right),$$

$$\max\left(0, \frac{\beta}{2} + \frac{1}{4}\right) \le \sigma \le \min\left(\frac{\beta}{2} + \frac{3}{4}, \beta + 1\right).$$

The proof of this theorem can be found in [278].

4.3.2 Lagrange Interpolation in Sobolev Spaces

Let $\mathcal{X} \subset [-1, 1]$ be an arbitrary array of nodes. We consider the Lagrange polynomial $L_n(\mathcal{X}, f) \in \mathcal{P}_{n-1}$ interpolating functions f belonging to the Sobolev spaces

$$W_r^\infty(u) := \left\{ f \in C_u \mid \|f\|_{W_\infty^r} := \|fu\|_\infty + \|f^{(r)} \varphi^r u\|_\infty < +\infty \right\},$$

where $u := v^{\alpha,\beta}$ is an arbitrary Jacobi weight and $\varphi(x) := \sqrt{1-x^2}$.

The interpolation error in this subspace of C_u is given by

$$\|f - L_n(\mathcal{X}, f)\|_{W_r^\infty(u)} = \|[f - L_n(\mathcal{X}, f)]u\|_\infty + \|[f - L_n(\mathcal{X}, f)]^{(r)} \varphi^r u\|_\infty. \quad (4.3.13)$$

First, it is easy to check that

$$\|[f - L_n(\mathcal{X}, f)]u\|_\infty \le C \|L_n(\mathcal{X})\|_{u,\infty} E_{n-1}(f)_{u,\infty}, \quad (4.3.14)$$

where $C \ne C(n, f)$ and

$$\|L_n(\mathcal{X})\|_{u,\infty} := \sup_{\|fu\|_\infty \le 1} \|L_n(\mathcal{X}, f)u\|_\infty$$

4.3 Weighted Interpolation

is the norm of the Lagrange operator considered as a map $L_n(\mathcal{X}) : C_u \to C_u$, and

$$E_n(f)_{u,\infty} := \inf_{P \in \mathcal{P}_n} \|(f-P)u\|_\infty$$

denotes the error of the best weighted uniform approximation.

The following result holds for the second term in (4.3.13):

Theorem 4.3.5 *For all $n \in \mathbb{N}$ and $f \in W_r^\infty(u)$, we have*

$$\|[f - L_n(\mathcal{X}, f)]^{(r)} \varphi^r u\|_\infty \leq C \|L_n(\mathcal{X})\|_{u,\infty} E_{n-1-r}(f^{(r)})_{\varphi^r u,\infty}, \quad (4.3.15)$$

where $C \neq C(n, f)$.

Proof The result easily follows from the Favard inequality

$$E_n(f)_{u,\infty} \leq \frac{C}{n} E_{n-1}(f')_{\varphi u,\infty}, \quad C \neq C(n, f), \quad (4.3.16)$$

and from the estimate [245]

$$\|(f - P)^{(r)} \varphi^r u\|_\infty \leq C n^r \|(f - P)u\|_\infty + C \sum_{k=1}^r n^{r-k} E_{n-k}(f^{(k)})_{\varphi^k u,\infty}, \quad (4.3.17)$$

that holds for all $P \in \mathcal{P}_n$ with $C \neq C(n, f, P)$.

In fact, using (4.3.16) in (4.3.17), we have for all $P \in \mathcal{P}_n$

$$\|(f - P)^{(r)} \varphi^r u\|_\infty \leq C \left[n^r \|(f-P)u\|_\infty + E_{n-r}(f^{(r)})_{\varphi^r u,\infty} \right].$$

Taking $P = L_n(\mathcal{X}, f) \in \mathcal{P}_{n-1}$, by (4.3.14) and (4.3.16), we get

$$\|[f - L_n(\mathcal{X}, f)]^{(r)} \varphi^r u\|_\infty$$
$$\leq C(n-1)^r \|(f - L_n(\mathcal{X}, f))u\|_\infty + C E_{n-1-r}(f^{(r)})_{\varphi^r u,\infty}$$
$$\leq C \|L_n(\mathcal{X})\|_{u,\infty} (n-1)^r E_{n-1}(f)_{u,\infty} + C E_{n-1-r}(f^{(r)})_{\varphi^r u,\infty}$$
$$\leq C \|L_n(\mathcal{X})\|_{u,\infty} E_{n-1-r}(f^{(r)})_{\varphi^r u,\infty}. \qquad \square$$

Theorem 4.3.5 and (4.3.14) give the following error estimate in Sobolev spaces:

Corollary 4.3.1 *If $f \in W_s^\infty(u)$, then for all $n \in \mathbb{N}$ and any positive integer $r \leq s$, we have*

$$\|f - L_n(\mathcal{X}, f)\|_{W_r^\infty(u)} \leq C \|L_n(\mathcal{X})\|_{u,\infty} \frac{\|f\|_{W_s^\infty(u)}}{n^{s-r}}, \quad (4.3.18)$$

where C is an absolute constant.

Proof By (4.3.13), (4.3.14) and (4.3.15), we get

$$\|f - L_n(\mathcal{X}, f)\|_{W_r^\infty(u)} \leq C \|L_n(\mathcal{X})\|_{u,\infty} \left[E_{n-1}(f)_{u,\infty} + E_{n-1-r}(f^{(r)})_{\varphi^r u, \infty} \right].$$

Using (4.3.16) success, we obtain

$$\|f - L_n(\mathcal{X}, f)\|_{W_r^\infty(u)}$$

$$\leq C\|L_n(\mathcal{X})\|_{u,\infty} \left[\frac{E_{n-1-s}(f^{(s)})_{\varphi^s u, \infty}}{n^s} + \frac{E_{n-1-s}(f^{(s)})_{\varphi^s u, \infty}}{n^{s-r}} \right]$$

$$\leq C\|L_n(\mathcal{X})\|_{u,\infty} \frac{E_{n-1-s}(f^{(s)})_{\varphi^s u, \infty}}{n^{s-r}}$$

$$\leq C\|L_n(\mathcal{X})\|_{u,\infty} \frac{\|f^{(s)} \varphi^s u\|_\infty}{n^{s-r}}$$

$$\leq C\|L_n(\mathcal{X})\|_{u,\infty} \frac{\|f\|_{W_s^\infty(u)}}{n^{s-r}}. \qquad \square$$

The estimate of type (4.3.18) also holds in the Besov spaces (cf. [305]).

4.3.3 Interpolation at Laguerre Zeros

We consider Lagrange interpolating polynomials at the zeros of the Laguerre polynomials $p_n(w_\alpha)$, where $w_\alpha(x) := x^\alpha e^{-x}$, $\alpha > -1$, is the Laguerre weight on $(0, +\infty)$.

With this kind of interpolation processes we want to approximate locally continuous functions on $(0, +\infty)$ which could be unbounded at the points 0 and $+\infty$. More precisely, we set $u(x) := x^\gamma e^{-x/2}$ with $\gamma \geq 0$, and consider the weighted functional space C_u defined as the set of all the functions f that are continuous on each compact subinterval $[a,b] \subset (0, +\infty)$ and

$$\lim_{x \to 0^+} u(x) f(x) = 0 \quad \text{and} \quad \lim_{x \to +\infty} u(x) f(x) = 0. \quad (4.3.19)$$

This means that f can tend to infinity with an algebraic growth as $x \to 0^+$, and with an exponential growth as $x \to +\infty$.

Obviously in the case $\gamma = 0$ we omit the first limiting condition in (4.3.19) and the definition of C_u is modified as follows

$$f \in C_u \iff f \text{ continuous on } [0, +\infty) \quad \text{and} \quad \lim_{x \to +\infty} e^{-x/2} f(x) = 0.$$

For all $\gamma \geq 0$, the space C_u is equipped with the following norm

$$\|f\|_{C_u} := \|fu\|_\infty = \max_{x \geq 0} |f(x) u(x)|$$

and it is a Banach space.

4.3 Weighted Interpolation

For each $f \in C_u$, the Lagrange polynomial that interpolates the function f at the zeros $\{x_k\}_{k=1,2,\ldots,n}$ of the Laguerre polynomial $p_n(w_\alpha)$ is denoted, as usual, by $L_n(w_\alpha, f)$. Using the Lagrange formula, we have

$$L_n(w_\alpha, f; x) = \sum_{k=1}^{n} \ell_{n,k}(w_\alpha; x) f(x_k),$$

where

$$\ell_{n,k}(w_\alpha; x) = \frac{p_n(w_\alpha; x)}{p_n'(w_\alpha; x_k)(x - x_k)}, \quad k = 1, \ldots, n,$$

are the fundamental Lagrange polynomials.

Starting from $L_n(w_\alpha, f)$, we consider the Lagrange polynomial $\mathcal{L}_{n+1}(w_\alpha, f)$ interpolating f at the Laguerre zeros $\{x_k\}_{k=1,2,\ldots,n}$ and at an additional special knot $x_{n+1} = 4n$. Such a polynomial can be written as

$$\mathcal{L}_{n+1}(w_\alpha, f; x) = \sum_{k=1}^{n+1} \widetilde{\ell}_{n,k}(w_\alpha; x) f(x_k), \tag{4.3.20}$$

where

$$\widetilde{\ell}_{n,k}(w_\alpha; x) = \begin{cases} \ell_{n,k}(w_\alpha; x) \dfrac{4n - x}{4n - x_k} & \text{if } k = 1, \ldots, n, \\ \dfrac{p_n(w_\alpha; x)}{p_n(w_\alpha; 4n)} & \text{if } k = n+1. \end{cases} \tag{4.3.21}$$

The Lebesgue constants of this interpolation process are defined in the usual way by

$$\|\mathcal{L}_{n+1}(w_\alpha)\|_{C_u} := \sup_{\|f\|_{C_u}=1} \|\mathcal{L}_{n+1}(w_\alpha, f)\|_{C_u} = \max_{x \geq 0} \sum_{k=1}^{n+1} \frac{u(x)}{u(x_k)} \left| \widetilde{\ell}_{n,k}(w_\alpha; x) \right| \tag{4.3.22}$$

and their behaviour determines the order of the interpolation error

$$\|(f - \mathcal{L}_{n+1}(w_\alpha, f))u\|_\infty \leq \left(1 + \|\mathcal{L}_{n+1}(w_\alpha)\|_{C_u}\right) E_n(f)_{u,\infty},$$

where

$$E_n(f)_{u,\infty} := \inf_{P \in \mathcal{P}_n} \|(f - P)u\|_\infty$$

is the error of the best approximation of f in C_u.

Similarly to the Lagrange interpolation on a compact interval, we have

$$\|\mathcal{L}_{n+1}(w_\alpha)\|_{C_u} \geq C \log n, \quad C \neq C(n).$$

This result is due to Vértesi [499] and holds more generally for any array of knots in $[0, 4n]$.

The following theorem states the necessary and sufficient for the Lebesgue constants to be optimal.

Theorem 4.3.6 *For all $\gamma \geq 0$ and any $\alpha > -1$, we have*

$$\|\mathcal{L}_{n+1}(w_\alpha)\|_{C_u} \sim \log n \qquad (4.3.23)$$

if and only if the following inequalities

$$\frac{\alpha}{2} + \frac{1}{4} \leq \gamma \leq \frac{\alpha}{2} + \frac{5}{4} \qquad (4.3.24)$$

are satisfied.

Proof We assume (4.3.24) holds and prove (4.3.23), stating that

$$\|\mathcal{L}_{n+1}(w_\alpha, f)u\|_\infty \leq C \max_{x_1 \leq x \leq 4n} |(fu)(x)| \log n$$

holds for all $f \in C_u$, with $C \neq C(n, f)$.

Taking into account that for $P \in \mathcal{P}_n$, the equality (cf. [293])

$$\|Pu\|_\infty = \max_{x \in [a/n, 4n+4\gamma]} |P(x)u(x)|$$

holds, where $a > 0$ is a fixed constant, then for all $f \in C_u$ we have

$$\|\mathcal{L}_{n+1}(w_\alpha, f)u\|_\infty \leq \max_{a/n \leq x \leq 4n+4\gamma} |\mathcal{L}_{n+1}(w_\alpha, f; x)u(x)|$$

$$\leq \max_{a/n \leq x \leq 4n+4\gamma} \sum_{k=1}^{n+1} |\tilde{\ell}_{n,k}(w_\alpha; x)| \frac{u(x)}{u(x_k)} |(fu)(x_k)|$$

$$\leq \left(\max_{x_1 \leq x \leq 4n} |(fu)(x)| \right) \max_{a/n \leq x \leq 4n+4\gamma} \sum_{k=1}^{n+1} |\tilde{\ell}_{n,k}(w_\alpha; x)| \frac{u(x)}{u(x_k)}.$$

Now, we set

$$\sum_{k=1}^{n+1} |\tilde{\ell}_{n,k}(w_\alpha; x)| \frac{u(x)}{u(x_k)} = \sum_{k=1}^{n} |\tilde{\ell}_{n,k}(w_\alpha; x)| \frac{u(x)}{u(x_k)} + |\tilde{\ell}_{n,n+1}(w_\alpha; x)| \frac{u(x)}{u(4n)}$$

$$=: I_1(x) + I_2(x).$$

In order to estimate the first term, we note that by (4.3.21) we have

$$I_1(x) = \sum_{k=1}^{n} |\tilde{\ell}_{n,k}(w_\alpha; x)| \frac{4n-x}{4n-x_k} \frac{u(x)}{u(x_k)}.$$

Recalling that if d is the index of a Laguerre zero x_d closest to $x \in [a/n, 4n+4\gamma]$, then we have

$$|\tilde{\ell}_{n,d}(w_\alpha; x)| \frac{u(x)}{u(x_d)} \sim 1. \qquad (4.3.25)$$

4.3 Weighted Interpolation

Moreover in [247] (see also [374]) it was proved that

$$|\ell_{n,k}(w_\alpha; x)| = \left|\frac{p_n(w_\alpha; x)}{x - x_k}\right| \sqrt{x_k \lambda_{n,k}(w_\alpha)}, \qquad (4.3.26)$$

where $\lambda_{n,k}(w_\alpha)$ are the Christoffel numbers which satisfy the following estimate (cf. (2.3.60))

$$\lambda_k(w_\alpha) \sim w_\alpha(x_k) \Delta x_k \sim w_\alpha(x_k) \sqrt{\frac{x_k}{4n - x_k}}. \qquad (4.3.27)$$

Thus, by (4.3.25), (4.3.26) and (4.3.27), we have

$$I_1(x) = \sum_{k=1}^{n} |\ell_{n,k}(w_\alpha; x)| \frac{4n - x}{4n - x_k} \frac{u(x)}{u(x_k)}$$

$$\leq C + C \left[|p_n(w_\alpha; x)| \sqrt{w_\alpha(x)} \sqrt[4]{x(4n - x + \sqrt[3]{4n})} \right.$$

$$\left. \times \left[\sum_{\substack{k=1 \\ k \neq d}}^{n} \left(\frac{x}{x_k}\right)^{\gamma - \alpha/2 - 1/4} \left(\frac{4n - x}{4n - x_k}\right)^{3/4} \frac{\Delta x_k}{|x - x_k|} \right] \right].$$

Using the following estimate of Laguerre polynomials (see also Sect. 2.3.5)

$$|p_n(w_\alpha; x)| \sqrt{w_\alpha(x)} \sqrt[4]{x(4n - x + \sqrt[3]{4n})} \sim \left|\frac{x - x_d}{x_d - x_{d\pm 1}}\right|, \qquad (4.3.28)$$

when $a/n \leq x \leq 4n + 4\gamma$, we get

$$I_1(x) \leq C \left[1 + \sum_{\substack{k=1 \\ k \neq d}}^{n} \left(\frac{x}{x_k}\right)^{\gamma - \alpha/2 - 1/4} \left(\frac{4n - x}{4n - x_k}\right)^{3/4} \frac{\Delta x_k}{|x - x_k|} \right] \leq C \log n,$$

where in the last inequality we applied Lemma 4.1.3, with $\gamma - \alpha/2 - 1/4 \in [0, 1]$.

Now let us estimate $I_2(x)$. Using (4.3.21) and (4.3.28), we have

$$I_2(x) = \left|\frac{p_n(w_\alpha; x)}{p_n(w_\alpha; 4n)}\right| \frac{u(x)}{u(4n)} = \left(\frac{x}{4n}\right)^{\gamma - \alpha/2} \frac{\sqrt{w_\alpha(x)}|p_n(w_\alpha; x)|}{\sqrt{w_\alpha(4n)}|p_n(w_\alpha; 4n)|}$$

$$\sim \left(\frac{x}{4n}\right)^{\gamma - \alpha/2 - 1/4} \left(\frac{\sqrt[3]{4n}}{4n - x + \sqrt[3]{4n}}\right)^{1/4} \leq C \neq C(n),$$

being with $\gamma - \alpha/2 - 1/4 \geq 0$.

Summing up, for all $f \in C_u$, we get

$$\|\mathcal{L}_{n+1}(w_\alpha, f)u\|_\infty \leq \|fu\|_\infty \max_{a/n \leq x \leq 4n + 4\gamma} [I_1(x) + I_2(x)] \leq C\|fu\|_\infty \log n$$

and (4.3.23) follows.

In conclusion we prove that the inequalities in (4.3.24) are necessary conditions for having (4.3.23). Because of that, we assume $\gamma \notin [\alpha/2 + 1/4, \alpha/2 + 5/4]$ and prove that in this case $\|\mathcal{L}_{n+1}(w_\alpha)\|_{C_u} \geq Cn^\mu$, with $\mu > 0$ and $C \neq C(n)$.

In the case $\gamma < \alpha/2 + 1/4$, setting $\overline{x} = x_1/2$, by (4.3.22) and (4.3.21), we have

$$\|\mathcal{L}_{n+1}(w_\alpha)\|_{C_u} = \max_{x \geq 0} \left(\sum_{k=1}^{n+1} \frac{u(x)}{u(x_k)} |\widetilde{\ell}_{n,k}(w_\alpha; x)| \right) \geq \sum_{k=2}^{n+1} \frac{u(\overline{x})}{u(x_k)} |\widetilde{\ell}_{n,k}(w_\alpha; \overline{x})|.$$

Then, using (4.3.26), (4.3.27) and (4.3.28), we get

$$\|\mathcal{L}_{n+1}(w_\alpha)\|_{C_u} \geq C \sum_{k=2}^{n+1} \left(\frac{\overline{x}}{x_k} \right)^{\gamma - \alpha/2 - 1/4} \left(\frac{4n - \overline{x}}{4n - x_k} \right)^{3/4} \frac{\Delta x_k}{x_k - \overline{x}}.$$

Taking into account that $4n - \overline{x} > 4n - x_k$ and $-\gamma + \alpha/2 + 1/4 > 0$, we obtain

$$\|\mathcal{L}_{n+1}(w_\alpha)\|_{C_u} \geq C \sum_{k=2}^{n+1} \left(\frac{\overline{x}}{x_k} \right)^{\gamma - \alpha/2 - 1/4} \left(\frac{4n - \overline{x}}{4n - x_k} \right)^{3/4} \frac{\Delta x_k}{x_k - \overline{x}}$$

$$\geq C \overline{x}^{\gamma - \alpha/2 - 1/4} \sum_{k=2}^{n+1} \frac{\Delta x_k}{x_k^{\gamma - \alpha/2 - 1/4} (x_k - \overline{x})}$$

$$\geq Cn^{-\gamma + \alpha/2 + 1/4} \sum_{k=2}^{n+1} \frac{\Delta x_k}{x_k^{\gamma - \alpha/2 + 3/4}}$$

$$\geq Cn^{-2\gamma + \alpha + 1/2}.$$

Finally, in the case $\gamma > \alpha/2 + 5/4$, if $a, b > 0$ are fixed numbers and x_d, x_{d+1} are two Laguerre zeros of $p_n(w_\alpha)$ belonging to $[a, b]$, then we set $\overline{x} = (x_d + x_{d+1})/2$ and, analogously to the previous case, by (4.3.26), (4.3.27) and (4.3.28), we get

$$\|\mathcal{L}_{n+1}(w_\alpha)\|_{C_u} \geq C \sum_{x_1 \leq x_k \leq \overline{x}/2} \left(\frac{\overline{x}}{x_k} \right)^{\gamma - \alpha/2 - 1/4} \left(\frac{4n - \overline{x}}{4n - x_k} \right)^{3/4} \frac{\Delta x_k}{\overline{x} - x_k}$$

$$\geq C \left(\frac{4n - \overline{x}}{4n} \right)^{3/4} \overline{x}^{\gamma - \alpha/2 - 5/4} \sum_{x_1 \leq x_k \leq \overline{x}/2} \frac{\Delta x_k}{x_k^{\gamma - \alpha/2 - 1/4}}$$

$$\geq C \sum_{x_1 \leq x_k \leq \overline{x}/2} \frac{\Delta x_k}{x_k^{\gamma - \alpha/2 - 1/4}} \geq Cn^{\gamma - \alpha/2 - 5/4},$$

where we also used $4n - x_k < 4n$ and $\overline{x} - x_k < \overline{x} \sim 1$. □

By Theorem 4.3.6, the sequence $\{\mathcal{L}_{n+1}(w_\alpha)\}$ defines a "good" interpolation process in C_u for a suitable choice of the parameters α and γ. This is not true for the classical Lagrange interpolation at the Laguerre zeros $\{L_n(w_\alpha)\}$. In fact, it is not difficult to prove.

4.3 Weighted Interpolation

Theorem 4.3.7 *For all $\gamma \geq 0$ and any $\alpha > -1$ we have*

$$\|L_n(w_\alpha)\|_{C_u} \geq n^{1/6}. \quad (4.3.29)$$

Theorems 4.3.7 and 4.3.6 justify our study of $\mathcal{L}_{n+1}(w_\alpha)$ instead of the classical Lagrange polynomial $L_n(w_\alpha)$. Namely, in the space C_u the Lagrange interpolation at the Laguerre zeros is a "bad" interpolation process, but it can be improved by adding one special knot $x_{n+1} = 4n$ to the set of the Laguerre zeros. Using Theorem 4.3.6, in the case when the function space C_u is fixed, we know how to choose the Laguerre weight w_α which gives the Lagrange polynomial $\mathcal{L}_{n+1}(w_\alpha)$. Conversely, if the Lagrange polynomial $\mathcal{L}_{n+1}(w_\alpha)$ is fixed, then by means of (4.3.24) we can determine the function space C_u, where the given interpolation process is optimal.

Further, as for the interpolation on bounded intervals, in case when both the space C_u and the Lagrange polynomial $\mathcal{L}_{n+1}(w_\alpha)$ are fixed and (4.3.24) are not satisfied, we can apply the "additional knots method" to $\{\mathcal{L}_{n+1}(w_\alpha)\}$ which gives an optimal interpolation process under same conditions.

Similarly to the additional knots method we have seen on $[-1, 1]$, we add to the interpolation nodes of $\mathcal{L}_{n+1}(w_\alpha, f)$

$$0 < x_1 < x_2 < \cdots < x_n < x_{n+1} = 4n$$

a fixed number s of equispaced points between the end 0 and the first knot x_1. Thus, we consider the following $n + 1 + s$ interpolation nodes

$$0 < t_1 < \cdots < t_s < x_1 < x_2 < \cdots < x_n < x_{n+1} = 4n,$$

where

$$t_i := \frac{i}{s+1} x_1, \quad i = 1, \ldots, s,$$

and x_k, $k = 1, \ldots, n$, are the Laguerre zeros of $p_n(w_\alpha)$. If we denote by $\mathcal{L}_{n+1,s}(w_\alpha, f)$ the Lagrange polynomial interpolating a function $f \in C_u$ at the previous $n + 1 + s$ knots, the following theorem holds.

Theorem 4.3.8 *Let the parameters $\gamma \geq 0$, $\alpha > -1$ and $s \in \mathbb{N}$ satisfy the inequalities*

$$\frac{\alpha}{2} + \frac{1}{4} \leq \gamma + s \leq \frac{\alpha}{2} + \frac{5}{4}. \quad (4.3.30)$$

Then we have

$$\|\mathcal{L}_{n+1,s}(w_\alpha)\|_{C_u} \sim \log n. \quad (4.3.31)$$

In conclusion, we give some error estimates in the Sobolev-type spaces $W_r^\infty = W_r^\infty(u)$, defined as

$$W_r^\infty := \left\{ f \in C_u \ \middle| \ \|f^{(r)} \varphi^r u\|_\infty < +\infty \right\}, \quad r \geq 1, \ \varphi(x) := \sqrt{x},$$

and equipped with the norm

$$\|f\|_{W_r^\infty} := \|fu\|_\infty + \|f^{(r)}\varphi^r u\|_\infty.$$

The following result holds (cf. [301]):

Theorem 4.3.9 *Let $f \in W_s^\infty$, $s \geq 1$, and $\mathcal{L}_{n+1}(w_\alpha, f)$ be the Lagrange polynomial defined in (4.3.20). If (4.3.24) holds, then for some positive constant C, independent of n and f, and for any positive integer $r \leq s$, we have*

$$\|f - \mathcal{L}_{n+1}(w_\alpha, f)\|_{W_r^\infty} \leq C \frac{\log n}{(\sqrt{n})^{s-r}} \|f\|_{W_s^\infty}. \tag{4.3.32}$$

Proof The proof is based on the Favard-type inequality

$$E_n(f)_{u,\infty} \leq \frac{C}{\sqrt{n}} E_{n-1}(f')_{u\varphi,\infty} \leq \frac{C}{\sqrt{n}} \|f'u\varphi\|_\infty$$

and on the following estimate [301]

$$\|(f - P)^{(r)}u\varphi^r\|_\infty \leq Cn^{r/2}\|(f - P)u\|_\infty + CE_{n-r}(f^{(r)})_{u\varphi^r,\infty},$$

which holds for all $P \in \mathcal{P}_n$ and $f \in W_r^\infty$. More precisely, the last inequalities and Theorem 4.3.6 give

$$\|f - \mathcal{L}_{n+1}(w_\alpha, f)\|_{W_r^\infty}$$
$$= \|(f - \mathcal{L}_{n+1}(w_\alpha, f))u\|_\infty + \|(f - \mathcal{L}_{n+1}(w_\alpha, f))^{(r)}u\varphi^r\|_\infty$$
$$\leq C \left[n^{r/2}\|(f - \mathcal{L}_{n+1}(w_\alpha, f))u\|_\infty + E_{n-r}(f^{(r)})_{u\varphi^r,\infty} \right]$$
$$\leq C \left[n^{r/2} \log n \, E_n(f)_{u,\infty} + E_{n-r}(f^{(r)})_{u\varphi^r,\infty} \right]$$
$$\leq C \frac{\log n}{(\sqrt{n})^{s-r}} E_{n-s}(f^{(s)})_{u\varphi^s,\infty} + \frac{C}{(\sqrt{n})^{s-r}} E_{n-s}(f^{(s)})_{u\varphi^s,\infty}$$
$$\leq C \frac{\log n}{(\sqrt{n})^{s-r}} \|f^{(s)}u\varphi^s\|_\infty \leq C \frac{\log n}{(\sqrt{n})^{s-r}} \|f\|_{W_s^\infty}. \qquad \square$$

Remark 4.3.1 Let $\theta \in (0, 1)$ and j be defined as

$$x_j = \min\{x_k \mid x_k \geq 4\theta n\},$$

where $x_k = x_{n,k}$ is the k-th zero of the Laguerre polynomial $p_n(w_\alpha)$. Denote by $\psi \in C^\infty(\mathbb{R})$ a non-decreasing function such that $\psi(x) = 0$ if $x \leq 0$ and $\psi(x) = 1$ if $x \geq 1$. Set $\psi_j(x) = \psi((x - x_j)/(x_{j+1} - x_j))$, for each function $f \in C_u$, we define $f_j(x) = f(x) - \psi_j(x)f(x)$. By definition, it follows that $f_j(x) = f(x)$ if $x \in [0, x_j]$ and $f_j(x) = 0$ for $x \geq x_{j+1}$. Then, $f_j \in C_u$ and we have the following statement:

4.3 Weighted Interpolation

Proposition 4.3.1 *Let* $M = [\theta n/(1+\theta)]$, $\theta \in (0,1)$ *fixed. Then, for each* $f \in C_u$, *we have*

$$\|(f-f_j)u\|_\infty \leq C\left(E_M(f)_{u,\infty} + e^{-An}\|fu\|_\infty\right), \tag{4.3.33}$$

where $C \neq C(n,f)$.

Before we prove (4.3.33), we observe that, by previous proposition, the norm of any function $f \in C_u$ can be decomposed as

$$\|fu\|_\infty \leq C\left(\|fu\|_{L^\infty[0,x_j]} + E_M(f)_{u,\infty}\right).$$

Therefore, in the polynomial approximation of functions $f \in C_u$, it is sufficient to consider only their finite sections.

Proof of Proposition 4.3.1 For each polynomial $P_M \in \mathcal{P}_M$ we can write

$$\|(f-f_j)u\|_\infty = \|\psi_j fu\|_\infty \leq \max_{x \geq 4\theta n} |(fu)(x)|$$

$$\leq \max_{x \geq 4\theta n} |(f-P_M)(x)u(x)| + \max_{x \geq 4\theta n} |P_M(x)u(x)|$$

$$\leq \|(f-P_M)u\|_\infty + \max_{x \geq 4\theta n} |P_M(x)u(x)|.$$

At this point we use the following inequality (see [293])

$$\max_{x \geq 4(1+\delta)n} |P(x)u(x)| \leq Ce^{-An} \max_{x \geq 0} |P(x)u(x)|,$$

which holds for any polynomial $P \in \mathcal{P}_n$ and any fixed $\delta > 0$, with positive constants C and A depending on δ (but, not on n and P). Taking a maximal M such that $4M(1+\theta) \leq 4n\theta$, i.e., $M = [\theta n/(1+\theta)]$, we have

$$\max_{x \geq 4\theta n} |P_M(x)u(x)| \leq \max_{x \geq 4(1+\theta)M} |P_M(x)u(x)| \leq Ce^{-An} \max_{x \geq 0} |P_M(x)u(x)|$$

$$\leq Ce^{-An}\left(\|(f-P_M)u\|_\infty + \|fu\|_\infty\right).$$

Therefore,

$$\|(f-f_j)u\|_\infty \leq C\left(\|(f-P_M)u\|_\infty + e^{-An}\|fu\|_\infty\right).$$

Assuming the infimum over $P_M \in \mathcal{P}_M$, the estimate (4.3.33) follows. $\square$

The previous observations suggest us to consider

$$\mathcal{L}_{n+1}(w_\alpha, f_j; x) = \sum_{k=1}^{j} \widetilde{\ell}_{n,k}(w_\alpha; x) f(x_k)$$

instead of (4.3.20) by neglecting $[cn]$, $c < 1$, terms. Obviously, all the previous theorems also hold for $\mathcal{L}_{n+1}(w_\alpha, f_j)$. In fact, it is sufficient to observe that

$$\|[f - \mathcal{L}_{n+1}(w_\alpha, f_j)]u\|_\infty \leq \|(f - f_j)u\|_\infty + \|[f_j - \mathcal{L}_{n+1}(w_\alpha, f_j)]u\|_2$$

and to apply the previous estimates.

More simply we can consider the following truncated interpolation process. Consider the sequence of functions $\{\Delta_j L_n(w_\alpha, \Delta_j f)\}_n$, where j is defined as in Remark 4.3.1, $\Delta_j(x)$ is the characteristic function of the interval $[0, x_j]$, and

$$L_n(w_\alpha, \Delta_j f; x) = \sum_{k=1}^{j} \ell_{n,k}(x) f(x_k),$$

with

$$\ell_{n,k}(x) = \frac{p_n(w_\alpha; x)}{p'_n(w_\alpha; x)(x - x_k)}.$$

Now we can state the following result:

Theorem 4.3.10 *If the parameters γ and α of the weight functions w_α and u satisfy the condition*

$$\frac{\alpha}{2} + \frac{1}{4} \leq \gamma \leq \frac{\alpha}{2} + \frac{5}{4},$$

then we have

$$\|\Delta_j L_n(w_\alpha, \Delta_j f) u\|_\infty \leq C \|f \Delta_j u\|_\infty \log n \qquad (4.3.34)$$

and

$$\|[f - \Delta_j L_n(w_\alpha, \Delta_j f)] u\|_\infty \leq C \left[E_M(f)_{u,\infty} \log n + e^{-An} \|fu\|_\infty \right], \qquad (4.3.35)$$

where $M = [\theta n/(1+\theta)]$ and the constants C and A are independent of n and f.

Proof Taking into account that $x_k \leq x_j \sim 4\theta n$, (4.3.34) can easily be deduced from (4.3.25)–(4.3.28). To prove (4.3.35), we use the following decomposition that is true for every $P \in \mathcal{P}_M$:

$$[f - \Delta_j L_n(w_\alpha, \Delta_j f)]u$$
$$= [f(1 - \Delta_j)u] + [f - L_n(w_\alpha, \Delta_j f)]u\Delta_j$$
$$= [f(1 - \Delta_j)u] + [f - P]u\Delta_j - [L_n(w_\alpha, \Delta_j(f - P))u\Delta_j]$$
$$- [L_n(w_\alpha, (1 - \Delta_j)P)u\Delta_j].$$

The estimate of the norm of the first term is given by (4.3.33), the norm of the second term is dominated by $E_M(f)_{u,\infty}$, and for the third term we use (4.3.34).

4.3 Weighted Interpolation

Finally, for some $M > 0$, we can prove

$$\|L_n(w_\alpha, (1-\Delta_j)P)u\Delta_j\|_\infty \leq Cn^r\|(1-\Delta_j)Pu\|_\infty \leq Cn^r\|Pu\|_{L^\infty[4\theta n, +\infty)}.$$

Continuing as in the proof of Proposition 4.3.1, we deduce

$$\|L_n(w_\alpha, (1-\Delta_j)P)u\Delta_j\|_\infty \leq Ce^{-An}\|fu\|_\infty. \qquad \square$$

The behaviour of the Fourier sum relative to the same class of functions is analogous. In fact, if for a function $f \in C_u$, we consider

$$S_n(w_\alpha, f; x) = \sum_{k=0}^{n-1} c_k p_k(w_\alpha; x), \quad c_k = \int_0^{+\infty} f p_k(w_\alpha) w_\alpha,$$

and Δ_n^* is the characteristic function of the interval $[0, 4\theta n]$, for the sequence $\{\Delta_n^* S_n(w_\alpha, \Delta_n f)\}_n$, the following result holds:

Theorem 4.3.11 *If the parameters γ (≥ 0) and α (> -1) of the weight functions $w_\alpha(x) = x^\alpha e^{-x}$ and $u(x) = x^\gamma e^{-x/2}$ satisfy the condition*

$$\frac{\alpha}{2} + \frac{1}{4} \leq \gamma \leq \frac{\alpha}{2} + \frac{3}{4},$$

then, for each $f \in C_u$, we have

$$\|\Delta_n^* S_n(w_\alpha, f\Delta_n^*)u\|_\infty \leq C\|fu\Delta_n^*\|_\infty (\log n)$$

and

$$\|[f - \Delta_n^* S_n(w_\alpha, f\Delta_n^*)]u\|_\infty \leq C\big[E_M(f)_{u,\infty} \log n + e^{-An}\|fu\|_\infty\big],$$

where the constants C and A are independent of n and f.

The proof of this theorem can be found in [296].

4.3.4 Interpolation at Hermite Zeros

In this subsection we consider the Hermite weight $w(x) := e^{-x^2}$ on $\mathbb{R}$ and interpolate functions f which are locally continuous on the real axis ($f \in C_{\text{loc}}(\mathbb{R})$) and have the following behaviour for $x \to \pm\infty$

$$\lim_{|x| \to +\infty} f(x)\sqrt{w(x)} = 0.$$

Denote by C_w the set of all such functions, i.e.,

$$C_w := \left\{ f \in C_{\text{loc}}(\mathbb{R}) \ \Big| \ \lim_{|x| \to +\infty} f(x)\sqrt{w(x)} = 0 \right\},$$

and equip C_w with the following norm

$$\|f\|_{C_w} := \|f\sqrt{w}\|_\infty = \max_{x \in \mathbb{R}} \left| f(x)\sqrt{w(x)} \right|$$

so that $(C_w, \|\cdot\|_{C_w})$ is a Banach space.

Let $\{x_k\}_{k=1,\ldots,n}$ be the zeros of the Hermite polynomial $p_n(w;x)$ and denote by $L_n(w,f)$ the Lagrange polynomial interpolating a function $f \in C_w$ at the Hermite zeros $\{x_k\}_{k=1,\ldots,n}$.

As an interpolation at the Laguerre zeros, also this interpolation process at the Hermite zeros $\{L_n(w,f)\}_n$ is not efficient to approximate functions $f \in C_w$. In fact, the corresponding Lebesgue constants satisfy the following estimate [463]:

$$\|L_n(w)\|_{C_w} := \sup_{\|f\sqrt{w}\|_\infty = 1} \|L_n(w,f)\sqrt{w}\|_\infty \sim n^{1/6}.$$

The same result holds for the more general weights $u_\alpha(x) := e^{-|x|^\alpha}$, with $\alpha > 1$. In [463] Szabados improved the interpolation process based on the zeros of $p_n(u_\alpha;x)$ by adding two special knots $\pm x_0$, defined by

$$\left| p_n(u_\alpha; x_0)\sqrt{u_\alpha(x_0)} \right| = \max_{x \in \mathbb{R}} \left| p_n(u_\alpha; x)\sqrt{u_\alpha(x)} \right|.$$

He proved that with this modification of the set of nodes, the corresponding Lebesgue constants have order $\log n$.

In the special case $w(x) = e^{-x^2}$, since $|x_k| < M_n = \sqrt{2n}$, $k = 1,\ldots,n$ (cf. Sects. 2.4.4 and 2.4.5), it seems more natural to replace $\pm x_0$ by $\pm\sqrt{2n}$, so that we avoid to compute the additional knot x_0 for any n.

More precisely, we consider the Lagrange polynomial $\mathcal{L}_{n+2}(w,f)$ interpolating $f \in C_w$ at the Hermite zeros $\{x_k\}_{k=1,\ldots,n}$ and at the two additional knots given by

$$x_0 := -\sqrt{2n}, \qquad x_{n+1} := \sqrt{2n}.$$

Using the Lagrange interpolation formula, we can write

$$\mathcal{L}_{n+2}(w,f;x) := \frac{(\sqrt{2n}-x)p_n(w;x)}{2\sqrt{2n}\,p_n(w;-\sqrt{2n})} f(-\sqrt{2n}) + \sum_{k=1}^n \frac{2n-x^2}{2n-x_k^2} \ell_{n,k}(w;x) f(x_k)$$

$$+ \frac{(\sqrt{2n}+x)p_n(w;x)}{2\sqrt{2n}\,p_n(w;\sqrt{2n})} f(\sqrt{2n}), \tag{4.3.36}$$

4.3 Weighted Interpolation

where $\ell_{n,k}(w; x)$ are the fundamental Lagrange polynomials corresponding to the Hermite zeros, i.e.,

$$\ell_{n,k}(w; x) = \frac{p_n(w; x)}{p'(w; x_k)(x - x_k)}, \quad k = 1, \ldots, n. \tag{4.3.37}$$

The following theorem holds:

Theorem 4.3.12 *For all $n \in \mathbb{N}$ and each $f \in C_w$, with $w(x) = e^{-x^2}$, we have*

$$\|\mathcal{L}_{n+2}(w, f)\sqrt{w}\|_\infty \leq C \log n \|f \sqrt{w}\|_\infty, \quad C \neq C(n, f). \tag{4.3.38}$$

Proof First we observe that from the Mhaskar-Rahmanov-Saff identity (2.4.22) for the Hermite weight (see also [423]), we have

$$\|\mathcal{L}_{n+2}(w, f)\sqrt{w}\|_\infty = \max_{x \in [-\sqrt{2n}, \sqrt{2n}]} \left| \mathcal{L}_{n+2}(w, f; x) \sqrt{w(x)} \right|.$$

Then we get by (4.3.36)

$$\|\mathcal{L}_{n+2}(w, f)\sqrt{w}\|_\infty$$

$$\leq \|f\sqrt{w}\|_\infty \max_{x \in [-\sqrt{2n}, \sqrt{2n}]} \left(\left| \frac{\sqrt{2n} - x}{2\sqrt{2n}} \right| \left| \frac{\sqrt{w(x)} p_n(w; x)}{\sqrt{w(-\sqrt{2n})} p_n(w; -\sqrt{2n})} \right| \right.$$

$$+ \sum_{k=1}^{n} \left| \frac{2n - x^2}{2n - x_k^2} \right| |\ell_{n,k}(w; x)| \sqrt{\frac{w(x)}{w(x_k)}}$$

$$+ \left. \left| \frac{\sqrt{2n} + x}{2\sqrt{2n}} \right| \left| \frac{\sqrt{w(x)} p_n(w; x)}{\sqrt{w(\sqrt{2n})} p_n(w; \sqrt{2n})} \right| \right),$$

i.e.,

$$\|\mathcal{L}_{n+2}(w, f)\sqrt{w}\|_\infty$$

$$\leq C \|f\sqrt{w}\|_\infty \max_{x \in [-\sqrt{2n}, \sqrt{2n}]} \left(\frac{|\sqrt{w(x)} p_n(w; x)|}{\sqrt{w(-\sqrt{2n})} |p_n(w; -\sqrt{2n})|} \right.$$

$$+ \sum_{k=1}^{n} \left| \frac{2n - x^2}{2n - x_k^2} \right| |\ell_{n,k}(w; x)| \sqrt{\frac{w(x)}{w(x_k)}} + \frac{|\sqrt{w(x)} p_n(w; x)|}{\sqrt{w(\sqrt{2n})} |p_n(w; \sqrt{2n})|} \right)$$

$$= C \|f\sqrt{w}\|_\infty \max_{x \in [-\sqrt{2n}, \sqrt{2n}]} [I_1(x) + I_2(x) + I_3(x)].$$

Now, we use the following estimate of Hermite polynomials (see Theorem 2.4.4 for $\beta = 0$ or [80])

$$|p_n(w;x)|\sqrt{w(x)}\sqrt[4]{2n-x^2+n^{1/3}} \sim \left|\frac{x-x_d}{x_{d\pm 1}-x_d}\right|, \qquad (4.3.39)$$

where, as usual, x_d is a Hermite zero closest to x, i.e., $|x-x_d| = \min_{1\le k\le n}|x-x_k|$. By (4.3.39) we deduce that

$$\left|p_n(w;x)\sqrt{w(x)}\right| \le Cn^{-1/12}, \qquad C \ne C(n,x),$$

and

$$\left|p_n(w;\pm\sqrt{2n})\sqrt{w(\pm\sqrt{2n})}\right| \ge Cn^{-1/12}, \qquad C \ne C(n),$$

hold. Hence we have

$$|I_1(x)| + |I_3(x)| := \left|\frac{\sqrt{w(x)}\,p_n(w;x)}{\sqrt{w(-\sqrt{2n})}\,p_n(w;-\sqrt{2n})}\right| + \left|\frac{\sqrt{w(x)}\,p_n(w;x)}{\sqrt{w(\sqrt{2n})}\,p_n(w;\sqrt{2n})}\right|$$

$$\le C \ne C(n,x). \qquad (4.3.40)$$

In order to estimate $I_2(x)$, we note that (cf. [257]) $|\ell_{n,d}(w;x)|\sqrt{w(x)/w(x_d)} \sim 1$, and then, using also (4.3.37), we can write

$$|I_2(x)| := \sum_{k=1}^{n}\left|\frac{2n-x^2}{2n-x_k^2}\right||\ell_{n,k}(w;x)|\sqrt{\frac{w(x)}{w(x_k)}}$$

$$\sim 1 + \sum_{\substack{k=1\\k\ne d}}^{n}\left|\frac{2n-x^2}{2n-x_k^2}\right|\left|\frac{p_n(w;x)}{(x-x_k)p_n'(w;x_k)}\right|\sqrt{\frac{w(x)}{w(x_k)}}$$

$$= 1 + \sum_{\substack{k=1\\k\ne d}}^{n}\left|\frac{2n-x^2}{2n-x_k^2}\right|^{3/4}\left|\frac{\sqrt[4]{2n-x^2}\sqrt{w(x)}\,p_n(w;x)}{\sqrt[4]{2n-x_k^2}\sqrt{w(x_k)}\,p_n'(w;x_k)}\right|\frac{1}{|x-x_k|}.$$

Thus, using (4.3.39) and the following estimates

$$\left|\sqrt{w(x_k)}\,p_n(w;x_k)\right| \sim \sqrt[4]{2n+x_k^2+n^{1/3}}, \qquad k=1,\ldots,n, \qquad (4.3.41)$$

$$\Delta x_k := x_{k+1}-x_k \sim (2n-x_k^2)^{-1/2}, \qquad (4.3.42)$$

4.3 Weighted Interpolation

we get

$$I_2(x) \leq C \left(1 + \sum_{\substack{k=1 \\ k \neq d}}^{n} \left| \frac{2n - x^2}{2n - x_k^2} \right|^{3/4} \frac{\Delta x_k}{|x - x_k|} \right).$$

Now, applying Lemma 4.1.4, we obtain $I_2(x) \leq C \log n$, $C \neq C(n, x)$, which together with (4.3.40) gives the statement. □

From (4.3.38), in a usual way, we deduce that

$$\|(\mathcal{L}_{n+2}(w, f) - f)\sqrt{w}\|_\infty \leq C E_{n+1}(f)_{\sqrt{w}, \infty} \log n, \quad C \neq C(n, f), \quad (4.3.43)$$

holds for all $f \in C_w$, where

$$E_n(f)_{\sqrt{w}, \infty} := \inf_{P \in \mathcal{P}_n} \|(f - P)\sqrt{w}\|_\infty$$

is the error of the best approximation in the C_w-norm.

Note that (4.3.38) gives the following result for the Lebesgue constants

$$\|\mathcal{L}_{n+2}(w)\|_{C_w} := \sup_{\|f\sqrt{w}\|_\infty} \|\mathcal{L}_{n+2}(w, f)\sqrt{w}\|_\infty \leq C \log n, \quad (4.3.44)$$

where $C \neq C(n)$. On the other hand, Szabados [463] proved that in C_w the Lebesgue constants corresponding to an arbitrary array $\mathcal{X}$ of nodes in $\mathbb{R}$, grow at least like $\log n$, as $n \to +\infty$, i.e.,

$$\|L_n(\mathcal{X})\|_{C_w} \geq C \log n, \quad C \neq C(n).$$

Hence, the estimate (4.3.44) is sharp and we can say that the interpolation process $\{\mathcal{L}_{n+2}(w)\}_n$ is "optimal".

As for the interpolation at the Laguerre zeros, it is possible to consider a "truncated interpolation" based on the Hermite zeros. The construction is quite similar to the Laguerre case. Denote by $x_1, x_2, \ldots, x_{[n/2]}$ the positive zeros of the n-th Hermite polynomial $p_n(w; x)$ and set $x_{-i} = -x_i$.

Let $j = j(n)$ be defined by $x_j = \min\{x_k : x_k \geq \theta\sqrt{2n}\}$, $\theta \in (0, 1)$. With ψ defined above, we put $\psi_j(x) = \psi((|x| - x_j)/(x_{j+1} - x_j))$ and $f_j(x) = f(x) - \psi_j(x) f(x)$. Then $f_j(x) = f(x)$ for $x \in [-x_j; x_j]$ and $f_j(x) = 0$ for $|x| \geq x_{j+1}$. Inequality (4.3.33) is replaced by the following

$$\|(f - f_j)\sqrt{w}\|_\infty \leq C\left[E_M(f)_{\sqrt{w}, \infty} + e^{-An}\|f\sqrt{w}\|_\infty\right],$$

with $M = [\theta n/(\theta + 1)]$ and C and A independent on f and n. Then the relation (4.3.36) can be replaced by

$$\mathcal{L}_{n+2}(w, f_j; x) = \sum_{k=-j}^{j} \frac{2n - x^2}{2n - x_k^2} \ell_{n,k}(x) f(x_k).$$

We complete this subsection with two results analogous to Theorems 4.3.10 and 4.3.11 for the Laguerre case. We use the previous definition of x_j, $j = j(n)$, and consider the sequence $\{\Delta_j L_n(w, \Delta_j f)\}$ in the space $C_{\sqrt{w}}$, where $\Delta_j(x)$ is the characteristic function of the interval $[-x_j, x_j]$.

Theorem 4.3.13 *For every $f \in C_{\sqrt{w}}$ we have*

$$\|\Delta_j L_n(w, \Delta_j f)\sqrt{w}\|_\infty \leq C \|\Delta_j f \sqrt{w}\|_\infty \log n$$

and

$$\|[f - \Delta_j L_n(w, \Delta_j f)]\sqrt{w}\|_\infty \leq C\big[E_M(f)_{\sqrt{w},\infty} \log n + e^{-An} \|f\sqrt{w}\|_\infty\big],$$

where $M = [\theta n/(1+\theta)]$ and the constants C and A are independent of n and f.

According to the Fourier sum, let Δ_n^* be the characteristic function of the interval $[-\theta\sqrt{2n}, \theta\sqrt{2n}]$ and consider the sequence $\{\Delta_n^* S_n(w, \Delta_n^* f)\}_n$, where

$$S_n(w, f; x) = \sum_{k=0}^{n-1} c_k p_k(w; x), \quad c_k = \int_{-\infty}^{+\infty} f p_k(w) w.$$

Then, the following result holds:

Theorem 4.3.14 *For every $f \in C_{\sqrt{w}}$ we have*

$$\|\Delta_n^* S_n(w, f\Delta_n^*)\sqrt{w}\|_\infty \leq C \|(\Delta_n^* f)\sqrt{w}\|_\infty \log n$$

and

$$\|[f - \Delta_n^* S_n(w, f\Delta_n^*)]\sqrt{w}\|_\infty \leq C\big[E_M(f)_{u,\infty} \log n + e^{-An} \|f\sqrt{w}\|_\infty\big],$$

where the constants C and A are independent of n and f and $M = [\theta n/(1+\theta)]$.

The proof of these two theorems can be found in [296].

4.3.5 Interpolation of Functions with Internal Isolated Singularities

The weighted polynomial approximation of continuous functions or of smooth functions with singular derivatives at some isolated points is of some theoretical interest and often proves to be useful in many applications. For example, such functions occur as solutions of integral equations with discontinuous right-hand sides. While there exists a wide literature about the polynomial approximation of functions with singularities at the endpoints, the case of singular functions with singularities at isolated points inside the interval have only recently been studied ([82, 308, 312]).

The inner singularities add new difficulties and require a more careful examination of the behaviour of the approximating polynomial around these singularities.

In this subsection we consider interpolation processes on bounded and unbounded intervals for functions with isolated singularities. First we need to prove a preliminary result.

Let $v^{\mu,\nu}(x) = (1-x)^\mu (1+x)^\nu$ and
$$-1 = x_0 < x_1 < x_2 < \cdots < x_n < x_{n+1} = 1$$
with $x_k = \cos\theta_k$ and $n(\theta_{k-1} - \theta_k) \sim 1$.

Set
$$\Gamma_n(x) := \sum_{k=1, k \neq d}^n \frac{v^{\mu,\nu}(x)\left(|x-t_0|+n^{-1}\right)^\rho}{v^{\mu,\nu}(x_k)\left(|x_k-t_0|+n^{-1}\right)^\rho} \frac{\Delta x_k}{|x-x_k|},$$
where $x_d = \min_k |x_k - x|$, $\Delta x_k = x_{k+1} - x_k$, $\mu, \nu, \rho \in \mathbb{R}$.

In a similar way, let $y_1, \ldots, y_n$ be the zeros of the n-th Laguerre polynomial $p_n(w_\alpha)$ orthogonal on $(0, +\infty)$ with respect to the weight $w_\alpha(x) = x^\alpha e^{-x}$. Set
$$A_n(x) := \sum_{k=1, k \neq d}^n \frac{x^\sigma \left(|t_0 - x| + 1/\sqrt{n}\right)^\tau}{y_k^\sigma \left(|t_0 - y_k| + 1/\sqrt{n}\right)^\tau} \frac{\Delta y_{k-1}}{|x - y_k|},$$
where $y_d = \min_k |x - y_k|$, $\Delta y_{k-1} = y_k - y_{k-1}$ and $\sigma, \tau \in \mathbb{R}$.

Lemma 4.3.1 *Let $a \in \mathbb{R}^+$ be a fixed number. We have*
$$\sup_{|x| \leq 1 - a/n^2} \Gamma_n(x) \sim \log n \tag{4.3.45}$$

if and only if $0 < \mu, \nu, \rho < 1$. Moreover,
$$\sup_{a/n \leq x \leq 4n} A_n(x) \sim \log n \tag{4.3.46}$$

if and only if $0 < \sigma, \tau < 1$.

Proof Let us prove (4.3.45). Since, for $x_k \neq x_d$, we have
$$\Gamma_n(x) = \sum_{x_k \leq 0} + \sum_{x_k > 0},$$
where in the first sum $1 - x_k \sim 1$ and in the second one $1 + x_k \sim 1$. Then, it will be sufficient to separately estimate
$$\Gamma_n' = \sum_{k=1, k \neq d}^n \left(\frac{1+x}{1+x_k}\right)^\nu \left(\frac{|x-t_0|+n^{-1}}{|x_k-t_0|+n^{-1}}\right)^\rho \frac{\Delta x_k}{|x-x_k|}$$

and

$$\Gamma_n'' = \sum_{k=1, k\neq d}^{n} \left(\frac{1-x}{1-x_k}\right)^{\nu} \left(\frac{|x-t_0|+n^{-1}}{|x_k-t_0|+n^{-1}}\right)^{\rho} \frac{\Delta x_k}{|x-x_k|}.$$

Let us consider Γ_n'. Let $\delta > 0$ be such that $\Delta = (t_0 - \delta, t_0 + \delta) \subset (-1, 1)$. Then, $\Gamma_n' = \sum_{x_k \in \Delta} + \sum_{x_k \notin \Delta}$, $x_k \neq x_d$. In the first sum $1 + x_k \sim 1$, and in the second one $|x_k - t_0| + n^{-1} \sim 1$. Since a similar decomposition also holds for Γ_n'', it is sufficient to separately estimate the next three sums

$$\sum_{k=1, k\neq d}^{n} \left(\frac{1-x}{1-x_k}\right)^{\mu} \frac{\Delta x_k}{|x-x_k|}, \quad \sum_{k=1, k\neq d}^{n} \left(\frac{|x-t_0|+n^{-1}}{|x_k-t_0|+n^{-1}}\right)^{\rho} \frac{\Delta x_k}{|x-x_k|},$$

$$\sum_{k=1, k\neq d}^{n} \left(\frac{1+x}{1+x_k}\right)^{\mu} \frac{\Delta x_k}{|x-x_k|},$$

since $\mu, \nu, \rho > 0$. But, all these sums are equivalent to $\log n$, when $0 \leq \mu, \nu, \rho \leq 1$ (see [375]).

Moreover, if $\nu > 1$ we have

$$\sup_{|x| \leq 1-a/n^2} \Gamma_n(x) \geq \Gamma_n(t_0/2) \geq \sum_{x_1 \leq x_k < 0} \frac{v^{\mu,\nu}(t_0/2)(t_0/2+n^{-1})^{\rho}}{v^{\mu,\nu}(x_k)(|t_0-x_k|+n^{-1})^{\rho}} \frac{\Delta x_k}{|t_0/2 - x_k|}$$

$$> \frac{v^{\mu,\nu}(t_0/2)}{2} \left(\frac{t_0/2+n^{-1}}{2}\right)^{\rho} \sum_{x_1 \leq x_k < 0} \frac{\Delta x_k}{(1+x_k)^{\mu}}$$

$$\geq C \int_{x_1}^{1/2} \frac{dt}{(1+t)^{\nu}} \sim n^{2(\nu-1)} > \log n,$$

and for $\nu < 0$,

$$\sup_{|x| \leq 1-a/n^2} \Gamma_n(x) \geq \Gamma_n(x_1) > (1-x_1)^{\mu}(1+x_1)^{\nu}\left(|t_0 - x_1| + n^{-1}\right)^{\rho}$$

$$\times \sum_{0 < x_k \leq t_0/2} \frac{1}{(1-x_k)^{\mu}(1+x_k)^{\nu}\left(|t_0 - x_k| + n^{-1}\right)^{\rho}} \frac{\Delta x_k}{|x_1 - x_k|}$$

$$\sim (1+x_1)^{\nu} \sum_{0 < x_k \leq t_0/2} \frac{\Delta x_k}{(1+x_k)^{\nu}}$$

$$\sim (1+x_1)^{\nu} \int_0^{t_0/2} \frac{dt}{(1+t)^{\nu}} \sim n^{-2\nu} > \log n.$$

4.3 Weighted Interpolation

One can proceed in a similar way if $\mu < 0$ or $\mu > 1$.

Now, if $\rho > 1$, it is sufficient to evaluate Γ_n at $t_0/2$ in order to get

$$\Gamma_n(t_0/2) \geq C \sum_{t_0-\delta < x_k < t_0+\delta} \frac{\Delta x_k}{\left(|t_0 - x_k| + n^{-1}\right)^\rho}$$

$$\geq C \int_{t_0-\delta}^{t_0} \frac{dt}{\left[(t_0-t) + n^{-1}\right]^\rho} \sim n^{\rho-1}.$$

Finally, if $\rho < 0$ one has

$$\Gamma_n\left(t_0 - n^{-1}\right) \geq n^{-\rho} \sum_{t_0+\delta/2 < x_k < t_0+\delta} \frac{\Delta x_k}{\left(|t_0 - x_k| + n^{-1}\right)^\rho} \sim n^{-\rho}$$

and the proof of (4.3.45) is complete.

We omit the proof of (4.3.46) because it is similar to the previous one (see [301, Lemma 4.1]). □

4.3.5.1 Interpolation Processes on Bounded Intervals

Let $f \in L_u^\infty$ and

$$u(x) = v^{\gamma,\delta}(x)|x - t_0|^\theta \quad (\gamma, \delta, \theta \geq 0, \ |t_0| < 1).$$

If we want to approximate the function f by a Lagrange interpolating polynomial, the point t_0 cannot be an interpolation knot, and therefore we use the following procedure.

Let $w(x) = v^{\alpha,\beta}(x)|x - t_0|^\eta$ be another generalized Jacobi weight and let $\{p_n(w)\}$ be the corresponding sequence of orthonormal polynomials with a positive leading coefficient. Denote by $x_1 < x_2 < \cdots < x_n$ the zeros of $p_n(w)$. Let x_c be the closest zero to t_0, i.e., $|x_c - t_0| = \min_k |x_k - t_0|$ and let $q_s \in \mathcal{P}_s$ be such that

$$\|(f - q_s) \cdot -t_0|^\theta\|_{[t_0-a/n, t_0+a/n]} \leq 2 \inf_{\deg q \leq s} \|(f - q_s) \cdot -t_0|^\theta\|_{[t_0-a/n, t_0+a/n]},$$

with a fixed $a > 0$. Let $\psi \in C^\infty(\mathbb{R})$ be a nondecreasing function, such that

$$\psi(x) = \begin{cases} 0, & x \leq 0, \\ 1, & x \geq 1. \end{cases}$$

Using ψ we define the functions

$$\psi_1(x) = \psi\left(\frac{x - x_{c-2}}{x_{c-1} - x_{c-2}}\right), \quad \psi_2(x) = \psi\left(\frac{x - x_{c+1}}{x_{c+2} - x_{c+1}}\right),$$

and
$$F = F_{t_0} = (1-\psi_1)f + (1-\psi_2)\psi_1 q_s + \psi_2 f.$$

It follows from the definition that

$$F = \begin{cases} f & \text{in } [-1, x_{c-2}] \cup [x_{c+2}, 1], \\ q_s & \text{in } [x_{c-1}, x_{c+1}], \\ (1-\psi_1)f + (1-\psi_2)\psi_1 q_s & \text{in } [x_{c-2}, x_{c-1}], \\ (1-\psi_2)q_s + \psi_2 f & \text{in } [x_{c+1}, x_{c+2}]. \end{cases} \quad (4.3.47)$$

Next we interpolate the function F at the zeros $x_1 < x_2 < \cdots < x_n$ of $p_n(w)$ and denote by $\tilde{L}_n(w, F)$ the corresponding Lagrange polynomial. Recalling (4.3.47) we have

$$\tilde{L}_n(w, F; x) = \sum_{k=1}^n \ell_k(x) F(x_k) = \sum_{k \neq c, c \pm 1}^n \ell_k(x) f(x_k) + \sum_{k=c-1}^{c+1} \ell_k(x) q_s(x_k), \quad (4.3.48)$$

where

$$\ell_k(x) = \frac{p_n(w; x)}{p_n'(w; x_k)(x - x_k)}.$$

Denoting the usual L^p norm ($1 \le p < +\infty$) by $\|\cdot\|_p$, we can state the following result for $\tilde{L}_n(w)$ (see [294]):

Theorem 4.3.15 Let $f \in L_u^\infty$, $u(x) = v^{\gamma,\delta}(x)|x - t_0|^\theta$ ($\gamma, \delta \ge 0, 0 \le \theta < 1$, $|t_0| < 1$), and $1 \le p < +\infty$. Then

$$\|u\tilde{L}_m(w, F)\|_p \le C\|uF\|_\infty, \quad C \neq C(m, F), \quad (4.3.49)$$

if and only if

$$\frac{u}{\sqrt{w\varphi}} \in L^p \quad \text{and} \quad \frac{\sqrt{w\varphi}}{u} \in L^1, \quad \varphi(x) = \sqrt{1-x^2}. \quad (4.3.50)$$

Moreover,

$$\|u\tilde{L}_n(w, F)\|_\infty \le C\|uF\|_\infty \log n, \quad C \neq C(n, F), \quad (4.3.51)$$

if and only if

$$\frac{u}{\sqrt{w\varphi}} \in L^\infty \quad \text{and} \quad \frac{\sqrt{w\varphi}}{u} \in L^1 \quad \text{or} \quad \begin{pmatrix} \gamma = \frac{\alpha}{2} + \frac{5}{4} \\ \delta = \frac{\beta}{2} + \frac{5}{4} \\ \theta = \frac{\eta}{2} + 1 \end{pmatrix}. \quad (4.3.52)$$

4.3 Weighted Interpolation

Remark 4.3.2 The conditions (4.3.50) and (4.3.52) can be expressed as follows

$$\frac{\alpha}{2} - \frac{1}{4} - \frac{1}{p} < \gamma < \frac{\alpha}{2} + \frac{5}{4},$$
$$\frac{\beta}{2} - \frac{1}{4} - \frac{1}{p} < \delta < \frac{\beta}{2} + \frac{5}{4}, \qquad (4.3.53)$$
$$\frac{\eta}{2} - \frac{1}{p} < \theta < \frac{\eta}{2} + 1$$

and

$$\frac{\alpha}{2} - \frac{1}{4} \leq \gamma \leq \frac{\alpha}{2} + \frac{5}{4},$$
$$\frac{\beta}{2} - \frac{1}{4} \leq \delta \leq \frac{\beta}{2} + \frac{5}{4}, \qquad (4.3.54)$$
$$\frac{\eta}{2} \leq \theta \leq \frac{\eta}{2} + 1,$$

respectively.

Proof of Theorem 4.3.15 Setting

$$A = \left(-1 + \frac{C}{n^2}, t_0 - \frac{C}{n}\right) \cup \left(t_0 + \frac{C}{n}, 1 - \frac{C}{n^2}\right),$$

for any fixed $C > 0$, we can write by the Remez inequality

$$\|u\widetilde{L}_n(w, F)\|_p \leq C\|u\widetilde{L}_n(w, F)\|_{L^p(A)}, \quad 1 \leq p \leq +\infty.$$

Putting $g(x) = \operatorname{sgn}(\widetilde{L}_n(w, F; x))|u(x)\widetilde{L}_n(w, F; x)|^{p-1}$ and

$$r(t) = \int_A \frac{p_n(w; x) - p_n(w; t)}{x - t} u(x) g(x) \, dx = H\left(p_n(w) u g, t\right) - p_n(w; t) H(ug, t),$$

where H denotes the Hilbert transform extended to A, we get

$$\|u\widetilde{L}_n(w, F)\|_{L^p(A)}^p = \sum_{k=1}^n \frac{F(x_k) u_n(x_k)}{p_n'(w; x_k) u_n(x_k)} r(x_k),$$

where $u_n(x) = v^{\gamma,\delta}(x)\left(|x - t_0| + n^{-1}\right)$.

Now, for $k \neq c$ $\left(|t_0 - x_c| = \min_k |t_0 - x_k|\right)$ we conclude that $u_n(x_k) \sim u(x_k)$. For $x = c$ we have

$$F(x_c) u_n(x_c) = q_s(x_c) u_n = q_s(x_{c-1}) u_n(x_c) + u_n(x_c) \int_{x_{c-1}}^{x_c} q_s'(t) dt.$$

Now, $u_n(x_c) \leq u_n(x_{c-1}) \leq cu(x_{c-1})$ and $|q_s(x_{c-1})u_n(x_c)| \leq c\|uF\|_{[x_{c-1},x_c],\infty}$. Moreover, for $0 < \theta < 1$ we get

$$\left|u_n(x_c)\int_{x_{c-1}}^{x_c} q_s'(t)dt\right| \leq \frac{C}{n^\theta}\int_{x_{c-1}}^{x_{c+1}} |q_s'(t)||t_0-t|^\theta \frac{dt}{|t_0-t|^\theta}$$

$$\leq \frac{C}{n^\theta}\|q_s'|t_0-\cdot|^\theta\|_{[x_{c-1},x_{c+1}]}\int_{x_{c-1}}^{x_{c+1}}\frac{dt}{|t_0-t|^\theta}$$

$$\sim \frac{C}{n^\theta}\cdot\frac{1}{n^{1-\theta}}\|q_s'|t_0-\cdot|^\theta\|_{[x_{c-1},x_{c+1}]}$$

$$\leq C\|uF\|,$$

by using the Markov inequality in $[x_{c-1}, x_{c+1}]$.

Then we have $|u_n(x_k)F(x_k)| \leq \|uF\|$ for any $k = 1, \ldots, n$. Furthermore (see [376])

$$\frac{1}{p_n'(w; x_k)} \sim \sqrt{w_n\varphi}(x_k)(x_{k+1}-x_k),$$

where $w_n(x) = v^{\alpha,\beta}(|x-t_0|+n^{-1})^\eta$. By using the Marcinkiewicz and Remez inequalities, we get for $p \in [1, +\infty)$

$$\|u\widetilde{L}_n(w, F)\|_p \leq C\|uF\|\int_A \frac{\sqrt{w\varphi(t)}}{u(t)}|r(t)|dt,$$

where the integral exists in view of (4.3.50).

Furthermore, recalling the definition of $r(t)$, we obtain

$$\int_A \frac{\sqrt{w\varphi(t)}}{u(t)}|r(t)|dt \leq \int_A \frac{\sqrt{w\varphi(t)}}{u(t)}|H(p_n(w)gu, t)|dt$$

$$+ \int_A \frac{\sqrt{w\varphi(t)}}{u(t)}|p_n(w;t)||H(gu, t)|dt =: I_1 + I_2.$$

Taking into account that $|p_n(w;t)| \leq c/\sqrt{w\varphi(t)}$ for $t \in A$, and the equality

$$\int_A fHg = -\int_A gHf \quad \text{for } f \in (L\log^+ L) \text{ and } g \in L^\infty,$$

we have

$$I_1 \leq \int_A \frac{u(t)}{\sqrt{w\varphi(t)}}g(t)\left|H\left(G_1\frac{\sqrt{w\varphi}}{u}, t\right)\right|dt,$$

$$I_2 \leq \int_A \frac{1}{u(t)}|H(gu, t)|dt \leq \int_A g(t)u(t)\left|H\left(\frac{G_2}{u}, t\right)\right|dt,$$

where $G_1 = \operatorname{sgn} H(\ldots)$ and $G_2 = \operatorname{sgn} H(\ldots)$.

4.3 Weighted Interpolation

If $1 < p < +\infty$ we use [376, Lemma] and

$$I_1 + I_2 \leq C \, \|L_n(w, f)u\|_p^{p-1}.$$

For $p = 1$, we have $|g(t)| \leq 1$ and

$$I_1 \leq \int_A H(G, t) dt + \int_A \frac{u(t)}{\sqrt{w\varphi(t)}} \left(\int_A \left| \frac{\frac{\sqrt{w\varphi(x)}}{u(x)} - \frac{\sqrt{w\varphi(t)}}{u(t)}}{x - t} \right| dx \right) dt.$$

Since $U(x) = \sqrt{w\varphi(x)}/u(x)$ is a generalized Jacobi weight, the inner integral is dominated by a constant times the product of the factors of U having negative exponents. Then, we have $I_1 \leq C$, since

$$\int_A H(G, t) dt \leq \sqrt{2} \left(\int_A H^2(G, t) dt \right)^{1/2} \leq 2.$$

For I_2 we use a similar argument. Thus (4.3.50) implies (4.3.49).

Let us prove that (4.3.49) implies (4.3.50). Now it easily follows from (4.3.49) and [313, Theorem 2.2] that

$$\frac{u}{\sqrt{w\varphi}} \in L^p, \qquad 1 \leq p < +\infty.$$

We have to prove that (4.3.49) implies $\sqrt{w\varphi} \in L^1$, i.e., $\alpha, \beta, \gamma, \delta, \theta, \eta$ satisfy (4.3.53). This can be done following [376, p. 688].

For θ and η, let us consider a function f with $|f(x)| \leq 1$ such that $f(\pm 1) = 0$, $f(x_k) = 0$ for $x_k \leq t_0 - \delta$ and $x_k \geq t_0 + \delta$ and $f(x_k) = \mathrm{sgn}\,(p'_n(w; x_k))$ if $x_k \in (t_0 - \delta, t_0 + \delta) \subset (-1, 1)$. Then

$$\widetilde{L}_n(w, f; x) = p_n(w; x) \sum_{t_0 - \delta \leq x_k \leq t_0 + \delta} \frac{1}{|p'_n(w; x_k)|\,(x - x_k)\,u_n(x_k)}$$

and

$$|u(x)\widetilde{L}_n(w, f; x)| \geq 2 |p_n(w; x) u(x)| \sum_{t_0 - \delta \leq x_k \leq t_0 + \delta} \frac{1}{|p'_n(w; x_k)|\,u_n(x_k)}$$

$$\sim |p_n(w; x) u(x)| \sum_{t_0 - \delta \leq x_k \leq t_0 + \delta} \frac{\sqrt{w\varphi}(x_k)}{u_n(x_k)} \Delta x_k$$

$$\sim |p_n(w; x) u(x)| \sum_{t_0 - \delta \leq x_k \leq t_0 + \delta} \left(|x_k - t_0| + n^{-1} \right)^{\eta/2 - \theta} \Delta x_k$$

$$\sim |p_n(w; x) u(x)| \int_{t_0 - \delta}^{t_0 + \delta} \left(|t - t_0| + n^{-1} \right)^{\eta/2 - \theta} dt.$$

Then
$$\sup_n \|p_n(w)u\|_p \int_{t_0-\delta}^{t_0+\delta} \left(|t-t_0|+n^{-1}\right)^{\eta/2-\theta} dt \leq C.$$

But,
$$\|p_n(w)u\|_p \sim \left\|\frac{u}{\sqrt{w\varphi}}\right\|_p$$

and $\eta/2 - \theta > -1$.

To prove the second part of Theorem 4.3.15 it is sufficient to show that

$$\Gamma_n = \max_{x \in A} \left\{ u(x) \sum_{k=1, k \neq d}^{n} \frac{|\ell_k(x)|}{u_n(x_k)} \right\} \leq C \log n$$

if and only if (4.3.54) holds. But, with $w_n(x) = v^{\alpha,\beta}(x)\left(|x-t_0|+n^{-1}\right)^{\eta}$ and $x \in A$

$$\Gamma(x) := |p_n(w;x)u(x)| \sum_{k=1, k\neq d}^{n} \frac{|\ell_k(x)|}{u_n(x_k)}$$

$$\sim |p_n(w;x)\sqrt{w_n\varphi}(x)| \left(\frac{u_n(x)}{\sqrt{w\varphi}(x)} \sum_{k=1, k\neq d}^{n} \frac{\sqrt{w_n\varphi}(x_k)}{u_n(x_k)} \frac{\Delta x_k}{|x-x_k|} \right).$$

Using Lemma 4.3.1, with

$$\mu = \gamma - \frac{\alpha}{2} - \frac{1}{4}, \quad \nu = \delta - \frac{\beta}{2} - \frac{1}{4}, \quad \rho = \theta - \frac{\eta}{2},$$

we deduce

$$\Gamma(x) \sim C \left| p_n(w;x)\sqrt{w\varphi}(x) \right| \log n$$

if and only if (4.3.54) are satisfied. Then, the second part of Theorem 4.3.15 follows by recalling that $\| p_n(w)\sqrt{w_n\varphi} \| \sim 1$. □

Corollary 4.3.2 *Suppose that f and u satisfy the conditions in Theorem 4.3.15 and let*

$$A_n(f) = E_{n-1}(f)_u + \inf_{\deg q \leq s} \|u(f-q)\|_{[t_0-a/n, t_0+a/n]}.$$

Then for $1 \leq p < +\infty$ we have

$$\|u[f - \tilde{L}_n(w, F)]\|_p \leq C A_n(f), \quad C \neq C(n, F), \quad (4.3.55)$$

if and only if (4.3.50) holds. Moreover,

$$\|u[f - \tilde{L}_n(w, F)]\|_\infty \leq C \log n\, A_n(f), \quad C \neq C(n, F), \quad (4.3.56)$$

if and only if (4.3.52) holds.

4.3 Weighted Interpolation

Proof First we prove inequality (4.3.55). We have

$$\|u[f - \tilde{L}_n(w, F)]\|_p \leq \|u[f - F]\|_p + \|u[F - \tilde{L}_n(w, F)]\|_p.$$

It follows from the definition of F that

$$\|u[f - F]\|_p \leq C \inf_{q_s \in \mathcal{P}_s} \|u[f - q_s]\|_{L^\infty(t_0 - C/n, t_0 + C/n)}.$$

Moreover, using Theorem 4.3.15, we get for all polynomials $P \in \mathcal{P}_{n-1}$

$$\|u[F - \tilde{L}_n(w, F)]\|_p \leq \|u(F - P)\|_p + \|u\tilde{L}_n(w, F - P)\|_p$$

$$\leq C\|u(F - P)\|_\infty \leq C\|u(F - f)\|_\infty + \|u(f - P)\|_\infty$$

$$\leq C \inf_{q_s \in \mathcal{P}_s} \|u(f - q_s)\|_\infty + \|u(f - P)\|_\infty.$$

Therefore, assuming that $P \in \mathcal{P}_{n-1}$ minimizes the last expression, we arrive at the estimate

$$\|u[f - \tilde{L}_n(w, F)]\|_p \leq C \left(E_{n-1}(f)_{u,\infty} + \inf_{q_s \in \mathcal{P}_s} \|u(f - q_s)\|_\infty \right),$$

i.e., (4.3.55).

The proof of (4.3.56) is similar and therefore Corollary 4.3.2 is proved. □

Now, we give some remarks:

For $\theta = 0$, Theorem 4.3.15 directly follows from [376, Theorem 1], but for $\theta > 0$ some nontrivial difficulties appear in the proof.

We deduce from the definition of $\omega_\varphi^r(f, t)_{u,\infty}^*$ (see (2.5.48)) that

$$A_n(f) \leq C \omega_\varphi^r (f, 1/n)_{u,\infty}^*.$$

If the function f is smooth "around" the singularity, $A_n(f)$ can be estimated by Theorem 2.5.4 and Corollary 4.3.2. For example, if $f(x) = \text{sgn}(x)$, we get $A_n(f) \leq Cn^{-\theta}$.

In particular, if $f^{(r-1)}(t_0)$ exists and $\|f^{(r)} \varphi^r u\| < +\infty$, with $r \geq 1$, then using Lemma 2.5.1, we can obtain an estimate for $A_n(f)$ in the following form

$$A_n(f) \leq \frac{C}{n^r} \left(\|f^{(r)} \varphi^r u\| + \|uf\| \right).$$

Moreover, if the above assumptions on f are satisfied, then we can set in (4.3.48) $q_s(x_k) = f(x_k)$.

It is not difficult to see that the (strong) assumption $0 < \theta < 1$ is not required from the conditions (4.3.50) and (4.3.52), but from the presence of the weight u in the norm of the function (see the proof of Theorem 4.3.15). However, if $\theta \geq 1$ in the weight u, a slight modification can be made in the previous Lagrange polynomial $\widetilde{L}_n(w, F)$. In fact, it is sufficient to interpolate the function $F = F_{t_0}$ at the zeros of $p_{n+1}(w; x)/(x - x_c)$, where $x_c = x_{n+1,c}$, by defining the following interpolation process

$$L_n^*(w, F; x) = \sum_{k=1, k \neq c}^{n} \ell_k(x) \frac{x_k - x_c}{x - x_c} F(x_k)$$

$$= \sum_{k \neq c, c \pm 1}^{n} \ell_k(x) f(x_k) \frac{x_k - x_c}{x - x_c} + \sum_{k=c-1, k \neq c}^{c+1} \ell_k(x) q_s(x_k) \frac{x_k - x_c}{x - x_c}$$

where ℓ_k is as in (4.3.48) with $n + 1$ instead of n. For this last polynomial the following theorem, which is complementary in some sense to Theorem 4.3.15, holds (see [294]).

Theorem 4.3.16 *Let f and u be as in Theorem 4.3.15 and $1 \leq p < +\infty$. Then there exists a positive constant $C \neq C(n, F)$ such that*

$$\|uL_n^*(w, F)\|_p \leq C\|uF\|_\infty \tag{4.3.57}$$

if and only if

$$\frac{u}{|\cdot - t_0|\sqrt{w\varphi}} \in L^p \quad \text{and} \quad \frac{|\cdot - t_0|\sqrt{w\varphi}}{u} \in L^1. \tag{4.3.58}$$

Moreover, for some positive constant $C \neq C(n, f)$ we have

$$\|uL_n^*(w, f)\|_\infty \leq C\|uf\|_\infty \log n \tag{4.3.59}$$

if and only if

$$\frac{u}{|\cdot - t_0|\sqrt{w\varphi}} \in L^\infty \quad \text{and} \quad \frac{|\cdot - t_0|\sqrt{w\varphi}}{u} \in L^1 \quad \text{or} \quad \begin{pmatrix} \gamma = \frac{\alpha}{2} + \frac{5}{4} \\ \delta = \frac{\beta}{2} + \frac{5}{4} \\ \theta = \frac{\eta}{2} + 1 \end{pmatrix}. \tag{4.3.60}$$

We omit the proof of this theorem since it is very similar to that of Theorem 4.3.15. Of course, (4.3.55) and (4.3.56) of Corollary 4.3.2 hold again, if we set $L_n^*(w)$ instead of $\widetilde{L}_n(w)$, and if (4.3.58) replaces (4.3.50), and (4.3.60) replaces (4.3.52).

It follows from (4.3.60) that

$$1 + \frac{\eta}{2} \leq \theta \leq 2 + \frac{\eta}{2},$$

4.3 Weighted Interpolation

i.e., $\theta > 1/2$ and Theorem 4.3.16 is not true for $\theta \leq 1/2$. Therefore, the interpolation processes $\{\widetilde{L}_n(w, F)\}$ and $\{L_n^*(w, F)\}$ are complementary and they can approximate every function in L_u^∞. However, $\widetilde{L}_n(w)$ and $L_n^*(w)$ use the zeros of the generalized Jacobi polynomial and their construction (except some special cases) requires a high computational cost, since until now only a few properties of these polynomials are known. To overcome this problem we propose a third procedure which uses the zeros of Jacobi polynomials and replaces $L_n^*(w)$ (not $\widetilde{L}_n(w)$!).

Indeed, following an idea from [214], let $v^{\alpha,\beta}$ be the Jacobi weight and let $\{p_n(v^{\alpha,\beta})\}$ be the corresponding sequence of orthonormal polynomials with positive leading coefficients. Given $v \in \mathbb{N}$, let $x_1 < x_2 < \cdots < x_{n+v}$ be the zeros of $p_{n+v}(v^{\alpha,\beta})$ and let us denote by x_c the zero of $p_{n+v}(v^{\alpha,\beta})$ which is closest to t_0, i.e., $|x_c - t_0| = \min_k |x_k - t_0|$. Moreover, let $y_i < \cdots < x_c < \cdots < y_v$ be v zeros of $p_{n+v}(v^{\alpha,\beta})$ of type $x_{c \pm (i-1)}$. We set $\pi(x) = \prod_{i=1}^{v}(x - y_i)$. Finally, let $L_n(v^{\alpha,\beta}, f)$ be the Lagrange polynomial interpolating $f \in L_u^\infty$ at the zeros of $p_{n+v}(v^{\alpha,\beta}; x)/\pi(x)$, i.e.,

$$L_n(v^{\alpha,\beta}, f; x) = \sum_{x_k \in \mathcal{B}} \frac{p_{n+v}(v^{\alpha,\beta}; x) f(x_k)}{\pi(x)} \frac{\pi(x_k)}{p'_{n+v}(v^{\alpha,\beta}; x_k)(x - x_k)},$$

where $\mathcal{B} = \{y_1, \ldots, y_v\}$.

Now, we are able to state the following theorem which is similar to the previous one (see [294]):

Theorem 4.3.17 *Let f and u be as in Theorem 4.3.15 and $1 \leq p < +\infty$. Then there exists a positive constant $C \neq C(n, F)$ such that*

$$\|uL_n(v^{\alpha,\beta}, f)\|_p \leq C \|uf\|_\infty \quad (4.3.61)$$

if and only if

$$\frac{u}{|\cdot - t_0|^v \sqrt{v^{\alpha,\beta}} \varphi} \in L^p \quad \text{and} \quad \frac{|\cdot - t_0|^v \sqrt{v^{\alpha,\beta}} \varphi}{u} \in L^1. \quad (4.3.62)$$

Moreover, for some positive constant $C \neq C(n, f)$ we have

$$\|uL_n(v^{\alpha,\beta}, f)\|_\infty \leq C \|uf\|_\infty \log n \quad (4.3.63)$$

if and only if

$$\frac{u}{|\cdot - t_0|^v \sqrt{v^{\alpha,\beta}} \varphi} \in L^\infty \quad \text{and} \quad \frac{|\cdot - t_0|^v \sqrt{v^{\alpha,\beta}} \varphi}{u} \in L^1 \quad \text{or} \quad \begin{pmatrix} \gamma = \frac{\alpha}{2} + \frac{5}{4} \\ \delta = \frac{\beta}{2} + \frac{5}{4} \\ v = \theta - 1 \end{pmatrix}. \quad (4.3.64)$$

Proof Let $d = \max(t_0 - y_1, y_r - t_0)$ and set

$$A = \left(-1 + \frac{a}{n^2}, t_0 - 2d\right) \cup \left(t_0 + 2d, 1 - \frac{a}{n^2}\right),$$

where $a > 0$ is fixed. Since the measure of $[t_0 - 2d, t_0 + 2d]$ is of order of n^{-1}, we use the Remez inequality to obtain

$$\|uL_n(v^{\alpha,\beta}, f)\|_p \leq C\|uL_n(v^{\alpha,\beta}, f)\|_{L^p(A)}, \quad 1 \leq p \leq +\infty.$$

Moreover, for $x \in A$, i.e., $|t_0 - x| > 2d$ and $n^{-1} \leq C|x - t_0|$, it follows that

$$\frac{|x - t_0|}{2} < |x - y_i| \leq C\frac{|x - t_0|}{2}$$

and $\pi(x) \sim |x - t_0|^\nu$, $x \in A$. Then, letting $q_n(x) = p_{n+\nu}(v^{\alpha,\beta}; x)/\pi(x)$,

$$g(x) = \mathrm{sgn}\left(L_n(v^{\alpha,\beta}, f; x)\right)|u(x)L_n(v^{\alpha,\beta}, f; x)|^{p-1}$$

and

$$r(t) = \int_A \frac{q_n(x) - q_n(t)}{x - t} u(x)g(x)dx \in \mathcal{P}_{n-1},$$

we can write

$$\|uL_n(v^{\alpha,\beta}, f)\|_{L^p(A)} \leq C \sum_{\substack{k=1 \\ x_k \notin B}}^n \frac{|f(x_k)\pi(x_k)|}{p'_{n+\nu}(v^{\alpha,\beta}; x_k)}|r(x_k)|.$$

Recalling the relation

$$\frac{1}{|p'_{n+\nu}(v^{\alpha,\beta}; x_k)|} \sim \sqrt{v^{\alpha,\beta}\varphi(x_k)}(x_{k+1} - x_k),$$

we get

$$\|uL_n(v^{\alpha,\beta}, f)\|_p \leq C\|uf\|_\infty \sum_{\substack{k=1 \\ x_k \notin B}}^n \frac{\sqrt{v^{\alpha,\beta}\varphi(x_k)}}{|x_k - t_0|^{\theta-\nu}}|r(x_k)|\Delta x_k, \quad 1 \leq p < +\infty.$$

Now, if we repeat the proof of Theorem 4.3.15 step by step and recall that

$$|q_n(x)| \leq \frac{C}{|x - t_0|^\nu \sqrt{v^{\alpha,\beta}\varphi(x)}}, \quad x \in A,$$

the equivalence of (4.3.61) and (4.3.62) follows easily.
  Finally, we consider the case $p = +\infty$. Let

$$\ell_k^*(x) = \frac{p_{n+\nu}(v^{\alpha,\beta}; x)}{\pi(x)} \frac{\pi(x_k)}{p'_{n+\nu}(v^{\alpha,\beta}; x_k)(x - x_k)}.$$

4.3 Weighted Interpolation

If we denote by x_d one of the zeros which is closest zeros to x, then we have

$$u(x)\frac{|\ell_d^*(x)|}{u(x_d)} \sim 1,$$

$$u(x)\frac{|\ell_k^*(x)|}{u(x_k)} \sim \frac{|u(x)p_{n+v}(v^{\alpha,\beta};x)|}{|x-t_0|^v} \frac{\Delta x_k}{v^{\gamma-\frac{\alpha}{2}-\frac{1}{4},\delta-\frac{\beta}{2}-\frac{1}{4}}(x_k)|t_0-x_k|^{\theta-v}|x-x_k|},$$

which implies

$$\max_{x\in A} \sum_{\substack{k=1 \\ x_k \notin \mathcal{B}}}^n u(x)\frac{|\ell_k^*(x)|}{u(x_k)} \sim 1 + \frac{|u(x)p_{n+v}(v^{\alpha,\beta};x)|}{|x-t_0|^v|x-\tau|^{\theta-v}v^{\sigma,\tau}(x)}$$

$$\times \sum_{\substack{k=1 \\ x_k \notin \mathcal{B}}}^n \frac{|x-\tau|^{\theta-v}v^{\sigma,\tau}(x)\Delta x_k}{|x_k-\tau|^{\theta-v}v^{\sigma,\tau}(x_k)|x-x_k|},$$

where

$$k \neq d, \quad \sigma = \gamma - \frac{\alpha}{2} - \frac{1}{4}, \quad \tau = \delta - \frac{\beta}{2} - \frac{1}{4}.$$

Moreover,

$$\frac{|u(x)p_{n+v}(v^{\alpha,\beta};x)|}{|x-t_0|^v|x-\tau|^{\theta-v}v^{\sigma,\tau}(x)} = |v^{\frac{\alpha}{2}+\frac{1}{4},\frac{\beta}{2}+\frac{1}{4}}(x)p_{n+v}(v^{\alpha,\beta};x)|$$

and

$$\max_{x\in A} |v^{\frac{\alpha}{2}+\frac{1}{4},\frac{\beta}{2}+\frac{1}{4}}(x)p_{n+v}(v^{\alpha,\beta};x)| \sim 1.$$

Then, using Lemma 4.3.1, we have

$$\sup_{\|uf\|_\infty=1} \|uL_n(v^{\alpha,\beta},f)\|_\infty \sim \max_{x\in A} \sum_{\substack{k=1 \\ x_k \notin \mathcal{B}}}^n u(x)\frac{|\ell_k^*(x)|}{u(x_k)} \sim \log n,$$

if and only if (4.3.64) holds. $\square$

It follows from (4.3.64) that $\theta - 1 \leq v \leq \theta$ and therefore, since $v \geq 1$, this implies $\theta \geq 1$. Theorem 4.3.16 can be replaced in numerical applications by the last theorem (but not by Theorem 4.3.15). Notice that (4.3.61) is equivalent to

$$\|u[f - L_n(v^{\alpha,\beta},f)]\|_p \leq CE_{n-1}(f)_{u,\infty}, \quad 1 \leq p < +\infty,$$

and (4.3.63) to

$$\|u[f - L_n(v^{\alpha,\beta},f)]\|_\infty \leq CE_{n-1}(f)_{u,\infty}\log n.$$

To simplify notations we have assumed that f has only one singular point (i.e., the weight u has only one internal zero). In the case of two or more points, for instance if $u(x) = v^{\gamma,\delta}(x)|x - t_0|^{\theta_0}|x - t_1|^{\theta_1}$, we use the zeros of the generalized Jacobi polynomials orthogonal with respect to the weight

$$w(x) = v^{\alpha,\beta}(x)|x - t_0|^{\eta_0}|x - t_1|^{\eta_1}$$

and construct a new function F by modifying the function f around the singularities t_0 and t_1. If we use the Jacobi zeros, then we consider the zeros of $p_{n+\nu_1+\nu_2}(v^{\alpha,\beta})$ and interpolate f at the zeros of

$$\frac{p_{n+\nu_1+\nu_2}(v^{\alpha,\beta}; x)}{\pi_{\nu_1}(x)\pi_{\nu_2}(x)},$$

where π_{ν_1} and π_{ν_2} are defined as before.

4.3.5.2 Interpolation Processes on Unbounded Intervals

Here, we consider functions $f \in L_v^\infty$, where

$$v(x) = \sqrt{w_{2\gamma}(x)}\,|x - t_0|^\eta, \quad w_{2\gamma}(x) = x^{2\gamma}e^{-x}, \quad t_0 > 0,\ \eta - \gamma > 0.$$

For such functions we are not able to establish the complete results obtained in the case of bounded intervals. In fact, very little is known about the orthogonal polynomials with respect to weights like $|x - t_0|^\lambda e^{-x}$ and, moreover, the behaviour of the weighted L_p-norm of the Lagrange polynomials based on the Laguerre zeros is settled.

Here we propose the following procedure:

Let w_α be the Laguerre weight, $w_\alpha(x) = x^\alpha e^{-x}$, $\alpha > -1$, $x > 0$. Let $\{p_n(w_\alpha)\}$ be the corresponding system of orthonormal polynomials with positive leading coefficients and let $x_1 < \cdots < x_{n+\nu}$, $\nu \geq 1$, be the zeros of $p_{n+\nu}(w_\alpha)$, where $x_c := x_{n+\nu,c}$ is one of the zeros which is closest to t_0. We denote by $y_1 < \cdots < x_c < \cdots < y_\nu$ the zeros of $p_{n+\nu}(w_\alpha)$ of the form $x_{c\pm(i-1)}$ and set $\pi(x) = \prod_{i=1}^\nu (x - y_i)$. Moreover, let $j := j(n)$ such that $x_j = \min\{x_k \geq 4\theta(n+\nu)\}$, $0 < \theta < 1$. Using the function ψ introduced above we define

$$\psi_j(x) := \psi\left(\frac{x - x_j}{x_{j+1} - x_j}\right) \quad \text{and} \quad f_j := (1 - \psi_j)f.$$

Finally, we denote by $L_{n+1}(w_\alpha, f_j)$ the Lagrange polynomial interpolating the function f_j at the zeros of the polynomial

$$\frac{4(n+\nu) - x}{\pi(x)} p_{n+\nu}(w_\alpha; x).$$

4.3 Weighted Interpolation

Since $f_j = f$ on $(0, x_j)$ and $f_j = 0$ on $[x_{j+1}, +\infty)$, we can write

$$L_{n+1}(w_\alpha, f_j; x) = \sum_{\substack{k=1 \\ x_k \notin \mathcal{B}}}^{j} \ell_k^*(x) f(x_k),$$

where $\mathcal{B} = \{y_1, \ldots, y_\nu\}$ and

$$\ell_k^*(x) = \frac{4(n+\nu) - x}{4(n+\nu) - x_k} \frac{p_{n+\nu}(w_\alpha; x)}{\pi(x)} \frac{\pi(x_k)}{p_{n+\nu}'(w_\alpha; x_k)(x - x_k)}.$$

Now, we state the following result [294]:

Theorem 4.3.18 Let $f \in L_v^\infty$, $v(x) = x^\gamma |x - t_0|^\eta e^{-x/2}$, with $\gamma > 0$ and $\eta \geq 1$. Then, with $M = [\frac{\theta}{1+\theta} n]$, $0 < \theta < 1$, we have

$$\|v[f - L_{n+1}(w_\alpha, f_j)]\|_\infty \leq C \left[E_M(f)_{v,\infty} \log n + e^{-An} \|vf\|_\infty \right]$$

if and only if

$$\frac{\alpha}{2} + \frac{1}{4} \leq \gamma \leq \frac{\alpha}{2} + \frac{5}{4} \quad \text{and} \quad \eta - 1 \leq \nu \leq \eta, \tag{4.3.65}$$

where C and A are positive constants independent of n and f.

Proof We first prove that

$$\sup_{\|vf_j\|_\infty = 1} \|v L_{n+1}(w_\alpha, f_j)\|_\infty \sim \log n \tag{4.3.66}$$

holds if and only if

$$\frac{\alpha}{2} + \frac{1}{4} \leq \gamma \leq \frac{\alpha}{2} + \frac{5}{4} \quad \text{and} \quad \eta - 1 \leq \nu \leq \eta.$$

To this end we set $d := \max(t_0 - y_1, y_\nu - t_0)$ and

$$A = \left(\frac{C}{n}, t_0 - 2d \right) \cup (t_0 + 2d, 4n).$$

Since the distance between the zeros in a neighborhood of t_0 is of order $1/\sqrt{n}$ (see [301]), we can use a Remez-type inequality [308] to obtain

$$\|v L_{n+1}(w_\alpha, f_j)\|_\infty \sim \|v L_{n+1}(w_\alpha, f_j)\|_{L^\infty(A)}$$

and $|\pi(x)| \sim |x - t_0|^\nu$, $x \in A$. Moreover (cf. [301]), by easy computations we get $v(x)|\ell_d^*(x)|/v(x_d) \sim 1$ and, for $j \geq k \neq d$,

$$v(x) \frac{|\ell_k^*(x)|}{v(x_k)} \sim \left| \sqrt{w_\alpha(x)} p_{n+\nu}(w_\alpha; x) \right| \sqrt[4]{x(4n - x)}$$

$$\times \left(\frac{x - t_0}{x_k - t_0} \right)^{\eta - \nu} \left(\frac{x}{x_k} \right)^{\gamma - \frac{\alpha}{2} - \frac{1}{4}} \frac{\Delta x_k}{|x - x_k|}.$$

Since
$$\max_{x \in A} |\sqrt{w_\alpha(x)} p_{n+\nu}(w_\alpha; x)| \sqrt[4]{x(4n-x)}| \sim 1,$$
we conclude that
$$\sup_{\|f_j v\|_\infty = 1} \|L_{n+1}(w_\alpha, f)v\|_\infty \sim \sup_{\|f_j v\|_\infty = 1} \|L_{n+1}(w_\alpha, f_j)v\|_{L^\infty(A)}$$

$$\sim \max_{x \in A} \sum_{\substack{k=1 \\ x_k \notin B}}^{j} v(x) \frac{|\ell_k^*(x)|}{v(x_k)}$$

$$\sim 1 + \sum_{\substack{k=1 \\ x_k \notin B}}^{j} \left|\frac{x-t_0}{x_k-t_0}\right|^{\eta-\nu} \left(\frac{x}{x_k}\right)^{\gamma-\frac{\alpha}{2}-\frac{1}{4}} \frac{\Delta x_k}{|x-x_k|}.$$

By Lemma 4.3.1, the last sum is equivalent to $\log n$ if and only if
$$\frac{\alpha}{2} + \frac{1}{4} \leq \gamma \leq \frac{\alpha}{2} + \frac{5}{4} \quad \text{and} \quad \eta - 1 \leq \nu \leq \eta.$$

Now, we have
$$\|v[f - L_{n+1}(w_\alpha, f_j)]\|_\infty \leq \|v[f - f_j]\|_\infty + \|v[f_j - L_n(w_\alpha, f_j)]\|_\infty.$$

Letting
$$M = \left[\frac{\theta}{1+\theta}(n+\nu)\right] \sim n,$$
we have (see [302])
$$\|v[f - f_j]\|_\infty \leq C\left(E_M(f)_{v,\infty} + e^{-An}\|vf\|_\infty\right).$$

Moreover, since for all polynomials $P \in \mathcal{P}_M$, $P = P_j + \psi_j P$, and
$$f_j - L_n(w_\alpha, f_j) = f_j - P - L_n(w_\alpha, f_j - P_j) + L_n(w_\alpha, \psi_j P)$$
$$= (f_j - f) + (f - P) - L_n(w_\alpha, (f-P)_j) + L_n(w_\alpha, \psi_j P),$$
we have
$$\|v[f_j - L_n(w_\alpha, f_j)]\|_\infty \leq \|v(f-P)\|_\infty + \|v(f-f_j)\|_\infty$$
$$+ \|vL_n(w_\alpha, (f-P)_j)\|_\infty + \|vL_n(w_\alpha, \psi_j P)\|_\infty.$$

Taking the infimum over $P \in \mathcal{P}_M$ and using (4.3.66), we see that the first three terms are dominated by
$$C\left(E_M(f)_{v,\infty} \log n + e^{-An}\|vf\|_\infty\right).$$

4.3 Weighted Interpolation

Now, it remains to estimate the last term. Thus,

$$|v(x)L_n(w_\alpha, \psi_j P; x)| = \left| \sum_{k>j} v(x) \frac{|\ell_k^*(x)|}{v(x_k)} P(x_k) v(x_k) \right|$$

$$\leq \|vP\|_{[4\theta n, 4n]} \sum_{\substack{k>j \\ x_k \notin B}} v(x) \frac{|\ell_k^*(x)|}{v(x_k)}.$$

Using Lemma 4.3.1 and recalling the conditions on $\alpha, \beta, \gamma, \delta, \nu$, and η, we see that the last sum is of order $\log n$. Finally, using an inequality proved in [308], we obtain

$$\|vP\|_{[4\theta n, \infty)} \leq Ce^{-An}\|vP\|_\infty \leq Ce^{-An}\|vf\|_\infty,$$

since P is the polynomial of best approximation of $f \in L_v^\infty$. □

Notice that Theorem 4.3.18 still holds true if $j = n$, but the "truncation" introduced by $L_{n+1}(w_\alpha, f_j)$ allows us to neglect the computation of $\mathcal{O}(n)$ terms of the sum and it can be useful in applications. Finally, we note that approximation of functions defined on the whole real axis and with some singular points can be obtained by using a similar argument, but we omit the details.

4.3.5.3 Numerical Examples

Now we consider now a few examples in order to illustrate the previous theoretical results, especially the ones given in Theorem 4.3.17 (for $p = +\infty$) and Theorem 4.3.18. All computations were performed in MATHEMATICA system, using standard machine precision known as double precision (m.p. $\approx 2.22 \times 10^{-16}$).

For the interval $[-1, 1]$ we take the weight u as in Theorem 4.3.17, i.e., $u(x) = v^{\gamma,\delta}(x)|x - t_0|^\theta$, where $v^{\gamma,\delta}(x) = (1-x)^\gamma (1+x)^\delta$ (Jacobi weight) and $\gamma, \delta \geq 0$, $\theta \geq 1$. The interpolation nodes are the zeros of the Jacobi polynomial $p_{n+\nu}(v^{\alpha,\beta}; x)$, excluding ν of them which are closest to the singular point $x = t_0$. We also give the corresponding *weighted Lebesgue function*,

$$\Lambda_n(u; x) = u(x) \sum_{k=1}^n \frac{|\ell_{n,k}(x)|}{u(x_k)}, \qquad (4.3.67)$$

where the interpolation nodes are denoted by x_k ($k = 1, \ldots, n$) and $\ell_{n,k}(x)$ are the fundamental Lagrange polynomials.

For the interval $[0, +\infty)$ we take the "space" weight

$$v(x) = x^\gamma e^{-x/2}|x - t_0|^\eta,$$

with $\gamma \geq 0$ and $\eta \geq 1$. The interpolation nodes are zeros of the generalized Laguerre polynomial $p_{n+\nu}(w_\alpha; x)$ ($w_\alpha(x) = x^\alpha e^{-x}$), excluding ν of them, which are the

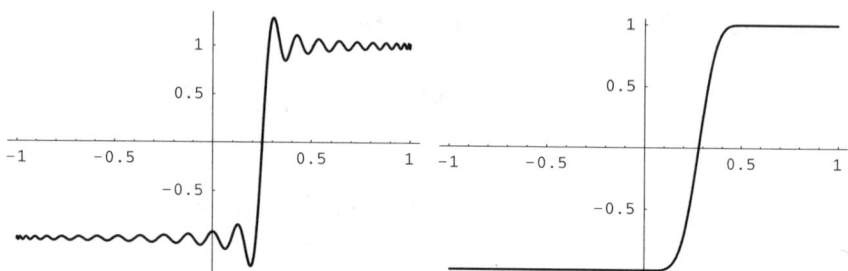

Fig. 4.3.1 Non-weighted (*left*) and weighted (*right*) Lagrange polynomial for the function $f(x) = \text{sgn}(x - 1/4)$ and $n = 50$ nodes

ones of the zeros that are closest to the singular point $x = t_0$, and adding the node $4(n+\nu)$. According to Theorem 4.3.18, a "truncation" of the Lagrange sum can be used, taking only j terms, where $j := j(n)$ is determined by $x_j = \min\{x_k \geq 4\theta(n+\nu)\}$ and $0 < \theta \leq 1$.

Example 4.3.3 We consider the simple function $f(x) = \text{sgn}(x - 1/4)$, which has a singularity at the point $x = t_0 = 1/4$. Because of this, a non-weighted Lagrange interpolation is bad. The case of such an interpolation at $n = 50$ Chebyshev nodes is displayed in Fig. 4.3.1 (left).

Since the function f is regular at ± 1, according to Theorem 4.3.17, we put $\gamma = \delta = 0$. As a weight function (Jacobi weight $v^{\alpha,\beta}$) we can take the Chebyshev weight of first kind,

$$w(x) = v^{-1/2,-1/2}(x) = \frac{1}{\sqrt{1-x^2}},$$

because $\alpha = \beta = -1/2$ satisfy the conditions

$$\frac{\alpha}{2} + \frac{1}{4} \leq \gamma \leq \frac{\alpha}{2} + \frac{5}{4} \quad \text{and} \quad \frac{\beta}{2} + \frac{1}{4} \leq \delta \leq \frac{\beta}{2} + \frac{5}{4}. \qquad (4.3.68)$$

Then, the interpolation nodes x_k ($k = 1, \ldots, n$) will be the zeros of $T_{n+\nu}(x)$, excluding ν ($0 - 1 \leq \nu \leq \theta$) of them, which are closest to the point $t_0 = 1/4$.

Taking $\theta = 8$ we extract $\nu = 7$ or $\nu = 8$ zeros of $T_{n+7}(x)$ or $T_{n+8}(x)$, respectively. The weighted Lebesgue functions in these cases are given in Fig. 4.3.2. We take $\nu = 7$ in our calculation, because this case gives slightly better results than the second one. The corresponding weighted Lagrange polynomial in this case for $n = 50$ is displayed in Fig. 4.3.1 (right). The cases for $n = 10$, $n = 20$, and $n = 100$ are shown in Fig. 4.3.3 (left).

The uniform norm of the weighted error, $\|u[f - L_n(v^{\alpha,\beta}, f)]\|_\infty$, for $n \leq 100$ is presented in Fig. 4.3.3 (right) as a linear-log plot.

4.3 Weighted Interpolation 311

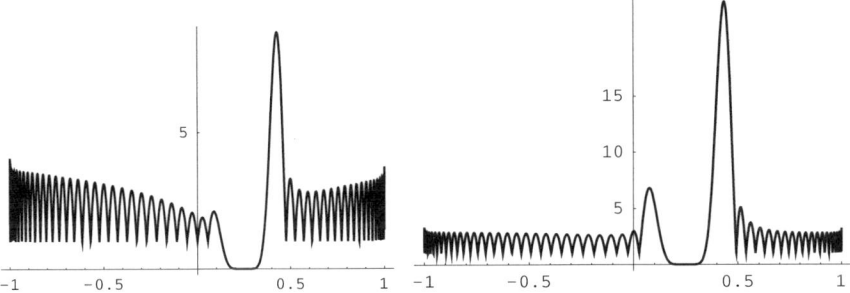

Fig. 4.3.2 The weighted Lebesgue function for $n = 50$ and $\nu = 7$ (*left*) and $\nu = 8$ (*right*)

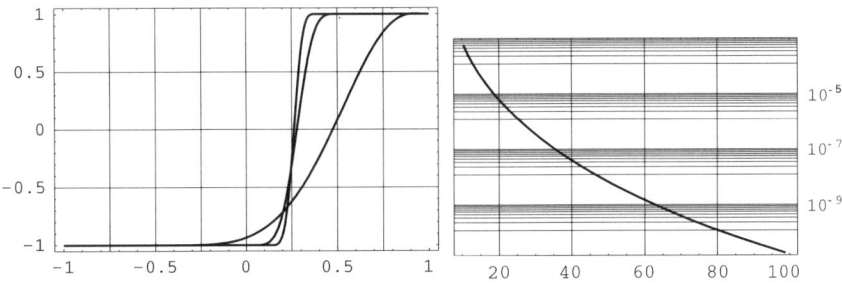

Fig. 4.3.3 The weighted Lagrange polynomial for $n = 10, 50$, and 100 nodes (*left*) and the uniform norm of the weighted error for $n \le 100$ (*right*)

Example 4.3.4 Consider the function f defined by

$$f(x) = \begin{cases} -\dfrac{e^{-x}}{\sqrt{1+x}}, & \text{for } x < 0, \\ \log \dfrac{1-x}{1+x}, & \text{for } x > 0. \end{cases}$$

Besides the end-point singularities, a singularity at $x = t_0 = 0$ exists with the jump equal to $\lim_{x \to 0+} f(x) - \lim_{x \to 0-} f(x) = 1$ (see Fig. 4.3.4 (left)).

We put $u(x) = (1-x)^{3/2}(1+x)^{5/2}|x|^{7/2}$ and $\alpha = 3/2$, $\beta = 7/2$, so that the conditions in (4.3.68) are satisfied. Taking $\theta = 7/2$, we must extract $\nu = 3$ nodes from the set of all zeros of the Jacobi polynomial $p_{n+\nu}(v^{3/2,7/2}; x)$. The corresponding weighted Lebesgue function (4.3.67) for $n = 50$ is presented in Fig. 4.3.4 (right).

The uniform norm of the weighted error for $n \le 100$ is displayed in Fig. 4.3.5 (right). The weighted Lagrange polynomial $L_{100}(v^{3/2,7/2}, f; x)$ is given on the left side of the same figure. Notice that for a small n, e.g., $n = 10$, this polynomial is a bad approximation to f (see Fig. 4.3.6 (left)). On the other hand, we can see that $u(x)L_{10}(v^{3/2,7/2}, f; x)$ is very close to $u(x)f(x)$, i.e., $\|u[f - L_{10}(v^{3/2,7/2}, f)]\|_\infty \approx 4.5 \times 10^{-3}$.

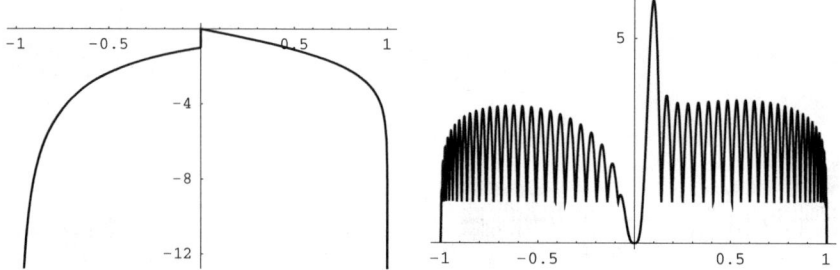

Fig. 4.3.4 The graphics $x \mapsto f(x)$ (*left*) and the weighted Lebesgue function $x \mapsto \Lambda_n(u; x)$ (*right*) for $n = 50$ nodes

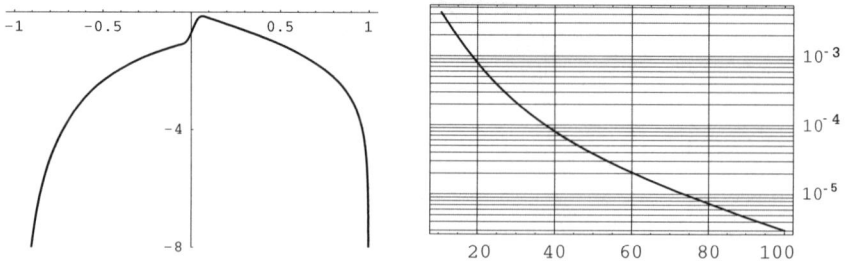

Fig. 4.3.5 The weighted Lagrange polynomial for $n = 100$ nodes (*left*) and the uniform norm of the weighted error for $n \leq 100$ (*right*)

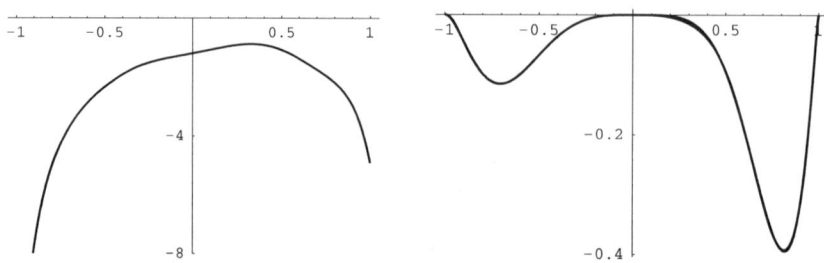

Fig. 4.3.6 The weighted Lagrange polynomial $x \mapsto L_{10}(v^{3/2,7/2}, f; x)$ (*left*) and the corresponding functions $x \mapsto u(x) L_{10}(v^{3/2,7/2}, f; x)$ and $x \mapsto u(x) f(x)$ (*right*)

Example 4.3.5 Let

$$f(x) = \frac{1}{\sqrt{|\sin(x - 1/2)|}} \log \frac{1}{1 - x^2}.$$

As we can see

$$\lim_{x \to \pm 1} f(x) = +\infty \quad \text{and} \quad \lim_{x \to 1/2} f(x) = +\infty.$$

4.3 Weighted Interpolation

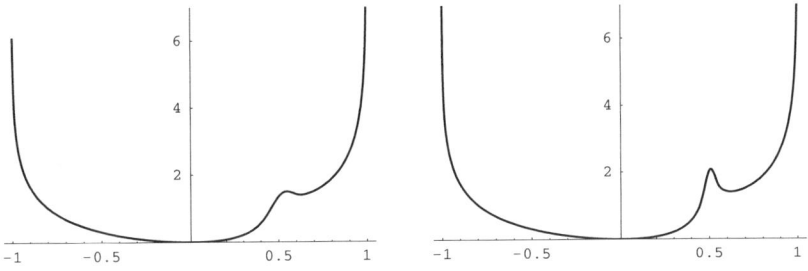

Fig. 4.3.7 The weighted Lagrange polynomial for $n = 50$ (*left*) and $n = 100$ nodes (*right*)

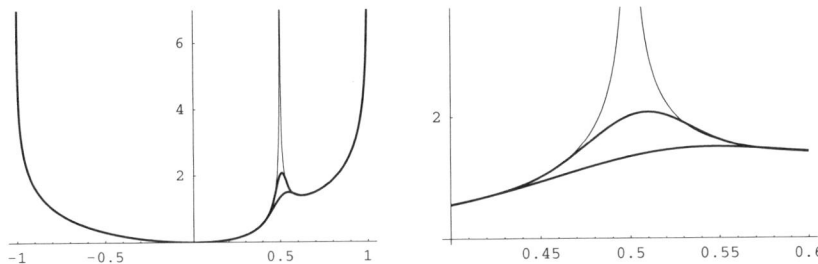

Fig. 4.3.8 The graphics $x \mapsto L_n(v^{3/2,3/2}, f; x)$ ($n = 50, 100$) and $x \mapsto f(x)$ in $(-1, 1)$ (*left*) and $(0.4, 0.6)$ (*right*)

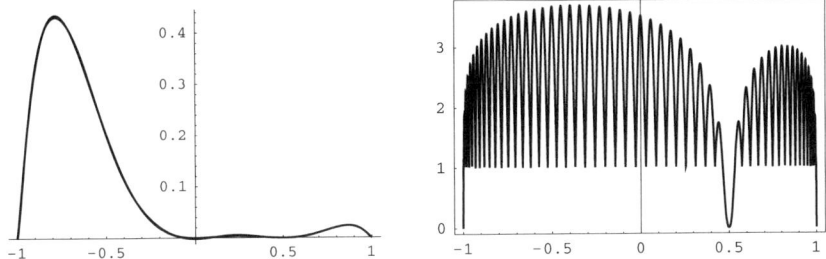

Fig. 4.3.9 Graphics of $x \mapsto u(x)L_{10}(v^{3/2,3/2}, f; x)$ and $x \mapsto u(x)f(x)$ (*left*) and the Lebesgue function $x \mapsto \Lambda_n(u; x)$ for $n = 50$ nodes (*right*)

We take $\gamma = \delta = 3/2$ and $\theta = 5/2$, i.e., $u(x) = (1 - x^2)^{3/2}|x - 1/2|^{5/2}$, and $\alpha = \beta = 3/2$. Notice that $\nu = 2$ in this case. The weighted Lagrange polynomials for $n = 50$ and $n = 100$ are displayed in Fig. 4.3.7. Figure 4.3.8 shows these polynomials and the original function $x \mapsto f(x)$ in the interval $(-1, 1)$ (left) and locally for $x \in (0.4, 0.6)$ (right).

In Fig. 4.3.9 we present the graphics of $x \mapsto u(x)L_{10}(v^{3/2,3/2}, f; x)$ and $x \mapsto u(x)f(x)$ (left), as well as the corresponding Lebesgue function for $n = 50$ nodes

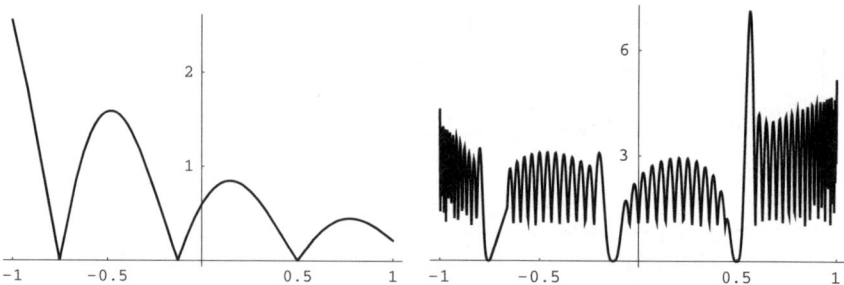

Fig. 4.3.10 The graphics $x \mapsto f(x)$ (*left*) and the weighted Lebesgue function $x \mapsto \Lambda_n(u; x)$ (*right*) for $n = 60$

Table 4.3.1 The uniform norm of errors in the Lagrange interpolation

Number of nodes n	Standard interpolation $\|\bar{e}_n\|_\infty$	$\|u\bar{e}_n\|_\infty$	Weighted interpolation $\|ue_n\|_\infty$
10	2.60(−1)	8.18(−2)	2.92(−5)
20	2.00(−1)	8.31(−3)	5.51(−6)
30	8.14(−2)	2.16(−3)	6.02(−7)
40	1.47(−1)	5.26(−3)	3.08(−7)
50	9.72(−2)	2.45(−3)	1.07(−7)
60	6.07(−2)	2.54(−3)	3.47(−8)
70	5.96(−2)	1.12(−3)	2.77(−8)
80	4.67(−2)	1.19(−4)	1.50(−8)
90	3.89(−2)	1.70(−4)	9.10(−9)
100	3.94(−2)	5.40(−4)	6.46(−9)

(right). We also mention that the uniform norm of the weighted error

$$\|u[f - L_n(v^{3/2,3/2}, f)]\|_\infty$$

is equal to 7.49×10^{-3}, 6.02×10^{-5}, and 7.73×10^{-6}, for $n = 10, 50$, and 100, respectively.

Example 4.3.6 Let $f(x) = e^{-x}|\sin 5(x - 1/2)|$. This function is continuous for $x \in [-1, 1]$, but there are three "critical points" in $(-1, 1)$:

$$t_0 = \frac{1}{2} - \frac{2\pi}{5}, \quad t_1 = \frac{1}{2} - \frac{\pi}{2}, \quad t_2 = \frac{1}{2},$$

in which the function f is not differentiable (see Fig. 4.3.10).

A direct application of the Lagrange interpolation with the Chebyshev nodes gives the results in Table 4.3.1. In the second column of this table we give the

4.3 Weighted Interpolation

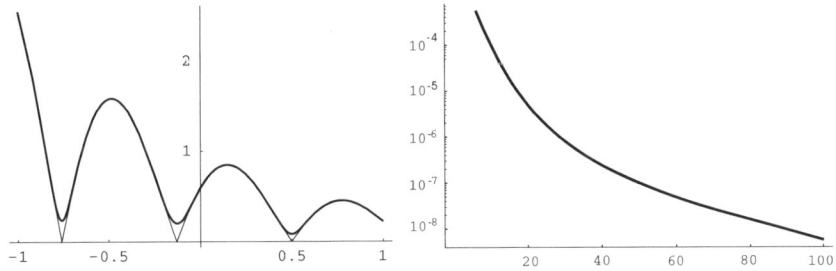

Fig. 4.3.11 The Lagrange polynomial $x \mapsto L_{60}(v^{-1/2,-1/2}; x)$ (*left*) and the uniform norm of the weighted error $x \mapsto u(x)[f(x) - L_n(v^{-1/2,-1/2}; x)]$ (*right*) for $n \leq 100$

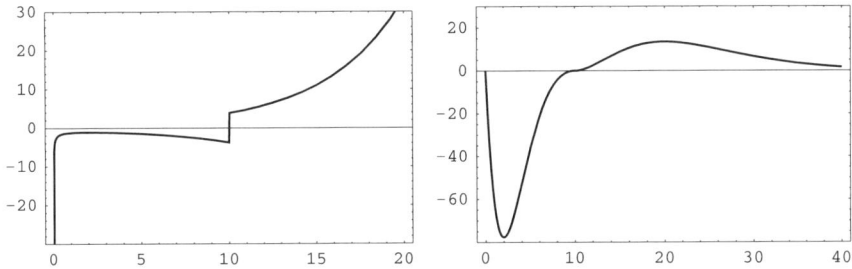

Fig. 4.3.12 The graphics $x \mapsto f(x)$ (*left*) and $x \mapsto v(x) f(x)$ (*right*)

uniform norm of the corresponding errors $\overline{e}_n(x) = f(x) - \overline{L}_n(v^{-1/2,-1/2}; x)$ for $n = 10(10)100$. Numbers in parentheses indicate decimal exponents.

According to Theorem 4.3.17 and the corresponding comments regarding this theorem, we put $\gamma = \delta = 0$, $\theta_0 = \theta_1 = \theta_2 = 7/2$, so that

$$u(x) = |x - t_0|^{7/2} |x - t_1|^{7/2} |x - t_2|^{7/2}.$$

This allows us to take the Chebyshev nodes ($\alpha = \beta = -1/2$) as the zeros of $T_{n+9}(x)$ and to extract nine points (three of them in the neighborhood of each point t_k, $k = 0, 1, 2$). The weighted Lebesgue function for such a distribution of nodes is displayed in Fig. 4.3.10 (right).

The uniform norm of the corresponding weighted error

$$u(x) e_n(x) = u(x)[f(x) - L_n(v^{-1/2,-1/2}; x)]$$

is presented in the last column in Table 4.3.1. In order to compare errors in non-weighted and weighted interpolation, we also introduce an additional column in this table, with the uniform norm of the previous error of standard interpolation $\overline{e}_n(x)$ multiplied by $u(x)$. As we can see, the advantage of weighted interpolation is evident.

In Fig. 4.3.11 we gave the graphics of the Lagrange interpolation polynomial $L_n(v^{-1/2,-1/2}; x)$ for $n = 60$ and the uniform norm $\|u e_n\|_\infty$ for $n \leq 100$ (see also Table 4.3.1).

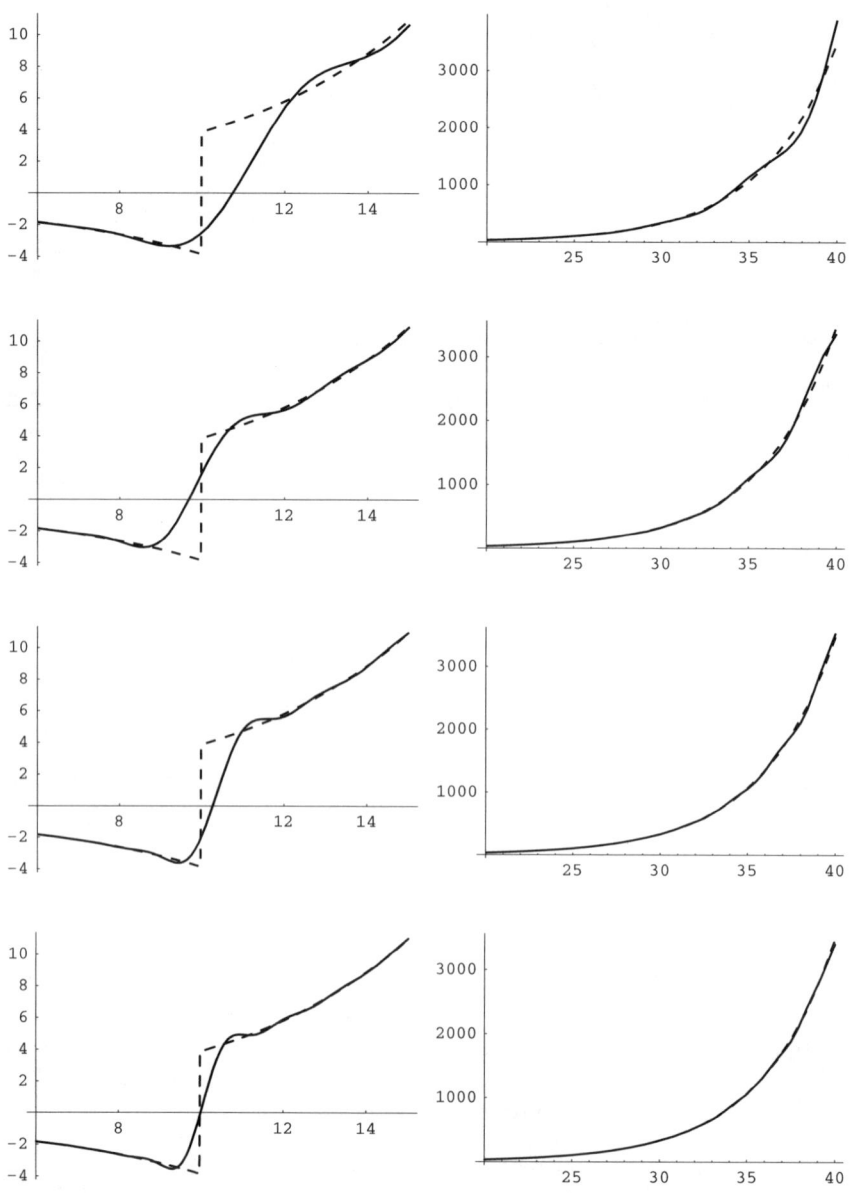

Fig. 4.3.13 The graphics $x \mapsto L_{n+1}(w_{5/2}; x)$ (*solid line*) and $x \mapsto f(x)$ (*broken line*) on $[6, 15]$ (*left*) and $[20, 40]$ (*right*) for $n = 50$, $n = 100$, $n = 200$, and $n = 300$

Example 4.3.7 Consider the function f defined on $(0, +\infty)$ by

$$f(x) = \frac{e^{x/4}}{\sqrt{x}} \operatorname{sgn}(x - 10).$$

4.3 Weighted Interpolation

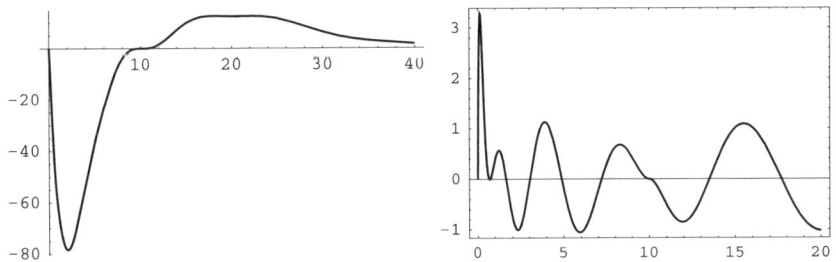

Fig. 4.3.14 The function $x \mapsto v(x)L_{n+1}(w_{5/2}; x)$ (*left*) and the weighted error $x \mapsto v(x)[f(x) - L_{n+1}(w_{5/2}; x])$ (*right*) for $n = 10$

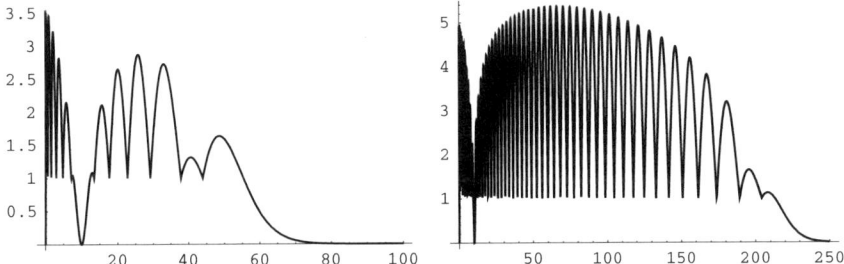

Fig. 4.3.15 The weighted Lebesgue function $x \mapsto \Lambda_{n+1}(x)$ for $n = 10$ (*left*) and $n = 50$ (*right*)

According to Theorem 4.3.18 we put $\gamma = 3/2$ and $\eta = 2$, i.e.,

$$v(x) = x^{3/2} e^{-x/2} |x - 10|^2.$$

The graphics of $x \mapsto f(x)$ and $x \mapsto v(x)f(x)$ are displayed in Fig. 4.3.12.

For the parameters α and ν which satisfy inequalities (4.3.65) we can take $\alpha = 5/2$ and $\nu = 1$. In this way, the weight w_α becomes the generalized Laguerre weight

$$w_{5/2}(x) = x^{5/2} e^{-x}, \qquad 0 \leq x < +\infty.$$

In Fig. 4.3.14 we present the graphic of the Lagrange polynomial $L_{n+1}(w_{5/2}; x)$ multiplied by the "space" weight $v(x)$ for $n = 10$, as well as the graphic of the corresponding weighted error.

Especially, it is interesting to consider the behaviour of the Lagrange polynomial $L_{n+1}(w_{5/2}; x)$ in some neighbourhood of the singular point $x = 10$. Figure 4.3.13 shows graphics of the Lagrange polynomial $x \mapsto L_{n+1}(w_{5/2}; x)$ and the function $x \mapsto f(x)$ for $x \in [6, 15]$, when $n = 50, 100, 200,$ and 300. The behaviour of the interpolation polynomial in the interval $[20, 40]$ is also presented.

The graphics of the weighted Lebesgue function $x \mapsto \Lambda_{n+1}(x)$ in this case for $n = 10$ and $n = 50$ are displayed in Fig. 4.3.15.

With a "truncation" of the Lagrange polynomial, i.e., by taking only j terms, determined by $x_j = \min\{x_k \geq 4\theta(n + \nu)\}$ and $0 < \theta \leq 1$, the computations can be

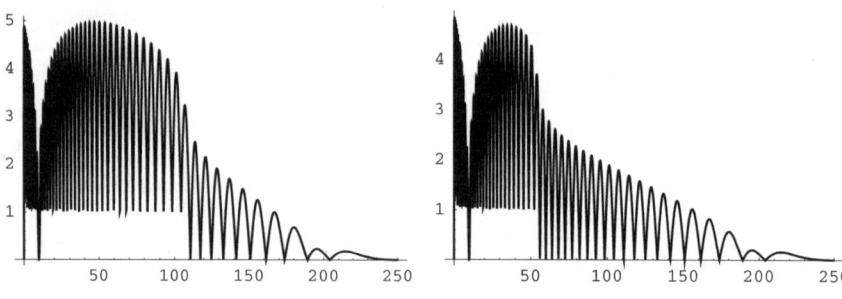

Fig. 4.3.16 The weighted Lebesgue function $x \mapsto \Lambda_{n+1}^{(\theta)}(x)$ for $n = 50$ with dropped nodes: $\theta = 1/2$ (*left*) and $\theta = 1/4$ (*right*)

significantly reduced. The corresponding weighted Lebesgue function is denoted by $\Lambda_{n+1}^{(\theta)}(x)$. The cases for $n = 50$ with dropped nodes when $\theta = 1/2$ and $\theta = 1/4$ are presented in Fig. 4.3.16.

As we can see, the corresponding weighted Lebesgue constants,

$$\Lambda_{n+1}^{(\theta)} = \max_{0 \leq x < +\infty} |\Lambda_{n+1}^{(\theta)}(x)|,$$

for $n = 50$ are almost the same when $\theta = 1$, $\theta = 1/2$, and $\theta = 1/4$. In other words, such a "truncation" in the weighted Lagrange polynomial does not change its numerical characteristics, but significantly reduces the computations.

Chapter 5
Applications

5.1 Quadrature Formulae

5.1.1 Introduction

As in Sect. 2.2.1, let $d\mu$ be a finite positive Borel measure on the real line such that its support is an infinite set, and all its moments $\mu_k = \int_{\mathbb{R}} x^k \, d\mu(x)$, $k = 0, 1, \ldots$, exist and are finite. Also, for real-valued functions $f, g \in L^2(d\mu)$ we define an inner product by (2.2.1), i.e.,

$$(f, g) = \int_{\mathbb{R}} f(x)g(x) \, d\mu(x). \qquad (5.1.1)$$

Now, we consider an *n-point quadrature formula*

$$\int_{\mathbb{R}} f(x) \, d\mu(x) = \sum_{k=1}^{n} A_k f(x_k) + R_n(f), \qquad (5.1.2)$$

where the sum

$$Q_n(f) = \sum_{k=1}^{n} A_k f(x_k) \qquad (5.1.3)$$

provides an approximation to the integral $I(f) = \int_{\mathbb{R}} f(x) \, d\mu(x)$ and $R_n(f)$ is the corresponding error. In the *quadrature sum* $Q_n(f)$, the points x_k are called the *nodes* and A_k are the *weights* of the quadrature formula (5.1.2). Here, we suppose the nodes are mutually distinct.

As before (cf. Sect. 2.2.1) we consider very often the case when $d\mu(x) = w(x) \, dx$, where the weight function $w(x)$ is non-negative and measurable in Lebesgue's sense for which all moments exists and $\mu_0 > 0$. In the case when $\mathrm{supp}(w) = [a, b]$, where $-\infty < a < b < +\infty$, we will always consider the standard interval $[-1, 1]$ and then we suppose that all nodes x_k belong to $[-1, 1]$.

In general, in order to approximate the integral $I(f) = \int_{-1}^{1} f(x)w(x) \, dx$ for continuous functions on $[-1, 1]$, we can consider Q_n as a linear functional from $C^0 = C[-1, 1]$ to $\mathbb{R}$. The norm of $Q_n: C^0 \to \mathbb{R}$,

$$\|Q_n\| = \|Q_n\|_{C^0 \to \mathbb{R}} = \sup_{\|f\|_{\infty}=1} |Q_n(f)| = \sum_{k=1}^{n} |A_k|, \qquad (5.1.4)$$

plays an important role in the convergence of $Q_n(f)$ to $I(f)$ for all continuous functions, as well as in the computation of $Q_n(f)$.

Definition 5.1.1 We say that the quadrature sum $Q_n(f)$ is *convergent* if and only if

$$\lim_{n\to+\infty} R_n(f) = \lim_{n\to+\infty} [I(f) - Q_n(f)] = 0$$

and $Q_n(f)$ is *stable* if and only if $\sup_n \|Q_n\| < +\infty$ (is *unstable* if and only if $\sup_n \|Q_n\| = +\infty$).

It is important to mention the following:

Remark 5.1.1 If $\sup_n \|Q_n\| = +\infty$, then, according to the Banach-Steinhaus theorem, there exists a function $f_0 \in C[-1, 1]$ such that $\lim_{n\to+\infty} R_n(f_0) = \pm\infty$, i.e., a unstable formula cannot be convergent for all continuous functions. However, we have the convergence of $Q_n(f)$ *if and only if* $Q_n(f)$ is stable and is convergent on a subset dense in C^0.

Remark 5.1.2 Let $\eta = \eta(x)$ be a perturbation function of f. Then we have

$$|Q_n(f+\eta) - Q_n(f)| \leq \|Q_n\| \|\eta\|.$$

Definition 5.1.2 The n-point quadrature formula (5.1.2) has degree of exactness d if for every $p \in \mathcal{P}_d$ we have $R_n(p) = 0$. In addition, if $R_n(p) \neq 0$ for some $p \in \mathcal{P}_{d+1}$, the formula (5.1.2) has precise degree of exactness d.

The convergence order of $Q_n(f)$ depends on the smoothness of the function f, as well as on its degree of exactness.

It is well known that for given n mutually different nodes x_k, $k = 1, \ldots, n$, we can always achieve a degree of exactness $d = n - 1$ by interpolating at these nodes and integrating the interpolation polynomial instead of f. Indeed, taking the *node polynomial*

$$q_n(x) = \prod_{k=1}^{n}(x - x_k), \qquad (5.1.5)$$

by integrating the Lagrange interpolation formula

$$f(x) = \sum_{k=1}^{n} \ell_k(x) f(x_k) + r_n(f; x),$$

where

$$\ell_k(x) = \frac{q_n(x)}{q_n'(x_k)(x - x_k)}, \quad k = 1, \ldots, n, \qquad (5.1.6)$$

we obtain (5.1.2), with

$$A_k = \frac{1}{q_n'(x_k)} \int_{\mathbb{R}} \frac{q_n(x)}{x - x_k} d\mu(x), \quad k = 1, \ldots, n, \qquad (5.1.7)$$

5.1 Quadrature Formulae

and

$$R_n(f) = \int_{\mathbb{R}} r_n(f;x)\,d\mu(x). \qquad (5.1.8)$$

Notice that for each $f \in \mathcal{P}_{n-1}$, we have $r_n(f;x) = 0$, and therefore $R_n(f) = 0$. Quadrature formulae obtained in this way are known as *interpolatory*. Usually, the interpolatory quadrature

$$\int_{-1}^{1} f(x)w(x)\,dx = \sum_{k=1}^{n} A_k f(x_k) + R_n(f), \qquad (5.1.9)$$

with given nodes $x_k \in [-1,1]$, is called the *weighted Newton-Cotes formula*. The classical Newton-Cotes formula is for $w(x) = 1$ and the equidistant nodes $x_k = -1 + 2(k-1)/(n-1)$, $k = 1,\ldots,n$.

According to (5.1.4) and Remark 5.1.1, for interpolatory quadratures the following result holds:

Theorem 5.1.1 *Any interpolatory quadrature* (5.1.9), *with* $A_k \geq 0$, $k = 1,\ldots,n$, *is convergent for all continuous functions.*

Proof First we conclude that the quadrature sum $Q_n(f) = \sum_{k=1}^{n} A_k f(x_k)$ is stable, because

$$\|Q_n\| = \sum_{k=1}^{n} |A_k| = \sum_{k=1}^{n} A_k = \int_{-1}^{1} w(x)\,dx = \mu_0 < +\infty.$$

Since $R_n(f) = 0$ for each $f \in \mathcal{P}_{n-1}$, denoting by P_{n-1} the polynomial of the best uniform approximation in $\mathcal{P}_{n-1}$, we have for each $f \in C^0$

$$|R_n(f)| = |R_n(f - P_{n-1})|$$

$$\leq \int_{-1}^{1} |f - P_{n-1}|(x)w(x)\,dx + \sum_{k=1}^{n} A_k |f - P_{n-1}|(x_k)$$

$$\leq 2E_{n-1}(f)_{\infty}\mu_0. \qquad \square$$

In the case of the classical Newton-Cotes quadratures the previous theorem cannot be applied. Namely, some of weight coefficients $A_k = A_k^{(n)}$ for $n \geq 8$ are negative. We mention here that this sequence $\{Q_n(f)\}_{n \in \mathbb{N}}$ indeed does not converge for each $f \in C^0$ (cf. Brass [49]). An account of the role played by moments and modified moments in the construction of interpolatory quadrature rules, especially weighted Newton-Cotes and Gaussian rules, is given by Gautschi [164].

One of the important uses of orthogonal polynomials is in the construction of quadrature formulas of the maximal, or nearly maximal, algebraic degree of exactness for integrals involving a positive measure $d\mu$. The following theorem is due to Jacobi [222] (see Gautschi [162, p. 48]):

Theorem 5.1.2 *Given a positive integer m ($\leq n$), the quadrature formula (5.1.2) has degree of exactness $d = n - 1 + m$ if and only if the following conditions are satisfied:*

1° *Formula (5.1.2) is interpolatory;*
2° *The node polynomial (5.1.5) satisfies*

$$(\forall p \in \mathcal{P}_{m-1}) \quad (p, q_n) = \int_{\mathbb{R}} p(x) q_n(x) \, d\mu(x) = 0.$$

Proof The necessity of the conditions 1° and 2° is trivial. In order to prove the sufficiency of these conditions we take an arbitrary $u \in \mathcal{P}_{n-1+m}$ and represent it in the form $u(x) = p(x) q_n(x) + r(x)$, where $p \in \mathcal{P}_{m-1}$ and $r \in \mathcal{P}_{n-1}$.

Since

$$\int_{\mathbb{R}} u(x) \, d\mu(x) = \int_{\mathbb{R}} p(x) q_n(x) \, d\mu(x) + \int_{\mathbb{R}} r(x) \, d\mu(x),$$

by the orthogonality condition 2° (with respect to the inner product (5.1.1)), the first integral on the right vanishes, and the second one, according to 1°, can be expressed exactly by the quadrature sum $Q_n(r)$, so that we have

$$\int_{\mathbb{R}} u(x) \, d\mu(x) = \sum_{k=1}^{n} A_k r(x_k) = \sum_{k=1}^{n} A_k u(x_k),$$

since $r(x_k) = u(x_k)$ for each $k = 1, \ldots, n$. Thus, $R_n(u) = 0$. □

5.1.2 Some Remarks on Newton-Cotes Rules with Jacobi Weights

In this subsection we consider the weighted Newton-Cotes rules with the Jacobi weight $w(x) = v^{\gamma,\delta}(x) = (1-x)^{\gamma}(1+x)^{\delta}$ on $[-1, 1]$, $\gamma > -1, \delta > -1$,

$$\int_{-1}^{1} f(x) v^{\gamma,\delta}(x) \, dx = \sum_{k=1}^{n} A_k f(x_k) + R_n(f), \quad (5.1.10)$$

where the nodes x_k are zeros of Jacobi polynomials (Jacobi abscissas) belonging to parameters other than γ, δ.

If the nodes x_k are zeros of $P_n^{(\alpha,\beta)}(x)$, then the weight coefficients in (5.1.10) are given by

$$A_k = \frac{1}{\frac{d}{dx} P_n^{(\alpha,\beta)}(x_k)} \int_{-1}^{1} \frac{P_n^{(\alpha,\beta)}(x)}{x - x_k} v^{\gamma,\delta}(x) \, dx, \quad k = 1, \ldots, n.$$

As we mentioned before (see Theorem 5.1.1) an important property of interpolatory quadratures is positivity of their weight coefficients (*positive quadratures*).

5.1 Quadrature Formulae

The earliest examples are the positive quadrature rules of Fejér (see Sect. 2.4.8) for $w(x) = v^{0,0}(x) = 1$ having as abscissas the Chebyshev points of the first and second kind. The case with the Chebyshev abscissas of the first kind ($\alpha = \beta = -1/2$) was given by (2.4.41) and (2.4.42). Subsequent work for the same weight dealt with ultraspherical and more general Jacobi abscissas, either for all n ([16–18]), or for selected fixed n ([450, 451]). Also, ultraspherical abscissas were considered in combination with the Chebyshev weight of the first kind $v^{-1/2,-1/2}$ and with ultraspherical weight functions in [243, 321], respectively.

Askey's conjecture on positivity of all Cotes numbers in (5.1.10), with $w(x) = v^{0,0}(x) = 1$ and Jacobi abscissas (see [17]) was recently revised by Gautschi [164]:

Conjecture 5.1.1 All Cotes coefficients A_k in (5.1.10), with $w(x) = v^{0,0}(x) = 1$ and Jacobi abscissas (zeros of $P_n^{(\alpha,\beta)}$), are positive only in the region

$$\{\alpha \leq \beta < \alpha + 2, \ -1 < \alpha < -1/2\} \cup \{-1/2 \leq \alpha \leq \beta \leq 3/2\},$$

as well as, by symmetry, in the companion region reflected along $\alpha = \beta$.

The quadrature formula (5.1.10) can be consider also taking as abscissas the zeros of two (related) Jacobi polynomials.

Conjecture 5.1.2 For any $\alpha > -1$, $\beta > -1$, let x_k be zeros of the polynomial $P_n^{(\alpha,\beta)}(x) P_{n-1}^{(\alpha+1,\beta+1)}(x)$. Then, for each $n \in \mathbb{N}$, the $(2n-1)$-point interpolatory quadrature

$$\int_{-1}^{1} f(x)(1-x)^{\alpha+1/2}(1+x)^{\beta+1/2} dx = \sum_{k=1}^{2n-1} A_k f(x_k) + R_{2n-1}(f)$$

has all nonnegative coefficients A_k.

Remark 5.1.3 Conjecture 5.1.2 was stated by Milovanović during the *Sixth Conference on Applied Mathematics* held on the Serbian mountain Tara in 1988. The conjecture was checked numerically by Marinković in her Master's thesis [283] and by Gautschi [164] for many parameters (α, β). For example, Gautschi investigated the cases $\alpha = -0.75(0.25)4.00$, $\beta = \alpha(0.25)4.00$, and $n = 5(5)40$, as well as $\alpha = -0.9(0.1)1.0$, $\beta = \alpha(0.1)1.0$, $n = 1(1)40$, and always confirmed the conjecture in these cases as well.

The Fejér $(2n-1)$-point formula with Chebyshev points of the second kind is evidently the special case $\alpha = \beta = -1/2$ of the previous conjecture, since $U_{2n-1}(x) = 2T_n(x)U_{n-1}(x)$. Namely, the nodes in this case are the zeros of $U_{2n-1}(x)$, i.e., $P_n^{(-1/2,-1/2)}(x) P_{n-1}^{(1/2,1/2)}(x)$, and the weight is $w(x) = 1$.

5.1.3 Gauss-Christoffel Quadrature Rules

According to Theorem 5.1.2, an n-point quadrature formula (5.1.2) with respect to the positive measure $d\mu(t)$ has the maximal algebraic degree of exactness $2n-1$, i.e., $m=n$ is optimal. The higher m ($>n$) is impossible. Indeed, according to 2°, the case $m=n+1$ requires the orthogonality $(p,q_n)=0$ for all $p \in \mathcal{P}_n$, which is impossible when $p=q_n$.

The quadrature formula (5.1.2) with the maximal algebraic degree of exactness $2n-1$ is called the *Gaussian quadrature formula* with respect to the measure $d\mu$ or sometimes the *Gauss-Christoffel quadrature formula*. This famous method of numerical integration was discovered in 1814 by Gauss [142], using his theory of continued fractions associated with hypergeometric series. It is interesting to mention that for $d\mu(x) = dx$, Gauss determined quadrature parameters, the nodes x_k and the weights A_k, $k=1,\ldots,n$, for all $n \leq 7$. An elegant alternative derivation of this method was provided by Jacobi, and a significant generalization to arbitrary measures was given by Christoffel. The error term and convergence were proved by Markov and Stieltjes, respectively. A nice survey of Gauss-Christoffel quadrature formulae was written by Gautschi [146].

Thus, in the case $m=n$, the orthogonality condition 2° from Theorem 5.1.2 evidently shows that the node polynomial q_n must be (monic) orthogonal polynomial with respect to the measure $d\mu$. Thus, the nodes x_k must be zeros of the polynomial $q_n(x) = \pi_n(d\mu; x)$. The corresponding weights A_k (Christoffel numbers) can be obtained from (5.1.7) and expressed in terms of orthogonal polynomials, which are completely done in Sect. 2.2.3 (in particular, see (2.2.43) and (2.2.44)). In the following statement we summarize these results.

Theorem 5.1.3 *The parameters of the n-point Gauss-Christoffel quadrature rule (5.1.2), with respect to a positive measure $d\mu$, are given by*

$$x_k = x_{n,k}, \quad A_k = \lambda_{n,k} = \lambda_n(d\mu; x_{n,k}) > 0, \quad k=1,\ldots,n,$$

i.e., the nodes are zeros of the orthogonal polynomial $\pi_n(d\mu; x)$ and the weights are values of the Christoffel function $\lambda_n(d\mu; x)$ at these zeros.

5.1.3.1 Gauss-Christoffel Quadratures for the Classical Weights

In the case of the classical weight functions (see Definition 2.3.1 for $w \in CW$), for the Christoffel numbers there exist analytic expressions in terms of orthogonal polynomials. For example, for $w(x) = v^{\alpha,\beta}(x) = (1-x)^\alpha(1+x)^\beta$, i.e., in the Gauss-Jacobi quadrature formula

$$\int_{-1}^{1} f(x) v^{\alpha,\beta}(x)\, dx = \sum_{k=1}^{n} A_k f(x_k) + R_n(f), \qquad (5.1.11)$$

the nodes x_k are zeros of the Jacobi polynomial $P_n^{(\alpha,\beta)}(x)$ and the weights are given by (2.3.44), i.e.,

$$A_k = \lambda_{n,k}^{(\alpha,\beta)} = \frac{\Gamma(n+\alpha+1)\Gamma(n+\beta+1)}{n!\Gamma(n+\alpha+\beta+1)} \cdot \frac{2^{\alpha+\beta+1}}{(1-x_k^2)\left[\frac{d}{dx}P_n^{(\alpha,\beta)}(x_k)\right]^2}.$$

In the Chebyshev case of the first kind ($\alpha = \beta = -1/2$), the weights become π/n for each k and the corresponding Gauss-Chebyshev quadrature formula has a simple form

$$\int_{-1}^{1} \frac{f(x)}{\sqrt{1-x^2}} dx = \frac{\pi}{n} \sum_{k=1}^{n} f\left(\cos\frac{(2k-1)\pi}{2n}\right) + R_n(f). \qquad (5.1.12)$$

For the generalized Gauss-Laguerre quadrature formula

$$\int_0^{+\infty} f(x) x^\alpha e^{-x} dx = \sum_{k=1}^{n} A_k f(x_k) + R_n(f), \qquad (5.1.13)$$

the nodes x_k are zeros of the generalized Laguerre polynomial $L_n^\alpha(x)$ and the weights are given by (2.3.58), i.e.,

$$A_k = \lambda_{n,k}^{(\alpha)} = \frac{\Gamma(n+\alpha+1)}{n!} \cdot \frac{1}{\left[\frac{d}{dx}L_n^\alpha(x_k)\right]^2}.$$

Similarly to (5.1.11) and (5.1.13) we can give the corresponding result for the Gauss-Hermite quadrature formula

$$\int_{-\infty}^{+\infty} f(x) e^{-x^2} dx = \sum_{k=1}^{n} A_k f(x_k) + R_n(f). \qquad (5.1.14)$$

The nodes x_k are zeros of the Hermite polynomial $H_n(x)$ and the weights are given by

$$A_k = \lambda_{n,k} = \frac{2^{n+1} n! \sqrt{\pi}}{H_n'(x_k)^2} = \frac{2^{n-1}(n-1)! \sqrt{\pi}}{n H_{n-1}(x_k)^2}.$$

5.1.3.2 Computation of Gauss-Christoffel Quadratures

For generating Gauss-Christoffel quadrature rules there are numerical methods, which are computationally much better than a computation of nodes by using Newton's method and then a direct application of the previous expressions for the weights (see e.g. Davis and Rabinowitz [77]). The characterization of the Gauss-Christoffel quadratures via an eigenvalue problem for the Jacobi matrix has become the basis of current methods for generating these quadratures. The most popular of them is one due to Golub and Welsch [185]. Their method is based on determining the eigenvalues and the first components of the eigenvectors of a symmetric tridiagonal Jacobi matrix.

Theorem 5.1.4 *The nodes x_k in the Gauss-Christoffel quadrature rule (5.1.2), with respect to a positive measure $d\mu$, are the eigenvalues of the n-th order Jacobi matrix*

$$J_n(d\mu) = \begin{bmatrix} \alpha_0 & \sqrt{\beta_1} & & & O \\ \sqrt{\beta_1} & \alpha_1 & \sqrt{\beta_2} & & \\ & \sqrt{\beta_2} & \alpha_2 & \ddots & \\ & & \ddots & \ddots & \sqrt{\beta_{n-1}} \\ O & & & \sqrt{\beta_{n-1}} & \alpha_{n-1} \end{bmatrix}, \qquad (5.1.15)$$

where α_ν and β_ν, $\nu = 0, 1, \ldots, n-1$, are the coefficients in the three-term recurrence relation (2.2.4) for the monic orthogonal polynomials $\pi_\nu(d\mu; \cdot)$. The weights A_k are given by

$$A_k = \beta_0 v_{k,1}^2, \qquad k = 1, \ldots, n,$$

where $\beta_0 = \mu_0 = \int_\mathbb{R} d\mu(x)$ and $v_{k,1}$ is the first component of the normalized eigenvector $\mathbf{v}_k$ corresponding to the eigenvalue x_k,

$$J_n(d\mu)\mathbf{v}_k = x_k \mathbf{v}_k, \qquad \mathbf{v}_k^\mathrm{T} \mathbf{v}_k = 1, \qquad k = 1, \ldots, n.$$

Proof First, we note that the Christoffel function $\lambda_n(d\mu; x) = 1/K_{n-1}(x,x)$ is defined by (2.1.30), so that, according to Theorem 5.1.3 (see also (2.2.6)), we have

$$A_k = \lambda_n(d\mu; x_k) = \frac{1}{K_{n-1}(x_k, x_k)} = \left(\sum_{\nu=1}^{n-1} p_\nu(x_k)^2 \right)^{-1}, \qquad (5.1.16)$$

where $p_\nu(x) = p_\nu(d\mu; x)$ are orthonormal polynomials, which satisfy the three-term recurrence relation (cf. Theorems 2.2.1 and 2.2.2)

$$xp_\nu(x) = \sqrt{\beta_{\nu+1}} p_{\nu+1}(x) + \alpha_\nu p_\nu(x) + \sqrt{\beta_\nu} p_{\nu-1}(x), \qquad \nu \geq 0, \qquad (5.1.17)$$

with $p_{-1}(x) = 0$ and $p_0(x) = 1/\sqrt{\mu_0}$.

Taking the first n equations from (5.1.17), we get (see (2.2.12))

$$J_n(d\mu)\mathbf{p}_n(x) = x\mathbf{p}_n(x) - \sqrt{\beta_n}\, p_n(x)\mathbf{e}_n, \qquad (5.1.18)$$

where $J_n(d\mu)$ is given by (5.1.15) and $\mathbf{p}_n(x) = [p_0(x)\ p_1(x)\ \ldots\ p_{n-1}(x)]^\mathrm{T}$.

Now, putting the zero x_k ($k = 1, \ldots, n$) of the polynomial $p_n(x)$ in (5.1.18) instead of x, it is clear that x_k is an eigenvalue of the Jacobi matrix $J_n(d\mu)$ and $\mathbf{p}_n(x_k)$ is the corresponding eigenvector, so that (5.1.16) can be expressed in the

5.1 Quadrature Formulae

form $A_k \|\mathbf{p}_n(x_k)\|_E^2 = 1$ $(k = 1, \ldots, n)$, where $\|\mathbf{a}\|_E^2 = \mathbf{a}^T\mathbf{a}$ $(\mathbf{a} \in \mathbb{R}^n)$. After a normalization of eigenvectors of the Jacobi matrix,

$$\frac{\mathbf{p}_n(x_k)}{\|\mathbf{p}_n(x_k)\|_E} =: \mathbf{v}_k = [v_{k,1}\ v_{k,2}\ \cdots\ v_{k,n}]^T,$$

and a fact that $\|\mathbf{p}_n(x_k)\|_E = \|\mathbf{p}_n(x_k)\|_E / \|\mathbf{v}_k\|_E = p_0(x_k)/v_{k,1} = (1/\sqrt{\mu_0})/v_{k,1}$, the Christoffel numbers become $A_k = \mu_0 v_{k,1}^2$, $k = 1, \ldots, n$. □

Simplifying QR algorithm so that only the first components of the eigenvectors are computed, Golub and Welsch [185] gave an efficient procedure for constructing the Gaussian quadrature rules. This procedure is implemented in several programming packages including the most known ORTHPOL given by Gautschi [161] (see also the Mathematica Package OrthogonalPolynomials [68]).

As we can see, the computation of Gauss-Christoffel quadrature formulas is connected to orthogonal polynomials. According to Theorem 5.1.4, we need the recursion coefficients α_k and β_k, $k \le m - 1$, for the monic polynomials $\pi_\nu(d\mu; \cdot)$, in order to construct the n-point Gauss-Christoffel quadrature formula (5.1.2), with respect to a positive measure $d\mu$, for each $n \le m$. In the case of the classical orthogonal polynomials (cf. Sect. 2.3), these coefficients are known explicitly[1] and the construction problem of Gaussian quadratures is completely solved by Theorem 5.1.4. However, in the case of strong non-classical polynomials (see Sect. 2.4.7), we need an additional numerical construction of recursion coefficients (see Sect. 2.4.8).

As an illustration, we consider now the construction of the n-point Gauss-Christoffel quadrature rules on $(0, +\infty)$, with respect to the weight function $w(x) = 1/\cosh^2 x$, using the discretized Stieltjes-Gautschi procedure described in Sect. 2.4.8.

Let $d\mu(x) = w(x)\,dx$ on $(0, +\infty)$. A natural discretization of the inner product $(p, q)_{d\mu} \approx (p, q)_{d\mu_N}$ can be obtained by writing the respective integrals in the form

$$\int_0^{+\infty} P(x)\,d\mu(x) = \int_0^{+\infty} P(x/2) \frac{2}{(1+e^{-x})^2} e^{-x}\,dx \quad (P \in \mathcal{P}),$$

(where $P = pq$) and applying N-point (classical) Gauss-Laguerre quadrature (5.1.13) (with $\alpha = 0$) to the integral on the right. The Gauss-Laguerre nodes x_k^L (zeros of the standard Laguerre polynomial $L_N(x)$) and the weights A_k^L $(= \lambda_{N,k}^{(0)})$ can be easily computed for an arbitrary N by the Golub-Welsch algorithm.

In this way, if $N \gg n$ we get an appropriate approximation of the inner product

$$(p, q)_{d\mu} \approx \sum_{k=1}^N \frac{2A_k^L}{\left(1 + e^{-x_k^L}\right)^2} P(x_k^L/2).$$

---

[1]Also, there are a few non-classical cases for which the recursion coefficients are known explicitly (see Sect. 2.4).

The first 40 recursion coefficients ($n = 40$) were obtained accurately to 30 decimal digits with $N = 520$, using the MICROVAX 3400 in Q-arithmetic with machine precision $\approx 1.93 \times 10^{-34}$ (see Milovanović [330]).

The same results were obtained also by a discretization procedure based on the composite Fejér quadrature rule, decomposing the interval of integration into four subintervals, $[0, +\infty] = [0, 10] \cup [10, 100] \cup [100, 500] \cup [500, +\infty]$ and using $N = 280$ points on each subinterval.

In the previous mentioned paper [330], the Gaussian quadrature rules on $(0, +\infty)$ with respect to the hyperbolic weight function $w(x) = \sinh x / \cosh^2 x$ were also constructed and applied to summation of the slowly convergent series (see Sect. 5.4).

As we mentioned on the end of Sect. 2.4.8, Gautschi and Milovanović [169] determined the recursion coefficients α_k and β_k, $k \leq 39$, for measures involving powers of Einstein's and Fermi's weights, $\varepsilon(x) = x/(e^x - 1)$ and $\varphi(x) = 1/(e^x + 1)$, respectively, and constructed the corresponding Gauss-Christoffel quadratures. Also, for the measure $d\mu(x) = [\varepsilon(x)]^r dx$ on $(0, +\infty)$, $r \geq 1$, the respective integrals in the discretization procedure were evaluated by the Gauss-Laguerre quadratures

$$\int_0^{+\infty} P(x) \, d\mu(x) = \frac{1}{r} \int_0^{+\infty} P(x/r) \left(\frac{x/r}{1 - e^{-x/r}} \right)^r e^{-x} dx$$

$$\approx \sum_{k=1}^N \frac{A_k^L}{r} \left(\frac{x_k^L/r}{1 - e^{-x_k^L/r}} \right)^r P(x_k^L/r),$$

where $P \in \mathcal{P}$.

Remark 5.1.4 There are several efficient algorithms for constructing some specific quadrature rules (e.g. see [459] for Gauss-Legendre quadratures). Some alternatives to the Golub and Welsch procedure can be found in Laurie [252]. In some special cases for calculating the weight coefficients A_k it is better to use the complete eigenvectors, instead of their first components, i.e., to use directly formula (5.1.16). An analysis of such cases is given in [342]. Notice that such a way was used in the period before an application of the QR-procedure (cf. Gautschi [143]).

5.1.4 Gauss-Radau and Gauss-Lobatto Quadrature Rules

In this section we consider quadratures which are very close to the Gaussian formulas. Suppose that the support interval $[a, b]$ of the measure $d\mu(x)$ is bounded from below, i.e., $a > -\infty$ and $b \leq +\infty$. Then we can include the end-point a in the set of quadrature nodes. Moreover, if $b < +\infty$ we can include both a and b to be nodes. It is sometimes convenient, especially when the function f vanishes at these points.

Now, we analyze these two cases.

5.1.4.1 Gauss-Radau Quadrature Formula

For the integrand f we introduce a function g by $f(x) = f(a) + (x-a)g(x)$, so that

$$\int_a^b f(x)\,d\mu(x) = f(a)\mu_0 + \int_a^b g(x)(x-a)\,d\mu(x),$$

where $\mu_0 = \int_a^b d\mu(x)$. If we define a new measure $d\mu_1(x) := (x-a)d\mu(x)$ and construct the corresponding n-point Gauss-Christoffel quadrature, we have

$$\int_a^b f(x)\,d\mu(x) = \mu_0 f(a) + \sum_{k=1}^n A_k^G g(x_k^G) + R_n^G(d\mu_1; g),$$

where, according to Theorem 5.1.3, $x_k^G = x_k^G(d\mu_1)$ are zeros of the orthogonal polynomial $\pi_n(d\mu_1; x)$ and $A_k^G = A_k^G(d\mu_1) = \lambda_n(d\mu_1; x_k^G) > 0$, $k = 1, \ldots, n$, and $R_n^G(d\mu_1; g)$ is the remainder in the corresponding Gaussian formula. In this way we obtain the so-called *Gauss-Radau* $(n+1)$-*point quadrature formula*

$$\int_a^b f(x)\,d\mu(x) = A_0^R f(a) + \sum_{k=1}^n A_k^R f(x_k^R) + R_{n,1}^R(d\mu; f), \qquad (5.1.19)$$

with nodes $x_0^R = a$, $x_k^R = x_k^G$, $k = 1, \ldots, n$, and weights A_k^R, given by

$$A_0^R = \mu_0 - \sum_{k=1}^n A_k^R, \quad A_k^R = \frac{A_k^G}{x_k^G - a}, \quad k = 1, \ldots, n.$$

The algebraic degree of exactness of the formula (5.1.19) is $d = 2n$.

Remark 5.1.5 The nodes and weights can be also obtained by a little modification of the Golub-Welsch Theorem 5.1.4. Namely, the matrix $J_n(d\mu)$ should be only changed by the following $(n+1)$-order matrix (see Golub [184] and Gautschi [166, pp. 155–156])

$$J_{n+1}^R(d\mu) = \begin{bmatrix} J_n(d\mu) & \sqrt{\beta_n}\,\mathbf{e}_n \\ \sqrt{\beta_n}\,\mathbf{e}_n^T & \alpha_n^R \end{bmatrix}, \quad \mathbf{e}_n^T = [0\ 0\ \cdots\ 1] \in \mathbb{R}^n,$$

where

$$\alpha_n^R = a - \beta_n(d\mu)\frac{\pi_{n-1}(d\mu; a)}{\pi_n(d\mu; a)}.$$

Taking more information on the function f at the node $x_0^R = a$ (e.g. on the first $r-1$ derivatives) we get the so-called *generalized Gauss-Radau quadrature formula*

$$\int_a^b f(x)\,d\mu(x) = \sum_{\nu=0}^{r-1} A_{0,\nu}^R f^{(\nu)}(a) + \sum_{k=1}^n A_k^R f(x_k^R) + R_{n,r}^R(d\mu;f), \qquad (5.1.20)$$

with degree of exactness $d = 2n + r - 1$. In order to get parameters, we start with

$$f(x) = \sum_{\nu=0}^{r-1} \frac{f^{(\nu)}(a)}{\nu!}(x-a)^\nu + (x-a)^r g(x)$$

and apply the same procedure as before, but now with the modified measure $d\mu_r(x) = (x-a)^r d\mu(x)$. Thus, with $x_k^G = x_k^G(d\mu_r)$ (zeros of the orthogonal polynomial $\pi_n(d\mu_r;x)$) and the Gaussian weights $A_k^G = A_k^G(d\mu_r) = \lambda_n(d\mu_r;x_k^G) > 0$, $k=1,\ldots,n$, we find

$$x_k^R = x_k^G, \quad A_k^R = \frac{A_k^G}{(x_k^G - a)^r}, \quad k=1,\ldots,n.$$

If we take successively $\pi_n(x)$, $(x-a)\pi_n(x)$, ..., $(x-a)^{r-1}\pi_n(x)$, instead of $f(x)$ in (5.1.20), where $\pi_n(x) = \prod_{\nu=1}^n (x - x_\nu^R)$, we get a upper triangular system of linear equations for determining the coefficients $A_{0,\nu}^R$, $\nu = 0, 1, \ldots, r-1$.

5.1.4.2 Gauss-Lobatto Quadrature Formula

Let $[a,b]$ be the support interval of the measure $d\mu(x)$. Introducing a function g by

$$f(x) - L_1(f;x) = (x-a)(b-x)g(x),$$

where

$$L_1(f;x) = \frac{x-b}{a-b}f(a) + \frac{x-a}{b-a}f(b),$$

and a new measure $d\mu_{1,1}(x) = (x-a)(b-x)d\mu(x)$, we have

$$\int_a^b f(x)\,d\mu(x) = \frac{f(a)}{b-a}\int_a^b (b-x)\,d\mu(x) + \frac{f(b)}{b-a}\int_a^b (x-a)\,d\mu(x)$$

$$+ \int_a^b g(x)\,d\mu_{1,1}(x).$$

Now, we construct the n-point Gauss-Christoffel rule with respect to the measure $d\mu_{1,1}(x)$,

$$\int_a^b g(x)\,d\mu_{1,1}(x) = \sum_{k=1}^n A_k^G g(x_k^G) + R_n^G(d\mu_{1,1};g),$$

5.1 Quadrature Formulae

where $x_k^G = x_k^G(d\mu_{1,1})$ are zeros of the orthogonal polynomial $\pi_n(d\mu_{1,1}; x)$ and $A_k^G = A_k^G(d\mu_{1,1}) = \lambda_n(d\mu_{1,1}; x_k^G) > 0$, $k = 1, \ldots, n$, and $R_n^G(d\mu_{1,1}; g)$ is the corresponding remainder. As in the Radau case, we obtain here the *Gauss-Lobatto quadrature formula*

$$\int_a^b f(x)\, d\mu(x) = A_0^L f(a) + \sum_{k=1}^n A_k^L f(x_k^L) + A_{n+1}^L f(b) + R_{n,1,1}^L(d\mu; f), \tag{5.1.21}$$

with nodes $x_0^L = a$, $x_k^L = x_k^G$, $k = 1, \ldots, n$, $x_{n+1}^L = b$, and weights A_k^L, given by

$$A_k^L = \frac{A_k^G}{(x_k^G - a)(b - x_k^G)}, \quad k = 1, \ldots, n,$$

and

$$A_0^L = \frac{1}{b-a}\left\{\int_a^b (b-x)\, d\mu(x) - \sum_{k=1}^n (b - x_k^G) A_k^L\right\},$$

$$A_{n+1}^L = \frac{1}{b-a}\left\{\int_a^b (x-a)\, d\mu(x) - \sum_{k=1}^n (x_k^G - a) A_k^L\right\}.$$

Remark 5.1.6 The nodes and weights in (5.1.21) can be obtained from the eigenvalue problem for the matrix of order $n + 2$ (see Golub [184] and Gautschi [166, pp. 159–160])

$$J_{n+2}^L(d\mu) = \begin{bmatrix} J_{n+1}(d\mu) & \sqrt{\beta_{n+1}^L}\, \mathbf{e}_{n+1} \\ \sqrt{\beta_{n+1}^L}\, \mathbf{e}_{n+1}^T & \alpha_{n+1}^L \end{bmatrix}, \quad \mathbf{e}_{n+1}^T = [0\ 0\ \cdots\ 1] \in \mathbb{R}^{n+1},$$

where α_{n+1}^L and β_{n+1}^L are given by the following system of equations

$$\begin{bmatrix} \pi_{n+1}(d\mu; a) & \pi_n(d\mu; a) \\ \pi_{n+1}(d\mu; b) & \pi_n(d\mu; b) \end{bmatrix} \begin{bmatrix} \alpha_{n+1}^L \\ \beta_{n+1}^L \end{bmatrix} = \begin{bmatrix} a\pi_{n+1}(d\mu; a) \\ b\pi_{n+1}(d\mu; b) \end{bmatrix}.$$

The algebraic degree of exactness of this $(n + 2)$-point formula is $d = 2n + 1$. The formula

$$\int_a^b f(x)\, d\mu(x) \approx \sum_{\nu=0}^{r-1} A_{0,\nu}^L f^{(\nu)}(a) + \sum_{k=1}^n A_k^L f(x_k^L) + \sum_{\nu=0}^{r-1} (-1)^\nu A_{n+1,\nu}^L f^{(\nu)}(b)$$

of the exactness $d = 2n + 2r - 1$, is known as the *generalized Gauss-Lobatto quadrature formula*. Taking the Gaussian parameters x_k^G and A_k^G for the measure

$d\mu_{r,r}(x) = (x-a)^r(b-x)^r d\mu(x)$, it is easy to see that

$$x_k^L = x_k^G, \quad A_k^L = \frac{A_k^G}{(x_k^G - a)^r(b - x_k^G)^r}, \quad k = 1, \ldots, n.$$

5.1.5 Error Estimates of Gaussian Rules for Some Classes of Functions

As we mentioned in Sect. 5.1.3, Markov [284] investigated the error term $R_n(f)$ in the Gauss quadrature formula. He considered the Hermite interpolation polynomial $h_{2n-1}(f; \cdot) \in \mathcal{P}_{2n-1}$ satisfying

$$h_{2n-1}(f; x_\nu) = f(x_\nu), \quad h'_{2n-1}(f; x_\nu) = f'(x_\nu), \quad \nu = 1, \ldots, n. \qquad (5.1.22)$$

(For the existence and uniqueness of the Hermite interpolation polynomial see Example 1.3.4 in Sect. 1.3.5.)

Using the Lagrange basis polynomials (5.1.6), with

$$q_n(x) = \pi_n(d\mu; x) = (x - x_1) \cdots (x - x_n),$$

and taking

$$U_\nu(x) = [1 - 2(x - x_\nu)\ell'_\nu(x_\nu)]\ell_\nu(x)^2, \quad V_\nu(x) = (x - x_\nu)\ell_\nu(x)^2, \quad \nu = 1, \ldots, n,$$

the polynomial $h_{2n-1}(f; \cdot)$ can be expressed in the form

$$h_{2n-1}(f; x) = \sum_{\nu=1}^n [U_\nu(x) f(x_\nu) + V_\nu(x) f'(x_\nu)]. \qquad (5.1.23)$$

Since $U_k, V_k \in \mathcal{P}_{2n-1}$ and $U_\nu(x_k) = \delta_{\nu,k}$, $V_\nu(x_k) = 0$, applying the Gauss quadrature rule we obtain

$$\int_{\mathbb{R}} U_\nu(x) d\mu(x) = \sum_{k=1}^n A_k U_\nu(x_k) = A_k, \quad \int_{\mathbb{R}} V_\nu(x) d\mu(x) = \sum_{k=1}^n A_k V_\nu(x_k) = 0,$$

so that

$$\int_{\mathbb{R}} h_{2n-1}(f; x) d\mu(x) = \sum_{k=1}^n A_k h_{2n-1}(f; x_k), \qquad (5.1.24)$$

where x_k and A_k are given in Theorem 5.1.3.

Remark 5.1.7 From the previous considerations, it is clear that we can omit the assumption on the existence of the first derivative of f at the points x_ν. Namely, in (5.1.22) it is enough to put $h'_{2n-1}(f; x_\nu) = \alpha_\nu$, $\nu = 1, \ldots, n$, where α_ν are arbitrary numbers.

5.1 Quadrature Formulae

Suppose that $[a, b] = \operatorname{supp}(d\mu)$. Using the previous facts, a classical result for the remainder term can be proved.

Theorem 5.1.5 *Let $f \in C^{2n}[a, b]$. Then, there exists $\xi \in (a, b)$ such that for the remainder $R_n(f)$ in the Gauss-Christoffel rule the following formula*

$$R_n(f) = \frac{\|\pi_n(d\mu; \cdot)\|^2}{(2n)!} f^{(2n)}(\xi)$$

holds.

Proof For functions $f \in C^{2n}[a, b]$, it is well-known that the error of the Hermite interpolation polynomial (5.1.23) can be expressed in the form

$$r_n(f; x) = f(x) - h_{2n-1}(f; x) = \frac{f^{(2n)}(\eta)}{(2n)!} q_n(x)^2, \quad (5.1.25)$$

where $q_n(x) = \pi_n(d\mu; x) = (x - x_1) \cdots (x - x_n)$ and $\eta = \eta(x) \in (a, b)$. According to (5.1.8), (5.1.24), and (5.1.25), we get

$$R_n(f) = \frac{1}{(2n)!} \int_{\mathbb{R}} f^{(2n)}(\eta(x)) \pi_n(d\mu; x)^2 \, d\mu(x).$$

Finally, an application of the mean value theorem of integration gives the desired result. $\square$

For example, the remainder term in the Gauss-Chebyshev quadrature (5.1.12) reduces to

$$R_n(f) = \frac{\pi}{2^{2n-1}(2n)!} f^{(2n)}(\xi), \quad -1 < \xi < 1,$$

when $f \in C^{2n}[-1, 1]$.

Let $d\mu(x) = w(x) \, dx$ and let $R_n(f)_w$ be the error in the corresponding n-point Gauss-Christoffel formula, with the nodes and weights, x_k and $A_k = \lambda_k(w)$, $k = 1, \ldots, n$, i.e.,

$$R_n(f)_w = \int_{-1}^{1} w(x) f(x) \, dx - \sum_{k=1}^{n} \lambda_k(w) f(x_k). \quad (5.1.26)$$

The estimate of $R_n(f)_w$ for different classes of functions is a very interesting and important problem. The general tools, useful in several contexts, are the following Posse-Markov-Stieltjes inequalities [134, p. 33]

$$\sum_{k=1}^{d-1} \lambda_k(w) g(x_k) \leq \int_{-\infty}^{x_d} g(x) w(x) \, dx \leq \sum_{k=1}^{d} \lambda_k(w) g(x_k),$$

where $d > 1$, g is such that $g^{(k)}(x) \geq 0$, $k = 0, 1, \ldots, 2n - 1$, $n > 1$, and

$$\sum_{k=d+1}^{n} \lambda_k(w) g(x_k) \leq \int_{x_{d+1}}^{+\infty} g(x) w(x) \, dx \leq \sum_{k=d}^{n} \lambda_k(w) g(x_k),$$

where $n - 1 \geq d \geq 1$, $(-1)^k g^{(k)}(x) \geq 0$, $k = 0, 1, \ldots, 2n - 1$, $n > 1$.

Depending on the class of functions, there are many methods for estimating the remainder term $R_n(f)_w$ in (5.1.26). Here, we analyze some of them.

5.1.5.1 Error Estimates for Analytic Functions

Let Γ be a simple closed curve in the complex plane surrounding the interval $[-1, 1]$ and $D = \text{int} \, \Gamma$ be its interior. If the integrand f is analytic in D and continuous on $\overline{D}$, then, according to (1.4.22), the remainder term $R_n(f)_w$ in (5.1.26) admits the contour integral representation

$$R_n(f)_w = \frac{1}{2\pi i} \oint_\Gamma K_n(z) f(z) \, dz, \tag{5.1.27}$$

where the kernel is given by

$$K_n(z) = \frac{1}{\pi_n(z)} \int_{-1}^{1} \frac{w(x) \pi_n(x)}{z - x} \, dx, \qquad z \notin [-1, 1],$$

and $\pi_n(z) = \prod_{k=1}^{n}(z - x_k)$ is the monic polynomial orthogonal with respect to the measure $d\mu(x) = w(x) \, dx$ on $(-1, 1)$.

An alternative representation for $K_n(z)$ is

$$K_n(z) = R_n\left(\frac{1}{z - \cdot}\right)_w = \int_{-1}^{1} \frac{w(x)}{z - x} \, dx - \sum_{k=1}^{n} \frac{\lambda_k(w)}{z - x_k}.$$

The integral representation (5.1.27) leads to the error estimate

$$|R_n(f)_w| \leq \frac{\ell(\Gamma)}{2\pi} \left(\max_{z \in \Gamma} |K_n(z)|\right) \left(\max_{z \in \Gamma} |f(z)|\right), \tag{5.1.28}$$

where $\ell(\Gamma)$ is the length of the contour Γ. In order to get the estimate (5.1.28), one has to study the magnitude of $|K_n(z)|$ on Γ.

More generally, if we apply the Hölder inequality to (5.1.27), we get

$$|R_n(f)_w| = \frac{1}{2\pi} \left| \oint_\Gamma K_n(z) f(z) \, dz \right|$$

$$\leq \frac{1}{2\pi} \left(\oint_\Gamma |K_n(z)|^r |dz| \right)^{1/r} \left(\oint_\Gamma |f(z)|^{r'} |dz| \right)^{1/r'}$$

5.1 Quadrature Formulae

$$= \frac{1}{2\pi} \|K_n\|_r \|f\|_{r'}, \qquad (5.1.29)$$

where $1 \le r \le +\infty$, $1/r + 1/r' = 1$, and

$$\|f\|_r := \begin{cases} \left(\oint_\Gamma |f(z)|^r |dz|\right)^{1/r}, & 1 \le r < +\infty, \\ \max_{z \in \Gamma} |f(z)|, & r = +\infty. \end{cases}$$

In the case $r = +\infty$ ($r' = 1$), this estimate becomes

$$|R_n(f)_w| \le \frac{1}{2\pi} \left(\max_{z \in \Gamma} |K_n(z)|\right)\left(\oint_\Gamma |f(z)||dz|\right). \qquad (5.1.30)$$

Evidently, from (5.1.30) it follows the estimate (5.1.28).
On the other hand for $r = 1$ ($r' = +\infty$), the estimate (5.1.29) reduces to

$$|R_n(f)_w| \le \frac{1}{2\pi} \left(\oint_\Gamma |K_n(z)||dz|\right)\left(\max_{z \in \Gamma} |f(z)|\right), \qquad (5.1.31)$$

which is evidently stronger than (5.1.28), because of the inequality

$$\oint_\Gamma |K_n(z)||dz| \le \ell(\Gamma)\left(\max_{z \in \Gamma} |K_n(z)|\right).$$

Many authors have used (5.1.28) to derive bounds of $|R_n(f)_w|$ (see [56, 57, 78, 103, 128, 146, 159, 174, 176, 256, 447, 457, 471], etc.), but the same technique has already been used much earlier by Hermite [210] and Heine [208, p. 16] to derive the error estimation for a polynomial interpolation. Two choices of the contour Γ have been widely used by these authors: a circle C_r with center at the origin and radius r (> 1), i.e., $C_r = \{z : |z| = r\}$, $r > 1$, and an ellipse with foci at the points ± 1 and sum of semiaxes $\varrho > 1$,

$$\mathcal{E}_\varrho = \left\{z \in \mathbb{C} \;\Big|\; z = \frac{1}{2}(\varrho e^{i\theta} + \varrho^{-1} e^{-i\theta}),\; 0 \le \theta < 2\pi\right\}.$$

When $\varrho \to 1$, then the ellipse shrinks to the interval $[-1, 1]$, while with increasing ϱ it becomes more and more circle-like. The advantage of the elliptical contours, compared to the circular ones, is that such a choice needs the analyticity of f in a smaller region of the complex plane, especially when ϱ is near 1 (see Fig. 5.1.1).

Since the ellipse $\mathcal{E}_\varrho$ has the length $\ell(\mathcal{E}_\varrho) = 4\varepsilon^{-1} E(\varepsilon)$, where ε is the eccentricity of $\mathcal{E}_\varrho$, i.e., $\varepsilon = 2/(\varrho + \varrho^{-1})$, and

$$E(\varepsilon) = \int_0^{\pi/2} \sqrt{1 - \varepsilon^2 \sin^2 \theta}\, d\theta$$

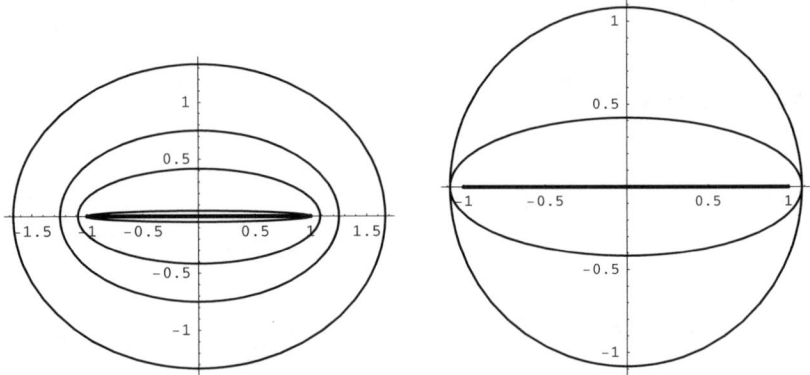

Fig. 5.1.1 Elliptical contours for $\varrho = 3, 2, 1.5$ and 1.05 (*left*) and a circular contour with $r = 13/12$ and an elliptical contour with $\varrho = 3/2$ (*right*)

is the complete elliptic integral of the second kind, the estimate (5.1.28) reduces to

$$|R_n(f)w| \leq \frac{2E(\varepsilon)}{\pi \varepsilon} \left(\max_{z \in \mathcal{E}_\varrho} |K_n(z)| \right) \|f\|_\varrho, \quad \varepsilon = \frac{2}{\varrho + \varrho^{-1}}, \quad (5.1.32)$$

where $\|f\|_\varrho = \max_{z \in \mathcal{E}_\varrho} |f(z)|$. As we can see, the bound on the right in (5.1.32) is a function of ϱ, so that it can be optimized with respect to $\varrho > 1$.

In [174] Gautschi and Varga studied error bounds of the form (5.1.28) for Gaussian quadratures of analytic functions, especially for four Chebyshev weights $w(t) = w_i(t)$ (i.e., Jacobi weights with parameters $\pm 1/2$):

(a) $w_1(t) = (1-t^2)^{-1/2}$,   (b) $w_2(t) = (1-t^2)^{1/2}$,
(c) $w_3(t) = (1-t)^{-1/2}(1+t)^{1/2}$, (d) $w_4(t) = (1-t)^{1/2}(1+t)^{-1/2}$.

For example, for the Chebyshev weight of the first kind w_1 they proved

$$\max_{z \in \mathcal{E}_\varrho} |K_n(z)| = K_n\left((\varrho + \varrho^{-1})/2\right) = \frac{4\pi}{\varrho^n} \frac{1}{(\varrho - \varrho^{-1})(\varrho^n + \varrho^{-n})}$$

for any $\varrho > 1$ and each $n \in \mathbb{N}$.

The cases of Gaussian rules with the Bernstein-Szegő weight functions and with some symmetric weights including especially the Gegenbauer weight were studied by Peherstorfer [390] and Schira [426], respectively. Some of these results have been extended to the Gauss-Radau and Gauss-Lobatto formulas (cf. Gautschi [158], Gautschi and Li [168], Schira [425], Hunter and Nikolov [218]).

The first approach in the sense (5.1.31) for Gaussian quadrature rules, using the elliptical contours, was given by Hunter [217]. According to (5.1.31) he studied the

5.1 Quadrature Formulae

quantity

$$L_n(E_\varrho) = \frac{1}{2\pi} \oint_{E_\varrho} |K_n(z)| \, |dz|.$$

Since $z = \frac{1}{2}(\xi + \xi^{-1})$, $\xi = \varrho e^{i\theta}$, and $|dz| = 2^{-1/2}\sqrt{a_2 - \cos 2\theta} \, d\theta$, where

$$a_j = a_j(\varrho) = \frac{1}{2}\left(\varrho^j + \varrho^{-j}\right), \quad j \in \mathbb{N}, \ \varrho > 1,$$

the quantity $L_n(E_\varrho)$ reduces to

$$L_n(E_\varrho) = \frac{1}{2\pi\sqrt{2}} \int_0^{2\pi} \frac{|f_n(z)|(a_2 - \cos 2\theta)^{1/2}}{|\pi_n(z)|} \, d\theta,$$

where f_n is the corresponding function on the second kind (see Sect. 2.2.4). This integral can be evaluated numerically by using a quadrature formula. However, if $w(t) = w_i(t)$ (one of the Chebyshev weights) it is possible to get an explicit expression for $L_n(E_\varrho)$. For example, in the case of $w = w_1$, Hunter [217] obtained

$$L_n(E_\varrho) = \frac{4}{\varrho^{2n}+1} K\left(\frac{2}{\varrho^n + \varrho^{-n}}\right),$$

where K is the complete elliptic integral of the first kind, i.e.,

$$K(k) = \int_0^{\pi/2} (1 - k^2 \sin^2\theta)^{-1/2} \, d\theta \quad (|k| < 1).$$

5.1.5.2 Error Estimates for Some Classes of Continuous Functions

We consider the error $R_n(f)_w$ in the Gauss-Christoffel quadrature formula (5.1.26) for three classes of continuous functions f.

Class C^0. For general weights w and continuous functions $f \in C^0 = C[-1, 1]$, the remainder term can be estimated in terms of the best approximation in the uniform norm. Following the proof of Theorem 5.1.1, we obtain the following result:

Theorem 5.1.6 *For a general weight function w and $f \in C^0$ we have*

$$|R_n(f)_w| \leq 2\|w\|_1 E_{2n-1}(f)_\infty, \tag{5.1.33}$$

where $\|w\|_1 = \int_{-1}^1 w(x) \, dx$.

Class C_w. For the Jacobi weight $w(x) = v^{\alpha,\beta}(x) = (1-x)^\alpha(1+x)^\beta$, with the parameters $\alpha, \beta > 0$ and functions from the weighted uniform space C_w (see (4.3.1)), we have the following estimate:

Theorem 5.1.7 *If $w = v^{\alpha,\beta}$, $\alpha, \beta > 0$ and $f \in C_w$, then*

$$|R_n(f)w| \leq C E_{2n-1}(f)_{w,\infty}, \qquad (5.1.34)$$

where $E_{2n-1}(f)_{w,\infty}$ denotes the error of the best weighted uniform approximation by polynomials of degree at most $2n - 1$ and $C \neq C(n, f)$ is a positive constant.

Proof Here, we use $\lambda_k(w) \sim w(x_k)\Delta x_k$, where $\Delta x_k = (1 - x_k^2)^{1/2}/n$ (cf. (2.3.47)). As in the proof of Theorem 5.1.6, for each $P \in \mathcal{P}_{2n-1}$, we find

$$|R_n(f)w| = |R_n(f - P)w|$$

$$\leq \int_{-1}^{1} w(x)|f(x) - P(x)|\,dx + C\sum_{k=1}^{n}|f(x_k) - P(x_k)|w(x_k)\Delta x_k$$

$$\leq 2\|(f - P)w\|_\infty + C\|(f - P)w\|_\infty \sum_{k=1}^{n} \Delta x_k \leq C\|(f - P)w\|_\infty.$$

Taking infimum with respect to $P \in \mathcal{P}_{2n-1}$ we get (5.1.34). □

Notice that in this case f can be singular at the endpoints, for example $f(x) = \log(1 - x^2)$ belongs to C_w.

Class $W_r^1(w)$. Now, we prove the error estimate for functions from the Sobolev space $W_r^1(w)$ $(r \geq 1)$.

Theorem 5.1.8 *Let $w = v^{\alpha,\beta}$ $(\alpha, \beta > -1)$ and $f \in W_r^1(w)$ $(r \geq 1)$. Then*

$$|R_n(f)w| \leq \frac{C}{2n-1} E_{2n-2}(f')_{w\varphi,1}, \qquad (5.1.35)$$

where $\varphi(x) = \sqrt{1 - x^2}$ and $C \neq C(n, f)$ is a positive constant.

Proof We use the inequalities

$$(b-a)\begin{cases}|f(a)| \\ |f(b)|\end{cases} \leq \int_a^b |f(t)|\,dt + (b-a)\int_a^b |f'(t)|\,dt \qquad (5.1.36)$$

where $-\infty < a < b < +\infty$, as well as $\lambda_k(w) \sim w(x_k)\Delta_k$, $\Delta x_k = (1-x_k^2)^{1/2}/n$, so that

$$\sum_{k=1}^{n} \lambda_k(w)|f(x_k)| \leq C\sum_{k=1}^{n} |f(x_k)|\Delta x_k\, w(x_k).$$

Now, for $k = 1, \ldots, n-1$, we apply the first inequality from (5.1.36), with $a = x_k$, $b = x_{k+1}$, to obtain

$$\Delta x_k|f(x_k)| \leq \int_{x_k}^{x_{k+1}} |f(t)|\,dt + \Delta x_k \int_{x_k}^{x_{k+1}} |f'(t)|\,dt.$$

5.1 Quadrature Formulae

Since $1 \pm x_k \sim 1 \pm t \sim 1 \pm x_{k+1}$, we get

$$\lambda_k(w)|f(x_k)| \sim |f(x_k)|\Delta x_k\, w(x_k)$$

$$\leq C \int_{x_k}^{x_{k+1}} |f(t)|w(t)\,dt + \frac{C_1}{n} \int_{x_k}^{x_{k+1}} |f'(t)|\varphi(t)w(t)\,dt.$$

For $k = n$ we use the second inequality of (5.1.36), $\Delta x_n \sim \Delta x_{n-1}$, as well as the previous consideration, so that we have

$$\lambda_n(w)|f(x_n)| \sim |f(x_n)|\Delta x_{n-1}\, w(x_n)$$

$$\leq C \int_{x_{n-1}}^{x_n} |f(t)|w(t)\,dt + \frac{C_1}{n} \int_{x_{n-1}}^{x_n} |f'(t)|\varphi(t)w(t)\,dt.$$

Then, taking the sum with respect to $k = 1, \ldots, n$, we obtain

$$\sum_{k=1}^{n} \lambda_k(w)|f(x_k)| \leq C \int_{-1}^{1} |f(t)|w(t)\,dt + \frac{C_1}{n} \int_{-1}^{1} |f'(t)|\varphi(t)w(t)\,dt. \quad (5.1.37)$$

Using (5.1.37), for each $P \in \mathcal{P}_{2n-1}$ we get

$$|R_n(f)_w| = |R_n(f - P)_w|$$

$$\leq \int_{-1}^{1} |(f - P)w|w(x)\,dx + \sum_{k=1}^{n} \lambda_k(w)|f(x_k) - P(x_k)|$$

$$\leq \|(f - P)w\|_1 + C\|(f - P)w\|_1 + \frac{C_1}{n}\|(f - P)'\varphi w\|_1$$

$$\leq C\|(f - P)w\|_1 + \frac{C_1}{n}\|(f - P)'\varphi w\|_1.$$

But, by a lemma of Ky [245], the second norm on the right can be estimated by

$$\|(f - P)'\varphi w\|_1 \leq C(2n - 1)\|(f - P)w\|_1 + C_1 E_{2n-2}(f')_{w\varphi,1},$$

so that

$$|R_n(f)_w| \leq C\|(f - P)w\|_1 + \frac{C_1}{n} E_{2n-2}(f')_{w\varphi,1}.$$

Taking the infimum over $P \in \mathcal{P}_{2n-1}$ we get

$$|R_n(f)_w| \leq C E_{2n-1}(f)_{w,1} + \frac{C_1}{n} E_{2n-2}(f')_{w\varphi,1}.$$

Finally, using the Favard inequality we obtain

$$|R_n(f)_w| \leq \frac{C}{2n - 1} E_{2n-2}(f')_{w\varphi,1}. \qquad \square$$

Example 5.1.1 We consider the error term in the Gauss-Christoffel formula (5.1.26) for $f(x) = (1+x)\log(1+x)$, $f(-1) = 0$, and the Jacobi weight $w = v^{\alpha,\beta}$ for some selected parameters α and β. Since this function f belongs to all of the previous classes of functions, we will analyze separately the estimates (5.1.33), (5.1.34), and (5.1.35).

1° Since
$$E_{2n-1}(f)_\infty \leq \int_0^{1/n} \frac{\Omega_\varphi^r(f,t)_\infty}{t} \, dt, \quad r \leq n,$$

we estimate the modulus by (2.5.13),

$$\Omega_\varphi^r(f,t)_\infty \leq \sup_{0<h\leq t} h^r \|f^{(r)}\varphi^r\|_{L^\infty(-1+h^2,1-h^2)}$$

$$\sim \sup_{0<h\leq t} \left\{ h^r \max_{|x|\leq 1-h^2} |(1+x)^{-r+1}(1-x^2)^{r/2}| \right\} \sim t^2.$$

Therefore, $E_{2n-1}(f)_\infty \leq C/n^2$ and (5.1.33) gives the estimate

$$|R_n(f)_w| \leq \frac{C}{n^2}. \tag{5.1.38}$$

2° Since
$$E_{2n-1}(f)_{v^{\alpha,\beta},\infty} \leq \int_0^{1/n} \frac{\Omega_\varphi^r(f,t)_{v^{\alpha,\beta},\infty}}{t} \, dt,$$

by a similar computation, we find $\Omega_\varphi^r(f,t)_{v^{\alpha,\beta},\infty} \sim t^{2+2\beta}$ and then (5.1.34) gives

$$|R_n(f)_w| \leq \frac{C}{n^{2+2\beta}}. \tag{5.1.39}$$

3° Since $f'(x) = 1 + \log(1+x)$ and $f \in W_1^1(v^{\alpha,\beta})$, $\alpha, \beta > -1$, we have

$$E_{2n-2}(f')_{v^{\alpha,\beta}\varphi,1} \leq C \int_0^{1/n} \frac{\Omega_\varphi^{r-1}(f',t)_{v^{\alpha,\beta}\varphi,1}}{t} \, dt$$

and

$$\Omega_\varphi^{r-1}(f',t)_{v^{\alpha,\beta}\varphi,1} \leq \sup_{0<h\leq t} h^{r-1} \|f^{(r)}\varphi^{r-1}\varphi v^{\alpha,\beta}\|_{L^1(|x|\leq 1-h^2)}$$

$$\sim \sup_{0<h\leq t} \left\{ h^{r-1} \int_{-1+h^2}^{1-h^2} (1+x)^{-r+1}(1-x^2)^{r/2} v^{\alpha,\beta}(x) \, dx \right\}$$

$$\leq \sup_{0<h\leq t} \left\{ h^{r-1} \int_{-1+h^2}^{1-h^2} (1+x)^{-r/2+1+\beta}(1-x)^\alpha \, dx \right\} \sim t^{3+2\beta}.$$

5.1 Quadrature Formulae

Table 5.1.1 The error term $|R_n(f)_w|$ in the Gaussian approximation of the integral $\int_{-1}^{1} w(x)f(x)\,dx$ for $f(x) = (1+x)\log(1+x)$, $w = v^{\alpha,\beta}$ and some selected parameters (α, β)

n	$(-1/2, -1/2)$	$(0, 0)$	$(1/2, 1/2)$	$(1, 2)$	$(1, 3)$
10	7.10(−4)	8.35(−5)	1.56(−5)	2.54(−7)	2.92(−8)
20	8.86(−5)	5.70(−6)	6.02(−7)	1.89(−9)	7.55(−11)
30	2.62(−5)	1.16(−6)	8.56(−8)	9.30(−11)	1.88(−12)
40	1.11(−5)	3.74(−7)	2.11(−8)	1.05(−11)	1.27(−13)
50	5.67(−6)	1.55(−7)	7.09(−9)	1.90(−12)	1.53(−14)

Then $E_{2n-2}(f')_{v^{\alpha,\beta}\varphi,1} \leq C/n^{3+2\beta}$ and (5.1.35) gives

$$|R_n(f)_w| \leq \frac{C}{n^{4+2\beta}}. \tag{5.1.40}$$

In Table 5.1.1 we give $|R_n(f)_w|$ for $w = v^{\alpha,\beta}$, when $(\alpha, \beta) = (-1/2, -1/2)$ (Chebyshev weight of the first kind), $(0, 0)$ (Legendre weight), $(1/2, 1/2)$ (Chebyshev weight of the second kind), and for two Jacobi weights: $(\alpha, \beta) = (1, 2)$ and $(1, 3)$. As we can see, the convergence order increases with β.

5.1.5.3 Error Estimates for Gauss-Laguerre Formula

For the generalized Laguerre weight $w_\alpha(x) = x^\alpha e^{-x}$, $\alpha > -1$, $x \geq 0$, we set

$$R_n(f)_{w_\alpha} = \int_0^{+\infty} f(x) w_\alpha(x)\,dx - \sum_{k=1}^{n} \lambda_k(w_\alpha) f(x_k),$$

where $x_1 < \cdots < x_n$ are zeros of the generalized Laguerre orthonormal polynomial $p_n(w_\alpha)$ and $\lambda_k(w_\alpha)$ are the corresponding Christoffel numbers.

The following simple proposition completes the Uspensky theorem (see [482]). First we define the function spaces

$$L^\infty_{w_{\alpha,\beta}} = \left\{ f \in C^0(0, +\infty) \,\Big|\, \lim_{\substack{x \to 0 \\ x \to +\infty}} f(x) w_{\alpha,\beta}(x) = 0 \right\},$$

for $w_{\alpha,\beta}(x) = (1+x)^\beta w_\alpha(x)$, $\alpha > 0$, and

$$L^\infty_{w_{0,\beta}} = \left\{ f \in C^0[0, +\infty) \,\Big|\, \lim_{x \to +\infty} f(x) e^{-x}(1+x)^\beta = 0 \right\},$$

for $\alpha \leq 0$.

Proposition 5.1.1 *For all $f \in L^\infty_{w_{\alpha,\beta}}$, $\alpha > 0$ and $\beta > 1$, we have*

$$|R_n(f)_{w_\alpha}| \leq C E_{2n-1}(f)_{w_{\alpha,\beta},\infty}. \tag{5.1.41}$$

For all $f \in L^\infty_{w_{0,\beta}}$, $\alpha \le 0$ and $-\alpha + \beta > 1$, we have

$$|R_n(f)_{w_\alpha}| \le C E_{2n-1}(f)_{w_{0,\beta},\infty}. \tag{5.1.42}$$

Here, $C \ne C(n, f)$ is positive constant.

Proof With $\alpha > 0$ and $\beta > 1$, for each $P \in \mathcal{P}_{2n-1}$, we have

$$|R_n(f)_{w_\alpha}| = |R_n(f - P)_{w_\alpha}|$$

$$\le \int_0^{+\infty} \frac{|(f-P)(x)|}{(1+x)^\beta} w_{\alpha,\beta}(x)\,dx$$

$$+ \sum_{k=1}^n \frac{\lambda_k(w_\alpha)}{(1+x_k)^\beta} |f(x_k) - P(x_k)|(1+x_k)^\beta$$

$$\le \|(f-P)w_{\alpha,\beta}\|_\infty \int_0^{+\infty} \frac{dx}{(1+x)^\beta}$$

$$+ \sum_{k=1}^n \frac{\Delta x_k}{(1+x_k)^\beta} |f(x_k) - P(x_k)| w_{\alpha,\beta}(x_k)$$

$$\le \|(f-P)w_{\alpha,\beta}\|_\infty \left(\int_0^{+\infty} \frac{dx}{(1+x)^\beta} + \sum_{k=1}^n \frac{\Delta x_k}{(1+x_k)^\beta} \right)$$

$$\le C \|(f-P)w_{\alpha,\beta}\|_\infty,$$

since $\lambda_k(w_\alpha) \sim w_\alpha(x_k)\Delta x_k$ (cf. (2.3.60)) and

$$\sum_{k=1}^n \frac{\Delta x_k}{(1+x_k)^\beta} < \sum_{k=1}^n \int_{x_{k-1}}^{x_k} \frac{dx}{(1+x)^\beta} < \int_0^{+\infty} \frac{dx}{(1+x)^\beta}, \quad x_0 = 0.$$

Taking the infimum over all $P \in \mathcal{P}_{2n-1}$, we get (5.1.41).
If $-1 < \alpha \le 0$ and $-\alpha + \beta > 1$, we get

$$|R_n(f)_{w_\alpha}| \le \|(f-P)w_{0,\beta}\|_\infty \left(\int_0^{+\infty} \frac{dx}{x^{-\alpha}(1+x)^\beta} \right)$$

and (5.1.42) follows directly. $\square$

In order to estimate the error $E_{2n-1}(f)_{w_{\alpha,\beta},\infty}$, we can use the results from Sect. 2.5.3 and, in particular, a comment given at the end of this section.
Assume now $f \in W_r^1(w_\alpha)$, where

$$W_r^1(w_\alpha) = \left\{ f \in L^1_{w_\alpha} \;\middle|\; f^{(r-1)} \in AC(\mathbb{R}^+) \text{ and } \|f^{(r)}\varphi^r w_\alpha\|_1 < +\infty \right\}$$

and $\varphi(x) = \sqrt{x}$.

5.1 Quadrature Formulae

For such a class of functions ($r=1$) the estimate (as in the finite interval case)

$$|R_n(f)_{w_\alpha}| \le \frac{C}{\sqrt{n}} \|f'\varphi w_\alpha\|_1 \qquad (f \in W_1^1(w_\alpha)) \tag{5.1.43}$$

is false. Here, we only can obtain

$$|R_n(f)_{w_\alpha}| \le \frac{C}{n^{1/6}} \|f'\varphi w_\alpha\|_1. \tag{5.1.44}$$

In fact, it was proved (see [298]) that for every fixed n there exists a function $f_n \in AC$, with $0 < \|f_n'\varphi w_\alpha\|_1 < +\infty$, such that

$$|R_n(f_n)_{w_\alpha}| \ge \frac{C}{n^{1/6}} \|f_n'\varphi w_\alpha\|_1.$$

Similar phenomenon appears in the Hermite, Sonin-Markov and Freud cases. The reason lies in the zero distribution of the corresponding orthogonal polynomials. However, we can obtain an estimate like (5.1.43) if we simplify the Gaussian rule. Namely, using the notation from Sect. 4.3.3, with a sufficiently large n and a fixed $\theta \in (0,1)$, we define $x_j = x_{j(n)} = \min\{x_k \mid x_k \ge 4\theta n\}$ and $M = [n\theta/(1+\theta)] \sim n$, and then we consider the truncated Gaussian rule

$$G_{n,j}(f) = \sum_{k=1}^{j} \lambda_k(w_\alpha) f(x_k).$$

The following result holds ([298]):

Theorem 5.1.9 *For all $f \in W_1^1(w_\alpha)$, $\alpha > -1$, we have*

$$\left| \int_0^{+\infty} f(x) w_\alpha(x) \, dx - G_{n,j}(f) \right| \le C \left[\frac{E_M(f')_{w_\alpha \varphi, 1}}{\sqrt{n}} + e^{-An} \|f w_\alpha\|_1 \right],$$

where C and A are constants independent of n and f.

Using the Favard theorem, for $f \in W_r^1(w_\alpha)$, as a consequence we get

$$\left| \int_0^{+\infty} f(x) w_\alpha(x) \, dx - G_{n,j}(f) \right| \le \frac{C}{n^{r/2}} \|f\|_{W_r^1(w_\alpha)},$$

with the same order of the best $L_{w_\alpha}^1$-approximation for such a class of functions.

5.1.5.4 Error Estimates for Freud-Gaussian Rules

By $w^\alpha(x) = e^{-|x|^\alpha}$, $\alpha > 1$, we denote a Freud weight and by $\{p_n(w^\alpha)\}$ the corresponding sequence of orthonormal polynomials with positive leading coefficients.

We denote by $x_1 < x_2 < \cdots < x_{[n/2]}$ the positive zeros of $p_n(w^\alpha)$ and put $x_{-k} = x_k$ (with $x_0 = 0$ for odd n). Then, we can write

$$R_n(f)_{w^\alpha} = \int_\mathbb{R} f(x) w^\alpha(x) \, dx - \sum_{k=-[n/2]}^{[n/2]} \lambda_k(w^\alpha) f(x_k),$$

with $\lambda_0(w^\alpha) = 0$ if n is even.

For continuous functions in $\mathbb{R}$ we can state the following result:

Proposition 5.1.2 *Let $f \in C^0(\mathbb{R})$ and assume $\lim_{|x| \to +\infty} |f(x)| \sigma_{\alpha,\beta}(x) = 0$, where $\sigma_{\alpha,\beta}(x) = w^\alpha(x)(1+|x|)^\beta$ with $\beta > 1$. Then, we have*

$$|R_n(f)_{w^\alpha}| \leq C E_{2n-1}(f)_{\sigma_{\alpha,\beta},\infty}, \tag{5.1.45}$$

with $C \neq C(n,\alpha)$.

For functions in the L_1-Sobolev spaces, i.e., if $f \in AC(\mathbb{R})$ and $\|f' w^\alpha\|_1 < +\infty$, we can repeat, mutatis mutandis, the comments from the Laguerre case. To be more precise, let $M_n \sim n^{1/\alpha}$ be the Mhaskar-Rakhmanov-Saff number with respect to the weight w^α (see Sect. 2.4.5). Then, for the previous class of functions we can only get the estimate

$$|R_n(f)_{w^\alpha}| \leq C \left(\frac{M_n}{n} \right)^{1/3} \|f' w^\alpha\|_1,$$

while the error of the best $L^1_{w^\alpha}$-approximation is

$$E_n(f)_{w^\alpha,1} \leq C \frac{M_n}{n} \|f' w^\alpha\|_1. \tag{5.1.46}$$

However, as before, we can also consider the "truncated" Gaussian sum

$$\sum_{k=-j}^{j} \lambda_k(w^\alpha) f(x_k),$$

with the corresponding remainder term

$$\widetilde{R}_n(f)_{w^\alpha} = \int_\mathbb{R} f(x) w^\alpha(x) \, dx - \sum_{k=-j}^{j} \lambda_k(w^\alpha) f(x_k),$$

where we take a fixed $\theta \in (0,1)$, n is sufficiently large, and $j = j(n)$ we define by $x_j = x_{j(n)} = \min\{x_k \mid x_k \geq \theta M_n\}$ and $M = [n(\theta/(1+\theta))^\alpha] \sim n$.

For smoother functions we introduce the Sobolev space

$$W^1_r(w^\alpha) = \left\{ f \in L^1_{w^\alpha} \mid f^{(r-1)} \in AC(\mathbb{R}) \text{ and } \|f^{(r)} w^\alpha\|_1 < +\infty \right\}, \quad r \geq 1,$$

5.1 Quadrature Formulae

equipped with the usual norm

$$\|f\|_{W_r^1(w^\alpha)} := \|fw^\alpha\|_1 + \|f^{(r)}w^\alpha\|_1.$$

Then, we can get the following result for the "truncated" Gaussian rule (see [93]):

Theorem 5.1.10 *For all $f \in W_1^1(w^\alpha)$ we have*

$$|\widetilde{R}_n(f)_{w^\alpha}| \leq C \left[\frac{M_n}{n} E_{2n-2}(f')_{w^\alpha,1} + e^{-An}\|fw^\alpha\|_1 \right],$$

where C and A depend on θ, but not of n and f.

Using this theorem and the Favard inequality, for all $f \in W_r^1(w^\alpha)$, we obtain

$$|\widetilde{R}_n(f)_{w^\alpha}| \leq C \left(\frac{M_n}{n} \right)^r \|f\|_{W_r^1(w^\alpha)}.$$

Remark 5.1.8 The previous error estimates of the Freud-Gaussian rules hold under the condition $\alpha > 1$. In the case $\alpha \leq 1$ the corresponding Freud-Gaussian rules have no any numerical interest. In fact, for $\alpha = 1$ the convergence of such a quadrature is very slow. It is not difficult to prove that

$$|\widetilde{R}_n(f)_{w^1}| \leq \frac{C}{\log n} \|f\|_{W_1^1(w^1)}.$$

If $\alpha < 1$, then $|\widetilde{R}_n(f)_{w^\alpha}|$ not converge to zero, because the polynomials are not dense in $L_{w^\alpha}^p$, $p \geq 1$ (cf. [98, p. 196]).

5.1.6 Product Integration Rules

We consider quadrature formulae, the so-called *product integration rules* for the numerical approximation of the integral

$$I(k, f) = \int_{-1}^{1} k(x) f(x) \, dx, \qquad (5.1.47)$$

where k is a Lebesgue integrable function on $[-1, 1]$ (not necessarily continuous, bounded, or of constant sign on $[-1, 1]$) and f is an arbitrary continuous function. Also, k can be a function of two variables $k(x, y)$, which is very often the case in integral equations. For a given set of distinct points $\{x_1, \ldots, x_n\}$ in $[-1, 1]$, such quadratures have the form

$$Q_n(f) = \sum_{v=1}^{n} A_v(k) f(x_v), \qquad (5.1.48)$$

where the weights $A_\nu(k)$ are uniquely determined by requiring that the formula be exact for any polynomial of degree at most $n-1$. In other words, the formula (5.1.48) should be interpolatory (cf. [78, pp. 84–90]), i.e.,

$$Q_n(f) = \int_{-1}^{1} k(x) L_n(f;x)\, dx,$$

where $L_n(f;x)$ is the Lagrange interpolation polynomial. Thus, the weights are given by

$$A_\nu(k) = \frac{1}{q_n'(x_\nu)} \int_{-1}^{1} \frac{k(x) q_n(x)}{x - x_\nu}\, dx, \quad k = 1, \ldots, n,$$

where $q_n(x) = (x - x_1) \cdots (x - x_n)$.

As we can see, the product integration rules are in fact a generalization of the weighted Newton-Cotes quadratures (5.1.9), in which the (nonnegative) weight function w is replaced by an arbitrary Lebesgue integrable function k.

Evidently, the choice of the nodes plays an essential role in the product integration rules. If the points are carefully chosen the rules have certain nice properties for a wide class of functions k and f (see [110, 111, 442–444, 449], etc.). Some popular sets of abscissas were used in the mentioned papers:

$$\mathcal{X}_1 = \left\{ -\cos\frac{(2\nu-1)\pi}{2n},\ \nu = 1, \ldots, n \right\}_n,$$

$$\mathcal{X}_2 = \left\{ \cos\frac{(n-\nu)\pi}{n-1},\ \nu = 1, \ldots, n \right\}_n,$$

$$\mathcal{X}_3 = \left\{ \cos\frac{2(n-\nu)\pi}{2n-1},\ \nu = 1, \ldots, n \right\}_n.$$

The case with nodes of the Jacobi polynomial $p_n(v^{\alpha,\beta};x)$, $\alpha, \beta > -1$, was investigated by Smith and Sloan [449] (for optimal systems of nodes see Sect. 4.2).

If k satisfies

$$\int_{-1}^{1} \left| k(x)(1-x)^{-\max[(2\alpha+1)/4,0]} (1+x)^{-\max[(2\beta+1)/4,0]} \right|^p dx < +\infty$$

for some $p > 1$, Smith and Sloan [449] proved that $Q_n(f)$ converges to $I(kf)$ as $n \to +\infty$ for any $f \in C^0$, and also

$$\lim_{n \to +\infty} \sum_{\nu=1}^{n} |A_\nu(k)| = \int_{-1}^{1} |k(x)|\, dx.$$

This is a generalization of an earlier result for Chebyshev nodes $\mathcal{X}_1$.

Because of the nice properties, the product integration rules are applied in several problems, especially in integral equations (see Sect. 5.2). Therefore, in the sequel

5.1 Quadrature Formulae

we suppose that k is a function of two variables, $f \in C_u$, $u = v^{\gamma,\delta}$, $\gamma, \delta \geq 0$, and consider the quadrature formula with the Jacobi nodes $x_\nu = x_\nu(v^{\alpha,\beta})$, $\nu = 1, \ldots, n$,

$$\int_{-1}^{1} k(x, y) f(x)\,dx = \int_{-1}^{1} k(x, y) L_n(v^{\alpha,\beta}, f; x)\,dx + e_n(f; y) =: G_n f + e_n(f; y),$$

where

$$G_n f = G_n(y) f = \sum_{\nu=1}^{n} A_\nu(y) f(x_\nu),$$

$e_n(f; y)$ is the remainder term and

$$A_\nu(y) = \int_{-1}^{1} k(x, y) \ell_\nu(x)\,dx.$$

Let $p_\nu(v^{\alpha,\beta}; \cdot)$ be orthonormal Jacobi polynomials, with parameters $\alpha, \beta > -1$ and, according to (2.2.6),

$$K_{n-1}(v^{\alpha,\beta}; x, y) = \sum_{i=0}^{n-1} p_i(v^{\alpha,\beta}; x) p_i(v^{\alpha,\beta}; y).$$

Since

$$\ell_\nu(x) = \lambda_\nu(v^{\alpha,\beta}) K_{n-1}(v^{\alpha,\beta}; x, x_\nu),$$

we have

$$A_\nu(y) = \lambda_\nu(v^{\alpha,\beta}) \int_{-1}^{1} k(x, y) K_{n-1}(v^{\alpha,\beta}; x, x_\nu)\,dx = \lambda_\nu(v^{\alpha,\beta}) \sum_{i=0}^{n-1} m_i p_i(v^{\alpha,\beta}; x_\nu),$$

where

$$m_i = \int_{-1}^{1} k(x, y) p_i(v^{\alpha,\beta}; x)\,dx \tag{5.1.49}$$

are the modified moments. There are several papers devoted to the evaluation and application of the modified moments. In particular, the modified moments for several kernels $k(x, y)$ with respect to the Chebyshev polynomials ($\alpha = \beta = -1/2$) and Gegenbauer polynomials were considered by Piessens and Branders [398] and Lewanowicz [260, 262], respectively. In these papers, certain linear recurrence relations for the modified moments were derived. Similar relations can be obtained for the moments (5.1.49) with respect to the Jacobi polynomials.

Since

$$e_n(f; y) = 0, \quad f \in \mathcal{P}_m, \quad m \leq n - 1,$$

for every fixed y, the stability of $G_n : C_u \to \mathbb{R}$ implies the convergence for all functions in C_u.

To prove the stability and convergence of the product formula G_n, we use the following result, which can be deduced from a theorem of Nevai [376].

Theorem 5.1.11 *Assume $u = v^{\gamma,\delta}$, $\gamma, \delta \geq 0$, and*

$$\sup_{|y|\leq 1} \int_{-1}^{1} \frac{|k(x,y)|}{u(x)} \log\left(2 + \frac{|k(x,y)|}{u(x)}\right) dx < +\infty. \tag{5.1.50}$$

Then, for all functions $f \in C_u$, we have

$$\sup_{|y|\leq 1} \int_{-1}^{1} |L_n(v^{\alpha,\beta}, f; x) k(x,y)| dx < C \|fu\|_\infty, \quad C \neq C(n, f),$$

if and only if

$$\sup_{|y|\leq 1} \int_{-1}^{1} \frac{|k(x,y)|}{\sqrt{v^{\alpha,\beta}(x)}\sqrt{1-x^2}} dx < +\infty \quad \text{and} \quad \int_{-1}^{1} \frac{\sqrt{v^{\alpha,\beta}(x)}\sqrt{1-x^2}}{u(x)} dx < +\infty. \tag{5.1.51}$$

Moreover, in general, we cannot replace (5.1.50) by a weaker condition.

Setting, for a fixed y,

$$\|G_n(y)\|_u = \sup_{f \in C_u} \frac{|G_n f|(y)}{\|fu\|_\infty},$$

we can now prove the following result:

Theorem 5.1.12 *Assume that $k(x,y)$ satisfies the condition (5.1.50). Then*

$$\sup_{|y|\leq 1} \sup_n \|G_n(y)\|_u < +\infty \tag{5.1.52}$$

if and only if (5.1.51) holds. Consequently,

$$\sup_{|y|\leq 1} |e_n(f,y)| \leq C E_{n-1}(f)_{u,\infty}. \tag{5.1.53}$$

Proof Since, for a fixed y,

$$\|G_n(y)\|_u = \sup_{f \in C_u} \frac{|G_n f|(y)}{\|fu\|_\infty} = \sup_{f \in C_u} \frac{\left|\int_{-1}^{1} k(x,y) L_n(v^{\alpha,\beta}, f; x) dx\right|}{\|fu\|_\infty}$$

$$= \sum_{\nu=1}^{n} \frac{|A_\nu(y)|}{u(x_\nu)},$$

5.1 Quadrature Formulae

the first part of this theorem is equivalent to Theorem 5.1.11. Consequently, $G_n f$ converges for all functions $f \subset C_u$.

Now, we prove the estimate (5.1.53). In fact, for all $y \in [-1, 1]$ and each $P_{n-1} \in \mathcal{P}_{n-1}$, we have

$$|e_n(f; y)| = |e_n(f - P_{n-1}; y)|$$

$$= \left| \int_{-1}^{1} \frac{k(x, y)}{u(x)} [f - P_{n-1}](x) u(x) \, dx \right.$$

$$\left. - \sum_{\nu=1}^{n} \frac{A_\nu(y)}{u(x_\nu)} [f - P_{n-1}](x_\nu) u(x_\nu) \right|$$

$$\leq \|(f - P_{n-1}) u\|_\infty \sup_{|y| \leq 1} \left(\int_{-1}^{1} \frac{|k(x, y)|}{u(x)} \, dx + \sum_{\nu=1}^{n} \frac{|A_\nu(y)|}{u(x_\nu)} \right).$$

Taking the infimum over all $P_{n-1} \in \mathcal{P}_{n-1}$, we get

$$\sup_{|y| \leq 1} |e_n(f, y)| \leq C E_{n-1}(f)_{u, \infty}.$$

In conclusion, if (5.1.50) and (5.1.51) hold, then the product rule is stable and convergent for each $f \in C_u$. $\square$

For very smooth functions we have:

Theorem 5.1.13 *Assume $f \in C^n[-1, 1]$ and*

$$\sup_{|y| \leq 1} \int_{-1}^{1} |k(x, y)| \left(1 + \frac{1}{\sqrt{v^{\alpha, \beta}(x) \sqrt{1 - x^2}}} \right) dx \leq C < +\infty.$$

Then

$$\sup_{|y| \leq 1} |e_n(f, y)| \leq C \frac{\|f^{(n)}\|_\infty}{2^n n!}, \quad C \neq C(n, f). \tag{5.1.54}$$

Proof Since $f \in C^n[-1, 1]$ and $q_n(x) = p_n(v^{\alpha, \beta}; x)/\gamma_n$ (γ_n is the leading coefficient in the orthonormal Jacobi polynomial of degree n), we use the interpolation error in the Cauchy form given by Theorem 1.4.4. Thus, for every fixed y, we obtain

$$|e_n(f, y)| \leq \frac{1}{\gamma_n n!} \int_{-1}^{1} |k(x, y)| |p_n(v^{\alpha, \beta}; x)| |f^{(n)}(\xi(x))| \, dx$$

$$\leq \frac{\|f^{(n)}\|_\infty}{\gamma_n n!} \int_{-1}^{1} |k(x, y)| |p_n(v^{\alpha, \beta}; x)| \, dx.$$

Since for $x \in [-1, 1]$ we have

$$|p_n(v^{\alpha,\beta}; x)| \leq C\left(\sqrt{1-x} + \frac{1}{n}\right)^{-\alpha-1/2}\left(\sqrt{1+x} + \frac{1}{n}\right)^{-\beta-1/2}$$

$$\leq C\left[1 + \frac{1}{\sqrt{v^{\alpha,\beta}(x)\varphi(x)}}\right]$$

and $\gamma_n \sim 2^n$, (5.1.54) follows immediately. □

5.1.7 Integration of Periodic Functions on the Real Line with Rational Weight

In this section we consider integrals of (2π)-periodic functions over the real line $\mathbb{R}$,

$$I(f) = \int_{\mathbb{R}} f(t)w(t)\,dt, \tag{5.1.55}$$

with a given even rational weight function of the form

$$w(t) = \frac{P(t^2)}{Q(t^2)}, \tag{5.1.56}$$

where

$$Q(t) = \prod_{k=1}^{n}(t + b_k^2), \qquad 0 < b_1 \leq b_2 \leq \cdots \leq b_n,$$

and $P(t)$ is a polynomial of degree at most $m < n$, which is nonnegative on the half line $[0, +\infty)$.

First the problem (5.1.55) can be simplified by obtaining the partial fraction decomposition of (5.1.56) in the form

$$w(t) = \sum_{j=1}^{m}\sum_{\nu=1}^{r_j} \frac{C_{j\nu}}{(t^2 + b_j^2)^\nu},$$

where the sum is over all pairs of conjugate complex poles $\pm ib_j$ of $Q(t^2)$, with corresponding multiplicities r_j ($j = 1, \ldots, m$). Here, $\sum_{j=1}^{m} r_j = n$.

Thus, without loss of generality, we can consider only weights of the form

$$w_\nu(t) = w_\nu(t; b) = \frac{1}{(t^2 + b^2)^\nu} \qquad (\nu \geq 1), \tag{5.1.57}$$

5.1 Quadrature Formulae

i.e., the integrals

$$I_\nu(f) = I_\nu(f;b) = \int_{\mathbb{R}} f(t) \frac{dt}{(t^2+b^2)^\nu} \qquad (b>0,\ \nu \geq 1). \tag{5.1.58}$$

First we show how to reduce the integral (5.1.58) to an integral on a finite interval. For this purpose we need the sum of the following series

$$W_\nu(\tau) = W_\nu(\tau;b) = \sum_{k=-\infty}^{+\infty} w_\nu(2k\pi + \tau) = \sum_{k=-\infty}^{+\infty} \frac{1}{\left[(2k\pi+\tau)^2 + b^2\right]^\nu}. \tag{5.1.59}$$

Since (cf. [403, p. 685])

$$\sum_{k=-\infty}^{+\infty} \frac{1}{(k+\alpha)^2 + \beta^2} = \frac{\pi}{\beta} \cdot \frac{\sinh 2\pi\beta}{\cosh 2\pi\beta - \cos 2\pi\alpha},$$

in the simplest but most important case $\nu = 1$, for $2\pi\alpha = \tau$ and $2\pi\beta = b$, we obtain

$$W_1(\tau) = W_1(\tau;b) = \frac{\sinh b}{2b} \cdot \frac{1}{\cosh b - \cos \tau}. \tag{5.1.60}$$

In the general case we can prove:

Lemma 5.1.1 *Let w_ν be given by (5.1.57), $\xi^{\pm} = -(\tau \pm ib)/(2\pi)$, and $\zeta = -\xi^+$. Then*

$$W_\nu(\tau) = -\frac{(2\pi)^{1-2\nu}}{2(\nu-1)!} \left\{ \lim_{z \to \xi^+} \frac{d^{\nu-1}}{dz^{\nu-1}} \left[\frac{\cot \pi z}{(z+\overline{\zeta})^\nu} \right] + \lim_{z \to \xi^-} \frac{d^{\nu-1}}{dz^{\nu-1}} \left[\frac{\cot \pi z}{(z+\zeta)^\nu} \right] \right\}.$$

The proof of this result can be done by an integration of the function $z \mapsto g(z) = \pi \cot(\pi z) w_\nu(2\pi z + \tau)$ over the rectangular contour C_N with vertices at the points $(N+\frac{1}{2})(\pm 1 \pm i)$, where $N \in \mathbb{N}$ is such that the poles $\xi^{\pm}$ of the function g are inside C_N. Then, taking $N \to +\infty$, the corresponding integral over C_N tends to zero, because $w_\nu(z) = O(1/z^{2\nu})$ when $z \to \infty$. Then, by Cauchy's residue theorem, we get

$$W_\nu(\tau) = \sum_{k=-\infty}^{+\infty} w_\nu(2k\pi + \tau) = -\left(\operatorname*{Res}_{z=\xi^+} g(z) + \operatorname*{Res}_{z=\xi^-} g(z) \right),$$

i.e., the desired result.

For $\nu = 1$, Lemma 5.1.1 gives (5.1.60). When $\nu = 2$ we obtain

$$W_2(\tau) = \frac{b \cosh b - \sinh b}{4b^3} \cdot \frac{\cos \tau + a}{(\cosh b - \cos \tau)^2},$$

where
$$a = \frac{\sinh 2b - 2b}{2b \cosh b - 2 \sinh b}.$$

In [295] the following result was proved:

Theorem 5.1.14 *Let* $x = \cos \tau$ *and* $c = \cosh b$. *Then*
$$W_\nu(\tau) = W_\nu(\tau; b) = \frac{p_\nu(x)}{(c-x)^\nu} \quad (\nu = 1, 2, \ldots), \tag{5.1.61}$$

where $p_\nu(x) = p_\nu(x; b)$ *is a nonnegative polynomial on* $[-1, 1]$ *of degree* $\nu - 1$. *These polynomials satisfy the recurrence relation*
$$p_{\nu+1}(x) = \frac{1}{2b\nu} \left\{ \nu\sqrt{c^2-1}\, p_\nu(x) - (c-x)\frac{\partial p_\nu(x)}{\partial b} \right\}, \tag{5.1.62}$$

where $p_1(x) = \sqrt{c^2 - 1}/(2b)$.

Proof We start with (5.1.60) written in the form $(c - x)W_1(\tau) = p_1(x)$, where
$$p_1(x) = \frac{\sinh b}{2b} = \frac{\sqrt{c^2-1}}{2b}, \quad x = \cos \tau, \ c = \cosh b.$$

Thus, the formula (5.1.61) is true for $\nu = 1$.

Suppose that (5.1.61) holds for some $\nu (\geq 1)$. Then, differentiating
$$(c-x)^\nu W_\nu(\tau) = p_\nu(x)$$

with respect to b, we get
$$\nu(c-x)^{\nu-1}\frac{dc}{db}W_\nu(\tau) + (c-x)^\nu \frac{\partial W_\nu(\tau)}{\partial b} = \frac{\partial p_\nu(x)}{\partial b},$$

from which it follows
$$\nu\sqrt{c^2-1}(c-x)^\nu W_\nu(\tau) - 2b\nu(c-x)^{\nu+1}W_{\nu+1}(\tau) = (c-x)\frac{\partial p_\nu(x)}{\partial b},$$

i.e.,
$$(c-x)^{\nu+1}W_{\nu+1}(\tau) = \frac{1}{2b\nu}\left\{\nu\sqrt{c^2-1}\,p_\nu(x) - (c-x)\frac{\partial p_\nu(x)}{\partial b}\right\} =: p_{\nu+1}(x).$$

Thus, the result is proved. □

We are ready now to give a transformation of the integral (5.1.57) to one on a finite interval. Putting $t = 2k\pi + \tau$ and using the periodicity of the function f,
$$f(t) = f(2k\pi + \tau) = f(\tau),$$

5.1 Quadrature Formulae

we have

$$I_\nu(f) = I_\nu(f;b) = \sum_{k=-\infty}^{+\infty} \int_{(2k-1)\pi}^{(2k+1)\pi} f(t)w_\nu(t)dt$$

$$= \sum_{k=-\infty}^{+\infty} \int_{-\pi}^{\pi} f(\tau)w_\nu(2k\pi+\tau)d\tau$$

$$= \int_{-\pi}^{\pi} f(\tau)\left(\sum_{k=-\infty}^{+\infty} w_\nu(2k\pi+\tau)\right)d\tau,$$

because of the uniform convergence of the series (5.1.59). Thus,

$$I_\nu(f) = \int_{-\pi}^{\pi} f(\tau)W_\nu(\tau)d\tau,$$

where $W_\nu(\tau)$ is defined by (5.1.59) and given by (5.1.61). We see that $W_\nu(-\tau) = W_\nu(\tau)$, i.e., W_ν is an even weight function.

Because of the last property of the weight function, we have

$$I_\nu(f) = I_\nu(f;b) = \int_{-\pi}^{0} f(\tau)W_\nu(\tau)d\tau + \int_{0}^{\pi} f(\tau)W_\nu(\tau)d\tau$$

$$= \int_{0}^{\pi} \big(f(\tau) + f(-\tau)\big)W_\nu(\tau)d\tau.$$

Changing the variables $\cos\tau = x$ and putting

$$f(\tau) + f(-\tau) = F(\cos\tau), \qquad (5.1.63)$$

we get the following result:

Theorem 5.1.15 *The integral* (5.1.58) *can be transformed into the form*

$$I_\nu(f) = I_\nu(f;b) = \int_{-1}^{1} F(x)\frac{p_\nu(x)}{(c-x)^\nu} \cdot \frac{dx}{\sqrt{1-x^2}}, \qquad (5.1.64)$$

where $c = \cosh b$, $p_\nu(x)$ is a polynomial determined by the recurrence relation (5.1.62), and F is defined by (5.1.63).

In order to evaluate the integral (5.1.64) it would seem more natural and simpler to apply the Gauss-Chebyshev quadrature formula. Therefore, we take $\Phi(x) = F(x)p_\nu(x)/(c-x)^\nu$ ($c > 1$) as an integrating function with respect to the Chebyshev weight $v_0(x) = (1-x^2)^{-1/2}$.

In this case, when for some $r \geq 1$ the function F satisfies the condition

$$\int_{-1}^{1} F^{(r)}(x)(\sqrt{1-x^2})^{(r-1)}dx < +\infty,$$

the error $R_n(\Phi)_{v_0}$ of the n-point Gauss-Chebyshev quadrature can be estimated in the form (see [290])

$$|R_n(\Phi)_{v_0}| \le \frac{A}{n^r} \int_{-1}^{1} \left| \frac{d^r}{dx^r} \left[\frac{F(x)p_v(x)}{(c-x)^v} \right] \right| (1-x^2)^{(r-1)/2} dx,$$

where $A > 0$ is a constant independent of Φ and n. Hence, when $c > 1$ is very close to 1, even if the integrand is bounded, it gives a very large bound.

On the contrary, if we take $v_\nu(x) = (1-x^2)^{-1/2}/(c-x)^\nu$ as a weight function (Szegő-Bernstein weight), then the error of the corresponding Gaussian formula is bounded as follows

$$|R_n(\Psi)| \le \frac{B}{n^r} \int_{-1}^{1} \left| \frac{d^r}{dx^r} [F(x)p_v(x)] \right| (1-x^2)^{(r-1)/2} \frac{dx}{(c-x)^v},$$

where $B > 0$ is a constant independent of Ψ ($\Psi(x) = F(x)p_v(x)$) and n. It is clear that the last integral is much smaller than the previous one. Also, some numerical evidences confirm this argument.

Thus, for evaluating the integral (5.1.64) it is more convenient to construct the Gaussian quadratures

$$\int_{-1}^{1} \Psi(x) d\lambda_\nu(x) = \sum_{k=1}^{n} A_k^{(n)} \Psi(x_k^{(n)}) + R_n(\Psi)_{v_\nu}, \quad R_n(\mathcal{P}_{2n-1})_{v_\nu} \equiv 0, \quad (5.1.65)$$

for the measure

$$d\lambda_\nu(x) = v_\nu(x) dx = \frac{dx}{(c-x)^\nu \sqrt{1-x^2}} \quad (\nu \ge 1), \quad (5.1.66)$$

where the function Ψ includes the algebraic polynomial $p_v(x)$ ($\Psi(x) = F(x)p_v(x)$).

It is well-known that the corresponding orthogonal polynomials $\pi_{n,\nu}(x)$ for the measure (5.1.66) can be explicitly calculated provided $\nu < 2n$ (cf. Szegő [470, p. 31]). On the other hand, there is a nonlinear algorithm to produce the recursion coefficients in the three-term recurrence relation for the monic polynomials $\pi_{n,\nu}(x)$,

$$\pi_{n+1,\nu}(x) = (x - \alpha_n^{(\nu)}) \pi_{n,\nu}(x) - \beta_n^{(\nu)} \pi_{n-1,\nu}(x), \quad n \ge 0,$$

$$\pi_{0,\nu}(x) = 1, \quad \pi_{-1,\nu}(x) = 0 \quad \left(\beta_0^{(\nu)} \triangleq m_0^{(\nu)} = \int_{-1}^{1} d\lambda_\nu(x) \right) \quad (5.1.67)$$

in terms of the ones for the polynomials $\pi_{n,\nu-1}(x)$ orthogonal with respect to the measure $d\lambda_{\nu-1}(x) = d\lambda_\nu(x)/(c-x)$. However, such an algorithm is numerically quite unstable unless c is very close to the support interval of the measure (see Gautschi [162, p. 102]). Two numerical algorithms for this purpose were also discussed in [127]. Our goal is to find analytic expressions for the recursion coefficients for some appropriate values of ν and then to apply a procedure for constructing the corresponding Gaussian formulae.

5.1 Quadrature Formulae

First we introduce the modified moments for $d\lambda_\nu(x)$ by the orthogonal polynomials $\pi_{n,\nu-1}(x)$,

$$m_n^{(\nu)} = \int_{-1}^{1} \pi_{n,\nu-1}(x) d\lambda_\nu(x) = \int_{-1}^{1} \frac{\pi_{n,\nu-1}(x) dx}{(c-x)^\nu \sqrt{1-x^2}} \quad (n \geq 0). \quad (5.1.68)$$

Notice that $\pi_{n,0}(x)$ are the monic Chebyshev polynomials of the first kind $\hat{T}_n(x)$ ($\hat{T}_0(x) = 1$, $\hat{T}_n(x) = 2^{1-n} \cos(n \arccos x)$, $n \geq 1$).

It is easy to prove the following auxiliary result:

Lemma 5.1.2 *For the first moment we have*

$$m_0^{(\nu)} = \int_{-1}^{1} \frac{dx}{(c-x)^\nu \sqrt{1-x^2}} = \frac{\pi Q_{\nu-1}(c)}{(c^2-1)^{\nu-1/2}},$$

where

$$Q_\nu(c) = \frac{1}{\nu} \left[(2\nu-1)c Q_{\nu-1}(c) - (c^2-1) Q'_{\nu-1}(c) \right], \quad Q_0(c) = 1.$$

Thus, we find

$$Q_1(c) = c, \quad Q_2(c) = c^2 + \frac{1}{2}, \quad Q_3(c) = c^3 + \frac{3}{2}c, \quad Q_4(c) = c^4 + 3c^2 + \frac{3}{8}, \text{ etc.}$$

According to [403, p. 415] we have

$$Q_\nu(c) = (c^2-1)^{\nu/2} P_\nu\left(\frac{c}{\sqrt{c^2-1}}\right),$$

where P_ν is the Legendre polynomial of order ν.

In order to get connection with the Chebyshev measure $d\lambda_0(x)$ it is convenient to put $Q_{-1}(c) = (c^2-1)^{-1/2}$. Then it gives $m_0^{(0)} = \pi$.

Now, we can prove ([295]):

Theorem 5.1.16 *The polynomials $\pi_{n,\nu}(x)$ can be expressed in terms of polynomials $\{\pi_{k,\nu-1}(x)\}$ in the form*

$$\pi_{n,\nu}(x) = \pi_{n,\nu-1}(x) - q_n^{(\nu)} \pi_{n-1,\nu-1}(x), \quad (5.1.69)$$

where $q_n^{(\nu)} = m_n^{(\nu)}/m_{n-1}^{(\nu)}$ and the moments $m_n^{(\nu)}$ are given by (5.1.68).

If $\alpha_n^{(\nu-1)}$ and $\beta_n^{(\nu-1)}$ are the recursion coefficients in (5.1.67) for polynomials $\{\pi_{n,\nu-1}(x)\}$, and

$$r_\nu = \frac{m_0^{(\nu-1)}}{m_0^{(\nu)}} = (c^2-1) \frac{Q_{\nu-2}(c)}{Q_{\nu-1}(c)}, \quad (5.1.70)$$

where the polynomials $Q_\nu(c)$ are defined in Lemma 5.1.2, then

$$q_1^{(\nu)} = c - \alpha_0^{(\nu-1)} - r_\nu, \qquad q_{n+1}^{(\nu)} = c - \alpha_n^{(\nu-1)} - \frac{\beta_n^{(\nu-1)}}{q_n^{(\nu)}} \qquad (n \geq 1). \qquad (5.1.71)$$

The coefficients in (5.1.67) are given by

$$\alpha_0^{(\nu)} = \alpha_0^{(\nu-1)} + q_1^{(\nu)}, \qquad \alpha_n^{(\nu)} = \alpha_n^{(\nu-1)} + q_{n+1}^{(\nu)} - q_n^{(\nu)} \qquad (n \geq 1)$$

and $\beta_0^{(\nu)} \triangleq m_0^{(\nu)} = \pi Q_{\nu-1}(c)/(c^2-1)^{\nu-1/2}$,

$$\beta_n^{(\nu)} = \beta_n^{(\nu-1)} + q_n^{(\nu)} \left[\alpha_n^{(\nu-1)} - \alpha_{n-1}^{(\nu-1)} + q_{n+1}^{(\nu)} - q_n^{(\nu)} \right] \qquad (n \geq 1).$$

Alternatively,

$$\beta_n^{(\nu)} = \beta_{n-1}^{(\nu-1)} \frac{q_n^{(\nu)}}{q_{n-1}^{(\nu)}} \qquad (n \geq 2).$$

Proof Putting

$$\pi_{n,\nu}(x) = \pi_{n,\nu-1}(x) - \sum_{k=0}^{n-1} q_{n,k}^{(\nu)} \pi_{k,\nu-1}(x)$$

and using the inner product

$$(f,g)_{\nu-1} = \int_{-1}^{1} f(x)g(x) d\lambda_{\nu-1}$$

with respect to the measure $d\lambda_{\nu-1}$, because of orthogonality, we obtain that for each $0 \leq i \leq n-2$,

$$(\pi_{n,\nu}, \pi_{i,\nu-1})_{\nu-1} = -q_{n,i}^{(\nu)} (\pi_{i,\nu-1}, \pi_{i,\nu-1})_{\nu-1}$$

and

$$(\pi_{n,\nu}, \pi_{i,\nu-1})_{\nu-1} = \int_{-1}^{1} (c-x) \pi_{n,\nu}(x) \pi_{i,\nu-1}(x) d\lambda_\nu(x)$$

$$= -\int_{-1}^{1} \pi_{n,\nu}(x) \left(x \pi_{i,\nu-1}(x) \right) d\lambda_\nu(x) = 0.$$

Thus, we conclude that $q_{n,i}^{(\nu)} = 0$ for such values of i and formula (5.1.69) holds, where we put $q_{n,n-1}^{(\nu)} \equiv q_n^{(\nu)}$.

From (5.1.69), because of orthogonality

$$0 = (\pi_{n,\nu}, 1)_\nu = (\pi_{n,\nu-1}, 1)_\nu - q_n^{(\nu)} (\pi_{n-1,\nu-1}, 1)_\nu,$$

5.1 Quadrature Formulae

we get $q_n^{(\nu)} = m_n^{(\nu)}/m_{n-1}^{(\nu)}$, where the modified moments are defined by (5.1.68). Using the recurrence relation for the polynomials $\{\pi_{n,\nu-1}(x)\}$ we find that

$$m_{n+1}^{(\nu)} = (c - \alpha_n^{(\nu-1)})m_n^{(\nu)} - \beta_n^{(\nu-1)} m_{n-1}^{(\nu)} - \int_{-1}^{1} \pi_{n,\nu-1}(x) d\lambda_{\nu-1},$$

which gives

$$m_1^{(\nu)} = (c - \alpha_0^{(\nu-1)})m_0^{(\nu)} - m_0^{(\nu-1)}$$

and

$$m_{n+1}^{(\nu)} = (c - \alpha_n^{(\nu-1)})m_n^{(\nu)} - \beta_n^{(\nu-1)} m_{n-1}^{(\nu)} \quad (n \geq 1).$$

These equalities give (5.1.71).

Finally, changing $\pi_{k,\nu}(x)$ ($k = n-1, n, n+1$) in the recurrence relation (5.1.67) by (5.1.69) and using the corresponding relation for polynomials $\{\pi_{k,\nu-1}(x)\}$ we get for $n \geq 2$

$$\pi_{n+1,\nu-1}(x) = \left(x - \alpha_n^{(\nu)} + q_{n+1}^{(\nu)} - q_n^{(\nu)}\right) \pi_{n,\nu-1}(x)$$
$$- \left(\beta_n^{(\nu)} - q_n^{(\nu)}\alpha_n^{(\nu)} + q_n^{(\nu)}\alpha_{n-1}^{(\nu-1)}\right) \pi_{n-1,\nu-1}(x)$$
$$+ \left(\beta_n^{(\nu)} q_{n-1}^{(\nu)} - \beta_{n-1}^{(\nu-1)} q_n^{(\nu)}\right) \pi_{n-2,\nu-1}(x).$$

Comparing with the recurrence relation for $\{\pi_{n,\nu-1}(x)\}$ we obtain formulas for the recursion coefficients. The case $n = 1$ should be separately considered. □

Notice that in Chebyshev case ($\nu = 0$) we have

$$\alpha_n^{(0)} = 0 \ (n \geq 0), \quad \beta_0^{(0)} \triangleq \pi, \ \beta_1^{(0)} = \frac{1}{2}, \ \beta_n^{(0)} = \frac{1}{4} \ (n \geq 2).$$

Also, from (5.1.70) it follows

$$r_1 = \sqrt{c^2 - 1}, \ r_2 = \frac{c^2 - 1}{c}, \ r_3 = \frac{c(c^2 - 1)}{c^2 + \frac{1}{2}}, \ r_4 = \frac{(c^2 + \frac{1}{2})(c^2 - 1)}{c^3 + \frac{3}{2}c}, \text{ etc.}$$

Now, using the previous theorem we give explicit expressions for the recursion coefficients for some important special cases, where $c = \cosh b$.

Case $\nu = 1$. Here we have $q_1^{(1)} = e^{-b}$, $q_n^{(1)} = \frac{1}{2} e^{-b}$ ($n \geq 2$), and the recursion coefficients

$$\alpha_0^{(1)} = e^{-b}, \ \alpha_1^{(1)} = -\frac{1}{2} e^{-b}, \ \alpha_n^{(1)} = 0 \ (n \geq 2);$$

$$\beta_0^{(1)} \triangleq \frac{\pi}{\sinh b}, \ \beta_1^{(1)} = \frac{1}{2}\left(1 - e^{-2b}\right), \ \beta_n^{(1)} = \frac{1}{4} \ (n \geq 2).$$

Case $\nu = 2$. Here, $q_1^{(2)} = e^{-b}\tanh b$, $q_n^{(2)} = \frac{1}{2}e^{-b}$ $(n \geq 2)$, and

$$\alpha_0^{(2)} = \frac{1}{\cosh b}, \quad \alpha_1^{(2)} = -e^{-b}\tanh b, \quad \alpha_n^{(2)} = 0 \ (n \geq 2);$$

$$\beta_0^{(2)} \triangleq \frac{\pi \cosh b}{\sinh^3 b}, \quad \beta_1^{(2)} = \frac{1}{2}\left(1 - e^{-2b}\right)\tanh^2 b, \quad \beta_2^{(2)} = \frac{1}{4}\left(1 + e^{-2b}\right),$$

and $\beta_n^{(2)} = \frac{1}{4}$ $(n \geq 3)$.

Case $\nu = 3$. Here, $q_1^{(3)} = \sinh b \tanh b/(2+\cosh(2b))$, $q_2^{(3)} = e^{-2b}\cosh b$, $q_n^{(3)} = \frac{1}{2}e^{-b}$ $(n \geq 3)$, and

$$\alpha_0^{(3)} = \frac{3\cosh b}{2+\cosh(2b)}, \quad \alpha_1^{(3)} = e^{-2b}\cosh b + \left(e^{-b} + \frac{\sinh b}{2+\cosh(2b)}\right)\tanh b,$$

$$\alpha_2^{(3)} = \frac{1}{2}e^{-3b}, \quad \alpha_n^{(3)} = 0 \ (n \geq 3);$$

$$\beta_0^{(3)} \triangleq \frac{\pi\left(\cosh^2 b + \frac{1}{2}\right)}{\sinh^5 b}, \quad \beta_1^{(3)} = \frac{(1-e^{-2b})^4}{2(1-4e^{-2b}+e^{-4b})^2},$$

$$\beta_2^{(3)} = \frac{1}{4}\left(1 + 3e^{-2b} - 3e^{-4b} - e^{-6b}\right), \quad \beta_n^{(3)} = \frac{1}{4} \ (n \geq 3).$$

As we can see the recurrence coefficients for the polynomials $\pi_{n,\nu}(x)$ reduce to the corresponding coefficients for the Chebyshev polynomials for $n \geq n_0$ $(n \in \mathbb{N})$. Precisely, the calculations show that

$$\alpha_n^{(\nu)} = \alpha_n^{(0)} = 0 \qquad n \geq \left[\frac{\nu+1}{2}\right] + 1$$

and

$$\beta_n^{(\nu)} = \beta_n^{(0)} = \frac{1}{4} \qquad n > \left[\frac{\nu}{2}\right] + 1.$$

In order to illustrate the presented transformation method, we consider a few numerical examples. All computations were done in D-arithmetic with machine precision $\approx 2.22 \times 10^{-16}$.

Example 5.1.2 Consider integrals of the form

$$I_\nu(f;b) = \int_{-\infty}^{+\infty} \frac{2\sin 2t - 1}{3 + 2\cos 3t} \cdot \frac{e^{-\cos 2t}}{(t^2+b^2)^\nu}\,dt \qquad (\nu \geq 1).$$

5.1 Quadrature Formulae

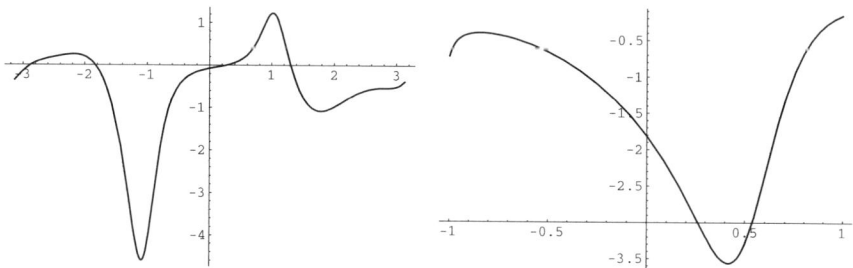

Fig. 5.1.2 The periodic function $f(t)$ (*left*); The function $F(x)$ obtained by the transformation (5.1.63) (*right*)

The function
$$f(t) = \frac{2\sin 2t - 1}{3 + 2\cos 3t} e^{-\cos 2t}$$
is a (2π)-periodic and its graph on the interval $[-\pi, \pi]$ is displayed in Fig. 5.1.2.

Since
$$f(\tau) + f(-\tau) = -\frac{2e^{-\cos 2\tau}}{3 + 2\cos 3\tau},$$
putting $x = \cos \tau$ and using (5.1.63), we find
$$F(x) = F(\cos \tau) = f(\tau) + f(-\tau) = \frac{-2e^{1-2x^2}}{3 - 6x + 8x^3}$$
and, according to (5.1.64),
$$I_\nu(f; b) = \int_{-1}^{1} F(x) \frac{\pi_\nu(x)}{(c-x)^\nu} \frac{dx}{\sqrt{1-x^2}},$$
where $c = \cosh b$.

Let $\nu = 1$. Applying Gaussian quadratures with the Chebyshev weight (ChW) for $n = 5(5)50$ and taking $b = 10^m$ ($m = -2, -1, 0$) we get approximations of $I_1(f; b)$ with relative errors given in Table 5.1.2. Numbers in parentheses indicate decimal exponents.

Taking Gaussian quadratures for $n = 5(5)50$, with the Szegő-Bernstein weight (SBW), $v_1(x) = (1-x^2)^{-1/2}/(c-x)$, the corresponding errors are also presented in the same table. The corresponding exact values of $I_1(f; b)$ are obtained using Gaussian quadratures with the SBW in Q-arithmetic (machine precision $\approx 1.93 \times 10^{-34}$):

$I_1(f; 0.01) = -0.2586588216241823127882\ldots \times 10^2 \quad (c = 1.0000500\ldots),$

$I_1(f; 0.10) = -0.4968012877996286228355\ldots \times 10^1 \quad (c = 1.0050041\ldots),$

$I_1(f; 1.00) = -0.1673215409745331112726\ldots \times 10^1 \quad (c = 1.5430806\ldots).$

Table 5.1.2 Relative errors in Gaussian approximations of the integral $I_1(f;b)$ with respect to the Chebyshev weight (ChW) and the Szegő-Bernstein weight (SBW)

b	$b=0.01$		$b=0.1$		$b=1.0$	
n	ChW	SBW	ChW	SBW	ChW	SBW
5	8.4(−1)	1.3(−2)	2.2(−1)	6.3(−2)	1.2(−2)	5.5(−2)
10	8.0(−1)	2.4(−4)	1.1(−1)	1.5(−3)	2.7(−3)	3.5(−3)
15	7.6(−1)	1.1(−5)	4.3(−2)	4.2(−5)	1.5(−4)	7.0(−5)
20	7.2(−1)	9.0(−7)	1.6(−2)	4.4(−6)	1.0(−6)	3.5(−6)
25	6.7(−1)	1.6(−8)	6.0(−3)	9.7(−8)	1.8(−7)	2.3(−7)
30	6.3(−1)	7.4(−10)	2.2(−3)	2.8(−9)	9.8(−9)	4.6(−9)
35	5.9(−1)	5.9(−11)	8.2(−4)	2.9(−10)	6.7(−11)	2.3(−10)
40	5.5(−1)	1.0(−12)	3.0(−4)	6.4(−12)	1.2(−11)	1.5(−11)
45	5.2(−1)	4.8(−14)	1.1(−4)	1.8(−13)	6.5(−13)	3.1(−13)
50	4.8(−1)	4.7(−15)	4.1(−5)	1.9(−14)	5.2(−15)	1.6(−14)

We see that for smaller values of b (c is close to 1) the Gauss-Chebyshev quadratures (ChW) cannot be used directly. When b increases both quadratures become comparable. However, by writing $I_1(f;b)$ in the form

$$I_1(f;b) = \frac{\pi}{2b} F(c) - \frac{\sinh b}{2b} \int_{-1}^{1} \frac{F(c)-F(x)}{c-x} \frac{dx}{\sqrt{1-x^2}},$$

the Gauss-Chebyshev quadratures can be applied directly.

Consider now the case $v=2$, with the functions ϕ_k and the corresponding weights v_k ($k=0,1,2$), where

$$\phi_k(x) = \frac{F(x)p_2(x)}{(c-x)^{2-k}}, \quad v_k(x) = \frac{1}{(c-x)^k \sqrt{1-x^2}}.$$

Applying the Gaussian quadratures with the Chebyshev weight ChW ($k=0$) and the Szegő-Bernstein weights SBW$_1$ ($k=1$) and SBW$_2$ ($k=2$) we get approximations of the integral $I_2(f;b)$. The exact values of this integral for some selected b are:

$$I_2(f;0.01) = -0.1156183821140487028202\ldots \times 10^6,$$
$$I_2(f;0.10) = -0.1214706913588412300593\ldots \times 10^3.$$

The relative errors in Gaussian approximations for $n = 5(5)50$ are presented in Table 5.1.3.

The advantage of quadrature formulas for $k=2$ (in this case $v=2$) is evident. When b increases all quadratures give similar results.

5.1 Quadrature Formulae

Table 5.1.3 Relative errors in Gaussian approximations of the integral $I_2(f;b)$ with respect to the Chebyshev weight (ChW) and to the Szegő-Bernstein weights (SBW$_1$ and SBW$_2$)

b	b = 0.01			b = 0.1		
n	ChW	SBW$_1$	SBW$_2$	ChW	SBW$_1$	SBW$_2$
5	1.0(0)	9.1(−1)	5.5(−7)	8.9(−1)	3.7(−1)	1.1(−3)
10	1.0(0)	8.3(−1)	1.0(−7)	6.2(−1)	1.4(−1)	6.7(−5)
15	1.0(0)	7.5(−1)	4.7(−9)	3.4(−1)	5.0(−2)	4.3(−6)
20	9.9(−1)	6.8(−1)	1.9(−11)	1.6(−1)	1.9(−2)	6.4(−8)
25	9.9(−1)	6.1(−1)	4.6(−12)	7.4(−2)	6.8(−3)	4.4(−9)
30	9.8(−1)	5.5(−1)	2.6(−12)	3.2(−2)	2.5(−3)	2.8(−10)
35	9.7(−1)	5.0(−1)	2.3(−12)	1.3(−2)	9.2(−4)	4.3(−12)
40	9.6(−1)	4.5(−1)	2.3(−12)	5.6(−3)	3.4(−4)	3.1(−13)
45	9.5(−1)	4.1(−1)	2.3(−12)	2.3(−3)	1.2(−4)	1.4(−15)
50	9.3(−1)	3.7(−1)	2.3(−12)	9.2(−4)	4.6(−5)	2.0(−14)

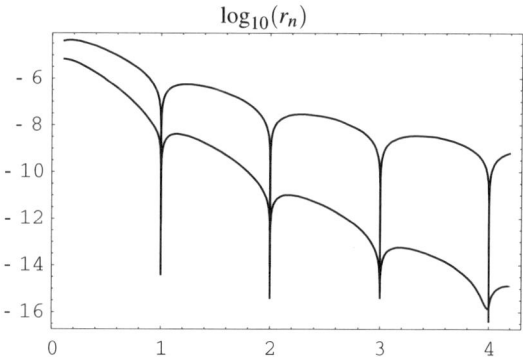

Fig. 5.1.3 Relative errors in Gaussian approximations with $n = 5$ (*upper curve*) and $n = 20$ nodes (*lower curve*) for $0 \leq \alpha < 4.5$

Example 5.1.3 Consider now the integral (5.1.58), with a nonanalytic function $f(t) = |\cos(t/2)|^{2\alpha}$ ($\alpha > 0$). After the transformation we obtain the integral

$$J(\alpha) = A \int_{-1}^{1} \left(\frac{1+x}{2}\right)^{\alpha} \frac{dx}{(c-x)\sqrt{1-x^2}},$$

where $A = 2p_1(x) = \sinh b/b$.

In order to evaluate this integral, we apply the Gaussian rule in n points with the Szegő-Bernstein weight ($\nu = 1$). A typical behaviour of the relative error r_n of Gaussian approximations with respect to the parameter α ($0 \leq \alpha < 4.5$) is displayed in Fig. 5.1.3 in the log-scale. Two cases for $n = 5$ and $n = 20$ are given, whereas $b = 0.01$. It is clear that the rapid increase of accuracy is achieved when the parameter α tends to an integer (i.e., when f becomes an analytic function).

5.2 Integral Equations

5.2.1 *Some Basic Facts*

In this section we introduce some numerical methods for computing approximate solutions of some classes of Fredholm integral equations of the second kind. Such methods are based on the so-called *Approximation and Polynomial Interpolation Theory* and lead to the construction of a polynomial sequence converging to the exact solution in some weighted uniform norm. However, the construction of such a sequence requires the solution of systems of linear equations that might be ill-conditioned. We devote our attention to this problem and obtain, as a useful result, that the systems of linear equations furnished by our methods are well conditioned (except for a log factor).

At the beginning we recall some basic fact on the linear functional analysis. The notion of a Banach space we assume as well-known, and we will denote such a space by $(X, \|\cdot\|)$, where $\|\cdot\|$ is the norm defined on X. Therefore, if $K : X \to X$ is a linear bounded map, its norm is defined by

$$\|K\| = \|K\|_{X \to X} = \sup_{\|f\|=1} \|Kf\|.$$

In the sequel we assume that X is a set of continuous functions defined on finite intervals and $\|\cdot\|$ is the sup-norm. In these spaces which we will specify, if an occasion requires, the Weierstrass theorem on the polynomial approximation holds. Consequently, a linear operator $K : X \to X$ is compact if and only if (cf. [475, p. 44])

$$\lim_{n \to +\infty} \sup_{\substack{f \in X \\ \|f\|=1}} E_n(Kf) = 0, \qquad (5.2.1)$$

where

$$E_n(F) = \inf_{p \in \mathcal{P}_n} \|F - p\|, \quad F \in X,$$

and $\mathcal{P}_n$ is the set of all algebraic polynomials of degree at most n.

We consider the Fredholm equation of second kind

$$(I - K)f = g, \qquad (5.2.2)$$

where I is the identity operator, K is a linear compact operator, the free term g is a known function and f is the unknown solution. The following theorems are sufficient for our aims.

Theorem 5.2.1 *Let $(X, \|\cdot\|)$ be a Banach space and let $K : X \to X$ be a linear operator. If $\|K\| \leq q < 1$ then $(I - K)^{-1}$ exists and*

$$\|(I - K)^{-1}\| \leq \frac{1}{1 - \|K\|} \leq \frac{1}{1 - q}.$$

5.2 Integral Equations

Moreover, let $K : X \to X$ *be a linear operator and* $\{K_n\}_n$ *be a sequence of linear operators, with* $K_n : X \to X$, *such that*

$$\lim_n \|K - K_n\| = 0.$$

If $(I - K)^{-1}$ *exists, then for n sufficiently large (say* $n > n_0$), $(I - K_n)^{-1}$ *exists and*

$$\|(I - K_n)^{-1}\| \le 2\|(I - K)^{-1}\| \tag{5.2.3}$$

holds.

Proof The first statement is the well-known J. Von Neuman theorem and we omit the proof of it. To prove (5.2.3), we first note that, according to our assumptions, $I - K$ has a bounded inverse and then

$$I - K_n = I - K + K - K_n = (I - K)[I - (I - K)^{-1}(K_n - K)]$$
$$:= (I - K)(I - D),$$

where

$$D := (I - K)^{-1}(K_n - K).$$

Moreover, since we assume that the sequence $\{K_n\}_n$ converges to K, there exists n_0 such that, for any $n > n_0$, we have

$$\|D\| \le \|(I - K)^{-1}\| \, \|K - K_n\| < \frac{1}{2}.$$

Now, by the Von Neuman theorem, $(I - D)^{-1}$ exists, as well as

$$(I - K_n)^{-1} = (I - D)^{-1}(I - K)^{-1},$$

with

$$\|(I - K_n)^{-1}\| \le \|(I - K)^{-1}\| \|(I - D)^{-1}\|$$
$$\le \frac{\|(I - K)^{-1}\|}{1 - \|(I - K)^{-1}\| \|(K - K_n)^{-1}\|} < 2\|(I - K)^{-1}\|$$

and (5.2.3) is proved. □

Theorem 5.2.2 *Let* K *be a linear bounded operator in a Banach space* X *and consider* (5.2.2). *Assume that* $(I - K)^{-1}$ *exists and denote by* f^* *the solution of* (5.2.2) *for a given function* g. *Moreover, consider the sequence of equations*

$$(I - K_n)f_n = g_n, \quad n = 1, 2, \ldots, \tag{5.2.4}$$

where $g_n, f_n \in X$, f_n is unknown and $\{K_n\}_n$, $K_n : X \to X$, is a sequence of linear bounded operators. If $\|g_n - g\| \to 0$ and $\|K - K_n\| \to 0$, then, for $n > n_0$, (5.2.4) has a unique solution f_n^* and we have

$$\|f^* - f_n^*\| \leq C\big(\|g - g_n\| + \|g\|\|K - K_n\|\big), \tag{5.2.5}$$

where $C \neq C(n, f^*, g)$.

Proof By Theorem 5.2.1 we conclude that $(I - K_n)^{-1}$ exists and (5.2.4) has a unique solution for all sufficiently large n, say $n > n_0$. Moreover, we have

$$\begin{aligned}(I - K_n)(f^* - f_n^*) &= (I - K_n)f^* - (I - K_n)f_n^* \\ &= (I - K)f^* - (K_n - K)f^* - (I - K_n)f_n^* \\ &= g - g_n - (K_n - K)f^*,\end{aligned}$$

from which we deduce

$$\begin{aligned}f^* - f_n^* &= (I - K_n)^{-1}[(g - g_n) - (K_n - K)f^*] \\ &= (I - K_n)^{-1}[(g - g_n) - (K_n - K)(I - K)^{-1}g]\end{aligned}$$

and, consequently,

$$\begin{aligned}\|f^* - f_n^*\| &\leq \|(I - K_n)^{-1}\|\|(g - g_n) + (K_n - K)(I - K)^{-1}g\| \\ &\leq \|(I - K_n)^{-1}\|\Big[\|g - g_n\| + \|(I - K)^{-1}\|\|K_n - K\|\|g\|\Big],\end{aligned}$$

i.e., (5.2.5). □

Now, we want to point out a numerical problem that appears in all procedures we are going to employ. We consider (5.2.2) with $g \in \mathcal{P}_{n-1}$ and a degenerate kernel,

$$(Kf)(y) = \lambda \int_{-1}^{1} k(x, y) f(x)\, dx, \quad k(x, y) = \sum_{i=1}^{n} a_i(x) b_i(y),$$

where a_i, $i = 1, \ldots, n$, are analytic functions and b_i are algebraic polynomials such that $\deg b_i = i - 1$, $i = 1, \ldots, n$ (e.g., monomials $b_i(y) = y^{i-1}$).
Since

$$(Kf)(y) = \lambda \sum_{i=1}^{n} b_i(y) \int_{-1}^{1} a_i(x) f(x)\, dx$$

is a polynomial of degree at most $n - 1$, if a solution f of (5.2.2) exists, then it is also a polynomial of degree at most $n - 1$. Choosing an arbitrary triangular infinity

5.2 Integral Equations

array of knots belonging to $[-1, 1]$

$$\mathcal{X} = \left\{ \begin{matrix} x_{1,1} \\ x_{2,1} \; x_{2,2} \\ \vdots \quad \ddots \\ x_{n,1} \; x_{n,2} \; \cdots \; x_{n,n} \\ \vdots \quad \quad \quad \ddots \end{matrix} \right\},$$

we denote by

$$\ell_k(x) = \ell_{n,k}(\mathcal{X}; x) := \frac{q_n(x)}{q_n'(x_{n,k})(x - x_{n,k})}, \quad k = 1, \ldots, n,$$

the fundamental Lagrange polynomials based on the knots $x_k := x_{n,k}$, $k = 1, \ldots, n$, of the n-th row of $\mathcal{X}$, where $q_n(x) = \prod_{k=1}^{n}(x - x_{n,k})$ (see Sect. 1.4.2).
Expanding g and f in the Lagrange basis $\{\ell_1, \ldots, \ell_n\}$, we can write

$$g(y) = \sum_{k=1}^{n} \ell_k(y) g(x_k), \quad f(y) = \sum_{k=1}^{n} \ell_k(y) f_k,$$

as well as

$$(Kf)(y) = \sum_{i=1}^{n} \ell_i(y)(Kf)(x_i)$$

with

$$(Kf)(x_i) = \lambda \int_{-1}^{1} k(x, x_i) f(x) \, dx = \lambda \sum_{k=1}^{n} f_k \int_{-1}^{1} k(x, x_i) \ell_k(x) \, dx,$$

i.e.,

$$(Kf)(x_i) = \lambda \sum_{k=1}^{n} f_k C_{i,k}.$$

Assuming that the coefficients $C_{i,k} = \int_{-1}^{1} k(x, x_i) \ell_k(x) \, dx$ can be computed exactly, (5.2.2) is equivalent to the following system of linear equations

$$\begin{bmatrix} 1 - \lambda C_{1,1} & -\lambda C_{1,2} & \cdots & -\lambda C_{1,n} \\ -\lambda C_{2,1} & 1 - \lambda C_{2,2} & & -\lambda C_{2,n} \\ \vdots & & & \\ -\lambda C_{n,1} & -\lambda C_{n,2} & & 1 - \lambda C_{n,n} \end{bmatrix} \begin{bmatrix} f_1 \\ f_2 \\ \vdots \\ f_n \end{bmatrix} = \begin{bmatrix} g(x_1) \\ g(x_2) \\ \vdots \\ g(x_n) \end{bmatrix}, \quad (5.2.6)$$

i.e., $A_n \mathbf{f} = \mathbf{g}$, where

$$A_n = [\delta_{i,k} - \lambda C_{i,k}]_{i,k=1}^{n}, \quad \mathbf{f} = [f_1 \; f_2 \; \cdots \; f_n]^\mathrm{T}, \quad \mathbf{g} = [g(x_1) \; g(x_2) \; \cdots \; g(x_n)]^\mathrm{T}.$$

If λ is not an eigenvalue of the matrix A_n, then

$$f^*(y) = \sum_{i=1}^{n} \ell_i(y) f_i^*$$

is the unique solution of (5.2.2), where $\mathbf{f}^* = [f_1^* \ \ldots \ f_n^*]^T$ is the unique solution of the system (5.2.6).

The following proposition is crucial for the computation of $f_1^*, \ldots, f_n^*$.

Proposition 5.2.1 *Denoting by A_n the matrix of the system of equations (5.2.6), we have*

$$\mathrm{cond}\,(A_n) \leq \mathrm{cond}\,(I - K)\, \|L_n(\mathcal{X})\|_\infty^2,$$

where $\mathrm{cond}\,(A_n) = \|A_n\|_\infty \|A_n^{-1}\|_\infty$ *and* $\|L_n(\mathcal{X})\|_\infty = \sup_{|y| \leq 1} \sum_{i=1}^{n} |\ell_i(y)|$ *is the n-th Lebesgue constant.*

Proof Letting $\mathbf{a} = [f_1 \ \ldots \ f_n]^T$ and $\mathbf{b} = [g(x_1) \ \ldots \ g(x_n)]^T$, we can write (5.2.6) as $A_n \mathbf{a} = \mathbf{b}$. If λ is not an eigenvalue of A_n, then (5.2.2) has a unique solution for every $g \in \mathcal{P}_{n-1}$ if and only if $A_n \mathbf{a} = \mathbf{b}$ has a unique solution for every $\mathbf{b} \in \mathbb{C}^n$. Therefore, for all $\theta = [\theta_1 \ \ldots \ \theta_n]^T$ there exists $\eta = [\eta_1 \ \ldots \ \eta_n]^T$ such that $A_n \theta = \eta$ if and only if $(I - K)\widetilde{\theta}(y) = \widetilde{\eta}(y)$, where

$$\widetilde{\theta}(y) = \sum_{i=1}^{n} \ell_i(y)\theta_i, \quad \theta_i = \widetilde{\theta}(x_i) \quad \text{and} \quad \widetilde{\eta}(y) = \sum_{i=1}^{n} \ell_i(y)\eta_i, \quad \eta_i = \widetilde{\eta}(x_i).$$

Then, for all θ,

$$\|A_n\theta\|_{l^\infty} = \|\eta\|_{l^\infty} = |\eta_\nu| = |\widetilde{\eta}(x_\nu)| \leq \|\widetilde{\eta}\|_\infty$$
$$= \|(I - K)\widetilde{\theta}\|_\infty \leq \|(I - K)\big|_{\mathcal{P}_{n-1}}\| \|\widetilde{\theta}\|_\infty$$
$$\leq \|(I - K)\big|_{\mathcal{P}_{n-1}}\| \|\theta\|_{l^\infty} \|L_n(\mathcal{X})\|_\infty,$$

where $|\eta_\nu| = \max_{1 \leq i \leq n} |\eta_i|$. Analogously, for all η we have

$$\|A_n^{-1}\eta\|_{l^\infty} \leq \|(I - K)^{-1}\big|_{\mathcal{P}_{n-1}}\| \|\eta\|_{l^\infty} \|L_n(\mathcal{X})\|_\infty$$

and, consequently, this proposition follows. $\square$

Regarding this result, if the entries of the array $\mathcal{X}$ are the equal-spaced points, we have (see (1.4.31))

$$\|L_n(\mathcal{X})\|_\infty \sim \frac{2^n}{e n \log n}.$$

In that case the serious problems in the computation of the solution $[f_1 \ \ldots \ f_n]^T$ of the system (5.2.6) can be appeared. Therefore, a choice of the array of knots $\mathcal{X}$ for

5.2 Integral Equations

which
$$\|L_n(\mathcal{X})\|_\infty \sim \log n$$
holds is recommended.

Now, we recall also some results in polynomial approximation. Taking a Jacobi weight $v^{\gamma,\delta}(x) := (1-x)^\gamma (1+x)^\delta$, with parameters $\gamma, \delta \geq 0$, in Sects. 2.5.1 and 2.5.2 we introduced the space
$$C^0_{v^{\gamma,\delta}} := \left\{ f \in C^0((-1,1)) \;\middle|\; \lim_{|x|\to 1} |(f v^{\gamma,\delta})(x)| = 0 \right\},$$
as well as the Zygmund space
$$Z_s = Z_s(v^{\gamma,\delta})$$
$$:= \left\{ f \in C^0_{v^{\gamma,\delta}} \;\middle|\; \|f\|_{Z_s(v^{\gamma,\delta})} := \|f v^{\gamma,\delta}\|_\infty + \sup_{t>0} \frac{\Omega^r_\varphi(f,t)_{v^{\gamma,\delta},\infty}}{t^s} < +\infty \right\},$$
where $r > s > 0$ and
$$\Omega^r_\varphi(f,t)_{v^{\gamma,\delta},\infty} := \sup_{0 < h \leq t} \|(\Delta^r_{h\varphi} f) v^{\gamma,\delta}\|_{C^0(I_{h,r})},$$
with $r \in \mathbb{N}$, $t \in \mathbb{R}^+$, $\varphi(x) := \sqrt{1-x^2}$, $I_{h,r} := [-1+4r^2h^2, 1-4r^2h^2]$ and
$$\Delta^r_{h\varphi} f(x) := \sum_{i=0}^{r} (-1)^i \binom{r}{i} f\left(x + \frac{rh}{2}\varphi(x) - ih\varphi(x) \right).$$

For $f \in C^0_{v^{\gamma,\delta}}$, we define the error of the best weighted approximation of this function by means of algebraic polynomials of degree at most n by
$$E_n(f)_{v^{\gamma,\delta},\infty} = \inf_{P \in \mathcal{P}_n} \|(f-P) v^{\gamma,\delta}\|_\infty.$$

In $C^0_{v^{\gamma,\delta}}$ a weak Jackson type theorem
$$E_n(f)_{v^{\gamma,\delta},\infty} \leq C \int_0^{1/n} \frac{\Omega^r_\varphi(f,t)_{v^{\gamma,\delta},\infty}}{t} \, dt, \quad C \neq C(n,f), \tag{5.2.7}$$
and a Stechkin type inequality
$$\Omega^r_\varphi\left(f, \frac{1}{n}\right)_{v^{\gamma,\delta},\infty} \leq \frac{C}{n^r} \sum_{j=0}^{n} (1+j)^{r-1} E_j(f)_{v^{\gamma,\delta},\infty}, \quad C = C(r),$$
hold. In particular, if $f \in Z_s(v^{\gamma,\delta})$, we have
$$E_n(f)_{v^{\gamma,\delta},\infty} \leq \frac{C}{n^s} \|f\|_{Z_s(v^{\gamma,\delta})}. \tag{5.2.8}$$

Let $\alpha, \beta > -1$ and $L_n(v^{\alpha,\beta}, f)$ be the Lagrange polynomial interpolating a continuous function f on $(-1, 1)$ at the zeros of $p_n(v^{\alpha,\beta})$, i.e.,

$$L_n(v^{\alpha,\beta}, f; x) = \sum_{k=1}^{n} \ell_k(v^{\alpha,\beta}; x) f(x_k),$$

where $\ell_k(v^{\alpha,\beta})$ is the k-th fundamental Lagrange polynomial. Setting

$$\|L_n(v^{\alpha,\beta})\|_{v^{\gamma,\delta},\infty} = \max_{|x| \leq 1} \left(v^{\gamma,\delta}(x) \sum_{k=1}^{n} \frac{|\ell_k(v^{\alpha,\beta}; x)|}{v^{\gamma,\delta}(x_k)} \right),$$

where x_k, $k = 1, \ldots, n$, are the zeros of $p_n(v^{\alpha,\beta})$, we recall the following results (see Theorem 4.3.1):

Theorem 5.2.3 *Let $v^{\alpha,\beta}$ and $v^{\gamma,\delta}$ be two Jacobi weights with parameters $\alpha, \beta > -1$ and $\gamma, \delta \geq 0$. Then, we have $\|L_n(v^{\alpha,\beta})\|_{v^{\gamma,\delta},\infty} \sim \log n$, or equivalently*

$$(\forall f \in C^0_{v^{\gamma,\delta}}) \quad \|[f - L_n(v^{\alpha,\beta}, f)]v^{\gamma,\delta}\|_\infty \leq C \log n \, E_{n-1}(f)_{v^{\gamma,\delta},\infty},$$

where $C \neq C(n, f)$, if and only if the following conditions

$$\frac{\alpha}{2} + \frac{1}{4} \leq \gamma \leq \frac{\alpha}{2} + \frac{5}{4}, \quad \frac{\beta}{2} + \frac{1}{4} \leq \delta \leq \frac{\beta}{2} + \frac{5}{4}$$

are satisfied.

Remark 5.2.1 This result allows us to find a space $C^0_{v^{\gamma,\delta}}$ in which the interpolation process $L_n(v^{\alpha,\beta})$ is "optimal", i.e., $\|L_n(v^{\alpha,\beta})\|_{v^{\gamma,\delta},\infty} \sim \log n$. In particular, if $\alpha, \beta \leq -1/2$ we can choose $\gamma = \delta = 0$.

Theorem 5.2.4 *Let u and w be two Jacobi weights and let $f \in C^0[-1, 1]$. Then there exists a constant $C \neq C(n, f)$ such that we have*

$$\|L_n(w, f)u\|_1 \leq C\|f\|_\infty,$$

or equivalently

$$\|[f - L_n(w, f)]u\|_1 \leq C E_{n-1}(f)_\infty,$$

if and only if

$$u, \frac{u}{\sqrt{w\varphi}} \in L^1.$$

In particular, if $w = v^{\alpha,\beta}$, $u = v^{\alpha-\gamma,\beta-\delta}$, $\alpha, \beta > -1$, $\gamma, \delta \geq 0$, then we have

$$\|L_n(v^{\alpha,\beta}, f)v^{\alpha-\gamma,\beta-\delta}\|_1 \leq C\|f\|_\infty$$

5.2 Integral Equations

if and only if

$$0 \leq \gamma < \min\left(\frac{\alpha}{2}+\frac{3}{4}, \alpha+1\right), \quad 0 \leq \delta < \min\left(\frac{\beta}{2}+\frac{3}{4}, \beta+1\right).$$

Finally, let

$$S_n(v^{\alpha,\beta}, g) = \sum_{k=0}^{n} c_k p_k(v^{\alpha,\beta}), \quad c_k = \int_{-1}^{1} g p_k(v^{\alpha,\beta}) v^{\alpha,\beta}$$

be the n-th Fourier sum of a function g with respect to the Jacobi polynomial system.

Lemma 5.2.1 *Let $\alpha, \beta > -1$. If the parameters γ and δ satisfy the following conditions*

$$\max\left\{0, \frac{\alpha}{2}+\frac{1}{4}\right\} \leq \gamma < \min\left\{\frac{\alpha}{2}+\frac{3}{4}, \alpha+1\right\},$$

(5.2.9)

$$\max\left\{0, \frac{\beta}{2}+\frac{1}{4}\right\} \leq \delta < \min\left\{\frac{\beta}{2}+\frac{3}{4}, \beta+1\right\},$$

then, for every function g such that

$$A(g) := \int_{-1}^{1} |g(x) v^{\alpha-\gamma, \beta-\delta}(x)| \log(2 + |g(x) v^{\alpha-\gamma, \beta-\delta}(x)|) \, dx < +\infty,$$

we have

$$\|S_n(v^{\alpha,\beta}, g) v^{\alpha-\gamma, \beta-\delta}\|_1 \leq C A(g),$$

where C is a positive constant independent of g and n.

5.2.2 Fredholm Integral Equations of the Second Kind

In this section we consider the Fredholm integral equations of the second kind

$$f(y) - \lambda \int_{-1}^{1} k(x, y) v^{\alpha,\beta}(x) f(x) \, dx = g(y), \quad |y| \leq 1, \qquad (5.2.10)$$

where $\lambda \neq 0$, $v^{\alpha,\beta}$ is the Jacobi weight, $k(x, y)$ and $g(y)$ are appropriate known functions, and f is the unknown function.

Letting

$$(Kf)(y) = \lambda \int_{-1}^{1} k(x, y) v^{\alpha,\beta}(x) f(x) \, dx,$$

we can write (5.2.10) in the usual form $(I - K)f = g$.

Depending on the smoothness of the kernel, we give two different numerical methods for approximating the solution of (5.2.10).

5.2.2.1 Locally Smooth Kernels

We consider the integral equation (5.2.10) in $C^0_{v^{\gamma,\delta}}$, assuming that $(I - K)^{-1}$ exists and is bounded on the weighted space $C^0_{v^{\gamma,\delta}}$, with parameters γ and δ such that the inequalities (5.2.9) are satisfied. Moreover, we suppose that the kernel $k(x, y)$ and the function g satisfy the following conditions

$$M_s := \sup_{|x|\leq 1} \sup_{t>0} \frac{\Omega^r_\varphi(k_x, t)_{v^{\gamma,\delta},\infty}}{t^s} < +\infty, \tag{5.2.11}$$

$$N_s := \sup_{|y|\leq 1} v^{\gamma,\delta}(y) \sup_{t>0} \frac{\Omega^r_\varphi(k_y, t)_\infty}{t^s} < +\infty, \tag{5.2.12}$$

$$g \in Z_s(v^{\gamma,\delta}), \tag{5.2.13}$$

with $r > s > 0$ and $k_x(y) = k(x, y) = k_y(x)$.

In the sequel, a positive constant C will include sometimes the absolute value of the parameter λ, which appears in (5.2.10).

The following theorem shows that $K : C^0_{v^{\gamma,\delta}} \to C^0_{v^{\gamma,\delta}}$ is compact.

Theorem 5.2.5 *Let $\alpha, \beta > -1$. If γ and δ satisfy (5.2.9) and $k(x, y)$ satisfies the condition (5.2.11), then*

$$\|Kf\|_{Z_s(v^{\gamma,\delta})} \leq A\|fv^{\gamma,\delta}\|_\infty, \tag{5.2.14}$$

where

$$A := |\lambda| \left(\sup_{-1 \leq x, y \leq 1} |v^{\gamma,\delta}(y)k(x, y)| + M_s \right) \left(\int_{-1}^1 v^{\alpha-\gamma, \beta-\delta}(x)dx \right).$$

Consequently, $K : C^0_{v^{\gamma,\delta}} \to C^0_{v^{\gamma,\delta}}$ is compact.

Proof We have

$$|(Kf)(y)v^{\gamma,\delta}(y)| = \left| \lambda \int_{-1}^1 v^{\gamma,\delta}(y)k(x, y)(fv^{\gamma,\delta})(x)v^{\alpha-\gamma,\beta-\delta}(x)dx \right|$$

$$\leq |\lambda| \, \|fv^{\gamma,\delta}\|_\infty \int_{-1}^1 \left| k(x, y)v^{\gamma,\delta}(y)v^{\alpha-\gamma,\beta-\delta}(x) \right| dx$$

$$\leq |\lambda| \|fv^{\gamma,\delta}\|_\infty \sup_{-1 \leq x, y \leq 1} \left| v^{\gamma,\delta}(y)k(x, y) \right| \int_{-1}^1 v^{\alpha-\gamma,\beta-\delta}(x)dx.$$

5.2 Integral Equations

Then

$$\|Kfv^{\gamma,\delta}\|_\infty \le |\lambda| \|fv^{\gamma,\delta}\|_\infty \sup_{-1 \le x, y \le 1} |v^{\gamma,\delta}(y)k(x,y)| \int_{-1}^{1} v^{\alpha-\gamma,\beta-\delta}(x)dx, \tag{5.2.15}$$

and, because of the assumptions (5.2.9), the integral at the right-hand side is bounded. Moreover, for $0 < h \le t$ and $y \in I_{r,h} = [-1 + 4r^2h^2, 1 - 4r^2h^2]$ (see Sect. 2.5.1)

$$\left|v^{\gamma,\delta}(y)\Delta^r_{h\varphi}(Kf)(y)\right| = \left|\lambda \int_{-1}^{1} v^{\gamma,\delta}(y)\Delta^r_{h\varphi}k(x,y)(fv^{\gamma,\delta})(x)v^{\alpha-\gamma,\beta-\delta}(x)\right|$$

$$\le |\lambda| \|fv^{\gamma,\delta}\|_\infty \int_{-1}^{1} \Omega^r_\varphi(k_x, t)_{v^{\gamma,\delta},\infty} v^{\alpha-\gamma,\beta-\delta}(x)dx$$

$$\le |\lambda| \|fv^{\gamma,\delta}\|_\infty \sup_{|x| \le 1} \Omega^r_\varphi(k_x, t)_{v^{\gamma,\delta},\infty} \int_{-1}^{1} v^{\alpha-\gamma,\beta-\delta}(x)dx,$$

and then

$$\sup_{t>0} \frac{\Omega^r_\varphi(Kf, t)_{v^{\gamma,\delta},\infty}}{t^s} \le |\lambda| \|fv^{\gamma,\delta}\|_\infty M_s \int_{-1}^{1} v^{\alpha-\gamma,\beta-\delta}(x)dx. \tag{5.2.16}$$

Combining (5.2.15) and (5.2.16), (5.2.14) follows. Now, we get by (5.2.8)

$$E_n(Kf)_{v^{\gamma,\delta},\infty} \le \frac{C}{n^s} \|Kf\|_{Z_s(v^{\gamma,\delta})} \le \frac{C}{n^s} \|fv^{\gamma,\delta}\|_\infty,$$

i.e.,

$$\lim_n \left(\sup_{\|fv^{\gamma,\delta}\|_\infty = 1} E_n(Kf)_{v^{\gamma,\delta},\infty} \right) = 0$$

and then K is a compact operator in view of (5.2.1). □

In order to introduce some numerical methods for an approximation of the solution of (5.2.10), we define the polynomial sequence $\{g_n\}_n$ and the sequence of operators $\{K_n\}_n$ as

$$g_n = L_n(v^{\alpha,\beta}, g) \quad \text{and} \quad K_n f = L_n(v^{\alpha,\beta}, K^* f),$$

respectively, where

$$(K^* f)(y) := (K_n^* f)(y) = \lambda \int_{-1}^{1} L_n(v^{\alpha,\beta}, k(\cdot, y); x) f(x) v^{\alpha,\beta}(x) dx.$$

We consider now the following finite dimensional equation

$$(I - K_n) f_n = g_n, \tag{5.2.17}$$

where $f_n \in \mathcal{P}_{n-1}$ is unknown.

In order to apply Theorem 5.2.2, we show that the sequence K_n converges to K in norm and that the sequence g_n converges to $g \in C^0_{v^{\gamma,\delta}}$.

Theorem 5.2.6 *Let $\alpha, \beta > -1$. If the parameters γ and δ satisfy the conditions (5.2.9), the kernel $k(x, y)$ and the free term g satisfy the conditions (5.2.11)–(5.2.13), then*

$$\|(g - g_n)v^{\gamma,\delta}\|_\infty \leq C \frac{\log n}{n^s} \|g\|_{Z_s(v^{\gamma,\delta})} \tag{5.2.18}$$

and

$$\|K - K_n\|_{C^0_{v^{\gamma,\delta}} \to C^0_{v^{\gamma,\delta}}} \leq C(M_s + N_s) \frac{\log n}{n^s}, \tag{5.2.19}$$

where $C \neq C(n)$.

Proof Since we assume $g \in Z_s(v^{\gamma,\delta})$, using Theorem 5.2.3 and (5.2.8), (5.2.18) easily follows.

Now, we prove (5.2.19). Subtracting and adding $K^* f$, we have

$$\|(Kf - K_n f)v^{\gamma,\delta}\|_\infty \leq \|(Kf - K^* f)v^{\gamma,\delta}\|_\infty + \|(K^* f - K_n f)v^{\gamma,\delta}\|_\infty$$

$$:= A + B. \tag{5.2.20}$$

For the first term A we get

$$|(Kf)(y) - (K^* f)(y)|v^{\gamma,\delta}(y)$$

$$= v^{\gamma,\delta}(y) \left| \lambda \int_{-1}^{1} [k_y(x) - L_n(v^{\alpha,\beta}, k_y, x)] v^{\alpha-\gamma,\beta-\delta}(x)(fv^{\gamma,\delta})(x) dx \right|$$

$$\leq C \|fv^{\gamma,\delta}\|_\infty v^{\gamma,\delta}(y) \int_{-1}^{1} \left| k_y(x) - L_n(v^{\alpha,\beta}, k_y; x) \right| v^{\alpha-\gamma,\beta-\delta}(x) dx.$$

By conditions (5.2.9), it means that $v^{\alpha-\gamma,\beta-\delta}/\sqrt{v^{\alpha,\beta}\varphi} \in L^1$, $v^{\alpha-\gamma,\beta-\delta} \in L^1$ and Theorem 5.2.4 can be applied. Thus, it gives

$$|(Kf)(y) - (K^* f)(y)|v^{\gamma,\delta}(y) \leq C \|fv^{\gamma,\delta}\|_\infty v^{\gamma,\delta}(y) E_{n-1}(k_y)_\infty.$$

Moreover, using the inequality (5.2.7) and (5.2.12), we find

$$A \leq C \|fv^{\gamma,\delta}\|_\infty v^{\gamma,\delta}(y) E_{n-1}(k_y)_\infty \leq C \|fv^{\gamma,\delta}\|_\infty v^{\gamma,\delta}(y) \int_0^{1/n} \frac{\Omega^r_\varphi(k_y, t)_\infty}{t} dt,$$

i.e.,

$$A \leq C \|fv^{\gamma,\delta}\|_\infty \frac{N_s}{n^s}. \tag{5.2.21}$$

5.2 Integral Equations

Now, using Theorem 5.2.3 and (5.2.9), we obtain

$$B \leq C(\log n) E_{n-1}(K^* f)_{v^{\gamma,\delta},\infty}. \tag{5.2.22}$$

In order to estimate $E_{n-1}(K^* f)_{v^{\gamma,\delta},\infty}$ by means of the inequality (5.2.7), we proceed to the estimation of $\Omega_\varphi^r(K^* f, t)_{v^{\gamma,\delta},\infty}$. Using Theorem 5.2.4, we get for $0 < h \leq t$

$$|v^{\gamma,\delta}(y) \Delta_{h\varphi}^r (K^* f)(y)|$$

$$= \left| \lambda \int_{-1}^{1} L_n(v^{\alpha,\beta}, v^{\gamma,\delta}(y) \Delta_{h\varphi}^r k_y; x) v^{\alpha-\gamma,\beta-\delta}(x)(f v^{\gamma,\delta})(x) dx \right|$$

$$\leq C \|f v^{\gamma,\delta}\|_\infty \int_{-1}^{1} |L_n(v^{\alpha,\beta}, v^{\gamma,\delta}(y) \Delta_{h\varphi}^r k_y; x)| v^{\alpha-\gamma,\beta-\delta}(x) dx$$

$$\leq C \|f v^{\gamma,\delta}\|_\infty v^{\gamma,\delta}(y) \sup_x |\Delta_{h\varphi}^r k_x(y)|$$

$$\leq C \|f v^{\gamma,\delta}\|_\infty \sup_x \Omega_\varphi^r(k_x, t)_{v^{\gamma,\delta},\infty},$$

and then

$$\Omega_\varphi^r(K^* f, t)_{v^{\gamma,\delta},\infty} \leq C t^s \|f v^{\gamma,\delta}\|_\infty M_s.$$

Thus, using inequality (5.2.7), (5.2.22) becomes

$$B \leq C \|f v^{\gamma,\delta}\|_\infty \frac{M_s}{n^s} \log n. \tag{5.2.23}$$

Finally, combining (5.2.21) and (5.2.23) with (5.2.20), the estimate (5.2.19) follows. □

Now, we can apply Theorem 5.2.2.

Theorem 5.2.7 *If the kernel $k(x, y)$ and the function g satisfy (5.2.11)–(5.2.13), α, β, γ and δ satisfy (5.2.9), and (5.2.10) has a unique solution f for every fixed g, then, for all sufficiently large n, (5.2.17) has a unique solution f_n^* and*

$$\|(f - f_n^*) v^{\gamma,\delta}\|_\infty = \mathcal{O}\left(\frac{\log n}{n^s}\right), \tag{5.2.24}$$

where the constants in "$\mathcal{O}$" are independent of n and f.

Moreover, if we represent f_n^* in the base $\varphi_i(y) := \ell_i(v^{\alpha,\beta}; y)/v^{\gamma,\delta}(x_i)$, $i = 1, \ldots, n$, i.e.,

$$f_n^*(y) = \sum_{i=1}^{n} \varphi_i(y) a_i, \tag{5.2.25}$$

then $\boldsymbol{a} = [a_1 \cdots a_n]^T$ is the solution of the following system of linear equations

$$\sum_{k=1}^{n}\left[\delta_{i,k} - \lambda v^{\gamma,\delta}(x_i)\frac{\lambda_k(v^{\alpha,\beta})}{v^{\gamma,\delta}(x_k)}k(x_k,x_i)\right]a_k = b_i := v^{\gamma,\delta}(x_i)g(x_i), \qquad (5.2.26)$$

for $i = 1, \ldots, n$, where $\lambda_k(v^{\alpha,\beta})$, $k = 1, \ldots, n$, are the Christoffel numbers. Finally, if we denote by A_n the matrix of the system (5.2.26) and by cond (A_n) its condition number in the uniform norm, then

$$\sup_n \frac{\text{cond}(A_n)}{\log n} < +\infty. \qquad (5.2.27)$$

Proof The estimate (5.2.24) is a direct consequence of Theorems 5.2.2 and 5.2.6. In order to obtain the system (5.2.26), we express $K_n f$, g_n and f_n using the base $\{\varphi_i\}_{i=1,\ldots,n}$. Since, for every $q \in \mathcal{P}_{n-1}$, we have

$$q(x) = \sum_{i=1}^{n} \varphi_i(x)\gamma_i, \quad \gamma_i = q(x_i)v^{\gamma,\delta}(x_i),$$

we can write

$$f_n(y) = \sum_{i=1}^{n} \varphi_i(y)a_i, \quad g_n(y) = \sum_{i=1}^{n} \varphi_i(y)b_i, \quad b_i = v^{\gamma,\delta}(x_i)g(x_i),$$

and

$$(K_n f)(y) = \sum_{i=1}^{n} \varphi_i(y) v^{\gamma,\delta}(x_i)(K^* f)(x_i).$$

Moreover, using the Gauss quadrature formula, we get

$$(K^* f_n)(x_i) = \lambda \int_{-1}^{1} L_n(v^{\alpha,\beta}, k(\cdot, x_i); x) f_n(x) v^{\alpha,\beta}(x) dx$$

$$= \lambda \sum_{k=1}^{n} \lambda_k(v^{\alpha,\beta}) k(x_k, x_i) \frac{a_k}{v^{\gamma,\delta}(x_k)}$$

and then it gives

$$(K_n f_n)(y) = \lambda \sum_{i=1}^{n} \varphi_i(y) v^{\gamma,\delta}(x_i) \sum_{k=1}^{n} \frac{\lambda_k(v^{\alpha,\beta})}{v^{\gamma,\delta}(x_k)} k(x_k, x_i) a_k,$$

where $\lambda_k(v^{\alpha,\beta})$, $k = 1, \ldots, n$, are the Christoffel numbers. Therefore, the finite dimensional equation $(I - K_n)f_n = g_n$ becomes

$$\sum_{i=1}^{n} \varphi_i(y) a_i - \lambda \sum_{i=1}^{n} \varphi_i(y) v^{\gamma,\delta}(x_i) \sum_{k=1}^{n} \frac{\lambda_k(v^{\alpha,\beta})}{v^{\gamma,\delta}(x_k)} k(x_k, x_i) a_k = \sum_{i=1}^{n} \varphi_i(y) b_i$$

5.2 Integral Equations

and then (5.2.26) follows.

In order to prove (5.2.27), we could use Proposition 5.2.1 to obtain

$$\sup_n \frac{\operatorname{cond}(A_n)}{\log^2 n} < C \operatorname{cond}(I - K),$$

where C is a positive constant. But, in this case, the matrix A_n is explicitly known. Therefore, denoting by $C_{i,k}$ ($i, k = 1, \ldots, n$) the entries of A_n, by (5.2.26) we get

$$\|A_n\|_{l^\infty} = \sup_i \sum_{k=1}^n |C_{i,k}|$$

$$\leq 1 + C \left(\max_{-1 \leq x, y \leq 1} |k(x, y)| v^{\gamma,\delta}(y) \right) \left(\int_{-1}^1 v^{\alpha-\gamma,\beta-\delta}(x) \, dx \right),$$

i.e., $\|A_n\|_{l^\infty} \leq C \neq C(n)$, being

$$\sum_{k=1}^n \frac{\lambda_k(v^{\alpha,\beta})}{v^{\gamma,\delta}(x_k)} \leq C \int_{-1}^1 v^{\alpha-\gamma,\beta-\delta}(x) \, dx$$

and then (5.2.27) follows. □

In conclusion of this section we want to observe that the proposed method is convenient from the computational point of view, because the system (5.2.26) is well-conditioned and its coefficients are given by elementary formulas. Moreover, the estimate (5.2.24) of the remainder term is optimal under the assumed conditions. Regarding the assumptions on the kernel $k(x, y)$ and the free term g, we note that, if the parameter s is an integer greater than or equal to 1, then (5.2.11)–(5.2.13) become

$$M_s := \sup_{|x| \leq 1} \left\| \frac{\partial^s}{\partial y^s} k(x, \cdot) \varphi^s v^{\gamma,\delta} \right\|_\infty < +\infty,$$

$$N_s := \sup_{|y| \leq 1} v^{\gamma,\delta}(y) \left\| \frac{\partial^s}{\partial x^s} k(\cdot, y) \varphi^s \right\|_\infty < +\infty,$$

$$\|g^{(s)} \varphi^s v^{\gamma,\delta}\| < +\infty.$$

Anyway, the modulus Ω_φ^r of a differentiable function F can be estimated by means of

$$\Omega_\varphi^r(F, t)_{v^{\gamma,\delta}, \infty} \leq C \sup_{0 < h \leq t} h^r \|F^{(r)} \varphi^r v^{\gamma,\delta}\|_{I_{r,h}}, \qquad (5.2.28)$$

where $I_{r,h} = [-1 + (2rh)^2, 1 - (2rh)^2]$.

It is also useful to observe that the proposed method can be used when the kernel $k_x(y)$ and the known term g have weak singularities at the end points of the interval $[-1,1]$. For example, let for the sake of simplicity, $\alpha = \beta$, $\gamma = \delta$, $v^\rho = v^{\rho,\rho}$,

$$v(y) = v^{-\rho}(y)\log\frac{e}{1-y^2}, \quad 0 < \rho < \frac{\alpha}{2} + \frac{3}{4}$$

and assume

$$g(y) = g^*(y)v(y), \quad k(x,y) = k^*(x,y)v(y),$$

where g^* and k^* are smooth functions. We can consider the equation (5.2.10) in C_{v^γ}, choosing γ such that

$$0 < \rho < \gamma < \min\left\{\frac{\alpha}{2} + \frac{3}{4}, \alpha + 1\right\}.$$

In fact, since by (5.2.28) we get

$$\Omega_\varphi^r(g,t)_{v^\gamma,\infty} \sim t^{2(\gamma-\rho)}\log t^{-1} \sim \Omega_\varphi^r(k_x,t)_{v^\gamma,\infty},$$

where t is sufficiently small, the assumptions (5.2.11)–(5.2.13) are satisfied, the system (5.2.26) can be used (mutatis mutandis) and the right-hand side of (5.2.24) has to be replaced by $O\left(\log^2 n/n^{2(\gamma-\rho)}\right)$. As an example, the equation

$$f(y) - \frac{1}{8\sqrt[8]{1-y^2}}\log\frac{e}{1-y^2}\int_{-1}^1 |x-y|^5(1-x^2)^{3/2}f(x)dx$$

$$= \frac{|y|^5}{\sqrt[8]{1-y^2}}\log\frac{e}{1-y^2} \tag{5.2.29}$$

has a unique solution in $C_{v^{5/4}}$.

Using the system (5.2.26) we construct the polynomial sequence $\{f_n^*\}_n$ that converges in $C_{v^{5/4}}$ to the exact solution f^* with order $O\left(n^{-9/4}\log^2 n\right)$.

Finally, it is also convenient to consider the equation (5.2.10) in $C_{v^{\gamma,\delta}}$ when the kernel and the known term are continuous but their derivative are unbounded around ± 1, like $g(x) = (1+x)\log(1+x)$.

5.2.2.2 Numerical Examples

We consider some examples of integral equations showing the behaviour of their approximate solutions $v^{\gamma,\delta}f_n^*$ ($f_n^* \in \mathcal{P}_{n-1}$). The all computation are performed in D arithmetic with the machine precision $\approx 2.22 \times 10^{-16}$.

Example 5.2.1 We consider the following integral equation

$$f(y) - \frac{4}{5}\int_{-1}^1 \sin(x+y)f(x)\frac{dx}{\sqrt{1-x^2}} = e^y.$$

5.2 Integral Equations

Table 5.2.1 Numerical results for Example 5.2.1

n	cond	$f_n^*(-0.5)$	$f_n^*(0.5)$
4	14.96	$-1.5179903\underline{1}8171035$	$-8.67508\underline{2}232783050$
8	15.02	$-1.5178323032\underline{6}4140$	$-8.6755881617\underline{5}9954$
16	16.53	$-1.51783230326380\underline{8}$	$-8.67558816176094\underline{2}$
32	18.48	-1.517832303263807	-8.675588161760939

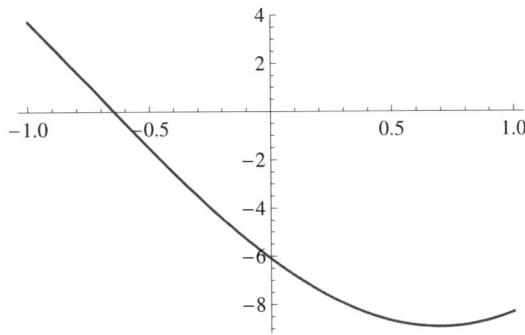

Fig. 5.2.1 The solution f_{32}^* in Example 5.2.1

Since the kernel $k(x, y) = \sin(x + y)$ and the function $g(y) = e^y$ are analytic functions, we obtain very accurate results. According to (5.2.9), we choose $\gamma = \delta = 0$.

The condition number (cond) of the linear system (5.2.26) (in uniform norm) is presented in Table 5.2.1 for $n = 4, 8, 16$, and 32, as well as the behaviour of the solution f_n^* at two points $y = \pm 0.5$. The first digit in error in these approximate solutions is underlined. As we can see, the error $\|f - f_{32}^*\|_\infty$ has the order of the machine precision.

Finally, in Fig. 5.2.1 we show the graph of the function f_{32}^*.

Example 5.2.2 We consider the following integral equation

$$f(y) - \frac{1}{2}\int_{-1}^{1} e^{x+y} f(x)\sqrt{1-x^2}\,dx = |y|^{9/2}.$$

Taking into account (5.2.9) we choose $\gamma = \delta = 0.9$. Since the kernel is an analytic function and $g \in Z_{9/2}$, according to (5.2.24), the theoretical error is of the order of $n^{-9/2}\log n$ and we have to increase n in order to obtain the correct digits.

In Table 5.2.2 we present the corresponding condition number of the linear system (5.2.26) and the weighted solutions $(v^{0.9,0.9} f_n^*)(y)$ at the points $y = 0.01$ and $y = 0.5$, with relative errors in the last row (for $n = 256$) less than 10^{-12}. The graph of the function $y \mapsto (v^{0.9,0.9} f_{256}^*)(y)$ is given in Fig. 5.2.2.

Table 5.2.2 Numerical results for Example 5.2.2

n	cond	$(v^{0.9,0.9}f_n^*)(0.01)$	$(v^{0.9,0.9}f_n^*)(0.5)$
8	16.38	−0.4681854456835951	−0.5558386462652317
16	22.07	−0.4681446429649453	−0.5557872315277880
32	26.60	−0.4681435187732979	−0.5557858149550620
64	29.01	−0.4681434902824499	−0.5557857790542810
128	30.26	−0.4681434896024169	−0.5557857781973842
256	30.89	−0.4681434895867625	−0.5557857781776584

Table 5.2.3 Numerical results for Example 5.2.3

n	cond	$(v^{5/4,5/4}f_n^*)(0.1)$	$(v^{5/4,5/4}f_n^*)(0.9)$
8	1.764	0.006774506685104795	0.29341826356237900
16	1.923	0.006956353848248328	0.29432937270107688
32	1.934	0.006966712560038358	0.29438316260645012
64	1.936	0.006967241715583831	0.29438593327674541
128	1.937	0.006967266128334941	0.29438606138448930
256	1.937	0.006967267182609122	0.29438606692011326

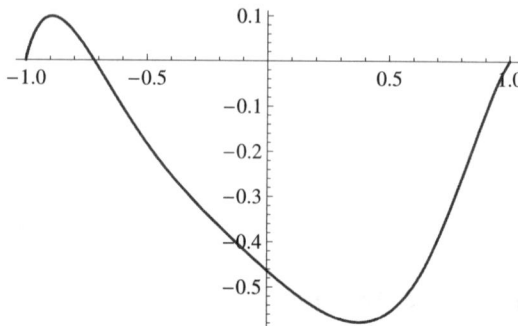

Fig. 5.2.2 The solution $(v^{0.9,0.9}f_{256}^*)(y)$ in Example 5.2.2

Example 5.2.3 We consider the equation (5.2.29) mentioned before. Choosing $\gamma = \delta = 5/4$, the kernel k and the function g belong to $Z_{9/4}$. According to (5.2.24), the theoretical error is of the order of $n^{-9/4}\log^2 n$.

In Table 5.2.3 we present the condition number of the linear system (5.2.26), as well as the behaviour of the weighted solution $(v^{5/4,5/4}f_n^*)(y)$ for $y = 0.1$ and $y = 0.9$. The results in the last row (for $n = 256$) are determined with the relative error less than 10^{-8}. The graph of the function $y \mapsto (v^{5/4,5/4}f_{256}^*)(y)$ is showed in Fig. 5.2.3.

5.2 Integral Equations

Fig. 5.2.3 The solution $(v^{5/4,5/4} f^*_{256})(y)$ in Example 5.2.3

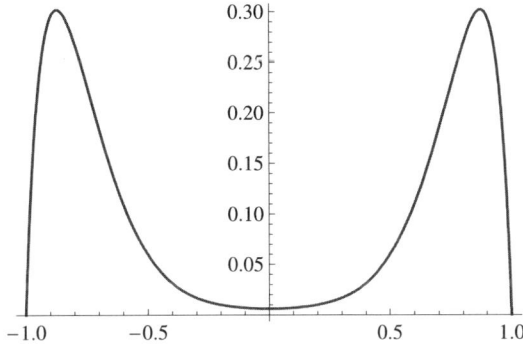

5.2.2.3 Weakly Singular Kernels

If $k(x, y)$ is weakly singular, we use a procedure that is theoretically simpler but numerically more expensive than the previous one.

With γ and δ satisfying (5.2.9), we consider $(I - K)f = g$ in $C^0_{v^{\gamma,\delta}}$ and we assume $g \in Z_s(v^{\gamma,\delta})$ and

$$\sup_{t>0} \frac{\Omega^r_\varphi(Kf, t)_{v^{\gamma,\delta},\infty}}{t^s} \leq C\|fv^{\gamma,\delta}\|_\infty. \tag{5.2.30}$$

Obviously the condition (5.2.11) implies (5.2.30), but the inverse implication is not true. For example, (5.2.30) is satisfied with $s = 1 + \mu$ if $k(x, y) = |x - y|^\mu$, $\mu > -1$.

Now, we introduce the sequences $\{K_n\}_n$ and $\{g_n\}_n$ by

$$(K_n f)(y) = L_n(v^{\alpha,\beta}, Kf, y)$$

and $g_n = L_n(v^{\alpha,\beta}, g)$, respectively.

As a consequence of Theorem 5.2.3 and relations (5.2.8) and (5.2.7), we have

$$\|(g - g_n)v^{\gamma,\delta}\|_\infty \leq C \frac{\log n}{n^s} \|g\|_{Z_s(v^{\gamma,\delta})}$$

and

$$\|K - K_n\|_{C^0_{v^{\gamma,\delta}} \to C^0_{v^{\gamma,\delta}}} = O\left(\frac{\log n}{n^s}\right).$$

Then we solve the approximate equation

$$f_n - K_n f_n = g_n, \tag{5.2.31}$$

where $f_n \in \mathcal{P}_{n-1}$ is the unknown sequence. Obviously, for all sufficiently large n, the last one has a unique solution f_n^*, if (5.2.10) is uniquely solvable in $C^0_{v^{\gamma,\delta}}$. By Theorem 5.2.2 the following estimate

$$\|(f - f_n^*)v^{\gamma,\delta}\|_\infty \leq C \frac{\log n}{n^s} \|g\|_{Z_s(v^{\gamma,\delta})} \tag{5.2.32}$$

holds. In order to compute f_n^* we write

$$f_n(x) = \sum_{k=1}^{n} \varphi_k(x) a_k, \qquad (5.2.33)$$

with the unknown coefficients $a_1, \ldots, a_n$, where $\varphi_k(y) = \ell_k(v^{\alpha,\beta}; y)/v^{\gamma,\delta}(x_k)$, $k = 1, \ldots, n$, and for (5.2.31) we have

$$\sum_{i=1}^{n} \varphi_i(y) a_i - \sum_{i=1}^{n} \varphi_i(y) v^{\gamma,\delta}(x_i)(K f_n)(x_i) = \sum_{i=1}^{n} \varphi_i(y)(g v^{\gamma,\delta})(x_i)$$

and therefore it gives

$$a_i - v^{\gamma,\delta}(x_i)(K f_n)(x_i) = (g v^{\gamma,\delta})(x_i), \quad i = 1, \ldots, n.$$

On the other hand

$$(K f_n)(x_i) = \lambda \int_{-1}^{1} k(x, x_i) f_n(x) v^{\alpha,\beta}(x) \, dx$$

$$= \lambda \sum_{k=1}^{n} \frac{a_k}{v^{\gamma,\delta}(x_k)} \int_{-1}^{1} k(x, x_i) \ell_k(v^{\alpha,\beta}; x) v^{\alpha,\beta}(x) \, dx.$$

Since

$$\ell_k(v^{\alpha,\beta}; x) = \lambda_k(v^{\alpha,\beta}) D_n(v^{\alpha,\beta}; x, x_k),$$

where $D_n(v^{\alpha,\beta}; x, x_k)$ is the Darboux-Christoffel kernel, we have

$$(K f_n)(x_i) = \lambda \sum_{k=1}^{n} \frac{a_k}{v^{\gamma,\delta}(x_k)} \lambda_k(v^{\alpha,\beta}) \int_{-1}^{1} D_n(v^{\alpha,\beta}; x, x_k) k(x, x_i) v^{\alpha,\beta}(x) dx$$

$$= \lambda \sum_{k=1}^{n} \frac{\lambda_k(v^{\alpha,\beta})}{v^{\gamma,\delta}(x_k)} S_n(v^{\alpha,\beta}, k(\cdot, x_i); x_k) a_k,$$

where $S_n(v^{\alpha,\beta}, k(\cdot, x_i); t)$ is the Fourier sum of the function $k(x, x_i)$ in the system $\{p_n(v^{\alpha,\beta})\}_n$, i.e.,

$$S_n(v^{\alpha,\beta}, k(\cdot, x_i); t) = \sum_{\nu=0}^{n-1} c_\nu p_\nu(v^{\alpha,\beta}; t),$$

with

$$c_\nu = \int_{-1}^{1} p_\nu(v^{\alpha,\beta}; x) k(x, x_i) v^{\alpha,\beta}(x) \, dx.$$

5.2 Integral Equations

We note that in the previous formula we use the values of this Fourier sum at the points $t = x_k$, $k = 1, \ldots, n$.

The computation of c_ν, $\nu = 0, 1, \ldots, n-1$, is the main computational effort. It can be performed by means of recurrence relations as many examples show.[2] In conclusion, (5.2.17) is equivalent to the following system of linear equations

$$\sum_{k=1}^{n} \left[\delta_{i,k} - \lambda v^{\gamma,\delta}(x_i) \frac{\lambda_k(v^{\alpha,\beta})}{v^{\gamma,\delta}(x_k)} S_n(v^{\alpha,\beta}, k(\cdot, x_i); x_k) \right] a_k = (g v^{\gamma,\delta})(x_i), \quad (5.2.34)$$

where $i = 1, \ldots, n$.

Using Proposition 5.2.1 we prove the bound

$$\sup_n \frac{\operatorname{cond}(A_n)}{\log^2 n} < +\infty,$$

where A_n denotes the matrix of the system (5.2.34) and $\operatorname{cond}(A_n)$ is its condition number in the uniform norm. Moreover, the following proposition holds:

Proposition 5.2.2 *Let α, β, γ and δ be as in (5.2.9). If the kernel $k(x, y)$ satisfy the condition*

$$\sup_{|y| \leq 1} \int_{-1}^{1} v^{\alpha-\gamma, \beta-\delta}(x) |k(x, y)| \log(1 + v^{\alpha-\gamma, \beta-\delta} |k(x, y)|) dx < +\infty,$$

then

$$\sup_n \frac{\operatorname{cond}(A_n)}{\log n} < +\infty.$$

Proof Following the proof of Proposition 5.2.1, we get

$$\|A_n^{-1}\|_{l^\infty} \leq C \|L_n(v^{\alpha,\beta})\|_{v^{\gamma,\delta}, \infty} \|(I - K_n)^{-1}\big|_{\mathcal{P}_{n-1}}\| \leq C \|(I - K_n)^{-1}\| \log n.$$

On the other hand, it follows by (5.2.34)

$$\|A_n\|_{l^\infty} \leq 1 + |\lambda| \max_i v^{\gamma,\delta}(x_i) \sum_{k=1}^{n} \frac{\lambda_k(v^{\alpha,\beta})}{v^{\gamma,\delta}(x_k)} |S_n(v^{\alpha,\beta}, k(\cdot, x_i); x_k)|$$

$$\leq 1 + C|\lambda| \max_i \int_{-1}^{1} |S_n(v^{\alpha,\beta}, v^{\gamma,\delta}(x_i) k(\cdot, x_i); t)| v^{\alpha-\gamma, \beta-\delta}(t) dt$$

$$\leq 1 + C \max_y \int_{-1}^{1} v^{\gamma,\delta}(y) \Big[|v^{\alpha-\gamma, \beta-\delta}(t) k(t, y)|$$

$$+ \log\big(2 + |v^{\alpha-\gamma, \beta-\delta}(t) k(t, y)|\big) \Big] dt,$$

---

[2]The reader can consult, for example, [398, 399].

using a Marcinkiewicz inequality [306] and applying Lemma 5.2.1. By our assumptions the statement follows. □

5.2.3 Nyström Method

In this section we want to approximate the solution of (5.2.10) by means of a modified version of the Nyström method. The idea of such a method is based on replacing (5.2.10) by

$$f(y) - \lambda(A_n f)(y) = g(y), \qquad (5.2.35)$$

where

$$(A_n f)(y) = \sum_{i=1}^{n} w_i(y) f(y_i)$$

is a suitable quadrature formula. Here, the functional $A_n f$ is characterized by the vector $[f(y_1) \ \ldots \ f(y_n)]^T$, which is the solution of the following system of linear equations

$$f(y_i) - \lambda \sum_{j=1}^{n} w_j(y_i) f(y_j) = g(y_i), \quad i = 1, \ldots, n, \qquad (5.2.36)$$

which is equivalent to (5.2.35). Of course, we have to prove that (5.2.35) or (5.2.36) admit a unique solution. If the system (5.2.36) has the unique solution $[f^*(y_1) \ \ldots \ f^*(y_n)]^T$, then

$$f_n^*(y) = \lambda \sum_{j=1}^{n} w_j(y) f^*(y_j) + g(y)$$

approximates the solution f^* of (5.2.10), with the following error

$$\|f^* - f_n^*\|_\infty \leq C \|Kf^* - A_n f^*\|_\infty.$$

Now, we consider two different cases of (5.2.10).

Case 1° The kernel $k(x, y)$ and the function g satisfy (5.2.11)–(5.2.13). Moreover the positive parameters γ and δ satisfy (5.2.9) and we assume that (5.2.10), for every function g, has a unique solution in $C_{v^{\gamma,\delta}}$. Then, we can write (5.2.10) in the following form

$$v^{\gamma,\delta}(y)f(y) - \lambda \int_{-1}^{1} v^{\gamma,\delta}(y) k(x, y) f(x) v^{\alpha,\beta}(x) dx = g(y) v^{\gamma,\delta}(y). \qquad (5.2.37)$$

5.2 Integral Equations

Using Nyström's idea we approximate the integral with the Gauss quadrature formula with respect to the weight function $v^{\alpha,\beta}$, so that

$$v^{\gamma,\delta}(y)f(y) - \lambda \sum_{k=1}^{n} \lambda_k(v^{\alpha,\beta}) \frac{v^{\gamma,\delta}(y)}{v^{\gamma,\delta}(x_k)} k(x_k, y)f(x_k)v^{\gamma,\delta}(x_k) = g(y)v^{\gamma,\delta}(y). \tag{5.2.38}$$

Therefore, (5.2.38) is equivalent to the following system of equations

$$v^{\gamma,\delta}(x_i)f(x_i) - \lambda \sum_{k=1}^{n} \lambda_k(v^{\alpha,\beta}) \frac{v^{\gamma,\delta}(x_i)}{v^{\gamma,\delta}(x_k)} k(x_k, x_i)f(x_k)v^{\gamma,\delta}(x_k) = g(x_i)v^{\gamma,\delta}(x_i), \tag{5.2.39}$$

for $i = 1, \ldots, n$, where $x_k = x_{n,k}(v^{\alpha,\beta})$, $k = 1, \ldots, n$, are the zeros of the Jacobi polynomial $p_n(v^{\alpha,\beta})$. Moreover, setting

$$a_k := (v^{\gamma,\delta}f)(x_k), \quad b_i := (gv^{\gamma,\delta})(x_i), \quad 1 \le i, k \le n,$$

(5.2.39) can be written in the form

$$\sum_{k=1}^{n} \left[\delta_{i,k} - \lambda \lambda_k(v^{\alpha,\beta}) \frac{v^{\gamma,\delta}(x_i)}{v^{\gamma,\delta}(x_k)} k(x_k, x_i) \right] a_k = b_i, \quad i = 1, \ldots, n. \tag{5.2.40}$$

But, we note that (5.2.40) is equal to (5.2.26), which admits a unique solution and its condition number satisfies (5.2.27). If $[a_1^* \ldots a_n^*]^T$ is the solution of (5.2.39), then the sequence $\{v^{\gamma,\delta}f_n^*\}_n$, defined by

$$(v^{\gamma,\delta}f_n^*)(y) = \lambda \sum_{k=1}^{n} \lambda_k(v^{\alpha,\beta}) \frac{v^{\gamma,\delta}(y)}{v^{\gamma,\delta}(x_k)} k(x_k, y)a_k^* + (v^{\gamma,\delta}g)(y), \tag{5.2.41}$$

approximates the solution of (5.2.10) and the weighted error is given by

$$\|(f^* - f_n^*)v^{\gamma,\delta}\|_\infty \le C \|(Kf^* - A_n f^*)v^{\gamma,\delta}\|_\infty$$

$$= C \max_{|y| \le 1} \left\{ \left| \int_{-1}^{1} k(x,y)f^*(x)v^{\alpha,\beta}(x)dx \right.\right.$$

$$\left.\left. - \sum_{k=1}^{n} \lambda_k(v^{\alpha,\beta}) k(x_k, y)f^*(x_k) \right| v^{\gamma,\delta}(y) \right\}$$

$$\le C \max_{|y| \le 1} \left\{ v^{\gamma,\delta}(y) \int_{-1}^{1} \left| k_y(x)f^*(x) - L_n(v^{\alpha,\beta}, k_y f^*; x) \right| v^{\alpha,\beta}(x)\, dx \right\}$$

$$\le C \max_{|y| \le 1} \left\{ v^{\gamma,\delta}(y) \max_{|x| \le 1} \left| k_y(x)f^*(x) - L_n(v^{\alpha,\beta}, k_y f^*; x) \right| v^{\gamma,\delta}(x) \right\}$$

$$\times \int_{-1}^{1} \frac{v^{\alpha,\beta}(x)}{v^{\gamma,\delta}(x)} dx$$

$$\leq C \log n \max_{|y|\leq 1} \{v^{\gamma,\delta}(y) E_{n-1}(k_y f^*)_{v^{\gamma,\delta},\infty}\}.$$

Since f^* belongs to the same class of functions, as well as g and k_x, and

$$E_{n-1}(k_y f^*)_{v^{\gamma,\delta},\infty} \leq \max_{|x|\leq 1} |k(x,y) v^{\gamma,\delta}(x)| E_{[\frac{n-1}{2}]}(f^*)_{v^{\gamma,\delta},\infty}$$

$$+ \|f^* v^{\gamma,\delta}\|_\infty E_{[\frac{n-1}{2}]}(k_y)_{v^{\gamma,\delta},\infty}$$

$$\leq \frac{C}{n^s} \left(\|k_y v^{\gamma,\delta}\|_\infty \|f^*\|_{Z_s(v^{\gamma,\delta})} + C_1 \|f^* v^{\gamma,\delta}\|_\infty \|k_y\|_{Z_s(v^{\gamma,\delta})} \right),$$

then we get

$$\|(f^* - f_n^*) v^{\gamma,\delta}\|_\infty \leq \frac{C \log n}{n^s} \left[\max_{1\leq x, y\leq 1} \left(v^{\gamma,\delta}(x) |k(x,y)| v^{\gamma,\delta}(y) \right) \|f^*\|_{Z_s(v^{\gamma,\delta})} \right.$$

$$\left. + N_s \|f^* v^{\gamma,\delta}\|_\infty \right],$$

which has the same rate of convergence as (5.2.24). Therefore, (5.2.41) can be taken instead of (5.2.25).

Case 2° The kernel $k(x, y)$ is weakly singular, the free term g belongs to $Z_s(v^{\gamma,\delta})$, γ and δ satisfy (5.2.9) and the linear operator K satisfies (5.2.30).

In such a case the equation (5.2.10), i.e., (5.2.37), is replaced by

$$v^{\gamma,\delta}(y) f(y) - \lambda \sum_{k=1}^{n} \lambda_k(v^{\alpha,\beta}) \frac{v^{\gamma,\delta}(y)}{v^{\gamma,\delta}(x_k)} S_n(v^{\alpha,\beta}, k(\cdot, y); x_k) f(x_k) v^{\alpha,\beta}(x_k)$$

$$= (g v^{\gamma,\delta})(y),$$

which is equivalent to the system of equations

$$a_i - \lambda \sum_{k=1}^{n} \lambda_k(v^{\alpha,\beta}) \frac{v^{\gamma,\delta}(x_i)}{v^{\gamma,\delta}(x_k)} S_n(v^{\alpha,\beta}, k(\cdot, x_i); x_k) a_k = b_i, \quad i=1,\ldots,n,$$

where a_j and b_i, $1 \leq i, j \leq n$, have the same mining as before. This system is equivalent to (5.2.34), with the unique solution $[a_1^* \ \ldots \ a_n^*]^T$. Therefore, the sequence $\{v^{\gamma,\delta} f_n^*\}_n$, defined by

$$(v^{\gamma,\delta} f_n^*)(y) = \lambda \sum_{k=1}^{n} \lambda_k(v^{\alpha,\beta}) \frac{v^{\gamma,\delta}(y)}{v^{\gamma,\delta}(x_k)} S_n(v^{\alpha,\beta}, k(\cdot, y), x_k) a_k^* + g(y) v^{\gamma,\delta}(y),$$

(5.2.42)

approximates $f^* v^{\gamma,\delta}$ and, because of $f^* \in Z_s(v^{\gamma,\delta})$, the error is given by

$$\|(f^* - f_n^*) v^{\gamma,\delta}\|_\infty$$

$$\leq C \sup_{|y| \leq 1} v^{\gamma,\delta}(y) \Bigg| \int_{-1}^{1} k_y(x) f^*(x) v^{\alpha,\beta}(x) dx$$

$$- \lambda \sum_{k=1}^{n} \lambda_k(v^{\alpha,\beta}) \frac{v^{\gamma,\delta}(y)}{v^{\gamma,\delta}(x_k)} S_n(v^{\alpha,\beta}, k(\cdot,y); x_k) a_k^* \Bigg|$$

$$\leq C \sup_{|y| \leq 1} v^{\gamma,\delta}(y) \int_{-1}^{1} |k_y(x)| \left| f^*(x) - L_n(v^{\alpha,\beta}, f^*; x) \right| v^{\alpha,\beta}(x) dx$$

$$\leq C \|[f^* - L_n(v^{\alpha,\beta}, f^*)] v^{\gamma,\delta}\|_\infty \sup_{|y| \leq 1} v^{\gamma,\delta}(y) \int_{-1}^{1} |k_y(x)| v^{\alpha-\gamma,\beta-\delta}(x) dx$$

$$\leq \frac{C \log n}{n^s} \|f^*\|_{Z_s(v^{\gamma,\delta})} \sup_{|y| \leq 1} v^{\gamma,\delta}(y) \int_{-1}^{1} |k_y(x)| v^{\alpha-\gamma,\beta-\delta}(x) dx.$$

Also in this case, the rate of convergence is the same as in (5.2.32). Moreover, (5.2.33) can be replaced by (5.2.42).

5.3 Moment-Preserving Approximation

The interpolation of a given function f by another one can be interpreted as a process of transferring the values of the function f (or values of its derivatives) at some selected points to another function, e.g. to its interpolation polynomial. This means that we have a value-preserving approximation. On the other hand, the moments of a function have often some very important physical meaning which should be preserved in an approximation, i.e., to have an approximation of f by a certain function s in such a way that as many of its moments as possible are the same as those of the function f. Finally, the value-preserving approximation (interpolation) and moment-preserving approximation can be combined. In this section we consider a few cases of such kind of approximation (cf. [44, 45, 139, 140, 150, 160], [166, pp. 227–239], [172, 188, 189, 237, 322, 335, 340, 352, 353]).

5.3.1 The Standard L^2-Approximation

Let $d\mu(t)$ be a given nonnegative measure on $\mathbb{R}$ and $U_n = \{g_0, g_1, g_2, \ldots, g_n\}$ be a given system of linearly independent functions in a real inner product space X, with

$$(f, g) = \int_\mathbb{R} f(t) g(t) d\mu(t) \quad (f, g \in X). \tag{5.3.1}$$

Supposing the moments of the function f, with respect to the measure $d\mu(t)$ and the system U_n, exist

$$\mu_k(f, U_n) = \int_{\mathbb{R}} f(t) g_k(t)\, d\mu(t) = (f, g_k), \quad k = 0, 1, \ldots, n, \qquad (5.3.2)$$

we want to approximate the function f by

$$s_n = \sum_{\nu=0}^{n} a_\nu g_\nu \; \big(\in X_n = \operatorname{span} U_n \big),$$

such that

$$\mu_k(s_n, U_n) = \mu_k(f, U_n), \quad k = 0, 1, \ldots, n. \qquad (5.3.3)$$

On the other hand, it is well-known that the standard L^2-approximation of $f \in X$ by $s_n \in X_n$, i.e.,

$$\min_{s \in X_n} \|f - s\|^2 = \min_{a_\nu \in \mathbb{R}} \left\| f - \sum_{\nu=0}^{n} a_\nu g_\nu \right\|^2 = \|f - s_n\|^2,$$

is attained *if and only if* the error $f - s_n$ is orthogonal to the subspace X_n, i.e., *if and only if*[3]

$$\left(f - \sum_{\nu=0}^{n} a_\nu g_\nu,\, g_k \right) = 0, \quad k = 0, 1, \ldots, n, \qquad (5.3.4)$$

which is equivalent to (5.3.3).

Thus, the standard L^2-approximation of a function is a moment-preserving approximation. Finally, we mention here that in the L^2-approximation we use an orthonormal (or orthogonal) system of functions, for example, $S_n = \{\phi_0, \phi_1, \phi_2, \ldots, \phi_n\}$, instead of an arbitrary system of linearly independent functions U_n. In that case, the conditions (5.3.4) directly give the Fourier coefficients $a_k = (f, \phi_k)$, $k = 0, 1, \ldots, n$ (see Sect. 2.1.2).

In the case of approximation with algebraic polynomials, S_n is a system of polynomials $\{p_k(d\mu; \cdot)\}$ orthonormal with respect to the measure $d\mu(t)$, and then the L^2-approximation is given by

$$s_n(t) = \sum_{k=0}^{n} a_k p_k(t), \quad a_k = \int_{\mathbb{R}} f(t) p_k(t)\, d\mu(t), \quad k = 0, 1, \ldots, n, \qquad (5.3.5)$$

---

[3] For these conditions see (2.1.7) in Chap. 2 and note that there X_n is generated by an orthonormal system of functions.

5.3 Moment-Preserving Approximation

and the corresponding error of the best L^2-approximation is given by (see Theorem 2.1.2)

$$E_n(d\mu, f)_2^2 = \|f - s_n\|^2 = \|f\|^2 - \sum_{k=0}^{n} a_k^2.$$

5.3.1.1 Generalization

If we introduce another system of linearly independent functions $V_n = \{\psi_0, \psi_1, \psi_2, \ldots, \psi_n\}$ besides U_n and, instead of (5.3.2), define the moments

$$\mu_k(f, V_n) = \int_{\mathbb{R}} f(t)\psi_k(t) \, d\mu(t) = (f, \psi_k), \quad k = 0, 1, \ldots, n,$$

with respect to this system V_n, then it is possible to consider a general problem of finding an approximation $s_n = \sum_{\nu=0}^{n} a_\nu g_\nu$ to a given function $f \in X$, such that it preserves the moments of f with respect to the system V_n. Since the measure $d\mu(t)$ is the same as before, the moments $\mu_k(f, V_n)$ are expressed again in terms of the inner product (5.3.1). Such an approximation problem,

$$\mu_k(s_n, V_n) = \mu_k(f, V_n), \quad k = 0, 1, \ldots, n, \tag{5.3.6}$$

was considered by Bojanov and Gori [44]. The system (5.3.6), i.e.,

$$\sum_{\nu=0}^{n} a_\nu (g_\nu, \psi_k) = (f, \psi_k), \quad k = 0, 1, \ldots, n,$$

has a unique solution *if and only if* the determinant

$$D_n = \begin{vmatrix} (g_0, \psi_0) & (g_1, \psi_0) & \cdots & (g_n, \psi_0) \\ (g_0, \psi_1) & (g_1, \psi_1) & & (g_n, \psi_1) \\ \vdots & & & \\ (g_0, \psi_n) & (g_1, \psi_n) & & (g_n, \psi_n) \end{vmatrix} \neq 0.$$

The construction of s_n is extremely simple when the system U_n is *biorthogonal* to V_n, i.e., when

$$(g_\nu, \psi_k) = \delta_{\nu k} \quad (\nu, k = 0, 1, \ldots, n).$$

In that case $D_n = 1$ and $a_\nu = (f, \psi_k)$, $k = 0, 1, \ldots, n$, so that

$$s_n(t) = \sum_{\nu=1}^{n} (f, \psi_\nu) g_\nu(t).$$

Remark 5.3.1 This approach can be used for finding moment-preserving spline approximations studied in [139, 140, 150, 172, 188, 189, 322].

5.3.2 The Constrained L^2-Polynomial Approximation

Sometimes in the L^2-approximation we want to transfer certain properties from the original function f to the approximant $\widetilde{s}_n$, e.g., to match certain number of data points exactly. Although this can be considered in a general fashion,[4] we restrict our consideration to the simplest case, given by the following constraints at m ($\leq n$) given points $\tau_1, \ldots, \tau_m$,

$$\widetilde{s}_n(\tau_\nu) = f(\tau_\nu), \quad \nu = 1, \ldots, m. \tag{5.3.7}$$

Thus, we need to minimize $\widetilde{E}_n(d\mu, f)_2^2 = \|f - \widetilde{s}\|^2$, when $\widetilde{s} \in \mathcal{P}_n$ and satisfies the conditions (5.3.7).

In order to find such a polynomial of the best L^2-approximation, with these constraints, we first construct the interpolation polynomial of degree at most $m - 1$, satisfying the conditions (5.3.7). Thus,

$$P_m(t) = L_m(f; t) = \sum_{\nu=1}^{m} f(\tau_\nu) \frac{q_m(t)}{(t - \tau_\nu) q'_m(\tau_\nu)},$$

where $q_m(t) = \prod_{\nu=1}^{m}(t - \tau_\nu)$. Now, we seek our least squares approximation in the form

$$\widetilde{s}(t) = P_m(t) + q_m(t) s(t), \quad s \in \mathcal{P}_{n-m},$$

where s is an arbitrary algebraic polynomial of degree at most $n - m$, which should be determined from the following minimization

$$\begin{aligned}
\widetilde{E}_n(d\mu, f)_2^2 &= \min_{s \in \mathcal{P}_{n-m}} \|f - P_m - q_m s\|^2 \\
&= \int_{\mathbb{R}} (f(t) - P_m(t) - q_m(t) s(t))^2 \, d\mu(t) \\
&= \int_{\mathbb{R}} \left(\frac{f(t) - P_m(t)}{q_m(t)} - s(t) \right)^2 q_m(t)^2 \, d\mu(t).
\end{aligned}$$

Introducing a new function $\widetilde{f}$ and a new measure $d\widetilde{\mu}$ given by

$$\widetilde{f}(t) = \frac{f(t) - P_m(t)}{q_m(t)} \quad \text{and} \quad d\widetilde{\mu}(t) = q_m(t)^2 d\mu(t),$$

respectively, we get the following new unconstrained L^2-approximation problem

$$\widetilde{E}_n(d\mu, f)_2^2 = E_{n-m}(d\widetilde{\mu}, \widetilde{f})_2^2 = \min_{s \in \mathcal{P}_{n-m}} \int_{\mathbb{R}} (\widetilde{f}(t) - s(t))^2 \, d\widetilde{\mu}(t).$$

---

[4]Constraints can be given by certain linear functionals, for example, $F_\nu(\widetilde{s}_n) = F_\nu(f)$ for $\nu = 1, \ldots, m$.

5.3 Moment-Preserving Approximation

If s_{n-m} is the solution of this L^2-problem, then

$$\widetilde{s}_n(t) = P_m(t) + q_m(t) s_{n-m}(t)$$

is the solution of the constrained L^2-approximation problem. According to (5.3.5), the solution s_{n-m} can be expressed in the form

$$s_{n-m}(t) = \sum_{k=0}^{n-m} \widetilde{a}_k p_k(d\widetilde{\mu}; t), \quad \widetilde{a}_k = \int_{\mathbb{R}} \widetilde{f}(t) p_k(d\widetilde{\mu}; t) \, d\widetilde{\mu}(t), \quad k = 0, 1, \ldots, n-m.$$

Some interesting examples are given in [166, pp. 223–225].

In [357, 358] Milovanović and Wrigge considered the L^2-approximation for functions $f \in L^2[-1, 1]$, with respect to the Gegenbauer weight $(1 - t^2)^{\lambda - 1/2}$, $\lambda > -1/2$, with the constraints $f(\pm 1) = 0$. The general theory was applied to the functions $f(t) = \cos(\pi t/2)$ and $f(t) = J_0(a_0 t)$, where a_0 is the smallest positive zero of the Bessel function J_0.

5.3.3 Moment-Preserving Spline Approximation

As we mentioned at the beginning of Sect. 5.3, because of certain physical meaning, the moment-preserving approximation can be applied in some problems in physics. Such kind of approximation with some classical methods, which are very sensitive to rounding errors, appeared in the physics literature by Laframboise and Stauffer [249] (approximation of the Maxwell velocity distribution by a linear combination of Dirac δ-functions) and Calder and Laframboise [55] (for the corresponding approximation by a linear combination of Heaviside step functions). In this section we describe a stable method for constructing such kind of approximations reducing the problem to quadratures (cf. [139, 140, 150, 172, 188, 189, 322]). As we noted in Remark 5.3.1 it can also be constructed by the approach given in [44].

5.3.3.1 Approximation on $[0, +\infty)$

Let f be a given function defined on the positive real line $\mathbb{R}_+ = [0, +\infty)$. The problem of an approximation of f by a piecewise constant function

$$s_n(t) = \sum_{\nu=1}^{n} a_\nu H(t_\nu - t), \quad a_\nu \in \mathbb{R}, \quad 0 < t_1 < \cdots < t_n < +\infty, \tag{5.3.8}$$

where H is the Heaviside step function, was solved by Gautschi [150], who also considered approximation by a linear combination of Dirac delta functions. The approximation was to preserve as many moments of f as possible. The method was extended to the general case of spline approximation of arbitrary degree m by Gautschi and Milovanović [172]. Now, we present this method of approximation.

Because of finite moments in this spline approximation on $\mathbb{R}_+$, the spline approximant must be the so-called *monospline*, i.e., it cannot have a purely polynomial part. Thus,

$$s_{n,m}(t) = \sum_{v=1}^{n} a_v (t_v - t)_+^m, \quad 0 \le t < +\infty, \qquad (5.3.9)$$

where the plus sign on the right is the cutoff symbol, $u_+ = u$ if $u > 0$ and $u_+ = 0$ if $u \le 0$, $0 < t_1 < \cdots < t_n$, $a_v \in \mathbb{R}$. Note that for $m = 0$, (5.3.9) reduces to (5.3.8).

We seek to determine $s_{n,m}$ such that

$$\int_0^{+\infty} s_{n,m}(t) t^j \, dV = \int_0^{+\infty} f(t) t^j \, dV, \quad j = 0, 1, \ldots, 2n-1, \qquad (5.3.10)$$

where dV is the volume element depending on the geometry of the problem. In some concrete applications in physics, up to unimportant numerical factors, $dV = t^{d-1} \, dt$, where $d = 1, 2$, and 3 for rectilinear, cylindric, and spherical geometry, respectively.

Theorem 5.3.1 *For fixed $n, m \in \mathbb{N}$ and $d \in \{1, 2, 3\}$, let f satisfy the following conditions:*

(a) $f \in C^{m+1}[0, +\infty)$;
(b) *the first $2n$ moments* $\mu_j = \int_0^{+\infty} t^{j+d-1} f(t) \, dt$, $j = 0, 1, \ldots, 2n-1$, *exist*;
(c) $f^{(v)}(t) = o(t^{-2n+1-d-v})$ *as* $t \to +\infty$, $v = 0, 1, \ldots, m$.

Then a spline function $s_{n,m}$ of the form (5.3.9) with positive knots t_v, that satisfies (5.3.10), exists and is unique if and only if the measure

$$d\lambda_m(t) = \frac{(-1)^{m+1}}{m!} t^{m+d} f^{(m+1)}(t) \, dt \quad \text{on} \quad [0, +\infty) \qquad (5.3.11)$$

admits an n-point Gauss-Christoffel quadrature formula

$$\int_0^{+\infty} g(x) \, d\lambda_m(x) = \sum_{v=1}^{n} \lambda_v^{(n)} g(\tau_v^{(n)}) + R_n(g; d\lambda_m), \qquad (5.3.12)$$

with distinct positive nodes $\tau_v^{(n)}$, where $R_n(g; d\lambda_m) = 0$ for all $g \in \mathcal{P}_{2n-1}$. In that event, the knots t_v and weights a_v in (5.3.9) are given by

$$t_v = \tau_v^{(n)}, \quad a_v = t_v^{-(m+d)} \lambda_v^{(n)}, \quad v = 1, \ldots, n. \qquad (5.3.13)$$

Proof Substituting (5.3.9) into (5.3.10) yields, since $t_v > 0$,

$$\sum_{v=1}^{n} a_v \int_0^{t_v} t^{j+d-1} (t_v - t)^m \, dt = \int_0^{+\infty} t^{j+d-1} f(t) \, dt, \quad j = 0, 1, \ldots, 2n-1.$$

5.3 Moment-Preserving Approximation

The left-hand side, through m integrations by parts, can be seen to be equal to

$$\frac{m!}{(j+d)(j+d+1)\cdots(j+d+m-1)} \sum_{\nu=1}^{n} a_\nu \int_0^{t_\nu} t^{j+d+m-1} dt$$

$$= \frac{m!}{(j+d)(j+d+1)\cdots(j+d+m)} \sum_{\nu=1}^{n} a_\nu t_\nu^{j+d+m}. \quad (5.3.14)$$

On the other hand, the integral on the right is transformed similarly by $m+1$ integration by parts. We carry out the first of them in detail to exhibit the reasonings involved. We have, for any $b > 0$,

$$\int_0^b t^{j+d-1} f(t) dt = \frac{1}{j+d} t^{j+d} f(t) \Big|_0^b - \frac{1}{j+d} \int_0^b t^{j+d} f'(t) dt.$$

The integrated term clearly vanishes at $t = 0$ and tends to zero as $t = b \to +\infty$ by assumption (c) with $\nu = 0$. Since $j \le 2n - 1$ and the integral on the left converges as $b \to +\infty$ by assumption (b), we conclude the convergence of the integral on the right as $b \to +\infty$. Therefore,

$$\int_0^{+\infty} t^{j+d-1} f(t) dt = -\frac{1}{j+d} \int_0^{+\infty} t^{j+d} f'(t) dt.$$

Continuing in this manner, using assumption (c) to show convergence to zero of the integrated term at the upper limit (its value at $t = 0$ always being zero) and the existence of $\int_0^{+\infty} t^{j+d-1+\nu} f^{(\nu)}(t) dt$ already established to infer the existence of the integrals $\int_0^{+\infty} t^{j+d+\nu} f^{(\nu+1)}(t) dt$, $\nu = 1, \ldots, m$, we arrive at

$$\int_0^{+\infty} t^{j+d-1} f(t) dt = \frac{(-1)^{m+1}}{(j+d)(j+d+1)\cdots(j+d+m)}$$

$$\times \int_0^{+\infty} t^{j+d+m} f^{(m+1)}(t) dt. \quad (5.3.15)$$

Comparing (5.3.15) with (5.3.14), we see that (5.3.10) are equivalent to

$$\sum_{\nu=1}^{n} (a_\nu t_\nu^{m+d}) t_\nu^j = \int_0^{+\infty} \left[\frac{(-1)^{m+1}}{m!} t^{m+d} f^{(m+1)}(t) \right] t^j dt, \quad j = 0, 1, \ldots, 2n-1.$$

These are precisely the conditions for t_ν to be the nodes of the Gauss-Christoffel formula (5.3.12) for the measure $d\lambda_m(t)$ given by (5.3.11) and $a_\nu t_\nu^{m+d}$ the corresponding weights. □

If f is *completely monotonic* on $[0, +\infty)$, i.e., such that $(-1)^m f^{(m)}(t) > 0$ for all $t \in [0, +\infty)$ and each $m \in \mathbb{N}_0$, then $d\lambda_m(t)$ in (5.3.11) is a positive measure for every m. Moreover, the first $2n$ moments of $d\lambda_m(t)$ exist, and therefore the Gauss-Christoffel quadrature formula (5.3.12) exists uniquely, with all distinct positive nodes $\tau_\nu^{(n)}$ and positive weights $\lambda_\nu^{(n)}$. The latter implies $a_\nu > 0$, $\nu = 1,\ldots,n$, in (5.3.9).

The next theorem gives the error in the moment-preserving spline approximation.

Theorem 5.3.2 *Given f as in Theorem 5.3.1, assume that the measure $d\lambda_m$ in (5.3.11) admits the n-point Gauss-Christoffel quadrature formula (5.3.12) with distinct positive nodes $\tau_\nu^{(n)}$ and the remainder term $R_n(g;d\lambda_m)$. Define*

$$\sigma_t(x) = x^{-(m+d)}(x-t)_+^m, \quad x,t > 0. \tag{5.3.16}$$

Then, for any $t > 0$, we have for the error of the spline approximation (5.3.9),

$$f(t) - s_{n,m}(t) = R_n(\sigma_t; d\lambda_m). \tag{5.3.17}$$

Proof By Taylor's formula, one has for any $b > 0$,

$$f(t) = f(b) + f'(b)(t-b) + \cdots + \frac{1}{m!}f^{(m)}(b)(t-b)^m$$

$$+ \frac{1}{m!}\int_b^t (t-x)^m f^{(m+1)}(x)\,dx. \tag{5.3.18}$$

Since by assumption (c) of Theorem 5.3.1, $\lim_{b\to+\infty} b^\nu f^{(\nu)}(b) = 0$ for $\nu = 0, 1, \ldots, m$, letting $b \to +\infty$ in (5.3.18), we obtain

$$f(t) = \frac{(-1)^{m+1}}{m!}\int_t^{+\infty}(x-t)^m f^{(m+1)}(x)\,dx$$

$$= \frac{(-1)^{m+1}}{m!}\int_0^{+\infty}(x-t)_+^m f^{(m+1)}(x)\,dx,$$

hence, by (5.3.11) and (5.3.16),

$$f(t) = \int_0^{+\infty}\sigma_t(x)\,d\lambda_m(x). \tag{5.3.19}$$

On the other hand, by (5.3.9) and (5.3.13),

$$s_{n,m}(t) = \sum_{\nu=1}^n t_\nu^{-(m+d)}\lambda_\nu^{(n)}(t_\nu - t)_+^m = \sum_{\nu=1}^n \lambda_\nu^{(n)}\sigma_t(\tau_\nu^{(n)}). \tag{5.3.20}$$

Subtracting (5.3.20) from (5.3.19) yields (5.3.17). □

Remark 5.3.2 To discuss convergence as $n \to +\infty$ (for fixed m), we assume f to satisfy the assumptions of Theorem 5.3.1 for all $n \in \mathbb{N}$. Then, by Theorem 5.3.2, the

5.3 Moment-Preserving Approximation

Table 5.3.1 Accuracy of the spline approximation for Example 5.3.1

d	$d=1$			$d=2$			$d=3$		
n	$m=1$	$m=2$	$m=3$	$m=1$	$m=2$	$m=3$	$m=1$	$m=2$	$m=3$
5	5.9(−2)	1.8(−2)	7.9(−3)	2.4(−2)	1.1(−2)	5.9(−3)	1.2(−2)	6.5(−3)	3.9(−3)
10	1.8(−2)	3.5(−3)	1.0(−3)	8.9(−3)	2.7(−3)	9.4(−4)	5.0(−3)	1.9(−3)	7.6(−4)
20	1.5(−2)	1.2(−3)	1.9(−4)	2.8(−3)	4.9(−4)	1.0(−4)	1.7(−3)	3.9(−4)	9.8(−5)
40	7.5(−3)	4.2(−4)	4.7(−5)	1.2(−3)	7.6(−5)	8.8(−6)	5.1(−4)	6.5(−5)	9.2(−6)

approximation process converges pointwise (at t), as $n \to +\infty$, if and only if the Gauss-Christoffel quadrature formula (5.3.12) converges when applied to the special function $g(x) = \sigma_t(x)$ in (5.3.16). Since σ_t is uniformly bounded on $\mathbb{R}$, this is true, for example, if $d\lambda_m$ is a positive measure and the moment problem for $d\lambda_m$ on $\mathbb{R}$ (with $d\lambda_m(x) = 0$ for $x < 0$) is determined (cf. Freud [134, Chap. 3, Theorem 1.1]).

Example 5.3.1 Let $f(t) = e^{-t}$ on $[0, +\infty)$ (exponential distribution). In this case the measure (5.3.11) becomes the generalized Laguerre measure

$$d\lambda_m(t) = \frac{1}{m!} t^{m+d} e^{-t} dt, \quad 0 \le t < +\infty.$$

The nodes t_ν of the spline function (5.3.9), therefore, are the zeros of the generalized Laguerre polynomial L_n^α with parameter $\alpha = m + d$, and the weights a_ν follow readily from (5.3.13) in terms of the corresponding Christoffel numbers $\lambda_\nu^{(n)}$.

Table 5.3.1 shows approximate values of the resulting maximum absolute errors $\|s_{n,m} - f\|_\infty = \max_{0 \le t \le t_n} |s_{n,m}(t) - f(t)|$, for $m = 1, 2, 3$; $d = 1, 2, 3$; and $n = 5, 10, 20, 40$. Numbers in parentheses indicate decimal exponents. Clearly, $|s_{n,m}(t) - f(t)| = f(t)$ for $t > t_n$. Since the moment problem for the generalized Laguerre measure is determined (cf. Freud [134, Chap. 2, Theorem 5.2]), it follows from Remark 5.3.2 that $s_{n,m}(t) \to f(t)$ as $n \to +\infty$, for any fixed $t > 0$.

It is likely that convergence also takes place if n is fixed and $m \to +\infty$. When $n = 1$, for example,

$$s_{1,m}(t) = \frac{(m+1)\cdots(m+d)}{(m+d+1)^d} \left(1 - \frac{t}{m+d+1}\right)_+^m,$$

which implies $s_{1,m}(t) = e^{-t} + O(m^{-1})$ as $m \to +\infty$.

Example 5.3.2 For the Maxwell distribution $f(t) = e^{-t^2}$ on $[0, +\infty)$, the measure (5.3.11) becomes

$$d\lambda_m(t) = \frac{1}{m!} t^{m+d} H_{m+1}(t) e^{-t^2} dt, \quad 0 \le t < +\infty,$$

Fig. 5.3.1 The functions w_0 (*dashed line*) and w_1 (*solid line*)

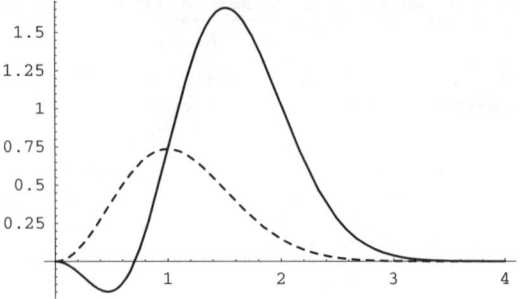

where H_{m+1} is the Hermite polynomial of degree $m+1$. If $m > 0$, as we assume, H_{m+1} changes sign at least once on $(0, +\infty)$, so that $d\lambda_m$ is no longer a positive measure. The existence of the Gauss-Christoffel quadrature formula (5.3.12) is therefore in doubt, and even if it exists, we cannot be sure that its nodes are all simple and positive as in Example 5.3.1. The matter depends on whether the nth degree orthogonal polynomial $\pi_n(d\lambda_m; \cdot)$ (relative to $d\lambda_m$) exists, and in addition, whether its zeros—the nodes $\tau_\nu^{(n)}$ in (5.3.12)—are distinct and positive. If so, the solution of this spline approximation problem is given by (5.3.13), where the Christoffel numbers $\lambda_\nu^{(n)}$ are uniquely determined by the nodes $\tau_\nu^{(n)}$; if not, the problem has no solution.

Let $d = 1$. The measures for $m = 0$ and $m = 1$ are

$$d\lambda_0(t) = w_0(t)\,dt = 2t^2 e^{-t^2}\,dt \quad \text{and} \quad d\lambda_1(t) = w_1(t)\,dt = 2t^2(2t^2 - 1)e^{-t^2}\,dt,$$

respectively. In Fig. 5.3.1 the functions w_m, $m = 0, 1$, are presented. As we can see the "weight" w_1 (solid line) changes sign at $1/\sqrt{2} \approx 0.707107$. Taking $n = 5$, the corresponding parameters (5.3.13) for splines $s_{5,m}$, $m = 0, 1$, are given in Table 5.3.2. The spline

$$s_{5,0}(t) = \sum_{\nu=1}^{5} a_\nu H(t_\nu - t), \quad 0 \leq t < +\infty,$$

is displayed in Fig. 5.3.2 (dashed line). The approximation $s_{5,1}$ also exists, but we can see that its coefficient a_1 is negative.

A complete investigation of this problem for $n \leq 20$, $m, d = 1, 2, 3$, was given by Gautschi and Milovanović [172].

Similarly, we can consider an approximation of a given function $t \mapsto f(t)$ on $[0, +\infty)$ by defective splines, e.g.,

$$s_{n,m}(t) = \sum_{\nu=1}^{n} \sum_{i=m-k+1}^{m} a_{i,\nu}(t_\nu - t)_+^i,$$

where $a_{i,\nu}$ are real numbers. Under suitable assumptions on f and $k = 2s + 1$, Milovanović and Kovačević [352, 353] showed that the approximation problem has

5.3 Moment-Preserving Approximation

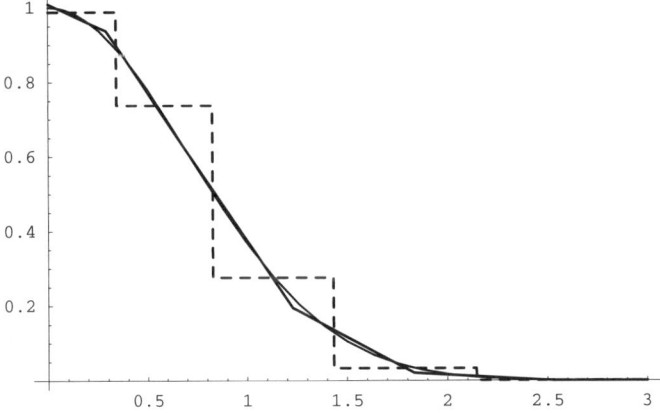

Fig. 5.3.2 Spline approximations $s_{5,0}$ (*dashed line*) and $s_{5,1}$ (*solid line*) for the Maxwell distribution $f(t) = e^{-t^2}$ on $[0, +\infty)$

Table 5.3.2 Parameters for the spline approximation in Example 5.3.2

ν	$m = 0$		$m = 1$	
	t_ν	a_ν	t_ν	a_ν
1	0.338409596069	0.250493551112	0.288714094070	−0.543359171726
2	0.826662543774	0.461399735580	1.228333755254	0.503506372966
3	1.432854372673	0.241856985893	1.833105588615	0.256795884026
4	2.145408222421	0.033293689727	2.528609144977	0.029655378867
5	3.014172467490	0.000683845871	3.372177757129	0.000486839143

the unique solution if and only if certain generalized Turán quadratures exist corresponding to a measure depending on f (for some details see Milovanović [335, 340]).

5.3.3.2 Approximation on a Compact Interval

Here we consider moment-preserving approximation of a function $t \mapsto f(t)$ given on a finite interval, which can be standardized to $[a, b] = [0, 1]$. A spline function of degree $m \geq 0$, with n (distinct) knots t_ν, $\nu = 1, \ldots, n$, in the interior of $[0, 1]$, can be written in terms of truncated powers in the form

$$s_{n,m}(t) = p(t) + \sum_{\nu=1}^{n} a_\nu (t_\nu - t)_+^m, \quad 0 \leq t \leq 1, \tag{5.3.21}$$

where a_ν are real numbers and p is an algebraic polynomial of degree at most m.

We consider two related problems:

Problem 1° Determine $s_{n,m}$ in (5.3.21) such that
$$\int_0^1 t^j s_{n,m}(t)\, dt = \int_0^1 t^j f(t)\, dt, \quad j = 0, 1, \ldots, 2n + m.$$

Problem 2° Determine $s_{n,m}$ in (5.3.21) such that
$$\int_0^1 t^j s_{n,m}(t)\, dt = \int_0^1 t^j f(t)\, dt, \quad j = 0, 1, \ldots, 2n - 1,$$

and
$$s_{n,m}^{(k)}(1) = f^{(k)}(1), \quad k = 0, 1, \ldots, m,$$

supposing that the first m derivatives of f at $t = 1$ exist and are known.

Note that $s_{n,m}(t) = p(t)$ for $t \geq 1$. If $f \in \mathcal{P}_m$, then we have a trivial solution $s_{n,m} = f$ for both problems.

Theorem 5.3.3 *Assume that $f \in C^{m+1}[0, 1]$. Then, Problem 1° has a unique solution if and only if the measure*
$$d\lambda_m(t) = \frac{(-1)^{m+1}}{m!} f^{(m+1)}(t)\, dt \quad \text{on } [0, 1] \tag{5.3.22}$$

admits a generalized Gauss-Lobatto quadrature formula
$$\int_0^1 g(x)\, d\lambda_m(x) = \sum_{k=0}^m [A_k g^{(k)}(0) + B_k g^{(k)}(1)] + \sum_{\nu=1}^n \lambda_\nu^L g(\tau_\nu^L) + R_{n,m}^L(d\lambda_m; g),$$

with $0 < \tau_1^L < \cdots < \tau_n^L < 1$, where $R_{n,m}^L(d\lambda_m; g) = 0$ for each $g \in \mathcal{P}_{2n+2m+1}$. If that is the case, the knots t_ν and coefficients a_ν in (5.3.21) are given by
$$t_\nu = \tau_\nu^L, \quad a_\nu = \lambda_\nu^L, \quad \nu = 1, \ldots, n,$$

and the polynomial p in (5.3.21) is determined by
$$p(t) = \sum_{k=0}^m \frac{1}{k!} \left[f^{(k)}(1) + (-1)^m m! B_{m-k} \right] (t - 1)^k.$$

Theorem 5.3.4 *Assume that $f \in C^{m+1}[0, 1]$. Then, Problem 2° has a unique solution if and only if the measure $d\lambda_m$ in (5.3.22) admits a generalized Gauss-Radau quadrature formula*
$$\int_0^1 g(x)\, d\lambda_m(x) = \sum_{k=0}^m A_k g^{(k)}(0) + \sum_{\nu=1}^n \lambda_\nu^R g(\tau_\nu^R) + R_{n,m}^R(d\lambda_m; g),$$

with $0 < \tau_1^R < \cdots < \tau_n^R < 1$, where $R_{n,m}^R(d\lambda_m; g) = 0$ for each $g \in \mathcal{P}_{2n+m}$. If that is the case, the knots t_ν and coefficients a_ν in (5.3.21) are given by

$$t_\nu = \tau_\nu^R, \quad a_\nu = \lambda_\nu^R, \quad \nu = 1, \ldots, n,$$

and the polynomial p in (5.3.21) is given by

$$p(t) = \sum_{k=0}^{m} \frac{1}{k!} f^{(k)}(1)(t-1)^k.$$

For the proofs of the previous theorems, error estimates in terms of the generalized Gauss-Lobatto and Gauss-Radau quadrature rules, convergence results, as well as numerical examples see Frontini, Gautschi and Milovanović [140]. Further extensions of the moment-preserving spline approximation on [0, 1] are given by Micchelli [322]. He relates this approximation to the theory of monosplines.

Using defective splines with odd defect $k = 2s + 1$, approximation problems reduce to certain generalized Gauss-Turán-Lobatto and Gauss-Turán-Radau quadrature formulas. A more general case with variable defects was considered in [189]. In that case, approximation problems reduce to the Gauss-Turán-Stancu type of quadratures and σ-orthogonal polynomials (cf. [146, 190, 355, 363, 431, 432]).

5.4 Summation of Slowly Convergent Series

Slowly convergent series appear in many problems in applied and computational sciences. There are several numerical methods based on linear and nonlinear transformations. In general, starting from the sequence of partial sums of the series, these transformations give other sequences with faster convergence to the same limit (the sum of the series). Some summation methods can be found in the books of Henrici [209], Lindelöf [265], and Mitrinović and Kečkić [365] (see also Jolley [223] for a collection of explicit expressions of some sums).

In this section we consider certain alternative methods of summation of slowly convergent series based on integral representations of series and an application of the Gaussian quadratures. Such summation/integration procedures for slowly convergent series have been recently developed in [69–71, 156, 157], [166, pp. 239–253], [167, 169, 330–332]. Here we describe some of these methods for convergent series of the type

$$T = \sum_{k=1}^{+\infty} a_k \quad \text{and} \quad S = \sum_{k=1}^{+\infty} (-1)^k a_k. \tag{5.4.1}$$

In the last subsection we give some remarks on methods for summation of slowly convergent power series.

5.4.1 Laplace Transform Method

Suppose that the general term of T (and S) is expressible in terms of the Laplace transform, or its derivative, of a known function. Here, we consider the both cases.

(a) Let $a_k = F(k)$, where

$$F(s) = \int_0^{+\infty} e^{-st} f(t)\, dt, \qquad \operatorname{Re} s \geq 1. \tag{5.4.2}$$

Then

$$T = \sum_{k=1}^{+\infty} F(k) = \sum_{k=1}^{+\infty} \int_0^{+\infty} e^{-kt} f(t)\, dt = \int_0^{+\infty} \left(\sum_{k=1}^{+\infty} e^{-kt} \right) f(t)\, dt,$$

i.e.,

$$T = \int_0^{+\infty} \frac{e^{-t}}{1 - e^{-t}} f(t)\, dt = \int_0^{+\infty} \frac{t}{e^t - 1} \frac{f(t)}{t}\, dt. \tag{5.4.3}$$

Thus, the summation of series is now transformed to an integration problem.

The first idea for numerical calculation of (5.4.3) is an application of the Gauss-Laguerre quadrature rule (5.1.13), with the weight $w(t) = e^{-t}$, to the function

$$\frac{f(t)}{1 - e^{-t}} = \frac{t}{1 - e^{-t}} \frac{f(t)}{t},$$

supposing that $f(t)/t$ is a smooth function. However, the convergence of these Gauss-Laguerre rules can be very slow, according to the presence of poles on the imaginary axis at $2k\pi i$ ($k = \pm 1, \pm 2, \ldots$).

Another approach was given in [169] by a construction of Gaussian quadrature formulas on $(0, +\infty)$,

$$\int_0^{+\infty} g(t) w(t)\, dt = \sum_{\nu=1}^n A_\nu g(\tau_\nu) + R_n(g), \tag{5.4.4}$$

with respect to the weight function $w(t) = \varepsilon(t) = t/(e^t - 1)$, and an application to (5.4.3). This function is widely used in solid state physics (e.g. in the Einstein-Bose distribution) and we called it as the *Einstein weight function*. It is also, incidentally, the generating function of the Bernoulli polynomials. If $g(t) = f(t)/t$ is a smooth function, the Gauss-Einstein formula (5.4.4) converges rapidly.

Similarly, for "alternating" series, we obtain

$$S = \sum_{k=1}^{+\infty} (-1)^k F(k) = \int_0^{+\infty} \left(\sum_{k=1}^{+\infty} (-1)^k e^{-kt} \right) f(t)\, dt,$$

5.4 Summation of Slowly Convergent Series

i.e.,

$$S = \int_0^{+\infty} \frac{-e^{-t}}{1+e^{-t}} f(t)\,dt = \int_0^{+\infty} \frac{1}{e^t+1}(-f(t))\,dt. \tag{5.4.5}$$

In this case a quadrature of Gaussian type (5.4.4) with respect to the *Fermi weight* $w(t) = \varphi(t) = 1/(e^t+1)$ is very convenient (see [169]).

(b) Let $a_k = -F'(k)$, where F is defined by (5.4.2).
Then, after a short calculation, we obtain

$$T = \sum_{k=1}^{+\infty}(-F'(k)) = \int_0^{+\infty} \frac{te^{-t}}{1-e^{-t}} f(t)\,dt = \int_0^{+\infty} \frac{t}{e^t-1} f(t)\,dt. \tag{5.4.6}$$

Also, for "alternating" series we get

$$S = \sum_{k=1}^{+\infty}(-1)^k(-F'(k)) = \int_0^{+\infty} \frac{-te^{-t}}{1+e^{-t}} f(t)\,dt = \int_0^{+\infty} \frac{1}{e^t+1}(-tf(t))\,dt. \tag{5.4.7}$$

As before, the Gauss-Einstein and Gauss-Fermi quadrature rules can be used for calculating the integrals in the obtained integral representations (5.4.6) and 5.4.7), respectively.

Example 5.4.1 We consider two simple examples

$$T = \sum_{k=1}^{+\infty} \frac{1}{(k+1)^2} = \frac{\pi^2}{6} - 1 \quad \text{and} \quad S = \sum_{k=1}^{+\infty} \frac{(-1)^k}{(k+1)^2} = \frac{\pi^2}{12} - 1.$$

According to the presented method in (a), we put $F(s) = (s+1)^{-2}$, which means that $f(t) = te^{-t}$. Then, (5.4.3) and (5.4.5) reduce to

$$T = \int_0^{+\infty} \frac{t}{e^t-1} e^{-t}\,dt \quad \text{and} \quad S = \int_0^{+\infty} \frac{1}{e^t+1}(-te^{-t})\,dt, \tag{5.4.8}$$

respectively. The same integral representations (5.4.8) can be obtained if we use (5.4.6) and (5.4.7), instead of (5.4.3) and (5.4.5), respectively. Namely, taking $-F'(s) = (s+1)^{-2}$, i.e., $F(s) = (s+1)^{-1}+C$, where $C = 0$, because of $F(s) \to 0$ as $s \to \infty$, we get in this case $f(t) = e^{-t}$. It is clear, that (5.4.6) and (5.4.7) give again (5.4.8).

In order to calculate the integrals T and S in (5.4.8) we apply the ordinary Gauss-Laguerre and Gauss-Einstein quadrature in n points. Table 5.4.1 shows the n-point approximations $T(n)$, together with the relative errors $r_n(T)$. Similarly, $S(n)$ and $r_n(S)$ are shown in Table 5.4.2. (The first digit in error is underlined and numbers in parentheses indicate decimal exponents.)

We can see the rapid convergence of the Gauss-Einstein quadrature in both cases. For example, in the case of the first integral T, 15-point quadrature already yields

Table 5.4.1 Gaussian approximations of the sum T and relative errors

q.f. n	Gauss-Laguerre $T(n)$	$r_n(T)$	Gauss-Einstein $T(n)$	$r_n(T)$
5	.6449244	1.5(−5)	.644742	3.0(−4)
10	.6449340525	2.2(−8)	.6449340594	1.1(−8)
15	.6449340668379	1.6(−11)	.644934066848017	3.2(−13)
20	.64493406684941	1.8(−12)	.644934066848226431 31	8.0(−18)
25	.644934066848194	4.9(−14)	.6449340668482264364772	1.8(−22)

Table 5.4.2 Gaussian approximations of the sum S and relative errors

q.f. n	Gauss-Laguerre $S(n)$	$r_n(S)$	Gauss-Einstein $S(n)$	$r_n(S)$
5	−.177330	1.1(−3)	−.177753	1.2(−3)
10	−.17753018	1.6(−5)	−.1775329780	6.5(−8)
15	−.17753307	6.0(−7)	−.17753296657625	2.1(−12)
20	−.17753296190	2.6(−8)	−.1775329665758867915	5.5(−17)
25	−.17753296671	7.6(−10)	−.17753296657588678176 4027	1.3(−21)

12 correct decimal digits. In contrast, 10 000 terms of the original series would give only 3-digit accuracy. Similarly for the second series S, on the basis of Leibniz' convergence criterion, the same accuracy as the one achieved for $n = 15$ would require the summation of approximately 690 000 terms.

As we mentioned before, because of the presence of poles on the imaginary axis, the convergence of the Gauss-Laguerre rules is slower. The influence of the poles in the second case is stronger, because the poles of the integrand are closer to the real line than the ones in the first case. Here, the poles are at the points $(2k + 1)\pi i$ ($k \in \mathbb{Z}$).

Example 5.4.2 Let $F(s) = s^{-1} \exp(-1/s)$, $-F'(s) = (s - 1)s^{-3} \exp(-1/s)$. The original function is here $f(t) = J_0(2\sqrt{t})$. This is an entire function, we expect the Gaussian quadratures (with respect to the Einstein and Fermi weights) to converge rapidly. This is confirmed in Table 5.4.3, which shows the relative errors for the n-point rule, $n = 2(2)12$. The exact sums (to 24 significant digits), as determined by Gaussian quadratures, are, respectively,

(a) $$\sum_{k=1}^{+\infty} (k-1)k^{-3} \exp(-1/k) = .342918943844609780961838,$$

(b) $$\sum_{k=1}^{+\infty} (-1)^{k-1}(k-1)k^{-3} \exp(-1/k) = -.044155938134083605273 6928,$$

5.4 Summation of Slowly Convergent Series

Table 5.4.3 Relative errors in Gaussian approximation of the sums (a), (b) and (c)

n	(a)	(b)	(c)
2	4.48(−2)	8.95(−1)	1.77(−2)
4	3.80(−6)	2.39(−4)	9.65(−7)
6	3.35(−11)	3.71(−9)	6.32(−12)
8	7.01(−17)	1.13(−14)	1.05(−17)
10	6.86(−23)	1.08(−20)	1.27(−23)
12		1.93(−23)	

(c) $\sum_{k=1}^{+\infty}(-1)^{k-1}k^{-1}\exp(-1/k) = .19710793639795065695 5672.$

The first 10 000 terms of the series yield, respectively, 3, 7 and 4 correct decimal digits. The Bessel function J_0 was evaluated by means of the rational approximations in [207] indexed as 5852, 6553 and 6953.

Thus, if the series T and S are slowly convergent and the respective functions in the integral representations are smooth, then the low-order Gaussian quadrature (5.4.4) applied to the integrals on the right, provides a possible summation procedure. Several numerical examples were analyzed in [169, §4].

A problem which arises with this procedure (*Laplace transform method*) is the determination of the original function f for a given series. For some other applications see [156, 157].

In [156], Gautschi treated the case when $a_k = k^{\nu-1}R(k)$, where $0 < \nu \le 1$ and $R(\cdot)$ is a rational function $R(s) = P(s)/Q(s)$, with P, Q real polynomials of degrees deg $P \le$ deg Q. By interpreting the terms in T and S again as Laplace transforms at integer values, Gautschi expressed the sum of the series as a weighted integral over $\mathbb{R}_+$ of certain special functions related to the incomplete gamma function. The weighting involves the product of a fractional power and either Einstein's function $\varepsilon(t)$ (for T) or Fermi's function $\varphi(t)$ (for S). The case $\nu = 1$ of purely rational series complements some traditional techniques of summations via quadratures (cf. [209, §7.2II]).

In particular, Gautschi [156] analyzed examples with $a_k = k^{-1/2}/(k+a)^m$, where $\operatorname{Re} a \ge 0$ and $m \ge 1$. The series T with $a = m = 1$ appeared in a study of spirals given by Davis [76].

5.4.2 Contour Integration Over a Rectangle

In this section we give an alternative summation/integration procedure for the series (5.4.1) when for $k \ge m$ we have $a_k = f(k)$, where $z \mapsto f(z)$ is an analytic function

Fig. 5.4.1 The contour of integration

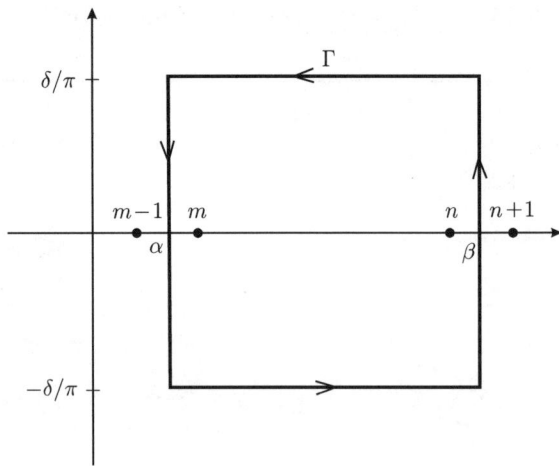

in the region

$$\{z \in \mathbb{C} \mid \operatorname{Re} z \geq \alpha, \ m-1 < \alpha < m\}. \tag{5.4.9}$$

In fact, we consider the series

$$T_m = \sum_{k=m}^{+\infty} a_k \quad \text{and} \quad S_m = \sum_{k=m}^{+\infty} (-1)^k a_k, \tag{5.4.10}$$

where $m \in \mathbb{Z}$.

The method requires the indefinite integral F of f chosen so as to satisfy certain decay properties ((C1)–(C3) below). Using contour integration over a rectangle in the complex plane we are able to reduce T_m and S_m to a problem of Gaussian quadrature rules on $(0, +\infty)$ with respect to the hyperbolic weight functions

$$w_1(t) = \frac{1}{\cosh^2 t} \quad \text{and} \quad w_2(t) = \frac{\sinh t}{\cosh^2 t}, \tag{5.4.11}$$

respectively (see Milovanović [330]).

Assume that f and g are analytic functions in a certain domain D of the complex plane with singularities $a_1, a_2, \ldots$ and $b_1, b_2, \ldots$, respectively, in a region $G = \operatorname{int} \Gamma \,(\subset D)$, where Γ is a closed contour. Then by Cauchy's residue theorem, we have

$$\frac{1}{2\pi i} \oint_\Gamma f(z) g(z)\, dz = \sum_\nu \operatorname*{Res}_{z=a_\nu} \Big(f(z) g(z)\Big) + \sum_\nu \operatorname*{Res}_{z=b_\nu} \Big(f(z) g(z)\Big). \tag{5.4.12}$$

Let

$$G = \left\{ z \in \mathbb{C} \mid \alpha \leq \operatorname{Re} z \leq \beta,\ |\operatorname{Im} z| \leq \frac{\delta}{\pi} \right\},$$

5.4 Summation of Slowly Convergent Series

where $m-1 < \alpha < m$, $n < \beta < n+1$ ($m, n \in \mathbb{Z}$, $m \leq n$), $\Gamma = \partial G$ (see Fig. 5.4.1), and $g(z) = \pi/\tan \pi z$. Then from (5.4.12) it immediately follows that (cf. [365, p. 212])

$$T_{m,n} = \sum_{\nu=m}^{n} f(\nu) = \frac{1}{2\pi i} \oint_\Gamma f(z) \frac{\pi}{\tan \pi z} dz - \sum_\nu \operatorname{Res}_{z=a_\nu} \left(f(z) \frac{\pi}{\tan \pi z} \right).$$

Similarly, for $g(z) = \pi/\sin \pi z$ we have

$$S_{m,n} = \sum_{\nu=m}^{n} (-1)^\nu f(\nu) = \frac{1}{2\pi i} \oint_\Gamma f(z) \frac{\pi}{\sin \pi z} dz - \sum_\nu \operatorname{Res}_{z=a_\nu} \left(f(z) \frac{\pi}{\sin \pi z} \right).$$

For a holomorphic function $z \mapsto f(z)$ in G, the last formulas become

$$T_{m,n} = \frac{1}{2\pi i} \oint_\Gamma f(z) \frac{\pi}{\tan \pi z} dz \quad \text{and} \quad S_{m,n} = \frac{1}{2\pi i} \oint_\Gamma f(z) \frac{\pi}{\sin \pi z} dz.$$

After integration by parts, these formulas reduce to

$$T_{m,n} = \frac{1}{2\pi i} \oint_\Gamma \left(\frac{\pi}{\sin \pi z} \right)^2 F(z) \, dz \tag{5.4.13}$$

and

$$S_{m,n} = \frac{1}{2\pi i} \oint_\Gamma \left(\frac{\pi}{\sin \pi z} \right)^2 \cos \pi z \, F(z) \, dz, \tag{5.4.14}$$

where F is an integral of f.

Assume now the following conditions for the function F (cf. [265, p. 57]):

(C1) F is a holomorphic function in the region (5.4.9);
(C2) $\lim_{|t| \to +\infty} e^{-c|t|} F(x + it/\pi) = 0$, uniformly for $x \geq \alpha$;
(C3) $\lim_{x \to +\infty} \int_{-\infty}^{+\infty} e^{-c|t|} |F(x + it/\pi)| \, dt = 0$,
where $c = 2$ or $c = 1$, when we consider $T_{m,n}$ or $S_{n,m}$, respectively.

Set $\alpha = m - 1/2$ and $\beta = n + 1/2$.

On the lines $z = x \pm i(\delta/\pi)$, we have that

$$\left| \frac{\pi}{\sin \pi z} \right| = \left| \frac{2i\pi}{e^{i\pi z} - e^{-i\pi z}} \right| = \frac{2\pi e^{-\delta}}{|1 - e^{-2\delta} e^{\pm i 2\pi x}|} \leq \frac{2\pi e^{-\delta}}{1 - e^{-2\delta}}$$

and also

$$|\cos \pi z| = \frac{1}{2} e^\delta |1 + e^{-2\delta} e^{\pm i 2\pi x}| \leq \frac{1}{2} e^\delta (1 + e^{-2\delta}).$$

Therefore, under condition (C2), the integrals on the lines $z = x \pm i(\delta/\pi)$, $\alpha \leq x \leq \beta$,

$$\frac{1}{2\pi i} \int_{\alpha \pm i(\delta/\pi)}^{\beta \pm i(\delta/\pi)} \left(\frac{\pi}{\sin \pi z} \right)^2 F(z) \, dz, \quad \frac{1}{2\pi i} \int_{\alpha \pm i(\delta/\pi)}^{\beta \pm i(\delta/\pi)} \left(\frac{\pi}{\sin \pi z} \right)^2 \cos \pi z \, F(z) \, dz,$$

tend to zero when $\delta \to +\infty$.

For $z = \beta + iy$ we have

$$\sin \pi z = (-1)^n \cosh \pi y, \qquad \cos \pi z = i(-1)^{n+1} \sinh \pi y,$$

$$\left|\frac{\pi}{\sin \pi z}\right|^2 = \frac{\pi^2}{\cosh^2 \pi y} = \frac{4\pi^2}{\left(e^{\pi y} + e^{-\pi y}\right)^2} \leq 4\pi^2 e^{-2\pi |y|},$$

and

$$\left|\frac{\pi}{\sin \pi z}\right|^2 |\cos \pi z| \leq 2\pi^2 e^{-\pi |y|},$$

so that

$$\left|\frac{1}{2\pi i} \int_{\beta - i(\delta/\pi)}^{\beta + i(\delta/\pi)} \left(\frac{\pi}{\sin \pi z}\right)^2 F(z)\, dz\right| \leq 2 \int_{-\delta}^{\delta} e^{-2|t|} \left|F(\beta + it/\pi)\right| dt$$

and

$$\left|\frac{1}{2\pi i} \int_{\beta - i(\delta/\pi)}^{\beta + i(\delta/\pi)} \left(\frac{\pi}{\sin \pi z}\right)^2 \cos \pi z\, F(z)\, dz\right| \leq \int_{-\delta}^{\delta} e^{-|t|} \left|F(\beta + it/\pi)\right| dt.$$

When $\delta \to +\infty$ and $n \to +\infty$ (i.e., $\beta \to +\infty$), because of (C3), the previous integrals tend to zero.

Thus, when $\delta \to +\infty$ and $n \to +\infty$, the integrals in (5.4.13) and (5.4.14) over Γ reduce to integrals along the line $z = \alpha + iy$ ($-\infty < y < +\infty$), so that

$$T_m = T_{m,\infty} = -\frac{1}{2\pi i} \int_{\alpha - i\infty}^{\alpha + i\infty} \left(\frac{\pi}{\sin \pi z}\right)^2 F(z)\, dz \qquad (5.4.15)$$

and

$$S_m = S_{m,\infty} = -\frac{1}{2\pi i} \int_{\alpha - i\infty}^{\alpha + i\infty} \left(\frac{\pi}{\sin \pi z}\right)^2 \cos \pi z\, F(z)\, dz.$$

Equality (5.4.15) can be reduced to

$$T_m = -\frac{1}{2} \int_{-\infty}^{+\infty} \frac{1}{\cosh^2 t} F(\alpha + it/\pi)\, dt,$$

i.e.,

$$T_m = \int_0^{+\infty} \Phi(\alpha, t/\pi)\, w_1(t)\, dt, \qquad (5.4.16)$$

where w_1 is defined in (5.4.11) and

$$\Phi(x, y) = -\frac{1}{2}\left[F(x+iy) + F(x-iy)\right].$$

5.4 Summation of Slowly Convergent Series

Similarly, (5.4.16) reduces to

$$S_m = \int_0^{+\infty} \Psi(\alpha, t/\pi) w_2(t) \, dt, \tag{5.4.17}$$

where w_2 is also defined in (5.4.11) and

$$\Psi(x, y) = \frac{(-1)^m}{2i} \left[F(x + iy) - F(x - iy) \right].$$

Here, $\alpha = m - 1/2$. Formulas (5.4.16) and (5.4.17) suggest to apply a Gaussian quadrature to the integrals on the right, using the weight functions w_1 and w_2, respectively (see [330]). The construction of such formulae is described in Sect. 5.1.3.

Remark 5.4.1 Instead of reducing integration to the positive half-line, one might keep integration over the whole real line and note that

$$T_m = \int_{-\infty}^{+\infty} \Phi(\alpha, t/(2\pi)) \frac{e^{-t}}{(1 + e^{-t})^2} \, dt$$

and

$$S_m = \int_{-\infty}^{+\infty} \Psi(\alpha, t/(2\pi)) \sinh(t/2) \frac{e^{-t}}{(1 + e^{-t})^2} \, dt.$$

Here, the weight function is $w(t) = e^{-t}/(1 + e^{-t})^2$, the logistic weight, for which the recursion coefficients for the respective orthogonal polynomials are explicitly known (see Sect. 2.4.6). Thus, no procedure is required to generate the recursion coefficients. Some comments on the convergence of the corresponding Gaussian quadrature will be given later.

Example 5.4.3 Consider again the series from Example 5.4.1, denoted now as T_1 and S_1, respectively.

Here, $f(z) = (z + 1)^{-2}$, and $F(z) = -(z + 1)^{-1}$, the integration constant being zero on account of the condition (C3). Thus,

$$\Phi(x, y) = \text{Re} \frac{1}{z + 1} = \frac{x + 1}{(x + 1)^2 + y^2} \quad \text{and} \quad \Psi(x, y) = \text{Im} \frac{1}{z + 1} = \frac{-y}{(x + 1)^2 + y^2}.$$

Now, we apply the Gaussian quadrature formulae with respect to the hyperbolic weights w_1 and w_2 given in (5.4.11) to T_1 and S_1, respectively. Table 5.4.4 shows the corresponding n-point Gaussian approximations $T_1(n)$ and $S_1(n)$ to T_1 and S_1, respectively, together with the relative errors $r_n(T_1)$ and $r_n(S_1)$, for $n = 5(5)40$.

The corresponding relative errors in the *Laplace transform method* (with Einstein weight) applied to T_1 are given in Table 5.4.1 (see also Table 5.4.7 for bigger values of n).

Table 5.4.4 Gaussian approximation of the sums T_1 and S_1 and relative errors

n	$T_1(n)$	$r_n(T_1)$	$S_1(n)$	$r_n(S_1)$
5	.644934149	1.3(−7)	−.1775520	1.1(−4)
10	.644934066776	1.1(−10)	−.17753303	3.5(−7)
15	.644934066848158	1.1(−13)	−.17753296569	5.0(−9)
20	.64493406684822733	1.4(−15)	−.1775329665917	8.9(−11)
25	.6449340668482264405	6.2(−18)	−.177532966575286	3.4(−12)
30	.6449340668482264363 07	2.6(−19)	−.177532966575929	2.4(−13)
35	.64493406684822643647604	5.6(−21)	−.1775329665758832	2.0(−14)
40	.64493406684822643647233	1.3(−22)	−.1775329665758870	1.5(−15)

Table 5.4.5 Relative errors in Gaussian approximation of the sum T_1 expressed in the form (5.4.18) for $m = 2(1)5$

n	$m=2$	$m=3$	$m=4$	$m=5$
5	5.4(−9)	1.9(−10)	8.6(−12)	3.7(−13)
10	1.1(−13)	1.7(−16)	7.9(−18)	2.0(−19)
15	3.8(−17)	3.7(−20)	1.1(−22)	3.8(−25)
20	4.0(−20)	1.2(−24)	1.9(−27)	2.3(−29)
25	1.1(−22)	2.0(−27)	2.6(−30)	2.5(−33)
30	1.4(−25)	1.1(−31)	2.2(−33)	
35	3.2(−27)	2.4(−32)		
40	3.6(−30)			

As we can see, for smaller values of n (≤ 15) we obtained better results than in the *Laplace transform method*. Furthermore, these results can be significantly improved if we apply this method to sum the series T_m, $m > 1$. That is, we use

$$T_1 = \sum_{k=1}^{m-1} \frac{1}{(k+1)^2} + T_m, \qquad T_m = \sum_{k=m}^{+\infty} \frac{1}{(k+1)^2}. \qquad (5.4.18)$$

Then, for $m = 2(1)5$ we obtain results whose relative errors are presented in Table 5.4.5. Also, in Table 5.4.6 we present the corresponding results for the sum S_1 expressed in a similar way

$$S_1 = \sum_{k=1}^{m-1} \frac{(-1)^k}{(k+1)^2} + S_m, \qquad S_m = \sum_{k=m}^{+\infty} \frac{(-1)^k}{(k+1)^2}. \qquad (5.4.19)$$

The rapidly increasing of convergence of the summation process as m increases is due to the poles $\pm i(m+1/2)\pi$ of $\Phi(m-1/2, t/\pi)$ moving away from the real line.

5.4 Summation of Slowly Convergent Series

Table 5.4.6 Relative errors in Gaussian approximation of the sum S_1 expressed in the form (5.4.19) for $m = 2(1)5$

n	m = 2	m = 3	m = 4	m = 5
5	1.9(−6)	2.2(−7)	1.5(−8)	4.5(−10)
10	1.9(−9)	2.3(−12)	1.0(−12)	1.1(−14)
15	8.1(−13)	1.9(−15)	3.2(−16)	9.2(−18)
20	6.6(−14)	1.1(−16)	6.2(−19)	7.6(−21)
25	6.2(−16)	6.8(−19)	2.5(−21)	1.3(−23)
30	2.4(−18)	9.7(−21)	1.2(−24)	1.3(−26)
35	1.1(−19)	4.3(−23)	1.3(−25)	2.1(−28)
40	5.6(−21)	1.3(−24)	1.1(−27)	9.8(−31)

Table 5.4.7 Relative errors in Gaussian approximation of the sum T_1 using the *Laplace transform method* (with Einstein weight) for $m = 1(1)3$

n	m = 1	m = 2	m = 3
5	3.0(−4)	8.4(−3)	3.0(−2)
10	1.1(−8)	1.8(−5)	3.8(−4)
15	3.2(−13)	2.8(−8)	3.7(−6)
20	8.0(−18)	3.9(−11)	3.1(−8)
25	1.8(−22)	5.1(−14)	2.5(−10)
30	3.9(−27)	6.3(−17)	1.9(−12)
35	8.7(−32)	7.6(−20)	1.4(−14)
40	4.6(−33)	8.8(−23)	1.0(−16)

It is interesting to note that a similar approach with the *Laplace transform method* does not lead to acceleration of convergence. For example, in the case of (5.4.18), we have that

$$T_m = \sum_{k=1}^{+\infty} \frac{1}{(k+m)^2} = \int_0^{+\infty} \varepsilon(t) e^{-mt}\, dt.$$

Then, applying the Gaussian quadrature to the integral on the right, using $w(t) = \varepsilon(t)$ as a weight function on $(0, +\infty)$, we can obtain approximations for the sum T_1 for different values of n and m. The corresponding relative errors for $n = 5(5)40$ and $m = 1(1)3$ are presented in Table 5.4.7. As we can see, the convergence of the process (as m increases) slows down considerably. The reason for this is the behavior of the function $t \mapsto e^{-mt}$, which tends to a discontinuous function when $m \to +\infty$. On the other hand, the function is entire, which explains the ultimately much better results in Table 5.4.7 when $m = 1$.

It is interesting to mention that the Gaussian quadrature over the whole real line with respect to the logistic function (see Remark 5.4.1) converges considerably more

Table 5.4.8 Relative errors in Gaussian approximation of the sum T_1 and S_1 with respect to the logistic weight for $m = 1(1)5$

n		m = 1	m = 2	m = 3	m = 4	m = 5
5	T_1	4.7(−5)	5.2(−7)	1.9(−8)	1.5(−9)	1.8(−10)
	S_1	1.1(−3)	1.1(−3)	8.2(−4)	6.3(−4)	4.8(−4)
10	T_1	1.1(−6)	1.2(−9)	6.2(−12)	8.0(−14)	2.0(−15)
	S_1	4.1(−6)	1.3(−7)	1.3(−7)	1.1(−7)	1.0(−7)
15	T_1	1.1(−7)	2.8(−11)	3.4(−14)	1.2(−16)	8.9(−19)
	S_1	4.0(−7)	1.2(−10)	1.7(−11)	1.6(−11)	1.5(−11)
20	T_1	2.1(−8)	1.8(−12)	7.5(−16)	9.4(−19)	2.7(−21)
	S_1	7.5(−8)	6.5(−12)	5.1(−15)	2.2(−15)	2.1(−15)
25	T_1	5.5(−9)	2.1(−13)	3.7(−17)	2.0(−20)	2.6(−23)
	S_1	2.0(−8)	7.5(−13)	1.4(−16)	3.8(−19)	2.9(−19)
30	T_1	1.9(−9)	3.5(−14)	3.1(−18)	8.6(−22)	5.6(−25)
	S_1	7.0(−9)	1.3(−13)	1.1(−17)	3.1(−21)	4.3(−23)
35	T_1	7.7(−10)	7.7(−15)	3.8(−19)	5.8(−23)	2.1(−26)
	S_1	2.8(−9)	2.8(−14)	1.4(−18)	2.1(−22)	7.1(−26)
40	T_1	3.5(−10)	2.1(−15)	6.1(−20)	5.5(−24)	1.2(−27)
	S_1	1.3(−9)	7.5(−15)	2.2(−19)	2.0(−23)	4.4(−27)

slowly than shown in Tables 5.4.4–5.4.6 for one-sided integration, even though the poles of the integrand have a distance twice as large from the real line. The reason, probably, is that these poles are now centered over the interval of integration, whereas in (5.4.16) and (5.4.17) they are located over the left endpoint of the interval. Numerical results for $n = 5(5)40$ and $m = 1(1)5$ are given in Table 5.4.8.

Example 5.4.4 The application of the *Laplace transform method* to the series

$$\sum_{k=1}^{+\infty}(k-1)k^{-3}\exp(-1/k) = .34291894384460978096183767790 2 \quad (5.4.20)$$

leads to an integration of the Bessel function $J_0(2\sqrt{t})$ (see Example 5.4.2). However, we work here with the exponential function $F(z) = -e^{-1/z}/z$, i.e.,

$$\Phi(x,y) = \frac{1}{r^2}e^{-x/r^2}\left(x\cos\frac{y}{r^2} + y\sin\frac{y}{r^2}\right), \quad r^2 = x^2 + y^2.$$

5.4 Summation of Slowly Convergent Series

Table 5.4.9 Relative errors in Gaussian approximation of the sum (5.4.20)

n	$m = 1$	$m = 2$	$m = 3$
2	2.9(−3)	1.2(−5)	2.1(−8)
6	1.3(−4)	3.7(−8)	1.2(−10)
10	1.8(−5)	3.7(−11)	9.9(−14)
14	1.2(−6)	1.2(−12)	1.2(−16)
18	1.3(−7)	8.5(−15)	6.6(−19)

Table 5.4.10 Relative errors in the *method of Laplace transform* for the series (5.4.21) with $a = 8$

$n = 5$	$n = 10$	$n = 15$	$n = 20$	$n = 25$	$n = 30$	$n = 35$	$n = 40$
1.4(−1)	2.3(−2)	1.5(−3)	1.9(−4)	2.5(−5)	2.1(−6)	2.5(−7)	2.6(−8)

As for accuracy, a similar situation prevails as in the previous example. Table 5.4.9 shows the relative errors in Gaussian approximations for $n = 2(4)18$ and $m = 1(1)3$.

Example 5.4.5 Consider now

$$T_1(a) = \sum_{k=1}^{+\infty} \frac{1}{\sqrt{k}(k+a)}. \quad (5.4.21)$$

This series with $a = 1$ appeared in a study of spirals (see Davis [76]) and defines the "Theodorus constant." The first 1 000 000 terms of the series $T_1(1)$ give the result 1.8580..., i.e., $T_1(1) \approx 1.86$ (only 3-digit accuracy). Using the *method of Laplace transform*, Gautschi (see [156, Example 5.1]) calculated (5.4.21) for $a = .5, 1, 2, 4, 8, 16$, and 32. As a increases, the convergence of the Gauss quadrature formula slows down considerably. For example, when $a = 8$, we have results with relative errors presented in Table 5.4.10.

In order to achieve better accuracy, when a is large, Gautschi [156] used "stratified" summation by letting $k = \lambda + \kappa a_0$ and summing over all $\kappa \geq 0$ for $\lambda = 1, 2, \ldots, a_0$, where $a_0 = \lfloor a \rfloor$ denotes the largest integer $\leq a$ ($a = a_0 + a_1$, $a_0 \geq 1$, $0 \leq a_1 < 1$).

Now, we directly apply the *method of contour integration over the rectangle* to (5.4.21) with

$$F(z) = \frac{2}{\sqrt{a}} \left(\arctan\sqrt{\frac{z}{a}} - \frac{\pi}{2} \right),$$

where the integration constant is taken so that $F(\infty) = 0$. For computing the arctan function in the complex plane ($z^2 \neq -1$) we use the formula

$$\arctan z = \frac{1}{2} \arg(u + iv) + \frac{i}{4} \log \frac{x^2 + (y+1)^2}{x^2 + (y-1)^2},$$

Table 5.4.11 Relative errors in Gaussian approximation of the sum (5.4.22) for $m=4$

n	$a=.5$	$a=1$	$a=2$	$a=4$
5	1.4(−11)	8.4(−12)	4.5(−12)	2.6(−12)
10	6.8(−18)	4.4(−18)	2.2(−18)	1.2(−18)
15	5.4(−22)	2.7(−22)	1.6(−22)	1.0(−22)
20	1.2(−25)	5.9(−26)	3.3(−26)	2.0(−26)
25	1.0(−28)	5.2(−29)	3.0(−29)	1.9(−29)
30	1.1(−31)	5.7(−32)	3.3(−32)	2.0(−32)

n	$a=8$	$a=16$	$a=32$	$a=64$
5	1.7(−12)	1.1(−12)	7.6(−13)	5.2(−13)
10	7.7(−19)	5.1(−19)	3.4(−19)	2.4(−19)
15	6.7(−23)	4.5(−23)	3.0(−23)	2.1(−23)
20	1.3(−26)	8.7(−27)	5.9(−27)	4.1(−27)
25	1.2(−29)	8.1(−30)	5.5(−30)	3.8(−30)
30	1.3(−32)	9.0(−33)	6.2(−33)	3.6(−33)

Table 5.4.12 The exact sums $T_1(a)$

a	$T_1(a)$
1/2	2.1344166429862372611014895 2804
1	1.8600250792211903071806959 1572
2	1.5396805123533020128750184 1998
4	1.2182740146698908458291597 6291
8	9.3137293400310387168575138 9665(−1)
16	6.9493171464104559016304607 1669(−1)
32	5.0992651702721134803613196 7602(−1)
64	3.6993169824967113220994236 4907(−1)

where $z = x+iy$, $u = 1-x^2-y^2$, $v = 2x$.

As before, we can represent (5.4.21) in the form

$$T_1(a) = \sum_{k=1}^{m-1} \frac{1}{\sqrt{k}(k+a)} + T_m(a), \quad T_m(a) = \sum_{k=m}^{+\infty} \frac{1}{\sqrt{k}(k+a)}, \quad (5.4.22)$$

and then use the Gaussian quadrature formula to calculate $T_m(a)$. Relative errors in approximations for $T_1(a)$, when $m=4$ and $a = p_\nu$, $\nu = 0(1)7$, where $p_0 = .5$ and $p_{\nu+1} = 2p_\nu$, are displayed in Table 5.4.11.

As we can see from Table 5.4.11, the method presented is very efficient. Moreover, its convergence is slightly faster if the parameter a is larger. The exact sums

5.4 Summation of Slowly Convergent Series

$T_1(a)$ (to 30 significant digits), as determined by Gaussian quadrature, are presented in Table 5.4.12.

Numerical experiments shows that is enough to use only the quadrature with respect to the first weight $w_1(t) = 1/\cosh^2 t$. Namely, in the series S_m we can include the hyperbolic sine as a factor in the corresponding integrand so that

$$S_m = \int_0^{+\infty} [\Psi(m - 1/2, t/\pi) \sinh(t)] w_1(t) \, dt.$$

Such an application was given in [332] to summation of slowly convergent series

$$T_m = T_m(\nu, a, p) = \sum_{k=m}^{+\infty} \frac{k^{\nu-1}}{(k+a)^p} \quad \text{and} \quad S_m = S_m(\nu, a, p) = \sum_{k=m}^{+\infty} (-1)^k \frac{k^{\nu-1}}{(k+a)^p},$$

where $m \in \mathbb{Z}$, $0 < \nu \le 1$, and a and p are such as to provide convergence of these series.

Remark 5.4.2 The Riemann zeta function $\zeta(z) = \sum_{k=1}^{+\infty} k^{-z}$ can be transformed to a weighted integral on $(0, +\infty)$ of the function

$$t \mapsto \exp(-(z/2) \log(1 + \beta_m^2 t^2)) \cos(z \arctan(\beta_m t)), \quad \beta = \frac{2}{(2m+1)\pi}, \quad m \in \mathbb{N}_0,$$

involving the hyperbolic weight $w(t) = 1/\cosh^2 t$. As an appropriate method for calculating values of $\zeta(z)$ the presented method can be used (see [332]).

Remark 5.4.3 Some methods for series with irrational terms were given in [331].

5.4.3 Remarks on Some Slowly Convergent Power Series

Numerical methods for summation of certain slowly convergent power series were considered by Gautschi [157], [166, pp. 249–253], Dassiè, Vianello, and Zanovello [73, 74], Dahlquist [71], etc. In this section we give a short account of these methods.

Gautschi [157] considered slowly convergent series occurring in plate contact problems

$$R_p(z) = \sum_{k=0}^{+\infty} \frac{z^{2k+1}}{(2k+1)^p} \quad \text{and} \quad S_p(z) = \sum_{k=0}^{+\infty} (-1)^k \frac{z^{2k+1}}{(2k+1)^p},$$

where $z \in \mathbb{C}$, $|z| \le 1$, and $p = 2$ or 3. The convergence of these series is very slowly, especially when $|z|$ is close or equal to 1. It is easy to see that $S_p(z) = i R_p(-iz)$, so that in the investigation it is sufficient to study only the first series $R_p(z)$. Note that for $z=1$ it can be expressed in terms of the Riemann zeta function,

$$R_p(1) = (1 - 2^{-p})\zeta(p).$$

For example, $R_2(1) = \pi^2/8$. Also, for z on the unit circle ($|z|=1$), some of these series can be summed explicitly as the Fourier series.

Using the idea of the *Laplace transform method*, we express the coefficient in the general term of the series $R_p(z)$ in terms of the Laplace transform, i.e., $(2k+1)^{-p} = F(k)$, where $F(s)$ defined by (5.4.2). Thus, in this case, we have

$$F(s) = \frac{1}{(2s+1)^p} \quad \text{and} \quad f(t) = \frac{t^{p-1}e^{-t/2}}{2^p(p-1)!}.$$

Then

$$R_p(z) = \sum_{k=0}^{+\infty} \frac{z^{2k+1}}{(2k+1)^p} = \frac{1}{2^p(p-1)!} \sum_{k=0}^{+\infty} z^{2k+1} \int_0^{+\infty} t^{p-1} e^{-t/2} e^{-kt} dt$$

$$= \frac{z}{2^p(p-1)!} \int_0^{+\infty} \frac{t^{p-1} e^{-t/2}}{1 - z^2 e^{-t}} dt.$$

For $z=1$ this integral becomes

$$R_p(1) = \frac{1}{2^p(p-1)!} \int_0^{+\infty} \frac{t^{p-1} e^{t/2}}{e^t - 1} dt,$$

and it can be evaluated by the quadrature formula (5.4.4), with respect to the Einstein weight function $\varepsilon(t) = t/(e^t - 1)$.

The case $z \neq 1$ is much more serious. Gautschi [157] reduces it to an integral over $(0, 1)$,

$$R_p(z) = \frac{1}{2^p(p-1)!z} \int_0^1 \frac{t^{-1/2}(\log(1/t))^{p-1}}{z^{-2} - t} dt,$$

and then he applies the Gaussian quadrature formulas with respect to the measure $d\lambda_p(t) = t^{-1/2}(\log(1/t))^{p-1} dt$.

Remark 5.4.4 In the same paper [157], Gautschi considered the series

$$T_p(x, b) = \sum_{k=0}^{+\infty} \frac{1}{(2k+1)^p} \frac{\cosh(2k+1)x}{\cosh(2k+1)b}$$

and

$$U_p(x, b) = \sum_{k=0}^{+\infty} \frac{1}{(2k+1)^p} \frac{\sinh(2k+1)x}{\cosh(2k+1)b}$$

5.4 Summation of Slowly Convergent Series

where $0 \le x \le b$, $b > 0$, and again $p = 2$ or 3, which are also of some interest in the plate contact problems.

The power series $S(z) = \sum_{k=1}^{+\infty} a_k z^k$ was considered recently by Dassiè, Vianello, and Zanovello [73, 74]. They gave an asymptotic expansion in powers of n^{-1} of the remainder $\sum_{k=n}^{+\infty} a_k z^k$, when the sequence a_k has a similar expansion. In the case of a numerical series ($z = 1$), the rigorous error estimates for the asymptotic approximations are provided. The results are applied to the evaluation of

$$S(z; m, a, b, \nu, p) = \sum_{k=m}^{+\infty} \frac{(k+b)^{\nu-1}}{(k+a)^p} z^k,$$

which generalizes various summation problems. In particular, in [74] the authors show that their method can be conveniently applied to the slowly convergent power series whose coefficients are rational functions of the summation index and provide several numerical examples.

For an interesting analysis of summation/integration procedures, with various questions concerning their construction and application, see [69–71]. In these papers Dahlquist gives a rigorous analysis of the summation formulas due to Plana, Lindelöf and Abel, and related Gauss-Christoffel rules. At the end, in order to stimulate further work on this subject, we only mention an interesting ill-conditioned power series $\sum_{k=0}^{+\infty} f(k; z)$, where

$$f(s; z) = \frac{g(s) z^s}{\Gamma(1 + s/2)^2}, \quad z = iy, \quad y \gg 1,$$

and $g(s)$ is analytic and bounded for $\mathrm{Re}(s) \ge 0$ (see [71]). This series converges for all z, but the moduli of the terms increase rapidly at the beginning. In the case $g(s) = 1$ and $z = iy$, the real part of the sum equals the Bessel function $J_0(2y)$. The largest term is easily estimated by means of Stirling's formula.

References

1. M. ABRAMOWITZ and I. A. STEGUN eds., *Handbook of Mathematical Functions with Formulas, Graphs, and Mathematical Tables*, Dover Publications, Inc., New York, 1972.
2. J. ACZÉL, Eine Bemerkung über die Charakterisierung der "klassischen" Orthogonalpolynome, *Acta Math. Acad. Sci. Hung.* **4** (1953), 315–321.
3. R. P. AGARWAL and G. V. MILOVANOVIĆ, One characterization of the classical orthogonal polynomials, In: *Progress in Approximation Theory* (P. Nevai, A. Pinkus, eds.), Academic Press, New York, 1991, pp. 1–4.
4. R. P. AGARWAL and G. V. MILOVANOVIĆ, Extremal problems, inequalities, and classical orthogonal polynomials, *Appl. Math. Comput.* **128** (2002), 151–166.
5. N. I. AHIEZER, On the weighted approximation of continuous functions by polynomials on the entire number axis, *AMS Translations*, Series 2, **22** (1962), 95–137.
6. N. I. AHIEZER, *Lectures in the Theory of Approximation*, Nauka, Moscow, 1965 (Russian).
7. N. I. AHIEZER and M. G. KREĬN, *On Some Problems in the Moment Theory*, GONTI, Har'kov, 1938.
8. G. ALEXITS, Sur l'ordre de grandeur de l'approximation d'une fonction par les moyennes de sa série de Fourier, *Mat. Fiz. Lapok* **48** (1941), 410–422 (Hungarian. French summary).
9. M. ALFARO and F. MARCELLÁN, Recent trends in orthogonal polynomials on the unit circle, In: *IMACS Annals on Computing and Applied Mathematics, Vol. 9: Orthogonal Polynomials and Their Applications* (C. Brezinski, L. Gori and A. Ronveaux, eds.), IMACS, Baltzer, Basel, 1991, pp. 3–14.
10. W. A. AL-SALAM, Characterization theorems for orthogonal polynomials, In: *Orthogonal Polynomials—Theory and Practice* (P. Nevai, ed.), NATO ASI Series, Series C; Mathematical and Physical Sciences, Vol. 294, Kluwer, Dordrecht, 1990, pp. 1–24.
11. C. R. R. ALVES and D. K. DIMITROV, Landau and Kolmogoroff type polynomial inequalities, *J. Ineq. Appl.* **4** (1999), 327–338.
12. G. E. ANDREWS and R. ASKEY, Classical orthogonal polynomials, In: *Polynômes Orthogonaux et Applications* (C. Brezinski, A. Draux, A. P. Magnus, P. Maroni, A. Ronveaux, eds.), Lect. Notes Math. No. 1171, Springer Verlag, Berlin, 1985, pp. 36–62.
13. G. E. ANDREWS, R. ASKEY, and R. ROY, *Special Functions*, Encyclopedia of Mathematics and Its Applications, The University Press, Cambridge, 1999.
14. J. R. ANGELOS, E. H. KAUFMAN JR., M. S. HENRY, and T. D. LENKER, Optimal nodes for polynomial interpolation, In: *Approximation Theory VI, Vol. 1 (College Station, TX, 1989)*, (C. K. Chui, L. L. Schumaker, J. D. Ward, eds.), Academic Press, Boston, 1989, pp. 17–20.
15. V. A. ANTONOV and K. V. HOLŠEVNIKOV, Estimation of a remainder of a Legendre polynomial generating function expansion generalization and refinement of the Bernšteĭn inequality, *Vestnik Leningrad. Univ. Mat. Mekh. Astronom.* 1980, vyp. 3, 5–7, 128 (Russian).
16. R. ASKEY, Positivity of the Cotes numbers for some Jacobi abscissas, *Numer. Math.* **19** (1972), 46–58.
17. R. ASKEY, Positivity of the Cotes numbers for some Jacobi abscissas II, *J. Inst. Math. Appl.* **24** (1979), 95–98.
18. R. ASKEY and J. FITCH, Positivity of the Cotes numbers for some ultraspherical abscissas, *SIAM J. Numer. Anal.* **5** (1968), 199–201.
19. R. ASKEY and S. WAINGER, Mean convergence of expansions in Laguerre and Hermite series, *Amer. J. Math.* **87** (1965), 695–708.
20. R. ASKEY and J. WILSON, Some basic hypergeometric orthogonal polynomials that generalize Jacobi polynomials, *Memoirs Amer. Mat. Soc.* **319**, Providence, RI, 1985.
21. R. ASKEY, G. GASPER, and L. A. HARRIS, An inequality for Tchebycheff polynomials and extensions, *J. Approx. Theory* **14** (1975), 1–11.
22. N. M. ATAKISHIYEV and S. K. SUSLOV, The Hahn and Meixner polynomials of an imaginary argument and some of their applications, *J. Phys. A: Math. Gen.* **18** (1985), 1583–1596.

23. G. V. BADALYAN, Generalisation of Legendre polynomials and some of their applications, *Akad. Nauk Armyan. SSR Izv. Ser. Fiz.-Mat. Estest. Tekhn. Nauk* **8**(5) (1955), 1–28 (Russian, Armanian summary).
24. V. M. BADKOV, Convergence in the mean and almost everywhere of Fourier series in polynomials orthogonal on an interval, *Mat. Sb.* **95** No. 137 (1974), 223–256.
25. V. M. BADKOV, Asymptotic and extremal properties of orthogonal polynomials with singularities in the weight, *Trudy Mat. Inst. Steklov* **198** (1992), 41–88 (Russian) [Engl. transl. *Proc. Steklov Inst. Math.* **198** (1994), 37–82].
26. G. I. BARKOV, Some systems of polynomials orthogonal in two symmetric intervals, *Izv. Vysš. Učebn. Zav. Matematika* No. 4 (1960), 3–16 (Russian).
27. P. BARRUCAND, Intégration numérique, Abscisses de Kronrod-Patterson et polynômes de Szegős, *C. R. Acad. Sci. Paris, Sér. A* **270** (1970), 147–158.
28. P. BARRUCAND and D. DICKINSON, On the associate Legendre polynomials, In: *Orthogonal Expansion and Their Continuous Analogues* (D. Haimo, ed.), Southern Illinois University Press, Carbondale, 1967, pp. 43–50.
29. H. BATEMAN and A. ERDÉLYI, *Higher Transcendental Functions*, Vol. 2, McGraw-Hill, New York, 1953.
30. W. C. BAULDRY, Estimates of Christoffel functions of generalized Freud-type-weights, *J. Approx. Theory* **46** (1986), 217–229.
31. C. BERG, Markov's theorem revisited, *J. Approx. Theory* **78** (1994), 260–275.
32. S. N. BERNSTEIN, Sur l'ordre de la meilleure approximation des fonctions continues par des polynômes de degré donné, *Mém. Acad. Roy. Belgique (2)* **4** (1912), 1–103.
33. S. BERNSTEIN, Quelques remarques sur l'interpolation, *Zap. Kharkov Mat. Ob-va (Comm. Kharkov Math. Soc.)* **15** (2) (1916), 49–61.
34. S. N. BERNSTEIN, Quelques remarques sur l'interpolation, *Math. Ann.* **79** (1918), 1–12.
35. S. N. BERNSTEIN, *Leçons sur les propriétés extrémales et la meilleure approximation des fonctions analytiques d'une variable réele*, Gauthier–Villars, Paris, 1926.
36. S. N. BERNSTEIN, Sur les polynomes ortogonaux relatifs à un segment fini. I, *J. Math. Pures Appl.* **9** (1930), 127–177.
37. S. N. BERNSTEIN, Sur les polynomes ortogonaux relatifs à un segment fini. II, *J. Math. Pures Appl.* **10** (1931), 219–286.
38. S. N. BERNSTEIN, Sur la limitation des valeurs d'un polynome $P_n(x)$ de degré n sur tout un segment par ses valeurs en $(n+1)$ points du segment, *Izv. Akad. Nauk SSSR* **8** (1931), 1025–1050.
39. J. P. BERRUT and L. N. TREFETHEN, Barycentric Lagrange interpolation, *SIAM Rev.* **46** (2004), 501–517.
40. D. BERTHOLD, W. HOPPE, and B. SILBERMANN, A fast alghoritm for solving the generalized airfol equation, *J. Comput. Appl. Math.* **43** (1992), 185–219.
41. O. V. BESOV, V. P. IL'IN, and S. M. NIKOL'SKIĬ, *Integral Representations of Functions and Imbedding Theorems*, Vol. II, V. H. Winston & Sons, Halsted Press Book, Wiley, New York, 1979.
42. H. F. BLICHFELDT, Note on the functions of the form $f(x) \equiv \varphi(x) + a_1 x^{n-1} + \cdots + a_n$, *Trans. Amer. Math. Soc.* **2** (1901), 100–102.
43. S. BOCHNER, Über Sturm-Liouvillesche Polynomsysteme, *Math. Z.* **29** (1929), 730–736.
44. B. D. BOJANOV and L. GORI, Moment preserving approximations, *Math. Balkanica (N. S.)* **13** (1999), 385–398.
45. B. D. BOJANOV and A. SRI RANGA, Some examples of moment preserving approximation, *Contemp. Math.* **239** (1999), 57–70.
46. B. D. BOJANOV and A. K. VARMA, On a polynomial inequality of Kolmogoroff's type, *Proc. Amer. Math. Soc.* **124** (1996), 491–496.
47. P. BORWEIN and T. ERDÉLYI, *Polynomials and Polynomial Inequalities*, Graduate Texts in Mathematics **161**, Springer-Verlag, New York, 1995.
48. P. BORWEIN, T. ERDÉLYI, and J. ZHANG, Müntz systems and orthogonal Müntz-Legendre polynomials, *Trans. Amer. Math. Soc.* **342** (1994), 523–542.

49. H. BRASS, Ein Gegenbeispiel zum Newton-Cotes-Verfahren, *Z. Angew. Math. Mech.* **57** (1977), 609.
50. M. G. DE BRUIN, Polynomials orthogonal on a circular arc, *J. Comput. Appl. Math.* **31** (1990), 253–266.
51. L. BRUTMAN, On the Lebesgue function for polynomial interpolation, *SIAM J. Numer. Anal.* **15** (1978), 694–704.
52. L. BRUTMAN, Lebesgue functions for polynomial interpolation—a survey, *Ann. Numer. Math.* **4** (1997), 111–127.
53. L. BRUTMAN, I. GOPENGAUZ, and D. TOLEDANO, On the integral of the Lebesgue function induced by interpolation at the Chebyshev nodes, *Acta Math. Hungar.* **90** (2001), 11–28.
54. G. J. BYRNE, T. M. MILLS, and S. J. SMITH, On Lagrange's interpolation with equidistant nodes, *Bull. Austral. Math. Soc.* **42** (1990), 81–89.
55. A. C. CALDER and J. G. LAFRAMBOISE, Multiple-water-bag simulation of inhomogeneous plasma motion near an electrode, *J. Comput. Phys.* **65** (1986), 18–45.
56. M. M. CHAWLA and M. K. JAIN, Error estimates for Gauss quadrature formulas for analytic functions, *Math. Comp.* **22** (1968), 82–90.
57. M. M. CHAWLA and M. K. JAIN, Asymptotic error estimates for the Gauss quadrature formula, *Math. Comp.* **26** (1972), 207–211.
58. P. L. CHEBYSHEV, *Théorie des mécanismes connus sous le nom de parallélogrammes*, Mém. Acad. Sci. St.-Pétersbourg **7** (1854), 539–564 [Œuvres, vol. 2, AN SSSR, Moscow–Leningrad, 1948, pp. 23–51].
59. P. L. CHEBYSHEV, *Sur les questions de minima qui se rattachent à la représentation approximative des fonctions*, Mém. Acad. Sci. St.-Pétersbourg, Sér. 7, **1** (1859), 1–81 [Œuvres, vol. 2, AN SSSR, Moscow–Leningrad, 1948, pp. 151–235].
60. T. S. CHIHARA, *An Introduction to Orthogonal Polynomials*, Gordon and Breach, New York, 1978.
61. Y. CHOW, L. GATTESCHI, and R. WONG, A Bernstein-type inequality for the Jacobi polynomial, *Proc. Amer. Math. Soc.* **121** (1994), 703–709.
62. G. CRISCUOLO and G. MASTROIANNI, Fourier and Lagrange operators in some weighted Sobolev type space, *Acta Sci. Math. (Szeged)* **60** (1995), 131–146.
63. G. CRISCUOLO, G. MASTROIANNI, and D. OCCORSIO, Convergence of extended Lagrange interpolation, *Math. Comp.* **55** (1990), 197–212.
64. G. CRISCUOLO, G. MASTROIANNI, and P. NEVAI, Associated generalized Jacobi functions and polynomials, *J. Math. Anal. Appl.* **158** (1991), 15–34.
65. G. CRISCUOLO, G. MASTROIANNI, and D. OCCORSIO, Uniform convergence of derivatives of extended Lagrange interpolation, *Numer. Math.* **60** (1991), 195–218.
66. G. CRISCUOLO, G. MASTROIANNI, and P. VÉRTESI, Pointwise simultaneous convergence of extended Lagrange interpolation with additional knots, *Math. Comp.* **59** (1992), 515–531.
67. G. CRISCUOLO, B. DELLA VECCHIA, D. S. LUBINSKY, and G. MASTROIANNI, Functions of the second kind for Freud weights and series expansions of Hilbert transforms, *J. Math. Anal. Appl.* **189** (1995), 256–296.
68. A. S. CVETKOVIĆ and G. V. MILOVANOVIĆ, The Mathematica Package "Orthogonal Polynomials", *Facta Univ. Ser. Math. Inform.* **19** (2004), 17–36.
69. G. DAHLQUIST, On summation formulas due to Plana, Lindelöf and Abel, and related Gauss-Christoffel rules, I, *BIT* **37** (1997), 256–295.
70. G. DAHLQUIST, On summation formulas due to Plana, Lindelöf and Abel, and related Gauss-Christoffel rules, II, *BIT* **37** (1997), 804–832.
71. G. DAHLQUIST, On summation formulas due to Plana, Lindelöf and Abel, and related Gauss-Christoffel rules, III, *BIT* **39** (1999), 51–78.
72. B. DANKOVIĆ, G. V. MILOVANOVIĆ, and S. LJ. RANČIĆ, Malmquist and Müntz orthogonal systems and applications, In: *Inner Product Spaces and Applications* (Th. M. Rassias, ed.), Pitman Res. Notes Math. Ser. 376, Longman, Harlow, 1997, pp. 22–41.
73. S. DASSIÈ, M. VIANELLO, and R. ZANOVELLO, Asymptotic summation of power series, *Numer. Math.* **80** (1998), 61–73.

74. S. Dassiè, M. Vianello, and R. Zanovello, A new summation method for power series with rational coefficients, *Math. Comp.* **69** (2000), 749–756.
75. P. J. Davis, *Interpolation and Approximation*, Dover Publications, Inc., New York, 1975.
76. P. J. Davis, *Spirals: from Theodorus to Chaos* (with contributions by Walter Gautschi and Arieh Iserles), A. K. Peters, Wellesley, 1993.
77. P. Davis and P. Rabinowitz, Abscissas and weights for Gaussian quadratures of high order, *J. Res. Nat. Bur. Standards* **56** (1956), 35–37.
78. P. J. Davis and P. Rabinowitz, *Methods of Numerical Integration* (2nd edn.), Computer Science and Applied Mathematics, Academic Press Inc., Orlando, 1984.
79. M. C. De Bonis, B. Della Vecchia, and G. Mastroianni, Approximation of the Hilbert transform on the real semiaxis using Laguerre zeros, *J. Comput. Appl. Math.* **140** (2002), 209–229.
80. M. C. De Bonis, B. Della Vecchia, and G. Mastroianni, Approximation of the Hilbert transform on the real line using Hermite zeros, *Math. Comp.* **71** (2002), 1169–1188.
81. M. C. De Bonis, G. Mastroianni, and M. Viggiano, K-functionals, moduli of smoothness and weighted best approximation on the semiaxis, In: *Functions, Series, Operators—Alexits Memorial Conference* (L. Leindler, F. Schipp, J. Szabados, eds.), János Bolyai Math. Soc., Budapest, 2002.
82. M. C. De Bonis, G. Mastroianni, and M. G. Russo, Polynomial approximation with special doubling weights, *Acta Sci. Math. (Szeged)* **69** (2003), 159–184.
83. C. De Boor and A. Pinkus, Proof of the conjectures of Bernstein and Erdős concerning the optimal nodes for polynomial interpolation, *J. Approx. Theory* **24** (1978), 289–303.
84. C. De Boor and E. B. Saff, Finite sequences of orthogonal polynomials connected by a Jacobi matrix, *Linear Algebra Appl.* **75** (1986), 43–55.
85. Ch.-J. De La Vallée Poussin, *Leçons sur l'approximation des fonctions d'une variable réelle*, Gautier-Villars, Paris, 1919.
86. P. A. Deift, *Orthogonal Polynomials and Random Matrices: A Riemann-Hilbert Approach*, Courant Lecture Notes in Mathematics, 3, New York University, Courant Institute of Mathematical Sciences, New York, American Mathematical Society, Providence, 1999.
87. P. Deift and X. Zhou, A steepest descent method for oscillatory Riemann-Hilbert problems, Asymptotics for the MKdV equation, *Ann. of Math. (2)* **137** (2) (1993), 295–368.
88. P. Deift, T. Kriecherbauer, K. T.-R. McLaughlin, S. Venakides, and X. Zhou, Asymptotics for polynomials orthogonal with respect to varying exponential weights, *Internat. Math. Res. Notices* (1997) 759–782.
89. P. Deift, T. Kriecherbauer, and K. T.-R. McLaughlin, New results on the equilibrium measure for logarithmic potentials in the presence of an external field, *J. Approx. Theory* **95** (1998), 388–475.
90. P. Deift, T. Kriecherbauer, K. T.-R. McLaughlin, S. Venakides, and X. Zhou, Uniform asymptotics for polynomials orthogonal with respect to varying exponential weights and applications to universality questions in random matrix theory, *Comm. Pure Appl. Math.* **52** (1999), 1335–1425.
91. P. Deift, T. Kriecherbauer, K. T.-R. McLaughlin, S. Venakides, and X. Zhou, Strong asymptotics of orthogonal polynomials with respect to exponential weights, *Comm. Pure Appl. Math.* **52** (1999), 1491–1552.
92. P. Deift, T. Kriecherbauer, K. T.-R. McLaughlin, S. Venakides, and X. Zhou, A Riemann-Hilbert approach to asymptotic questions for orthogonal polynomials, *J. Comput. Appl. Math.* **133** (2001), 47–63.
93. D. Della Vecchia and G. Mastroianni, Gaussian rules on unbounded intervals, *J. Complexity* **19** (2003), 247–258.
94. H. Dette and W. J. Studden, On a new characterization of the classical orthogonal polynomials, *J. Approx. Theory* **71** (1992), 3–17.
95. R. A. DeVore and G. G. Lorentz, *Constructive Approximation*, Grundlehren der mathematischen Wissenschaften, Vol. 303, Springer-Verlag, Berlin, 1993.
96. Z. Ditzian, On interpolation of $L_p[a, b]$ and weighted Sobolev spaces, *Pacific J. Math.* **90** (1980), 307–323.

97. Z. DITZIAN and D. S. LUBINSKY, Jackson and smoothness theorems for Freud weights in L_p $(0 < p \leq \infty)$, Constr. Approx. **13** (1997), 99–152.
98. Z. DITZIAN and V. TOTIK, *Moduli of Smoothness*, Springer Series in Computational Mathematics, Vol. 9, Springer, New York, 1987.
99. Z. DITZIAN and V. TOTIK, Remarks on Besov spaces and best polynomial approximation, Proc. Amer. Math. Soc. **104** (1988), 1059–1066.
100. Z. DITZIAN, V. H. HRISTOV, and K. G. IVANOV, Moduli of smoothness and K-functionals in L_p, $0 < p < 1$, Constr. Approx. **11** (1995), 67–83.
101. R. Ž. DJORDJEVIĆ and G. V. MILOVANOVIĆ, A generalization of E. Landau's theorem, Univ. Beograd. Publ. Elektrotehn. Fak. Ser. Mat. Fiz. **498–541** (1975), 91–96.
102. M. M. DJRBASHIAN, A survey on the theory of orthogonal systems and some open problems. In: *Orthogonal Polynomials: Theory and Practice* (P. Nevai, ed.), NATO ASI Series, Series C: Mathematical and Physical Sciences, Vol. **294**, Kluwer, Dordrecht, 1990, pp. 135–146.
103. J. D. DONALDSON and D. ELLIOTT, A unified approach to quadrature rules with asymptotic estimates of their remainders, SIAM J. Numer. Anal. **9** (1972), 573–602.
104. B. R. DRAGANOV and K. G. IVANOV, A new characterization of weighted Peetre K-functionals, Constr. Approx. **21** (2005), 113–148.
105. V. K. DZYADYK and V. V. IVANOV, On asymptotics and estimates for the uniform norms of the Lagrange interpolation polynomials corresponding to the Chebyshev nodal points, Anal. Math. **9** (1983), 85–97.
106. E. EGERVÁRY and P. TURÁN, Notes on interpolation. V, Acta Math. Acad. Sci. Hungar. **9** (1958), 259–267.
107. S. EHRICH, Asymptotic properties of Stieltjes polynomials and Gauss-Kronrod quadrature formulae, J. Approx. Theory **82** (1995), 287–303.
108. S. EHRICH and G. MASTROIANNI, Stieltjes polynomials and Lagrange interpolation, Math. Comp. **66** (1997), 311–331.
109. S. EHRICH and G. MASTROIANNI, Marcinkiewicz inequalities based on Stieltjes zeros, J. Comput. Appl. Math. **99** (1998), 129–141.
110. D. ELLIOTT and D. F. PAGET, Product integration rules and their convergence, BIT **16** (1976), 32–40.
111. D. ELLIOTT and D. F. PAGET, The convergence of product integration rules, BIT **18** (1978), 137–141.
112. T. ERDÉLYI and P. VÉRTESI, In memoriam Paul Erdős, J. Approx. Theory **94** (1998), 1–41.
113. P. ERDŐS, Some remarks on polynomials, Bull. Amer. Math. Soc. **53** (1947), 1169–1176.
114. P. ERDŐS, Problems and results on the theory of interpolation, I, Acta Math. Acad. Sci. Hungar. **9** (1958), 381–388.
115. P. ERDŐS, Problems and results on the theory of interpolation, II, Acta Math. Acad. Sci. Hungar. **12** (1961), 235–244.
116. P. ERDŐS, Problems and results on the convergence and divergence properties of the Lagrange interpolation polynomials and some extremal problems, Mathematica (Cluj) **10** (1968), 65–73.
117. P. ERDŐS and J. SZABADOS, On the integral of the Lebesgue function of interpolation, Acta Math. Acad. Sci. Hungar. **32** (1978), 191–195.
118. P. ERDŐS, J. SZABADOS, and P. VÉRTESI, On the integral of the Lebesgue function of interpolation. II, Acta Math. Hungar. **68** (1995), 1–6.
119. P. ERDŐS and P. TURÁN, On interpolation, III, Ann. of Math. **41** (1940), 510–553.
120. P. ERDŐS and P. VÉRTESI, On the almost everywhere divergence of Lagrange interpolatory polynomials for arbitrary system of nodes, Acta Math. Acad. Sci. Hungar. **36** (1980), 71–98 and **38** (1981), 263.
121. W. N. EVERITT and L. L. LITTLEJOHN, Orthogonal polynomials and spectral theory: a survey, In: *IMACS Annals on Computing and Applied Mathematics, Vol. 9, Orthogonal Polynomials and Their Applications* (C. Brezinski, L. Gori, and A. Ronveaux, eds.), J. C. Baltzer AG, Scientific Publ. Co., Basel, 1991, pp. 21–55.

122. W. N. EVERITT, K. H. KWON, L. L. LITTLEJOHN, and R. WELLMAN, Orthogonal polynomial solutions of linear ordinary differential equations, *J. Comput. Appl. Math.* **133** (2001), 85–109.
123. G. FABER, Über die interpolatorische Darstellung stetiger Funktionen, *Jahresber. der deutschen Math. Verein.* **23** (1914), 190–210.
124. R. P. FEINERMAN and D. J. NEWMAN, *Polynomial Approximation*, The Williams & Wilkins Company, Baltimore, 1974.
125. L. FEJÉR, Untersuchungen über Fouriersche Reihen, *Math. Ann.* **58** (1904), 51–69.
126. L. FEJÉR, Mechanische Quadraturen mit positiven Cotesschen Zahle, *Math. Z.* **37** (1933), 287–309.
127. B. FISCHER and G. GOLUB, How to generate unknown orthogonal polynomials out of known orthogonal polynomials, *J. Comput. Appl. Math.* **43** (1992), 99–115.
128. V. FOCK, On the remainder term of certain quadrature formulae, *Bull. Acad. Sci. Leningrad* **7** (1932), 419–448 (Russian).
129. A. S. FOKAS, A. R. ITS, and A. V. KITAEV, Isomonodromic approach in the theory of two-dimensional quantum gravity, *Uspekhi Mat. Nauk* **45** (1990), 135–136.
130. A. S. FOKAS, A. R. ITS, and A. V. KITAEV, Discrete Painleve equations and their appearance in quantum gravity, *Comm. Math. Phys.* **142** (1991), 313–344.
131. A. FRANSÉN, Accurate determination of the inverse gamma integral, *BIT* **19** (1979), 137–138.
132. G. FREUD, Über eine Klasse Lagrangescher Interpolationsverfahren, *Studia Sci. Math. Hungar.* **3** (1968), 249–255.
133. G. FREUD, Ein Beitrag zur Theorie des Lagrangeschen Interpolations-verfahrens, *Studia Sci. Math. Hungar.* **4** (1969), 379–384.
134. G. FREUD, *Orthogonal Polynomials*, Akadémiai Kiadó/Pergamon Press, Budapest, 1971.
135. G. FREUD, A certain class of orthogonal polynomials, *Mat. Zametki* **9** (1971), 511–520 (Russian).
136. G. FREUD, On two polynomials. II, *Acta Math. Sci. Hungar.* **23** (1972), 137–145.
137. G. FREUD and H. N. MHASKAR, Weighted polynomial approximation in rearrangement invariant Banach function spaces on the whole real line, *Indian J. Math.* **22** (3) (1980), 209–224.
138. G. FREUD and H. N. MHASKAR, K-Functionals and moduli of continuity in weighted polynomial approximation, *Ark. Mat.* **21** (1983), 145–161.
139. M. FRONTINI and G. V. MILOVANOVIĆ, Moment-preserving spline approximation on finite intervals and Turán quadratures, *Facta Univ. Ser. Math. Inform.* **4** (1989), 45–56.
140. M. FRONTINI, W. GAUTSCHI, and G. V. MILOVANOVIĆ, Discrete approximations to spherically symmetric distributions, *Numer. Math.* **50** (1987), 503–518.
141. M. I. GANZBURG, Strong asymptotics in Lagrange interpolation with equidistant nodes, *J. Approx. Theory* **122** (2003), 224–240.
142. C. F. GAUSS, Methodus nova integralium valores per approximationem inveniendi, Commentationes Societatis Regiae Scientarium Recentiores 3 (1814) [Werke III, pp. 123–162].
143. W. GAUTSCHI, Construction of Gauss-Christoffel quadrature formulas, *Math. Comp.* **22** (1968), 251–270.
144. W. GAUTSCHI, On generating Gaussian quadrature rules, In: *Numerische Integration* (G. Hämmerlin, ed.), ISNM, Vol. 45, Birkhäuser, Basel, 1979, pp. 147–154.
145. W. GAUTSCHI, Minimal solutions of three-term recurrence relations and orthogonal polynomials, *Math. Comp.* **36** (1981), 547–554.
146. W. GAUTSCHI, A survey of Gauss-Christoffel quadrature formulae, In: *E. B. Christoffel—The Influence of his Work on Mathematics and the Physical Sciences* (P. L. Butzer, F. Fehér, eds.), Birkhäuser, Basel, 1981, pp. 72–147.
147. W. GAUTSCHI, On generating orthogonal polynomials, *SIAM J. Sci. Statist. Comput.* **3** (1982), 289–317.
148. W. GAUTSCHI, Polynomials orthogonal with respect to the reciprocal gamma function, *BIT* **22** (1982), 387–389.

149. W. GAUTSCHI, How and how not to check Gaussian quadrature formulae, *BIT* **23** (1983), 209–216.
150. W. GAUTSCHI, Discrete approximations to spherically symmetric distributions, *Numer. Math.* **44** (1984), 53–60.
151. W. GAUTSCHI, On some orthogonal polynomials of interest in theoretical chemistry, *BIT* **24** (1984), 473–483.
152. W. GAUTSCHI, Gauss-Kronrod Quadrature—A Survey, In: *Numerical Methods and Approximation Theory III* (G. V. Milovanović, ed.), University of Niš, Niš, 1988, pp. 39–66.
153. W. GAUTSCHI, On the zeros of polynomials orthogonal on the semicircle, *SIAM J. Numer. Anal.* **20** (1989), 738–743.
154. W. GAUTSCHI, Computational aspects of orthogonal polynomials, In: *Orthogonal Polynomials (Columbus, OH, 1989)*, NATO Adv. Sci. Inst. Ser. C: Math. Phys. Sci., 294 (P. Nevai, ed.), Kluwer, Dordrecht, 1990, 181–216.
155. W. GAUTSCHI, Computational problems and applications of orthogonal polynomials, In: *IMACS Annals on Computing and Applied Mathematics, Vol. 9: Orthogonal Polynomials and Their Applications* (C. Brezinski, L. Gori and A. Ronveaux, eds.), IMACS, Baltzer, Basel, 1991, pp. 61–71.
156. W. GAUTSCHI, A class of slowly convergent series and their summation by Gaussian quadrature, *Math. Comp.* **57** (1991), 309–324.
157. W. GAUTSCHI, On certain slowly convergent series occurring in plate contact problems, *Math. Comp.* **57** (1991), 325–338.
158. W. GAUTSCHI, On the remainder term for analytic functions of Gauss-Lobatto and Gauss-Radau quadratures, *Rocky Mountain J. Math.* **21** (1991), 209–226.
159. W. GAUTSCHI, Remainder estimates for analytic functions, In: *Numerical Integration* (T. O. Espelid and A. Genz, eds.), Kluwer, Dordrecht, 1992, pp. 133–145.
160. W. GAUTSCHI, Spline approximation and quadrature formulae, *Atti Sem. Mat. Fis. Univ. Modena* **40** (1992), 169–182.
161. W. GAUTSCHI, Algorithm 726: ORTHPOL—a package of routines for generating orthogonal polynomials and Gauss-type quadrature rules, *ACM Trans. Math. Software* **20** (1994), 21–62.
162. W. GAUTSCHI, Orthogonal polynomials: applications and computation, *Acta Numerica* (1996), 45–119.
163. W. GAUTSCHI, *Numerical Analysis: An Introduction*, Birkhäuser, Basel, 1997.
164. W. GAUTSCHI, Moments in quadrature problems, *Comput. Math. Appl.* **33** (1997), 105–118.
165. W. GAUTSCHI, Computation of Bessel and Airy functions and of related Gaussian quadrature formulae, *BIT* **42** (2002), 110–118.
166. W. GAUTSCHI, *Orthogonal Polynomials: Computation and Approximation*, Clarendon Press, Oxford, 2004.
167. W. GAUTSCHI, The Hardy–Littlewood function: an exercise in slowly convergent series, *J. Comput. Appl. Math.* **179** (2005), 249–254.
168. W. GAUTSCHI and S. LI, The remainder term for analytic functions of Gauss-Radau and Gauss-Lobatto quadrature rules with multiple points, *J. Comput. Appl. Math.* **33** (1990), 315–329.
169. W. GAUTSCHI and G. V. MILOVANOVIĆ, Gaussian quadrature involving Einstein and Fermi functions with an application to summation of series, *Math. Comp.* **44** (1985), 177–190.
170. W. GAUTSCHI and G. V. MILOVANOVIĆ, Polynomials orthogonal on the semicircle, In: *International conference on special functions: Theory and Computation (Turin, 1984), Rend. Sem. Mat. Univ. Politec. Torino* **1985**, Special Issue, 179–185.
171. W. GAUTSCHI and G. V. MILOVANOVIĆ, Polynomials orthogonal on the semicircle, *J. Approx. Theory* **46** (1986), 230–250.
172. W. GAUTSCHI and G. V. MILOVANOVIĆ, Spline approximations to spherically symmetric distributions, *Numer. Math.* **49** (1986), 111–121.
173. W. GAUTSCHI and S. E. NOTARIS, An algebraic and numerical study of Gauss-Kronrod quadrature formulae for Jacobi weight functions, *Math. Comp.* **51** (1988), 321–348.

174. W. GAUTSCHI and R. S. VARGA, Error bounds for Gaussian quadrature of analytic functions, *SIAM J. Numer. Anal.* **20** (1983), 1170–1186.
175. W. GAUTSCHI, H. LANDAU, and G. V. MILOVANOVIĆ, Polynomials orthogonal on the semicircle, II, *Constr. Approx.* **3** (1987), 389–404.
176. W. GAUTSCHI, E. TYCHOPOULOS, and R. S. VARGA, A note on the contour integral representation of the remainder term for a Gauss-Chebyshev quadrature rule, *SIAM J. Numer. Anal.* **27** (1990), 219–224.
177. YA. L. GERONIMUS, On some properties of generalized orthogonal polynomials, *Mat. Sb.* **9** (51) (1941), 121–135 (Russian).
178. YA. L. GERONIMUS, Polynomials orthogonal on a circle and their applications, *Zap. Nauč.-issled. Inst. Mat. Mech. HMO* **19** (1948), 35–120 (Russian).
179. A. GHIZZETTI and A. OSSICINI, Su un nuovo tipo di sviluppo di una funzione in serie di polinomi, *Atti Accad. Naz. Lincei Rend. Cl. Sci. Fis. Mat. Natur.* (8) **43** (1967) 21–29.
180. A. GHIZZETTI and A. OSSICINI, *Quadrature Formulae*, Akademie Verlag, Berlin, 1970.
181. J. GILEWICZ and E. LEOPOLD, Location of the zeros of polynomials satisfying three-term recurrence relations. I. General case with complex coefficients, *J. Approx. Theory* **43** (1985), 1–14.
182. J. GILEWICZ and E. LEOPOLD, Location of the zeros of polynomials satisfying three-term recurrence relations with complex coefficients, *Integral Transform. Spec. Funct.* **2** (1994), 267–278.
183. J. GILEWICZ and E. LEOPOLD, Zeros of polynomials and recurrence relations with periodic coefficients, *J. Comput. Appl. Math.* **107** (1999), 241–255.
184. G. GOLUB, Some modified matrix eigenvalue problems, *SIAM Rev.* **15** (1973), 318–334.
185. G. GOLUB and J. H. WELSCH, Calculation of Gauss quadrature rules, *Math. Comp.* **23** (1969), 221–230.
186. V. L. GONČAROV, *Theory of Interpolation and Approximation of Functions*, GITTL, Moscow, 1954 (Russian).
187. I. E. GOPENGAUZ, On a theorem of A. F. Timan on approximation of functions by polynomials on a finite interval, *Mat. Zametki* **1** (1967), 163–172 (Russian).
188. L. GORI and E. SANTI, Moment-preserving approximations: a monospline approach, *Rend. Mat. Appl.* (7) **12** (1992), 1031–1044.
189. L. GORI, N. AMATI, and E. SANTI, On a method of approximation by means of spline functions, In: *Approximation, Optimization and Computing—Theory and Application* (A. G. Law and C. L. Wang, eds.), IMACS, Dalian, 1990, pp. 41–46.
190. L. GORI, M. L. LO CASCIO, and G. V. MILOVANOVIĆ, The σ-orthogonal polynomials: a method of construction, In: *IMACS Annals on Computing and Applied Mathematics, Vol. 9, Orthogonal Polynomials and Their Applications* (C. Brezinski, L. Gori, and A. Ronveaux, eds.), J. C. Baltzer AG, Scientific Publ. Co., Basel, 1991, pp. 281–285.
191. A. GORNY, Contribution à l'étude des fonctions dérivables d'une variable réelle, *Acta Math.* **71** (1939), 317–358.
192. Z. S. GRINSHPUN, Characteristic properties of orthogonal polynomials in terms of functions of the second kind, In: *Functional analysis, differential equations and their applications* 167, Kazakh. Gos. Univ., Alma-Ata, 1982, pp. 38–44 (Russian).
193. Z. GRINSHPUN, Special linear combinations of orthogonal polynomials, *J. Math. Anal. Appl.* **299** (2004), 1–18.
194. C. C. GROSJEAN, Theory of recursive generation of systems of orthogonal polynomials: An illustrative example, *J. Comput. Appl. Math.* **12&13** (1985), 299–318.
195. C. C. GROSJEAN, The weight functions, generating functions and miscellaneous properties of the sequences of orthogonal polynomials of the second kind associated with the Jacobi and the Gegenbauer polynomials, *J. Comput. Appl. Math.* **16** (1986), 259–307.
196. G. GRÜNWALD, Über Divergenzerscheinungen der Lagrangeschen Interpolationspolynome, *Acta Sci. Math. (Szeged)* **7** (1935), 207–221.
197. G. GRÜNWALD, Über Divergenzerscheinungen der Lagrangeschen Interpolationspolynome stetiger Funktionen, *Ann. of Math.* **37** (1936), 908–918.

198. A. GUESSAB, Some weighted polynomial inequalities in L^2-norm, *J. Approx. Theory* **79** (1994), 125–133.
199. A. GUESSAB, Weighted L^2 Markoff type inequality for classical weights, *Acta Math. Hung.* **66** (1995), 155–162.
200. A. GUESSAB and G. V. MILOVANOVIĆ, Weighted L^2-analogues of Bernstein's inequality and classical orthogonal polynomials, *J. Math. Anal. Appl.* **182** (1994), 244–249.
201. A. GUESSAB and G. V. MILOVANOVIĆ, Extremal problems of Markov's type for some differential operators, *Rocky Mountain J. Math.* **24** (1994), 1431–1438.
202. R. GÜNTTNER, Evaluation of Lebesgue constants, *SIAM J. Numer. Anal.* **17** (1980), 512–520.
203. R. GÜNTTNER, On asymptotics for the uniform norms of the Lagrange interpolation polynomials corresponding to extended Chebyshev nodes, *SIAM J. Numer. Anal.* **25** (1988), 461–469.
204. R. GÜNTTNER, Note on the lower estimate of optimal Lebesgue constants, *Acta Math. Hungar.* **65** (1994), 313–317.
205. G. HALÁSZ, The "coarse and fine theory of interpolation" of Erdős and Turán in a broader view, *Constr. Approx.* **8** (1992), 169–185.
206. G. H. HARDY, J. E. LITTLEWOOD, and G. POLYA, *Inequalities*, 2nd edn., Cambridge University Press, Cambridge, 1952.
207. J. F. HART et al., *Computer Approximations*, Wiley, New York, 1968.
208. E. HEINE, *Anwendungen der Kugelfunctionen und der verwandten Functionen*, 2nd edn., Reiner, Berlin, 1881.
209. P. HENRICI, *Applied and Computational Complex Analysis*, Vol. 1, Wiley, New York, 1984.
210. C. HERMITE, Sur la formule d'interpolation de Lagrange, *J. Reine Angew. Math.* **84** (1878), 70–79.
211. N. J. HIGHAM, The numerical stability of barycentric Lagrange interpolation, *IMA J. Numer. Anal.* **24** (2004), 547–556.
212. E. HILLE, On the analytical theory of semi-groups, *Proc. Nat. Acad. Sci. U.S.A.* **28** (1942), 421–424.
213. E. HILLE, Remark on the Landau-Kallman-Rota inequality, *Aequationes Math.* **4** (1970), 239–240.
214. A. HORVATH and J. SZABADOS, Polynomial approximation and interpolation on the real line with respect to general classes of weights, *Results Math.* **34** (1998), 120–131.
215. V. H. HRISTOV, Space of functions by the averaged moduli of functions of many variables, In: *Constructive Theory of Functions '84 (Varna, 1981)*, Publ. House Bulgar. Acad. Sci., Sofia, 1984, pp. 97–101.
216. R. HUNT, B. MUCKENHOUPT, and R. WHEEDEN, Weighted norm inequalities for the conjugate function and Hilbert transform, *Trans. Amer. Math. Soc.* **176** (1973), 227–251.
217. D. B. HUNTER, Some error expansions for Gaussian quadrature, *BIT* **35** (1995), 64–82.
218. D. B. HUNTER and G. NIKOLOV, On the error term of symmetric Gauss–Lobatto quadrature formulae for analytic functions, *Math. Comp.* **69** (2000), 269–282.
219. K. G. IVANOV, On the behaviour of two moduli of functions II, *Serdica* **12** (1986), 196–203.
220. D. JACKSON, *Über die Genauigkeit der Annäherung stetiger Funktionen durch ganze rationale Funktionen gegebenen Grades und trigonometrischen Summen gegebener Ordnung*, Diss., Göttingen, 1911.
221. D. JACKSON, *The Theory of Approximation*, Amer. Math. Soc. Colloq. Publ., **11**, Amer. Math. Soc., Providence, 1930.
222. C. G. J. JACOBI, Über Gaußs neue Methode, die Werte der Integrale näherungsweise zu finden, *J. Reine Angew. Math.* **30** (1826), 301–308.
223. L. B. W. JOLLEY, *Summation of Series*, Dover Publications, Inc., New York, 1961.
224. I. JOÓ, On some problems of M. Horváth, *Annales Univ. Sci. Budapest., Sect. Math.* **31** (1988), 243–260.
225. S. KARLIN and W. J. STUDDEN, *Tchebycheff Systems with Applications in Analysis and Statistics*, Pure and Applied Mathematics, Vol. XV, John Wiley Interscience, New York, 1966.

226. T. KASUGA and R. SAKAI, Orthonormal polynomials with generalized Freud-type weights, *J. Approx. Theory* **121** (2003), 13–53.
227. K. S. KAZARYAN and P. I. LIZORKIN, Multipliers, bases and unconditional bases of the weighted spaces B and SB. *Proc. Steklov Inst. Math.* **1990**, no 3, 111–130. (Russian) Studies in the theory of differentiable functions of several variables and its applications, 13. *Trudy Mat. Inst. Steklov.* **187** (1989), 98–115 (Russian).
228. T. KILGORE, A characterization of the Lagrange interpolating projection with minimal Tchebycheff norm, *J. Approx. Theory* **24** (1978), 273–288.
229. P. KIRCHBERGER, *Über Tschebysheff'sche Annäherungsmethoden*, Dissertation, Göttingen, 1902.
230. O. KIS, Lagrange interpolation with nodes at the roots of Sonin-Markov polynomials, *Acta Math. Acad. Sci. Hungar.* **23** (1972), 389–417 (Russian).
231. O. KIS and J. SZABADOS, On some de la Vallée Poussin type discrete linear operators, *Acta Math. Hungar.* **47** (1986), 239–260.
232. R. KOEKOEK and R. S. SWARTTOUW, The Askey-scheme of hypergeometric orthogonal polynomials and its q-analogue, *Reports of the Faculty of Technical Mathematics and Informatics* 98-17, Delft University of Technology, 1998, 120 pp.
233. A. KOLMOGOROV, On inequalities between upper bounds of the successive derivatives of an arbitrary function on an infinite interval, *Uchen. Zap. Moskov. Gos. Univ. Mat.* **30** (1939), 3–16 (Russian).
234. A. N. KORKIN and E. I. ZOLOTAREV, Sur un certain minimum, *Nouv. Ann. Math. Sér. 2* **12** (1873), 337–355.
235. N. KORNEICHUK, *Exact Constants in Approximation Theory*, Encyclopedia of Mathematics and its Applications, Vol. 38, Cambridge University Press, Cambridge, 1991.
236. M. A. KOVAČEVIĆ and G. V. MILOVANOVIĆ, Lobatto quadrature formulas for generalized Gegenbauer weight, In: *5th Conference on Applied Mathematics* (Z. Bohte, ed.), University of Ljubljana, Ljubljana, 1986, pp. 81–88.
237. M. A. KOVAČEVIĆ and G. V. MILOVANOVIĆ, Spline approximation and generalized Turán quadratures, *Portugal. Math.* **53** (1996), 355–366.
238. I. KRASIKOV, On the maximum of Jacobi polynomials, *J. Approx. Theory* **136** (2005), 1–20.
239. I. KRASIKOV, On the Erdélyi-Magnus-Nevai conjecture for Jacobi polynomials, *Constr. Approx.* **28** (2008), 113–125.
240. T. KRIECHERBAUER and K. T.-R. MCLAUGHLIN, Strong asymptotics of polynomials orthogonal with respect to Freud weights, *Internat. Math. Res. Notices*, no. **6**, 299–333.
241. D. G. KUBAYI and D. S. LUBINSKY, A Hilbert transform representation of the error in Lagrange interpolation, *J. Approx. Theory* **129** (2004), 94–100.
242. A. B. J. KUIJLAARS, K. T.-R. MCLAUGHLIN, W. VAN ASSCHE, and M. VANLESSEN, The Riemann-Hilbert approach to strong asymptotics for orthogonal polynomials on $[-1,1]$, *Adv. Math.* **188** (2004), 337–398.
243. M. KÜTZ, On the positivity of certain Cotes numbers, *Aequationes Math.* **24** (1982), 110–118.
244. K. H. KWON and D. W. LEE, Characterizations of Bochner-Krall orthogonal polynomials of Jacobi type, *Constr. Approx.* **19** (2003), 599–619.
245. N. X. KY, On simultaneous approximation by polynomials with weight, In: *Alfréd Haar Memorial Conference (Budapest, 1984)*, Colloq. Math. Soc. János Bolyai, 49, North-Holland, Amsterdam, 1985, pp. 661–665.
246. N. X. KY, On approximation by trigonometric polynomials in L_u^p-spaces, *Studia Sci. Math. Hungar.* **28** (1993), 183–188.
247. H. N. LADEN, An application of the classical orthogonal polynomials to the theory of interpolation. *Duke Math. J.* **8** (1941), 591–610.
248. H. N. LADEN, Fundamental polynomials of Lagrange interpolation and coefficients of mechanical quadrature, *Duke Math. J.* **10** (1943), 145–151.
249. J. G. LAFRAMBOISE and A. D. STAUFFER, Optimum discrete approximation of the Maxwell distribution, *AIAA J.* **7** (1969), 520–523.

250. E. LANDAU, Einige Ungleichungen für zweimal differenzierbare Funktionen, *Proc. London Math. Soc. (2)* **13** (1913), 43–49.
251. K. V. LAŠČENOV, On a class of orthogonal polynomials, *Učen. Zap. Leningrad. Gos. Ped. Inst.* **89** (1953), 167–189 (Russian).
252. D. P. LAURIE, Computation of Gauss-type quadrature formulas, In: *Numerical Analysis 2000, Vol. V, Quadrature and Orthogonal Polynomials* (W. Gautschi, F. Marcellán, and L. Reichel, eds.), *J. Comput. Appl. Math.* **127** (2001), 201–217.
253. S.-Y. LEE, The inhomogeneous Airy functions Hi(z) and Gi(z), *J. Chem. Phys.* **72** (1980), 332–336.
254. E. LEOPOLD, Location of the zeros of polynomials satisfying three-term recurrence relations. III. Positive coefficients case, *J. Approx. Theory* **43** (1985), 15–24.
255. P. LESKY, Die Charakterisierung der klassischen orthogonalen Polynome durch Sturm-Liouvillesche Differentialgleichungen, *Arch. Rat. Mech. Anal.* **10** (1962), 341–351.
256. F. G. LETHER, Error estimates for Gaussian quadrature, *Appl. Math. Comput.* **7** (1980), 237–246.
257. A. L. LEVIN and D. S. LUBINSKY, Christoffel functions, orthogonal polynomials and Nevai's conjecture for Freud weights, *Const. Approx.* **8** (1992), 463–535.
258. A. L. LEVIN and D. S. LUBINSKY, *Orthogonal Polynomials for Exponential Weights*, CMS Books in Mathematics/Ouvrages de Mathématiques de la SMC, 4. Springer-Verlag, New York, 2001.
259. A. L. LEVIN and D. S. LUBINSKY, Orthogonal polynomials for weights $x^{2\rho}e^{-2Q(x)}$ on $[0, d)$, *J. Approx. Theory* **134** (2005), 199–256.
260. S. LEWANOWICZ, Construction of a recurrence relation for modified moments, *J. Comput. Appl. Math.* **5** (1979), 193–206.
261. S. LEWANOWICZ, Properties of the polynomials associated with the Jacobi polynomials, *Math. Comp.* **47** (1986), 669–682.
262. S. LEWANOWICZ, A fast algorithm for the construction of recurrence relations for modified moments, *Appl. Math. (Warsaw)* **22** (1994), 359–372.
263. Z. LEWANDOWSKI and J. SZYNAL, An upper bound for the Laguerre polynomials, *J. Comput. Appl. Math.* **99** (1998), 529–533.
264. X. LI and R. N. MOHAPATRA, On the divergence of Lagrange interpolation with equidistant nodes, *Proc. Amer. Math. Soc.* **118** (1993), 1205–1212.
265. E. LINDELÖF, *Le Calcul des Résidus*, Gauthier-Villars, Paris, 1905.
266. L. LORCH, Alternative proof of a sharpened form of Bernstein's inequality for Legendre polynomials, *Applicable Anal.* **14** (1982/83), 237–240.
267. L. LORCH, Inequalities for ultraspherical polynomials and the gamma function, *J. Approx. Theory* **40** (1984), 115–120.
268. G. G. LORENTZ, K. JETTER, and S. D. RIEMENSCHNEIDER, *Birkhoff Interpolation*, Addison–Wessley, Reading, 1983.
269. D. S. LUBINSKY, A survey of general orthogonal polynomials for weights on finite and infinite intervals, *Acta Appl. Math.* **10** (1987), 237–296.
270. D. S. LUBINSKY, An update on orthogonal polynomials and weighted approximation on the real line, *Acta Appl. Math.* **33** (1993), 121–164.
271. D. S. LUBINSKY, Weierstrass' theorem in the twentieth century: a selection, *Quaestiones Mathematicae* **18** (1995), 91–130.
272. D. S. LUBINSKY, A taste of Erdős on interpolation, In: *Paul Erdős and his mathematics, I (Budapest, 1999)*, Bolyai Soc. Math. Stud., 11, János Bolyai Math. Soc., Budapest, 2002, pp. 423–454.
273. D. S. LUBINSKY, Asymptotics of orthogonal polynomials: Some old, some new, some identities, *Acta Appl. Math.* **61** (2000), 207–256.
274. D. S. LUBINSKY, Best approximation and interpolation of $(1 + (ax)^2)^{-1}$ and its transforms, *J. Approx. Theeory* **125** (2003), 106–115.
275. D. LUBINSKY and E. SAFF, *Strong Asymptotics for Extremal Polynomials Associated with Exponential Weights*, Springer Lecture Notes in Mathematics, Vol. 1305, Springer, Berlin, 1988.

276. D. LUBINSKY, H. MHASKAR, and E. SAFF, A proof of Freud's conjecture for exponential weights, *Constr. Approx.* **4** (1988), 65–83.
277. A. L. LUKASHOV and F. PEHERSTORFER, Zeros of polynomials orthogonal on two arcs of the unit circle, *J. Approx. Theory* **132** (2005), 42–71.
278. U. LUTHER and G. MASTROIANNI, Fourier projection in weighted L^∞ spaces, In: *Problems and methods in mathematical physics (Chemnitz, 1999)*, Oper. Theory Adv. Appl., 121, Birkhäuser, Basel, 2001, pp. 327–351.
279. F. W. LUTTMANN and T. J. RIVLIN, Some numerical experiments in the theory of polynomial interpolation, *IBM J. Res. Develop.* **9** (1965), 187–191.
280. A. P. MAGNUS, On Freud's equations for exponential weights, *J. Approx. Theory* **46** (1986), 65–99.
281. F. MALMQUIST, Sur la détermination d'une classe de fonctions analytiques par leur valeur dans un ensemble donné de points, In: *VI Skand. Matematikerkongres (Copenhagen, 1925)*, Gjellerups, Copenhagen, 1926, pp. 253–259.
282. J. MARCINKIEWICZ, Sur la divergence des pôlynoms d'interpolation, *Acta Sci. Math. (Szeged)* **8** (1937), 131–135.
283. S. D. MARINKOVIĆ, *Polynomials of Jacobi Type and Applications*, MS Thesis, University of Niš, Niš, 1995.
284. A. MARKOV, Sur la m'ethode de Gauss pour le calcul approch'e des int'egrales, *Math. Ann.* **25** (1885), 427–432.
285. A. MARKOV, *Differenzenrechnung*, Leipzig, 1895.
286. P. MARONI, Une caractérisation des polynômes orthogonaux semi-classiques, *C. R. Acad. Sci. Paris* **301** (1) (1985), 269–272.
287. P. MARONI, Prolégomènes à l'étude des polynômes orthogonaux semi-classiques, *Ann. Mat. Pura Appl.* **149** (4) (1987), 165–184.
288. P. MARONI, Une théorie algébrique des polynômes orthogonaux. Application aux polynômes orthogonaux semi-classiques, In: *IMACS Annals on Computing and Applied Mathematics, Vol. 9: Orthogonal Polynomials and Their Applications* (C. Brezinski, L. Gori and A. Ronveaux, eds.), IMACS, Baltzer, Basel, 1991, pp. 95–130.
289. G. MASTROIANNI, Uniform convergence of derivatives of Lagrange interpolation, *J. Comput. Appl. Math.* **43** (1992), 1–15.
290. G. MASTROIANNI, Generalized Christoffel functions and error of positive quadrature, *Numer. Algorithms* **10** (1995), 113–126.
291. G. MASTROIANNI, Some weighted polynomial inequalities, *J. Comput. Appl. Math.* **65** (1995), 279–292.
292. G. MASTROIANNI, Boundedness of Lagrange operator in some functional spaces. A survey. In: *Approximation Theory and Function Series (Budapest, 1995)*, Bolyai Soc. Math. Stud., 5, János Bolyai Math. Soc., Budapest, 1996, pp. 117–139.
293. G. MASTROIANNI, Polynomial inequalities, functional spaces and best approximation on the real semiaxis with Laguerre weights. In: *Orthogonal Polynomials, Approximation Theory and Harmonic Analysis (Inzel, 2000)*, *Electron. Trans. Numer. Anal.* **14** (2002), 125–134.
294. G. MASTROIANNI and G. V. MILOVANOVIĆ, Weighted interpolation of functions with isolated singularities. In: *Approximation Theory: A volume dedicated to Blagovest Sendov* (B. Bojanov, ed.), Darba, Sofia, 2002, pp. 310–341.
295. G. MASTROIANNI and G. V. MILOVANOVIĆ, Weighted integration of periodic functions on the real line, *Appl. Math. Comput.* **128** (2002), 365–378.
296. G. MASTROIANNI and G. V. MILOVANOVIĆ, Polynomial approximation on unbounded intervals by Fourier sums, *Facta Univ. Ser. Math. Inform.* **22** (2007), 155–168.
297. G. MASTROIANNI and G. MONEGATO, Nyström interpolants based on zeros of Laguerre polynomials for some Weiner-Hopf equations, *IMA J. Numer. Anal.* **17** (1997), 621–642.
298. G. MASTROIANNI and G. MONEGATO, Truncated quadrature rules over $(0, \infty)$ and Nyström-type methods, *SIAM J. Numer. Anal.* **41** (2003), 1870–1892.
299. G. MASTROIANNI and D. OCCORSIO, Legendre polynomials of the second kind, Fourier series and Lagrange interpolation, *J. Comput. Appl. Math.* **75** (1996), 305–327.

300. G. MASTROIANNI and D. OCCORSIO, Optimal systems of nodes for Lagrange interpolation on bounded intervals. A survey, *J. Comput. Appl. Math.* **134** (2001), 325–341.
301. G. MASTROIANNI and D. OCCORSIO, Lagrange interpolation at Laguerre zeros in some weighted uniform spaces, *Acta Math. Hungar.* **91** (2001), 27–52.
302. G. MASTROIANNI and D. OCCORSIO, Numerical approximation of weakly singular integrals on the half line, *J. Comput. Appl. Math.* **140** (2002), 587–598.
303. G. MASTROIANNI and D. OCCORSIO, Lagrange interpolation based at Sonin-Markov zeros, *Rend. Circ. Mat. Palermo, Ser. II, Suppl.* **68** (2002), 683–697.
304. G. MASTROIANNI and S. PRÖSSDORF, Some nodes matrices appearing in the numerical analysis for singular integral equations, *BIT* **34** (1994), 120–128.
305. G. MASTROIANNI and M. G. RUSSO, Lagrange interpolation in some weighted uniform spaces, *Facta Univ. Ser. Math. Inform.* **12** (1997), 185–201.
306. G. MASTROIANNI and M. G. RUSSO, Lagrange interpolation in weighted Besov spaces, *Constr. Approx.* **15** (1999), 257–289.
307. G. MASTROIANNI and M. G. RUSSO, Weighted Marcinkiewicz inequalities and boundedness of the Lagrange operator, In: *Mathematical Analysis and Applications*, Hadronic Press, Palm Harbor, 2000, pp. 149–182.
308. G. MASTROIANNI and J. SZABADOS, Polynomial approximation on infinite intervals with weights having inner singularities, *Acta Math. Hungar.* **96** (2002), 221–258.
309. G. MASTROIANNI and J. SZABADOS, Direct and converse polynomial approximation theorems on infinite intervals with weights having zeros, In: *Frontiers in Interpolation and Approximation* (N. K. Govil, H. N. Mhaskar, R. N. Mohapatra, Z. Nashed, and J. Szabados, eds.), Chapman & Hall/CRC, Boca Raton, 2007.
310. G. MASTROIANNI and V. TOTIK, Jackson type inequalities for doubling weights, II, *East J. Approx.* **5** (1999), 101–116.
311. G. MASTROIANNI and V. TOTIK, Weighted polynomial inequalities with doubling and A_∞ weights, *Constr. Approx.* **16** (2000), 37–71.
312. G. MASTROIANNI and V. TOTIK, Best approximation and moduli of smoothness for doubling weights, *J. Approx. Theory* **110** (2001), 180–199.
313. G. MASTROIANNI and P. VÉRTESI, Mean convergence of Lagrange interpolation on arbitrary system of nodes, *Acta Sci. Math. (Szeged)* **57** (1993), 429–441.
314. G. MASTROIANNI and P. VÉRTESI, Some applications of generalized Jacobi weights, *Acta Math. Hungar.* **77** (1997), 323–357.
315. G. MEINARDUS, *Approximation of Functions: Theory and Numerical Methods*, Springer Verlag, Berlin, 1967.
316. H. N. MHASKAR, *Introduction to the Theory of Weighted Polynomial Approximation*, World Scientific, Singapore, 1996.
317. H. N. MHASKAR, A tribute to Géza Freud, *J. Approx. Theory* **126** (2004), 1–15.
318. H. N. MHASKAR and E. B. SAFF, Extremal problems for polynomials with exponential weights, *Trans. Amer. Math. Soc.* **285** (1984), 203–234.
319. H. N. MHASKAR and E. B. SAFF, Where does the sup norm of a weighted polynomial live? *Constr. Approx.* **1** (1985), 71–91.
320. H. N. MHASKAR and E. B. SAFF, Where does the L_p norm of a weighted polynomial live? *Trans. Amer. Math. Soc.* **303** (1987), 109–124.
321. C. A. MICCHELLI, Some positive Cotes numbers for the Chebyshev weight function, *Aequationes Math.* **21** (1980), 105–109.
322. C. A. MICCHELLI, Monosplines and moment preserving spline approximation, In: *Numerical Integration III* (H. Brass and G. Hämmerlin, eds.), ISNM 85, Birkhäuser, Basel, 1988, pp. 130–139.
323. M. MICHALSKA and J. SZYNAL, A new bound for the Laguerre polynomials, *J. Comput. Appl. Math.* **133** (2001), 489–493.
324. T. M. MILLS and S. J. SMITH, The Lebesgue constant for Lagrange interpolation on equidistant nodes, *Numer. Math.* **61** (1992), 111–115.
325. G. V. MILOVANOVIĆ, On some functional inequalities, *Univ. Beograd. Publ. Elektrotehn. Fak. Ser. Mat. Fiz.* No. **599** (1977), 1–59 (Serbian).

326. G. V. MILOVANOVIĆ, Complex orthogonality on the semicircle with respect to Gegenbauer weight: theory and applications, In: *Topics in Mathematical Analysis, A Volume Dedicated to the Memory of A. L. Cauchy* (Th. M. Rassias, ed.), World Scientific, Singapore, 1989, pp. 695–722.
327. G. V. MILOVANOVIĆ, Some applications of the polynomials orthogonal on the semicircle, In: *Numerical methods (Miskolc, 1986)*, Colloq. Math. Soc. János Bolyai, **50**, North-Holland, Amsterdam, 1988, pp. 625–634.
328. G. V. MILOVANOVIĆ, *Numerical Analysis, I*, 3rd ed., Naučna Knjiga, Belgrade, 1991 (Serbian).
329. G. V. MILOVANOVIĆ, On polynomials orthogonal on the semicircle and applications, *J. Comput. Appl. Math.* **49** (1993), 193–199.
330. G. V. MILOVANOVIĆ, Summation of series and Gaussian quadratures, In: *Approximation and Computation* (R. V. M. Zahar, ed.), ISNM, Vol. 119, Birkhäuser Verlag, Basel, 1994, pp. 459–475.
331. G. V. MILOVANOVIĆ, Summation of slowly convergent series via quadratures, In: *Advances in Numerical Methods and Applications—$O(h^3)$* (I. T. Dimov, Bl. Sendov, and P. S. Vassilevski, eds.), World Scientific, Singapore, 1994, pp. 154–161.
332. G. V. MILOVANOVIĆ, Summation of series and Gaussian quadratures, II, *Numer. Algorithms* **10** (1995), 127–136.
333. G. V. MILOVANOVIĆ, Generalized Hermite polynomials on the radial rays in the complex plane, In: *Theory of Functions and Applications, Collection of Works Dedicated to the Memory of M. M. Djrbashian* (ed. H. B. Nersessian), Louys Publishing House, Yerevan, 1995, pp. 125–129.
334. G. V. MILOVANOVIĆ, Orthogonal polynomial systems and some applications, In: *Inner Product Spaces and Applications* (Th. M. Rassias, ed.), Pitman Res. Notes Math. Ser. **376**, Longman, Harlow, 1997, pp. 115–182.
335. G. V. MILOVANOVIĆ, S-orthogonality and generalized Turán quadratures: Construction and applications, In: *Approximation and Optimization, Vol. I* (Cluj-Napoca, 1996) (D. D. Stancu, Gh. Coman, W. W. Breckner, P. Blaga, eds.), Transilvania Press, Cluj-Napoca, 1997, pp. 91–106.
336. G. V. MILOVANOVIĆ, A class of orthogonal polynomials on the radial rays in the complex plane, *J. Math. Anal. Appl.* **206** (1997), 121–139.
337. G. V. MILOVANOVIĆ, Orthogonal polynomials on the radial rays and an electrostatic interpretation of zeros, *Publ. Inst. Math. (Beograd) (N. S.)* **64** (78) (1998), 53–68.
338. G. V. MILOVANOVIĆ, Müntz orthogonal polynomials and their numerical evaluation, In: *Applications and computation of orthogonal polynomials* (W. Gautschi, G. H. Golub, G. Opfer, eds.), ISNM, Vol. **131**, Birkhäuser, Basel, 1999, pp. 179–202.
339. G. V. MILOVANOVIĆ, Some generalized orthogonal systems and their connections, In: *Proceedings of the Symposium "Contemporary Mathematics" (Belgrade, 1998)* (N. Bokan, ed.), Faculty of Mathematics, University of Belgrade, 2000, pp. 181–200.
340. G. V. MILOVANOVIĆ, Quadrature with multiple nodes, power orthogonality, and moment-preserving spline approximation, In: *Numerical Analysis 2000, Vol. V, Quadrature and Orthogonal Polynomials* (W. Gautschi, F. Marcellán, and L. Reichel, eds.), *J. Comput. Appl. Math.* **127** (2001), 267–286.
341. G. V. MILOVANOVIĆ, Orthogonal polynomials on the radial rays in the complex plane and applications, *Rend. Circ. Mat. Palermo, Serie II, Suppl.* **68** (2002), 65–94.
342. G. V. MILOVANOVIĆ and A. S. CVETKOVIĆ, Note on a construction of weights in Gauss-type quadrature rule, *Facta Univ. Ser. Math. Inform.* **15** (2000), 69–83.
343. G. V. MILOVANOVIĆ and A. S. CVETKOVIĆ, Numerical integration of functions with logarithmic end point singularity, *Facta Univ. Ser. Math. Inform.* **17** (2002), 57–74.
344. G. V. MILOVANOVIĆ and A. S. CVETKOVIĆ, Uniqueness and computation of Gaussian interval quadrature formula for Jacobi weight function, *Numer. Math.* **99** (2004), 141–162.
345. G. V. MILOVANOVIĆ and A. S. CVETKOVIĆ, Remarks on "Orthogonality of some sequences of the rational functions and Müntz polynomials", *J. Comput. Appl. Math.* **173** (2005), 383–388.

References

346. G. V. MILOVANOVIĆ and A. S. CVETKOVIĆ, Orthogonal polynomials and Gaussian quadrature rules related to oscillatory weight functions, *J. Comput. Appl. Math.* **179** (2005), 263–287.
347. G. V. MILOVANOVIĆ and A. S. CVETKOVIĆ, Orthogonal polynomials related to the oscillatory-Chebyshev weight function, *Bull. Cl. Sci. Math. Nat. Sci. Math.* **30** (2005), 47–60.
348. G. V. MILOVANOVIĆ and A. S. CVETKOVIĆ, Gauss-Laguerre interval quadrature rule, *J. Comput. Appl. Math.* **182** (2005), 433–446.
349. G. V. MILOVANOVIĆ and A. S. CVETKOVIĆ, Gaussian type quadrature rules for Müntz systems, *SIAM J. Sci. Comput.* **27** (2005), 893–913.
350. G. V. MILOVANOVIĆ and A. S. CVETKOVIĆ, Gauss-Radau and Gauss-Lobatto interval quadrature rules for Jacobi weight function, *Numer. Math.* **102** (2006), 523–542.
351. G. V. MILOVANOVIĆ and M. A. KOVAČEVIĆ, Least squares approximation with constraint: generalized Gegenbauer case, *Facta Univ. Ser. Math. Inform.* **1** (1986), 73–81.
352. G. V. MILOVANOVIĆ and M. A. KOVAČEVIĆ, Moment-preserving spline approximation and Turán quadratures, In: *Numerical Mathematics (Singapore, 1988)* (R. P. Agarwal, Y. M. Chow and S. J. Wilson, eds.), INSM, Vol. 86, Birkhäuser, Basel, 1988, pp. 357–365.
353. G. V. MILOVANOVIĆ and M. A. KOVAČEVIĆ, Moment-preserving spline approximation and quadratures, *Facta Univ. Ser. Math. Inform.* **7** (1992), 85–98.
354. G. V. MILOVANOVIĆ and P. M. RAJKOVIĆ, On polynomials orthogonal on a circular arc, *J. Comput. Appl. Math.* **51** (1994), 1–13.
355. G. V. MILOVANOVIĆ and M. M. SPALEVIĆ, Quadrature formulae connected to σ-orthogonal polynomials, *J. Comput. Appl. Math.* **140** (2002), 619–637.
356. G. V. MILOVANOVIĆ and M. STANIĆ, Construction of multiple orthogonal polynomials by discretized Stieltjes-Gautschi procedure and corresponding Gaussian quadratures, *Facta Univ. Ser. Math. Inform.* **18** (2003), 9–29.
357. G. V. MILOVANOVIĆ and S. WRIGGE, On the least squares approximation with constraints, In: *IV Conference on Applied Mathematics (Split, 1984)*, Univ. Split, Split, 1985, pp. 103–108.
358. G. V. MILOVANOVIĆ and S. WRIGGE, Least squares approximation with constraints, *Math. Comp.* **46** (1986), 551–565.
359. G. V. MILOVANOVIĆ, B. DANKOVIĆ, and S. LJ. RANČIĆ, Some Müntz orthogonal systems, *J. Comput. Appl. Math.* **99** (1998), 299–310.
360. G. V. MILOVANOVIĆ, D. S. MITRINOVIĆ, and TH. M. RASSIAS, *Topics in Polynomials: Extremal Problems, Inequalities, Zeros*, World Scientific, Singapore, 1994.
361. G. V. MILOVANOVIĆ, P. M. RAJKOVIĆ, and Z. M. MARJANOVIĆ, A class of orthogonal polynomials on the radial rays in the complex plane, II, *Facta Univ. Ser. Math. Inform.* **11** (1996), 29–47.
362. G. V. MILOVANOVIĆ, P. M. RAJKOVIĆ, and Z. M. MARJANOVIĆ, Zero distribution of polynomials orthogonal on the radial rays in the complex plane, *Facta Univ. Ser. Math. Inform.* **12** (1997), 127–142.
363. G. V. MILOVANOVIĆ, M. M. SPALEVIĆ, and A. S. CVETKOVIĆ, Calculation of Gaussian type quadratures with multiple nodes, *Math. Comput. Modelling* **39** (2004), 325–347.
364. D. S. MITRINOVIĆ, *Analytic Inequalities*, Grundlehren der mathematischen Wissenschaften, Vol. 165, Springer-Verlag, Berlin, 1970.
365. D. S. MITRINOVIĆ and J. D. KEČKIĆ, *The Cauchy Method of Residues—Theory and Applications*, Reidel, Dordrecht, 1984.
366. G. MONEGATO, Positivity of weights of extended Gauss-Legendre quadrature rules, *Math. Comp.* **32** (1978), 243–245.
367. G. MONEGATO, An overview of results and questions related to Kronrod schemes, In: *Numerische Integration (Tagung, Math. Forschungsinst., Oberwolfach, 1978)* (G. Hämmerlin, ed.), ISNM, **45**, Birkhäuser, Basel, 1979, pp. 231–240.
368. G. MONEGATO, Stieltjes polynomials and related quadrature rules, *SIAM Review* **24** (1982), 137–158.

369. G. MONEGATO, An overview of the computational aspects of Kronrod quadrature rules, *Numer. Algorithms* **26** (2001), 173–196.
370. B. MUCKENHOUPT, Weighted norm inequalities for the Hardy maximal function, *Trans. Amer. Math. Soc.* **165** (1972), 207–226.
371. F. D. MURNAGHAN, The approximation of differentiable functions by polynomials, *An. Acad. Brasil. Ci.* **31** (1959), 25–29.
372. F. D. MURNAGHAN and J. W. WRENCH, JR., The determination of the Chebyshev approximating polynomial for a differentiable function, *Math. Tables Aids Comput.* **13** (1959), 185–193.
373. N. I. MUSKHELISHVILI, *Singular Integral Equations. Boundary problems of function theory and their application to mathematical physics*, Dover Publications, Inc., New York, 1992.
374. P. NEVAI, Laguerre interpolation based on the zeros of Laguerre polynomials, *Mat. Lapok* **22** (1971), 149–164 (Hungarian).
375. P. NEVAI, *Orthogonal Polynomials*, Mem. Amer. Math. Soc., Providence, 1979.
376. P. NEVAI, Mean convergence of Lagrange interpolation III, *Trans. Amer. Math. Soc.* **282** (1984), 669–698.
377. P. NEVAI, A new class of orthogonal polynomials, *Proc. Amer. Math. Soc.* **91** (1984), 409–415.
378. P. NEVAI, Géza Freud, orthogonal polynomials and Christoffel functions, A case study, *J. Approx. Theory* **48** (1986), 3–167.
379. P. NEVAI, Orthogonal polynomials, measures and recurrence relations on the unit circle, *Trans. Amer. Math. Soc.* **300** (1987), 175–189.
380. P. NEVAI, T. ERDÉLYI, and A. P. MAGNUS, Generalized Jacobi weights, Christoffel functions, and Jacobi polynomials, *SIAM J. Math. Anal.* **25** (2) (1994), 602–614.
381. A. F. NIKIFOROV and V. B. UVAROV, *Foundations of the Theory of Special Functions*, Nauka, Moscow, 1974 (Russian).
382. A. F. NIKIFOROV, S. K. SUSLOV, and V. B. UVAROV, *Classical Orthogonal Polynomials of a Discrete Variable*, Springer Series in Computational Physics, Springer-Verlag, Berlin, 1991 (Translated from the Russian).
383. E. M. NIKISHIN and V. N. SOROKIN, *Rational Approximations and Orthogonality*, Translation of Mathematical Monographs, Vol. 92, American Mathematical Society, Providence, 1991.
384. S. M. NIKOL'SKIĬ, *Approximation of Functions of Several Variables and Imbedding Theorems*, Nauka, Moscow, 1977 (Russian).
385. S. NOSCHESE and L. PASQUINI, On the nonnegative solution of a Freud three-term recurrence, *J. Approx. Theory* **99** (1999), 54–67.
386. B. OSILENKER, *Fourier Series in Orthogonal Polynomials*, World Scientific, Singapore, 1999.
387. A. OSSICINI, Costruzione di formule di quadratura di tipo Gaussiano, *Ann. Mat. Pura Appl.* (4) **72** (1966) 213–237.
388. J. PEETRE, On the connection between the theory of interpolation spaces and approximation theory, In: *Proc. Conf. Constr. Theory of Functions (Budapest, 1969)* (G. Alexits and S. B. Stechkin, eds.), Akad. Kiadó, Budapest, 1969, pp. 351–363.
389. F. PEHERSTORFER, On the asymptotic behaviour of functions of the second kind and Stieltjes polynomials and the Gauss-Kronrod quadrature formulas, *J. Approx. Theory* **70** (1992), 156–190.
390. F. PEHERSTORFER, On the remainder of Gaussian quadrature formulas for Bernstein-Szegő weight functions, *Math. Comp.* **60** (1993), 317–325.
391. F. PEHERSTORFER, Stieltjes polynomials and functions of the second kind, *J. Comput. Appl. Math.* **65** (1995), 319–338.
392. F. PEHERSTORFER, Minimal polynomials on several intervals with respect to the maximum norm—a survey, In: *Complex Methods in Approximation Theory (Almería, 1995)*, Monogr. Cienc. Tecnol., 2, Univ. Almería, Almería, 1997, pp. 137–159.
393. F. PEHERSTORFER and K. PETRAS, Ultraspherical Gauss-Kronrod quadrature is not possible for $\lambda > 3$, *SIAM J. Numer. Anal.* **37** (2000), 927–948.

394. F. PEHERSTORFER and R. STEINBAUER, Orthogonal polynomials on arcs of the unit circle, I, *J. Approx. Theory* **85** (1996), 140–184.
395. F. PEHERSTORFER and R. STEINBAUER, Orthogonal polynomials on arcs of the unit circle, II. Orthogonal polynomials with periodic reflection coefficients, *J. Approx. Theory* **87** (1996), 60–102.
396. F. PEHERSTORFER and R. STEINBAUER, Orthogonal polynomials on the circumference and arcs of the circumference, *J. Approx. Theory* **102** (2000), 96–119.
397. P. P. PETRUSHEV and V. A. POPOV, *Rational Approximation of Real Function*, Encyclopedia of Mathematics and its Applications, Vol. 28, Cambridge University Press, Cambridge, 1987.
398. R. PIESSENS and M. BRANDERS, The evaluation and application of some modified moments, *Nordisk Tidskr. Informationsbehandling (BIT)* **13** (1973), 443–450.
399. R. PIESSENS and M. BRANDERS, Tables of Gaussian quadrature formulas, *Appl. Math. Progr. Div. University of Leuven*, Leuven, 1975.
400. A. PINKUS, Weierstrass and approximation theory, *J. Approx. Theory* **107** (2000), 1–66.
401. T. POPOVICIU, Sur le reste dans certaines formules linéaires d'approximation de l'analyse, *Mathematica (Cluj)* **1** (24) (1959), 95–142.
402. S. PRÖSSDORF and B. SILBERMANN, *Numerical analysis for integral and related operator equations*, Akadem. Verlag, 1991.
403. A. P. PRUDNIKOV, YU. A. BRYCHKOV, and O. I. MARICHEV, *Integrals and Series. Elementary Functions*, Nauka, Moscow, 1981 (Russian).
404. J. RADON, Restausdrüche bei Interpolations und Quadraturformeln durch bestimmte Integrale, *Monatsh. Math. Phys.* **42** (1935), 389–396.
405. E. A. RAKHMANOV, On asymptotic properties of polynomials orthogonal on the real axis, *Mat. Sbornik* **119** (161) (1982), 163–203 (Russian) [Engl. transl. *Math. USSR Sb.* **47** (1984), 155–193].
406. E. A. RAKHMANOV, Strong asymptotics for orthogonal polynomials associated with exponential weights on $\mathbb{R}$, In: *Methods of Approximation Theory in Complex Analysis and Mathematical Physics* (A. A. Gonchar and E. B. Saff, eds.), Nauka, Moscow, 1992, pp. 71–97.
407. R. REEMTSEN, Modifications of the first Remez algorithm, *SIAM J. Numer. Anal.* **27** (1990), 507–518.
408. YA. E. REMEZ, Sur le calcul effectif des polynômes d'approximation de Tschebyscheff, *C. R. Acad. Sci. Paris* **199** (1934), 337–339.
409. YA. E. REMEZ, *Fundamentals of Numerical Methods of Chebyshev Approximation*, Naukova Dumka, Kiev, 1969 (Russian).
410. M. REVERS, On Lagrange interpolation with equally spaced nodes, *Bull. Austral. Math. Soc.* **62** (2000), 357–368.
411. M. REVERS, The divergence of Lagrange interpolation for x^α at equidistant nodes, *J. Approx. Theory* **103** (2000), 269–280.
412. M. REVERS, Approximation constants in equidistant Lagrange interpolation, *Period. Math. Hungar.* **40** (2) (2000), 167–175.
413. M. REVERS, On Lagrange interpolatory parabolas to x^α at equally spaced nodes, *Arch. Math.* **74** (2000), 385–391.
414. T. J. RIVLIN, *An Introduction to the Approximation of Functions*, Dover Publications, Inc., New York, 1969.
415. T. J. RIVLIN, *The Chebyshev Polynomials*, John Wiley & Sons, New York, 1974.
416. T. J. RIVLIN, The Lebsgue constants for polynomial interpolation, In: *Functional analysis and its applications, Internat. Conf., Eleventh Anniversary of Matscience (Madras, 1973)*, Lecture Notes in Math., Vol. 399, Springer, Berlin, 1974, pp. 422–437, dedicated to Alladi Ramakrishnan.
417. A. RONVEAUX, Sur l'équation différentielle du second ordre satisfaite par une classe de polynômes orthogonaux semi-classiques, *C. R. Acad. Sci. Paris* **305** (1) (1987), 163–166.
418. A. RONVEAUX and F. MARCELLÁN, Differential equation for classical-type orthogonal polynomials, *Canad. Math. Bull.* **32** (1989), 404–411.

419. A. RONVEAUX and G. THIRY, Differential equations of some orthogonal families in REDUCE, *J. Symb. Comp.* **8** (1989), 537–541.
420. P. G. ROONEY, Further inequalities for generalized Laguerre polynomials, *C. R. Math. Rep. Acad. Sci. Canada* **7** (1985), 273–275.
421. P. O. RUNCK and P. VÉRTESI, Some good point systems for derivatives of Lagrange interpolatory operators, *Acta Math. Hungar.* **56** (1990), 337–342.
422. E. B. SAFF, Orthogonal polynomials from a complex perspective, In: *Orthogonal Polynomials—Theory and Practice* (P. Nevai, ed.), NATO ASI Series, Series C; Mathematical and Physical Sciences, Vol. 294, Kluwer, Dordrecht, 1990, pp. 363–393.
423. E. B. SAFF and V. TOTIK, *Logarithmic Potentials with External Fields*, Grundlehren der mathematischen Wissenschaften, Vol. 316, Springer-Verlag, Berlin, 1997.
424. R. SALEM, *Essais sur les séries trigonometriques: Actualités scientifiques et industrialles*, N 862, Herman, Paris, 1940.
425. T. SCHIRA, The remander term for analytic functions of Gauss-Lobatto quadratures, *J. Comput. Appl. Math.* **76** (1996), 171–193.
426. T. SCHIRA, The remainder term for analytic functions of symmetric Gaussian quadratures, *Math. Comp.* **66** (1997), 297–310.
427. I. J. SCHOENBERG, The elementary cases of Landau's problem of inequalities between derivatives, *Amer. Math. Monthly* **80** (1973), 121–158.
428. A. SCHÖNHAGE, Fehlerfortpflanzung bei Interpolation, *Numer. Math.* **3** (1961), 62–71.
429. B. SENDOV and V. A. POPOV, *The Averaged Moduli of Smoothness. Applications in Numerical Methods and Approximation*, Pure and Applied Mathematics, A Wiley-Interscience Publication, John Wiley & Sons, Ltd., Chichester, 1988.
430. J. SHERMAN, On the numerators of the convergents of the Stieltjes continued fractions, *Trans. Amer. Math. Soc.* **35** (1933), 64–87.
431. Y. G. SHI, *Theory of Birkhoff Interpolation*, Nova Science Publishers, Inc., Hauppauge, 2003.
432. Y. G. SHI and G. XU, Construction of σ-orthogonal polynomials and Gaussian quadrature formulas, *Adv. Comput. Math.* **27** (2007), 79–94.
433. P. N. SHIVAKUMAR and R. WONG, Asymptotic expansion for the Lebesgue constants associated with polynomial interpolation, *Math. Comp.* **39** (1982), 195–200.
434. J. SHOHAT and J. SHERMAN, On the numerators of the continued fraction $\frac{\lambda_1|}{|x-c_1} - \frac{\lambda_2|}{|x-c_2} - \cdots$, *Proc. Nat. Acad. Sci. U.S.A.* **18** (1932), 283–287.
435. B. SIMON, Ratio asymptotics and weak asymptotic measures for orthogonal polynomials on the real line, *J. Approx. Theory* **126** (2004), 198–217.
436. B. SIMON, Orthogonal polynomials on the unit circle: new results, *Int. Math. Res. Not.*, **53** (2004), 2837–2880.
437. B. SIMON, *Orthogonal Polynomials on the Unit Circle. Part 1: Classical Theory*, Amer. Math. Soc. Colloq. Publ., **54**, Amer. Math. Soc., Providence, 2005.
438. B. SIMON, *Orthogonal Polynomials on the Unit Circle. Part 2: Spectral Theory*, Amer. Math. Soc. Colloq. Publ., **54**, Amer. Math. Soc., Providence, 2005.
439. B. SIMON, Fine structure of the zeros of orthogonal polynomials, II. OPUC with competing exponential decay, *J. Approx. Theory* **135** (2005), 125–139.
440. B. SIMON, Fine Structure of the zeros of orthogonal polynomials III: Periodic Recursion Coefficients, *Comm. Pure Appl. Math.* **59** (2006), 1–21.
441. M.-R. SKRZIPEK, Generalized associated polynomials and functions of second kind, *J. Comput. Appl. Math.* **178** (2005), 425–436.
442. I. H. SLOAN and W. E. SMITH, Product-integration with the Clenshaw-Curtis and related points: convergence properties, *Numer. Math.* **30** (1978), 514–428.
443. I. H. SLOAN and W. E. SMITH, Product-integration with the Clenshaw-Curtis points: implementation and error estimates, *Numer. Math.* **34** (1980), 378–401.
444. I. H. SLOAN and W. E. SMITH, Properties of interpolatory product integration rules, *SIAM J. Numer. Anal.* **19** (1982), 427–442.

References

445. I. SMIRNOV, Sur la théorie des polynomes orthogonaux à une variable complèxe, *Zh. Leningrad. Fiz.-Mat. Ob.* **2** (1928), 155–179.
446. I. SMIRNOV, Sur les valeurs limites des fonctions regulières à l'intérieur d'un cercle, *Zh. Leningrad. Fiz.-Mat. Ob.* **2** (1928), 22–37.
447. H. V. SMITH, Global error bounds for Gauss-Christoffel quadrature, *BIT* **21** (1981), 481–499.
448. R. SMITH, Similarity solutions of a non-linear differential equation, *IMA J. Appl. Math.* **28** (1982), 149–160.
449. W. E. SMITH and I. H. SLOAN, Product-integration rules based on the zeros of Jacobi polynomials, *SIAM J. Numer. Anal.* **17** (1980), 1–13.
450. G. SOTTAS, On the positivity of quadrature formulas with Jacobi abscissas, *Computing* **29** (1982), 83–88.
451. G. SOTTAS, Positivity domain of ultraspherical type quadrature formulas with Jacobi abscissas: Numerical investigations, In: Numerical Integration III (H. Brass and G. Hämmerlin, eds.), ISNM 85, Birkhäuser, Basel, 1988, pp. 285–294.
452. H. STAHL and V. TOTIK, *General Orthogonal Polynomials* Encyclopedia of Mathematics, vol. 43, Cambridge University Press, New York, 1992.
453. S. B. STECHKIN, On the order of the best approximation of continuous functions, *Izv. Akad. Nauk SSSR, Serie Math.* **15** (1951), 219–241 (Russian).
454. S. B. STECHKIN, Inequalities between the upper bounds of the derivatives of an arbitrary function on the half-line, *Mat. Zametki* **1** (1967), 665–574 (Russian).
455. N. M. STEEN, G. D. BYRNE, and E. M. GELBARD, Gaussian quadratures for the integrals $\int_0^\infty \exp(-x^2) f(x)\,dx$ and $\int_0^b \exp(-x^2) f(x)\,dx$, *Math. Comp.* **23** (1969), 661–671.
456. E. M. STEIN, *Harmonic Analysis*, Princeton University Press, Princeton, 1993.
457. F. STENGER, Bounds of the error of Gauss-type quadratures, *Numer. Math.* **8** (1966), 150–160.
458. P. K. SUETIN, *Classical Orthogonal Polynomials*, Nauka, Moscow, 1976 (Russian).
459. P. N. SWARZTRAUBER, On computing the points and weights for Gauss-Legendre quadrature, *SIAM J. Sci. Comput.* **24** (2002), 945–954.
460. J. SZABADOS, On an interpolatory analogon of the de la Vallée Poussin operator, *Studia Sci. Math. Hungar.* **9** (1974), 187–190.
461. J. SZABADOS, On the convergence of the derivatives of projection operators, *Analysis* **7** (1987), 349–357.
462. J. SZABADOS, On the norm of certain interpolating operators, *Acta Math. Hungar.* **55** (1990), 179–183.
463. J. SZABADOS, Weighted Lagrange and Hermite-Fejér interpolation on the real line, *J. Inequal. Appl.* **1** (1997), 99–123.
464. J. SZABADOS and P. VÉRTESI, On simultaneous optimization of norms of derivatives of Lagrange interpolation polynomials, *Bull. London Math. Soc.* **21** (1989), 475–481.
465. J. SZABADOS and P. VÉRTESI, *Interpolation of Functions*, World Scientific, Singapore, 1990.
466. F. H. SZAFRANIEC, Orthogonality of analytic polynomials: a little step further, *J. Comput. Appl. Math.* **179** (2005), 343–353.
467. P. SZÁSZ, On quasi-Hermite-Fejér interpolation, *Acta Math. Acad. Sci. Hungar.* **10** (1959), 413–439.
468. G. SZEGŐ, Beiträge zur Theorie der Toeplitzschen Formen, *Math. Z.* **6** (1920), 167–202.
469. G. SZEGŐ, Über die Entwicklung einer analytischen Funktion nach den Polynomen eines Orthogonalsystems, *Math. Ann.* **82** (1921), 188–212.
470. G. SZEGŐ, *Orthogonal Polynomials*, Amer. Math. Soc. Colloq. Publ., **23**, 4th ed., Amer. Math. Soc., Providence, 1975.
471. H. TAKAHASI and M. MORI, Estimation of errors in the numerical quadrature of analytic functions, *Appl. Anal.* **1** (1971), 201–229.
472. S. TAKENAKA, On the orthogonal functions and a new formula of interpolation, *Japan. J. Math.* **2** (1925), 129–145.

473. A. K. TASLAKYAN, Some properties of Legendre quasi-polynomials with respect to a Müntz system, *Mathematics, Èrevan University, Èrevan* **2** (1984), 179–189 (Russian, Armenian summary).
474. H. TIETZE, Eine Bemerkung zur Interpolation, *Z. Angew. Math. Phys.* **64** (1917), 74–90.
475. A. F. TIMAN, *Theory of Approximation of Functions of a Real Variable*, Dover Publications, Inc., New York, 1994.
476. V. TOTIK, *Weighted Approximation with Varying Weights*, Springer Lecture Notes in Mathematics, Vol. 1569, Springer, Berlin, 1994.
477. V. TOTIK, Orthogonal polynomials with respect to varying weights, *J. Comput. Appl. Math.* **99** (1998), 373–385.
478. L. N. TREFETHEN and J. A. C. WEIDEMAN, Two results on polynomial interpolation in equally spaced points, *J. Approx. Theory* **65** (1991), 247–260.
479. F. TRICOMI, *Serie orthogonali di Funzioni*, Torino, 1948.
480. A. H. TURECKIĬ, The bounding of polynomials prescribed at equally distributed points, *Proc. Pedag. Inst. Vitebsk* **3** (1940), 117–127 (Russian).
481. A. H. TURECKIĬ, *Theory of Interpolation in Problem Form*, Izdat. "Vyssh. Skola", Minsk, 1968 (Russian).
482. J. V. USPENSKY, On the convergence of quadrature formulas related to an infinite interval, *Trans. Amer. Math. Soc.* **30** (1928), 542–559.
483. W. VAN ASSCHE, Weighted zero distribution for polynomials orthogonal on an infinite interval, *SIAM J. Math. Anal.* **16** (1985), 1317–1334.
484. W. VAN ASSCHE, Orthogonal polynomials, associated polynomials and functions of the second kind, *J. Comput. Math. Appl.* **37** (1991), 237–249.
485. W. VAN ASSCHE, Orthogonal polynomials in the complex plane and on the real line, In: *Special Functions, q-Series and Related Topics* (M. E. H. Ismail et al., eds.), Fields Institute Communications **14** (1997), 211–245.
486. W. VAN ASSCHE, Approximation theory and analytic number theory, In: *Special Functions and Differential Equations* (K. Srinivasa Rao et al., eds.), Allied Publishers, New Delhi, 1998, pp. 336–355.
487. J. G. VAN DER CORPUT and C. VISSER, Inequalities concerning polynomials and trigonometric polynomials, *Nederl. Akad. Wetensch. Proc.* **49** (1946), 383–392 [= *Indag. Math.* **8** (1946), 238–247].
488. E. A. VAN DOORN, Representations and bounds for zeros of orthogonal polynomials and eigenvalues of sign-symmetric tri-diagonal matrices, *J. Approx. Theory* **51** (1987), 254–266.
489. E. A. VAN DOORN, On associated polynomials and decay rates for birth–death processes, *J. Math. Anal. Appl.* **278** (2003), 500–511.
490. E. A. VAN DOORN, Birth–death processes and associated polynomials, *J. Comput. Appl. Math.* **153** (2003), 497–506.
491. S. L. L. VAN EIJNDHOVEN and J. L. H. MEYERS, New orthogonality relations for the Hermite polynomials and related Hilbert spaces, *J. Math. Ann. Appl.* **146** (1990), 89–98.
492. M. VANLESSEN, Strong asymptotics of the recurrence coefficients of orthogonal polynomials associated to the generalized Jacobi weight, *J. Approx. Theory* **125** (2003), 198–237.
493. A. K. VARMA, A new characterization of Hermite polynomials, *Acta Math. Hungar.* **49** (1987), 169–172.
494. P. VÉRTESI, Remark on the Lagrange interpolation, *Studia. Sci. Math. Hungar.* **15** (1980), 277–281.
495. P. VÉRTESI, On Lagrange interpolation, *Periodica Math. Hungar.* **12**, 103–112.
496. P. VÉRTESI, On the zeros of Jacobi polynomials, *Studia Sci. Math. Hungar.* **25** (1990), 401–405.
497. P. VÉRTESI, Optimal Lebesgue constant for Lagrange interpolation, *SIAM J. Numer. Anal.* **27** (1990), 1322–1331.
498. P. VÉRTESI, On classical (unweighted) and weighted interpolation, *Rend. Circ. Mat. Palermo, Serie II, Suppl.* **68** (2002), 185–202.
499. P. VÉRTESI, Oral communication.

References

500. L. VINET and A. ZHEDANOV, A characterization of classical and semiclassical orthogonal polynomials from their dual polynomials, *J. Comput. Appl. Math.* **172** (2004), 41–48.
501. J. L. WALSH, *Interpolation and Approximation by Rational Functions in the Complex Domain*, 5th ed., Amer. Math. Soc. Colloq. Publ., **20**, Amer. Math. Soc., Providence, 1969.
502. K. WEIERSTRASS, *Mathematische Werke*, Berlin, 1915.
503. W. WERNER, Polynomial interpolation: Lagrange versus Newton, *Math. Comp.* **43** (1984), 205–217.
504. J. C. WHEELER, Modified moments and continued fraction coefficients for the diatomic linear chain, *J. Chem. Phys.* **80** (1984), 472–476.
505. H. WIDOM, Polynomials associated with measures in the complex plane, *J. Math. Mech.* **16** (1967), 997–1013.
506. H. WIDOM, Extremal polynomials associated with a system of curves in the complex plane, *Adv. Math.* **3** (1969), 127–232.
507. M. ZAMANSKY, Classes de saturation de certains procédés d'approximation des séries de Fourier des fonctions continues et applications à quelques problèmes d'approximation, *Ann. Sci. École. Norm. Sup.* (3) **66** (1949), 19–93.
508. A. ZYGMUND, *Trigonometric series*, Cambridge Univ. Press, London, 1959.

Index

Abel, N.H., 166, 413
Abramowitz, M., 142
Aczél, J., 126
Agarwal, R.P., 123
Ahieser, N.I., 20, 80
Al-Salam, W.A., 121
Alexits, G., 196
Algorithm
– first Remez, 23
– for finding optimal nodes for polynomial interpolation, 68
– Lanczos, 160
– modified Chebyshev, 159, 160
– QR, 327
– Remez, 20
– second Remez, 20
Alves, C.R.R., 123
Andrews, G.E., 121, 132
Antonov, V.A., 136
Approximant, 1
Approximation, 1
– moment-preserving, 385
– moment-preserving spline, 387
– standard L^2, 385
– the constrained L^2, 389
– unconstrained L^2, 388
– value-preserving, 385
– weighted, 166
Askey, R., 121, 132, 139, 323
Askey table, 121
– q-extension of, 121
Associated continued fraction, 114
– n-th convergent of, 114
Asymptotic properties of orthogonal polynomials, 103
Atakishiyev, N.M., 121

Badkov, V.M., 149
Barkov, G.I., 148
Bateman, H., 126
Bauldry, W.C., 155
Berg, C., 116, 117
Berman, D.L., 61
Bernoulli numbers, 65
Bernstein, S.N., 9, 19, 61, 64, 65, 67, 104, 136
Berrut, J.P., 50

Best approximation
– by polynomials, 7
– Chebyshev, 8
– element of, 1
– error of, 1
– in the uniform (supremum) norm, 3
– L^p, 7
– smoothness of a function and rate of convergence of, 35
– uniform, 7
Best weighted approximation
– in L^p by trigonometric polynomials, 230
Blichfeldt, H.F., 17
Bochner, S., 126
Bojanov, B.D., ix, 123, 387
Branders, M., 162, 165, 347
Brass, H., 321
Brutman, L., 64–67
Byrne, G.D., 165
Byrne, G.J., 62

Calder, A.C., 389
Canonical arrays, 67
Carleman's condition, 97
Cauchy, A.L., 54
Cauchy formula, 126
Cesàro mean, 143
Chebyshev, P.L., 8, 14, 17, 19, 86, 160
Chebyshev alternation, 18
Chebyshev array of nodes, 53
Chebyshev extremal problems, 14
Chebyshev nodes, 55
– transformed, 66
Chebyshev polynomials
– differential equation of, 10
– discrete, 86
– distribution of zeros of, 11
– extremal points of, 12
– limit distribution of zeros of, 12
– of the first kind, 6, 9, 122
– of the fourth kind, 122
– of the second kind, 6, 9, 122
– of the third kind, 122
– on the complex plane, 12
– three-term recurrence relation for, 9
– zeros of, 11
Chihara, T.S., 87, 97, 147

437

Chow, Y., 137
Christoffel, E.B., 324
Christoffel numbers, 115
Christoffel's formula, 98
Classical orthogonal polynomials, 121
– differential equation for, 122
– monic, 127
– Rodrigues' type formula, 126
– three-term recurrence relation for, 127
Condition
– Lipschitz, 109
– Lipschitz-Dini, 105
Condition number, 377
conjecture
– Askey, 323
– Bernstein-Erdős, 67
– Milovanović, 323
Constant
– doubling, 224
– Euler, 28, 65
– Lebesgue, 53, 63
– Lebesgue of the Fourier operator, 193
– optimal Lebesgue, 67
– weighted Lebesgue, 271
Convex hull, 93
– polynomial, 93
Convex set, 93
Cotes numbers, 115
Cotes-Christoffel coefficients, 115
Criscuolo, G., 157
Cvetković, A.S., ix

δ Dirac distribution, 146
Dahlquist, G., 166, 411, 413
Darboux formula, 137
Darboux's formulae, 163
Dassiè, S., 411, 413
Davis, P.J., 325, 401, 409
De Bonis, M.C., ix
De Boor, C., 67, 124
De Bruin, M.G., 88
De la Vallée Poussin, Ch.-J., 18, 28
De la Vallée Poussin interpolating polynomial, 204
De la Vallée Poussin means, 202
De la Vallée Poussin operators, 203
– discrete, 203
De la Vallée Poussin sums, 32, 196
Defective spline, 394
Deift, P.A., 150
Della Vecchia, B., 157
Determinant
– Hankel, 87, 95
– Vandermonde, 39

Dette, H., 124
DeVore, R.A., 18
Dimitrov, D.K., 123
Dirichlet-Mehler formula, 135
Discretized Stieltjes-Gautschi procedure, 160, 162
Divided differences, 49
Djrbashian, M.M., 81
Dzyadyk, V.K., 65

Egerváry, E., 264
Einstein-Bose distribution, 398
Equation
– Gauss hypergeometric, 132
– Schrödinger, 126
– Sturm-Liouville form, 125
Erdélyi, A., 126
Erdélyi, T., 138
Erdős, P., 53, 67, 68, 105
Error estimates for Freud-Gaussian rules, 343
Error estimates for Gauss-Laguerre formula, 341
Error in the moment-preserving spline approximation, 392
Euler's formulas, 4
Everitt, W.N., 147

Faber, G., 53, 67
Feinerman, R.P., 18
Fejér, L., 94, 164, 323
Fejér mean, 195
Fejér sums, 32
Finite forward differences, 32
Formula of Mehler-Heine type, 135
Fourier
– coefficients, 31, 77
– discrete operator, 203
– expansion, 77
– operator, 193
– projector, 194
– series, 30
– sums, 31, 193
Fourier partial sum
– integral form of the, 236
Fransén, A., 165
Freud, G., 106, 141, 154, 264, 393
Freud conjecture, 158
Frontini, M., 397
Function
– absolutely continuous, 95, 96
– Airy, 166
– Bessel, 135
– Christoffel, 91
– Dirac delta, 86

Index   439

- gamma, 127
- generalized Christoffel, 93
- generalized Steklov, 34
- generating, 128
- Heaviside step, 389
- hypergeometric, 132
- incomplete gamma, 401
- inhomogeneous Airy, 165
- jump, 96
- Lebesgue, 52
- measurable in Lebesgue's sense, 95
- modified Bessel, 166
- of the second kind, 117
- piecewise continuous, 182
- reciprocal gamma, 165
- Riemann zeta, 411
- Runge, 63
- singular, 96
- spline, 390
- Szegő, 103
- weight, 95
- weighted Lebesgue, 309
Functional
- moment, 86
- quasi-definite linear, 87

Ganzburg, M.I., 63
Gasper, G., 139
Gasper's Mehler-type integral, 137
Gatteschi, L., 137
Gauss, C.F., 324
Gauss-Christoffel quadrature formula, 324
- computation of, 325
- error estimates for analytic functions, 334
- error estimates for some classes of continuous functions, 337
- error (remainder) term in, 332
- for the classical weights, 324
- parameters of, 324
- truncated, 345
Gautschi, W., ix, 49, 88, 148, 160, 161, 164–166, 321, 323, 324, 327–329, 331, 336, 354, 389, 394, 397, 401, 409, 411, 412
Gelbard, E.M., 165
Generalized Laguerre polynomials, 140
- Christoffel function for, 144
- Christoffel numbers for, 144
- differential equation for, 140
- integral representation of, 141
- Rodrigues type formula for, 140
- three-term recurrence relation for, 140
Geronimus, Ya.L., 80, 104
Ghizzetti, A., 93
Golub, G., 326–329, 331

Gopengauz, I. E., 67
Gopengauz, I.E., 261
Gori, L., 387
Grinshpun, Z.S., 118
Grosjean, C.C., 112
Grünwald, G., 53, 212
Günttner, R., 65, 66

Haar property, 4
Halász, G., 264
Hamburger moment problem, 97
Harris, L.A., 139
Heine, E., 335
Henrici, P., 397
Hermite, C., 335
Hermite polynomials, 145
- differential equation for, 145
- Rodrigues type formula for, 145
Higham, N.J., 50
Hille, E., 196
Holševnikov, K.V., 136
Hristov, V.H., 214
Hunter, D.B., 336, 337

Identity
- Bernstein-Szegő, 108
- Christoffel-Darboux, 98
- Fokas-Its-Kitaev, 108
- Korous', 107
- Mhaskar-Rahmanov-Saff, 289
- Parseval, 37
- Rakhmanov's projection, 108, 111
- Riemann-Hilbert, 108
Inequality
- Bernstein, 35, 170
- Bessel, 78
- Cauchy, 91
- Cauchy-Schwarz-Buniakowsky, 75
- Chebyshev, 15
- Faber, 238
- Favard, 35, 171
- generalized Minkowski, 215
- Hölder, 206
- Jackson type, 170
- Marcinkiewicz, 205
- Markov, 298
- Markov-Bernstein type, 124
- Minkowski, 206
- Minkowski integral, 232
- Posse-Markov-Stieltjes, 333
- Remez, 297
- Salem-Stechkin, 35
- Stechkin type, 170
- Varma's, 123

– weak Marcinkiewicz, 229
Inner product, 75
– Hermitian symmetry of, 75
– homogeneity of, 75
– linearity of, 75
– positivity of, 75
– symmetry of, 75
Integral equations, 346
– Fredholm of the second kind, 362
Interlacing property, 100
Interpolation
– algebraic Lagrange, 39
– array, 51
– at Clenshaw's abscissas, 249
– at Hermite zeros, 287
– at Jacobi abscissas, 248
– at Laguerre zeros, 278
– at Stieltjes zeros, 266
– at the practical abscissas, 249
– at zeros of orthogonal polynomials, 235
– barycentric Lagrange, 50
– Birkhoff, 48
– Chebyshev, 55
– extended, 266
– Hermite, 46
– lacunary, 48
– Lagrange in Sobolev spaces, 276
– Lagrange-Hermite, 258
– nodes, 39
– of functions, 3
– of functions with internal isolated singularities, 292
– polynomial, 39
– Taylor, 47
– trigonometric, 40
– truncated based on the Hermite zeros, 291
– truncated based on the Laguerre zeros, 286
– weighted, 271
– with associated polynomials, 264
Interpolation error, 52, 54
– actual, 237
– in Cauchy form, 349
– in the class of analytic functions, 55
– in the class of continuous-differentiable functions, 54
– numerical, 237
– of the De la Vallée Poussin operator, 210
– of the discrete Fourier operator, 210
– of the Hermite interpolation polynomial, 333
– theoretical, 237
Interpolation formula
– Lagrange, 40, 51
– Newton, 49

– Riesz, 45
– trigonometric (in the Lagrange form), 42
Interpolation nodes
– arc sine distribution of, 59, 249
– optimal system of, 238
– system of, 51
– uniformly distributed, 58
Interpolation problem
– basic, 47
– general, 46
Interpolatory process, 52
– for locally continuous functions, 271
– uniform convergence of, 57
Ivanov, V.V., 65

Jacobi, C.G.J., 321, 324
Jacobi polynomials, 131
– Christoffel function for, 139
– Christoffel numbers for, 139
– differential equation for, 131
– Rodrigues' formula for, 131
– three-term recurrence relation for, 132
Jetter, K., vii
Jolley, L.B.W., 397
Joó, I., 141
Junghanns, P., ix

K-functional, 33, 169
– main part of, 169
– weighted, 230
Kasuga, T., 155
Kečkić, J.D., 397
Kernel
– Darboux, 236
– Darboux-Christoffel, 380
– De la Vallée Poussin, 29, 197
– degenerate, 364
– Dirichlet, 24
– Fejér, 24
– Jackson, 26
– locally smooth, 370
– reproducing, 91
– weakly singular, 379
Kilgore, T., 67
Kirchberger, P., 17
Kis, O., 152
Koekoek, R., 121
Kolmogorov, A., 123, 194
Korkin, A.N., 16
Korneichuk, N., 18
Kovačević, M.A., 394
Krasikov, I., 138
Kreĭn, M.G., 80
Kriecherbauer, T., 157

Kronrod, A.S., 266
Kubayi, D.G., 63
Kwon, K.H., 147
Ky, N.X., 339

Laframboise, J.G., 389
Lagrange, J.L., 49
Lagrange interpolation error
– in the L^p-norm, 212
Lagrange operator, 203
– convergence estimate in the Sobolev norm, 218
– uniform boundedness in the Sobolev spaces, 218
Landau, E., 123
Landau, H., 88
Laplace integral formula, 135
Laščenov, K.V., 147
Laurie, D.P., 328
Lee, S.-Y., 165
Leibniz' convergence criterion, 400
Lesky, P., 126
Levin, A.L., 106, 154, 155, 157
Lewandowski, Z., 143
Lewanowicz, S., 347
Li, S., 336
Li, X., 62
Lindelöf, E., 166, 397, 413
Littlejohn, L.L., 147
Lorch, L., 136
Lorentz, G.G., vii, 18
Lozinskiĭ, S.M., 61, 194
Lubinsky, D.S., ix, 3, 63, 106–108, 154, 155, 157, 158
Lukashov, A.L., 80

Magnus, A.P., 138, 158
Malkowsky, E., ix
Malmquist, F., 81
Malmquist-Takenaka basis, 81
Marcellán, F., 146
Marcinkiewicz, J., 53, 210
Marinković, S.D., 323
Markov, A., 116, 324, 332
Maroni, P., 147
Mastroianni, G., 153, 157
Matrix
– Gram, 76, 95
– Hankel, 95
– Hermitian, 76
– Jacobi, 99
– Jacobian, 71
Maxwell (velocity) distribution, 165

McLaughlin, K.T.-R., 157
Measure
– Borel, 89
– discrete, 96, 162
– generalized Laguerre, 393
Meinardus, G., 18, 20, 21
Méray, Ch., 56
Method
– additional nodes, 264
– bisection, 70
– Darboux's, 137
– Laplace transform, 398
– Newton-Kantorovič, 72
– Nyström, 382
– of contour integration over the rectangle, 402
– of moments, 159
– standard Newton, 70
Mhaskar, H.N., 106, 155, 158
Mhaskar-Rakhmanov-Saff number, 154
Micchelli, C.A., 397
Michalska, M., 143
Mills, T.M., 62, 64
Milovanović, G.V., 82, 88, 93, 123, 165, 166, 323, 328, 389, 394, 395, 397, 402
Mitrinović, D.S., 91, 397
Modulus of continuity
– main part, 167
Modulus of smoothness, 33
– φ-, 172
– global, 168
– main properties of, 33
Mohapatra, R.N., 62
Moment problem, 116
– determined, 116
– indeterminate, 116
– Markov's, 116
Monegato, G., 120
Monospline, 390
Muckenhoupt, B., 225
Murnaghan, F.D., 21

Nevai, P., 80, 93, 105, 106, 138, 141, 148, 149, 157, 158, 348
Nevai class, 106
Newman, D.J., 18
Newton, I., 49
Newton formula, 259
Nikiforov, A.F., 121
Nikolov, G., 336
Nonlinear Remez search, 73
Norm
– L^p, 3
– $L^r(d\mu)$, 92

– supremum, 3
– trigonometric Sobolev, 37
– uniform, 3

Occorsio, D., 153
Orthogonality
– interval of, 96
– on the semicircle, 88
– power, 93
– with respect to an oscillatory weight, 87
Orthogonalizing process
– Gram-Schmidt, 75
Ossicini, A., 93

Padé approximation, 113
Peetre, J., 33
Peherstorfer, F., 80, 105, 120, 336
Petras, K., 120
Petrushev, P.P., 18
Piessens, R., 162, 165, 347
Pinkus, A., 3, 67
Plana, G.A.A., 166, 413
Pochhammer's symbol, 127
Polynomials
– σ-orthogonal, 397
– algebraic, 2
– associated, 111, 265
– Chebyshev, 6, 9
– classical orthogonal, 121
– Clenshaw, 114
– cosine, 2
– discrete Chebyshev orthogonal, 86
– formal orthogonal with respect to a moment functional, 86
– Freud, 146, 155
– fundamental Lagrange, 39, 51
– Gegenbauer, 122, 133
– generalized exponential, 85
– generalized Freud, 155
– generalized Gegenbauer, 147
– generalized Hermite, 152
– generalized Jacobi, 148
– generalized Laguerre, 122, 140
– Hermite, 122, 145
– Hermite algebraic interpolation, 48
– Hermite interpolation, 47
– Hermite trigonometric interpolation, 48
– Hermite-Fejér, 264
– Jacobi, 14, 122, 131
– Lagrange algebraic interpolation, 39
– Legendre, 122
– monic, 2
– Müntz, 40, 82
– Müntz of the second kind, 84

– Müntz-Legendre, 82
– node, 54
– orthogonal on the ellipse, 80
– orthogonal on the radial rays, 81
– orthogonal on the real line, 95
– orthogonal on the unit circle, 80
– orthogonal on the unit disk, 80
– orthogonal sequences of, 14
– real trigonometric, 5
– s-orthogonal, 93
– self-inversive, 6
– semi-classical orthogonal, 147
– sine, 2
– Sonin-Markov, 146, 152
– standard Laguerre, 123
– Stieltjes, 119
– strong non-classical orthogonal, 159
– Szegő's orthogonal, 81
– Taylor, 48
– trigonometric, 2
– trigonometric interpolation, 42
– ultraspherical, 133
– very classical orthogonal, 121
Popov, V.A., 18
Product integration rules, 345
Programm package
– OrthogonalPolynomials, 327
– ORTHPOL, 327

Quadrature formula
– classical Newton-Cotes, 321
– Fejér, 164
– Gauss-Chebyshev, 325
– Gauss-Christoffel, 115, 324
– Gauss-Einstein, 398
– Gauss-Fermi, 399
– Gauss-Hermite, 325
– Gauss-Jacobi, 324
– Gauss-Laguerre, 325
– Gauss-Lobatto, 330
– Gauss-Radau, 329
– Gaussian, 324
– generalized Gauss-Lobatto, 331
– generalized Gauss-Radau, 330
– generalized Gauss-Turán-Lobatto, 397
– generalized Gauss-Turán-Radau, 397
– generalized Gauss-Turán-Stancu, 397
– generalized Turán, 395
– interpolatory, 321
– n-point, 319
– Newton-Cotes with Jacobi weight, 322
– nodes of, 319
– positive, 322

- weighted Newton-Cotes, 321
- weights of, 319
Quadrature sum, 319
- convergent, 320
- stable, 320
- unstable, 320
Quasi-projector, 196

Rabinowitz, P., 325
Rajković, P.M., 88
Rakhmanov, E.A., 106, 111, 155, 158
Rassias, Th.M., ix
Reemtsen, R., 23
Remainder term, 54
Revers, M., 62
Riemann-Hilbert problems, 109
Riemenschneider, S.D., vii
Riesz, M., 194
Rivlin, T.J., 15, 16, 65
Ronveaux, A., 146
Rooney, P.G., 142
Roy, R., 121, 132
Runge, 60, 64
Russo, M.G., ix

Saff, E.B., 59, 94, 124, 154, 155, 158
Sakai, R., 155
Schira, T., 336
Schönhage, A., 64
Sherman, J., 117
Shi, Y.G., vii
Shivakumar, P.N., 65
Shohat, J., 117
Simon, B., 80, 106
Skrzipek, M.-R., 114
Sloan, I.H., 346
Slowly convergent series, 397
Smirnov, I., 80
Smith, S.J., 62, 64
Smith, W.E., 346
Sokhotski-Plemelj formulae, 109
Spaces
- Besov, 36, 170
- Chebyshev, 38
- Haar, 38
- Hausdorff topological, 38
- inner product, 75
- L_1-Sobolev, 344
- L^p-Zygmund-Hölder, 36
- Sobolev, 33
- Sobolev-type, 167
- strictly normed, 2
- weighted Besov, 230
- weighted L^p, 225

- weighted uniform, 271
- Zygmund, 367
Stahl, H., 105
Stanić, M., ix
Stauffer, A.D., 389
Steen, N.M., 165
Stegun, I.A., 142
Steinbauer, R., 80
Stieltjes, T.J., 324
Stieltjes inversion formula, 116
Stieltjes procedure, 159
Stirling's formula, 413
Studden, W.J., 124
Suetin, P.K., 121, 126
Suslov, S.K., 121
Swarttouw, R.S., 121
Systems
- biorthogonal, 387
- Chebyshev, 38
- Haar, 38
- Malmquist-Takenaka of rational functions, 81
- Müntz, 40
- orthogonal, 75
- orthonormal, 75
- trigonometric, 79
Szabados, J., vii, ix, 54, 68, 205, 249, 264, 288, 291
Szász, P., 264
Szegő, G., 80, 86, 94, 98, 101, 103–105, 135–137, 140–142, 248, 354
Szegő's
- asymptotic, 104
- theory, 105
Szynal, J., 143

Takenaka, S., 81
Taylor's formula, 392
Themistoclakis, W., ix
Theodorus constant, 409
Theorem
- Banach-Steinhaus, 320
- Cauchy residue, 55
- Chebyshev alternation, 8, 17
- equioscillation, 18
- Favard's, 97
- fundamental of algebra, 4
- Gershgorin's, 100
- Gopengauz, 269
- Jackson, 34
- Markov's, 116
- Rolle, 54
- Szegő, 248
- Uspensky, 341

– Von Neuman, 363
– Weierstrass, 3, 167
Thiry, G., 146
Three-term recurrence relations, 96
Tietze, H., 64
Toledano, D., 67
Totik, V., ix, 59, 105, 154, 158
Transform
– Cauchy, 109
– finite Hilbert, 265
– Hilbert, 63, 109
– Laplace, 398
– Stieltjes, 114
Transformation
– Joukowski, 12
Trapezoidal rule, 198
Trefethen, L.N., 50, 65
Tricomi, F., 126
Turán, P., 105, 264
Tureckiĭ, A.H., 64

Uvarov, V.B., 121

Van Assche, W., 107, 118, 121
Van der Corput, J.G., 15
Van Doorn, E.A., 117
Vanlessen, M., 150
Varga, R.S., 336
Varma, A.K., 123
Vértesi, P., vii, ix, 53, 54, 67, 135, 249, 264, 279
Vianello, M., 411, 413
Vinet, L., 124
Visser, C., 15

Walsh, J.L., 81
Weak Jackson estimate, 171
Weideman, J.A.C., 65
Weierstrass, K., 3
Weight
– "A_p-class" of the doubling, 225
– Abel, 159
– Airy, 165
– Bernstein-Szegő, 108
– Chebyshev, 336

– Ditzian-Totik, 149
– doubling, 224
– Einstein's, 165
– even rational, 350
– Fermi's, 165
– Freud, 107, 154
– Gegenbauer, 336
– generalized Freud-type, 155
– generalized Jacobi, 148
– generalized Laguerre, 140
– Hermite, 145
– hyperbolic, 402
– Jacobi, 131, 166
– Lindelöf, 159
– logarithmic, 165
– logistic, 159
– modified Jacobi, 151
– one-sided Hermite, 165
– singular part of the, 229
– Szegő's class of, 103
– the hyperbolic, 166
Weight coefficients, 115
Weighted approximation, 166
– of functions having isolated interior singularities, 182
– on the real line, 178
– on the semi-axis, 174
– polynomial on $[-1, 1]$, 170
Weighted interpolation, 271
– at Jacobi zeros, 271
Wellman, R., 147
Welsch, J.H., 326–328
Werner, W., 50
Widom, H., 94, 105
Wilson, J., 121
Wong, R., 65, 137
Wrench, J.W., Jr., 21
Wrigge, S., 389
Wronskian-type relation, 113

Zanovello, R., 411, 413
Zhedanov, A., 124
Zhou, X., 150
Zolotarev, E.I., 16

Printing: Krips bv, Meppel, The Netherlands
Binding: Stürtz, Würzburg, Germany